# PRINCIPLES OF
# ENGINEERING ECONOMIC ANALYSIS

# PRINCIPLES OF ENGINEERING ECONOMIC ANALYSIS

SECOND EDITION                    REVISED PRINTING

## JOHN A. WHITE

School of Industrial and Systems Engineering
Georgia Institute of Technology
Atlanta, Georgia

## MARVIN H. AGEE

Department of Industrial Engineering and Operations Research
Virginia Polytechnic Institute and State University
Blacksburg, Virginia

## KENNETH E. CASE

School of Industrial Engineering and Management
Oklahoma State University
Stillwater, Oklahoma

JOHN WILEY & SONS
New York   Chichester
Brisbane   Toronto
Singapore

*Library of Congress Cataloging in Publication Data:*

White, John A., 1939–
   Principles of engineering economic analysis.

  Includes index.
  1. Engineering economy.  I. Agee, Marvin H.,
1931–    . II. Case, Kenneth E.  III. Title.
TA177.4.W48  1984    658.1'55    83-21586
ISBN 0-471-08607-X

Printed in the United States of America

10 9 8 7 6

# PREFACE

This textbook is for those who are interested in learning the principles involved in analyzing economic investment alternatives. Although the material is useful for accounting, economics, finance, and management students, its primary audience will be engineering and engineering technology students. Courses for which the book is intended are variously titled engineering economy, economic analysis, industrial economy, economic decision analysis, engineering decision analysis, and engineering economic analysis.

Over the past 24 years we have taught engineering economic analysis to more than 4000 persons, including undergraduate and graduate students from engineering and nonengineering academic programs. Additionally, we have taught refresher courses for professional engineering registration examinations and continuing education short courses, consulted, and worked in industrial positions that required the application of the material here. Based on that experience, we perceived a need for a textbook that presented a unified treatment of economic analysis principles. We also felt the need to address the subject from a *cash flow viewpoint*.

The following are some of the important features of this textbook.

1. A cash flow approach is taken throughout.
2. A systematic six-step approach is given for performing a comparison of investment alternatives.
3. The study period must be specified explicitly instead of unknowingly employing a least common multiple of lives study period.

4. Comprehensive treatments are provided of:
   a. Cost concepts, including cost terminology, cost estimation, accounting principles, and nonmonetary considerations.
   b. Inflation, as it affects the economic comparison of investment alternatives.
   c. Both post-1980 ACRS and pre-1981 methods of depreciation.
   d. Income taxes, including capital gains and losses, depreciation recapture, depletion, and investment tax credit considerations.
   e. Economic analyses in the public sector, emphasizing benefit-cost analysis procedures.
   f. Break-even, sensitivity, and risk analyses, with the latter including analytical and simulation approaches.
   g. Decision models, including decisions under risk and uncertainty and decision tree analyses.
5. Over 190 worked-out examples and approximately 400 problems are provided, with answers to even-numbered problems given at the end of the text.
6. Section numbers are provided for material covered by each end-of-chapter problem to facilitate a self-study program.
7. Modeling uniform, gradient, and geometric series, modeling changing interest rates, and modeling equity amounts in loan payments are presented to emphasize the importance of *modeling* cash flow series instead of using more conventional cookbook approaches.
8. New and expanded tables of compound interest factors are provided, including both discrete and continuous compounding, discrete and continuous flows, and gradient and geometric cash flow series.
9. Separate chapters on measuring investment worth and the comparison of alternatives are provided to permit a detailed understanding of the measures of investment worth prior to their use in comparing alternatives.
10. Methods of measuring investment worth include present worth, annual worth, future worth, capitalized worth, payback period, savings/investment ratio, internal rate of return, and external rate of return methods.
11. Methods of comparing investment alternatives include both total cash flow and incremental cash flow approaches utilizing each of the measures of investment worth.
12. Depreciation is covered in conjunction with the discussion of income taxes, since depreciation is not a cash flow and it influences cash flows through taxation.
13. Replacement analysis is covered in the discussion of comparison of alternatives, since a cash flow analysis of replacement problems is essentially no different than any other type of investment alternative.

14. The data have been packaged in a nine-chapter format in the sequence most natural to the coverage of the material.

15. Only material that is required in the performance of engineering economic analyses is presented, since it is believed that including tangential material not normally covered by the instructor has a deleterious effect on the student.

16. A detailed Instructor's Manual has been prepared that includes detailed solutions to all exercises.

The manuscript has been tested in the classroom. The orientation of the material and its organization have been influenced by the recommendations of numerous colleagues and undergraduate and graduate students.

The first seven chapters have been developed from our collective experience of teaching the subject to engineering sophomores from all disciplines; the material in the remaining two chapters has been covered in a senior-level elective course. Although elementary calculus is a sufficient prerequisite for the first seven chapters, an introductory course in probability theory is suggested as a prerequisite for the final two chapters.

Many people influenced the development of this textbook. However, one individual, Carl B. Estes, deserves special recognition. He has devoted considerable time and energy in providing detailed recommendations for improvement of the manuscript; he and his daughter, Lou Scott, have contributed immensely to the Second Edition. Additionally, we have benefited from either taking coursework from or teaching with John R. Canada, Wolter J. Fabrycky, Richard S. Leavenworth, William T. Morris, G. T. Stevens and Gerald J. Thuesen, who have authored or coauthored one or more texts on the subject. Our approach to the subject has also been influenced by our associations with Richard H. Bernhard, William E. Biles, Leland T. Blank, James R. Buck, Lynn E. Bussey, Claus Christiansen, Thomas P. Cullinane, Gerald A. Fleischer, David R. Freeman, Lynwood A. Johnson, Raymond P. Lutz, Leon F. McGinnis, M. Wayne Parker, Gunter P. Sharp, Donald R. Smith, Joe Tanchoco, and H. G. Thuesen. Therefore, the influence of others will be evident in various places in the book. Because it is impossible to credit the many individual contributions, we simply acknowledge gratefully their influence on our understanding of the subject.

Finally, we thank our wives Mary Lib, Barbara, and Lynn, for their patience and understanding during the project. They constantly demonstrated considerable insight on the subject by the gentle reminder, "If you know so much about investments, why aren't we rich?"

*John A. White*
*Marvin H. Agee*
*Kenneth E. Case*

# CONTENTS

CHAPTER SEVEN
## ECONOMIC ANALYSIS OF PROJECTS IN THE PUBLIC SECTOR

CHAPTER EIGHT
## BREAK-EVEN, SENSITIVITY, AND RISK ANALYSES

CHAPTER NINE
## DECISION MODELS

# PRINCIPLES OF
# ENGINEERING ECONOMIC ANALYSIS

# CHAPTER ONE
# **INTRODUCTION**

## 1.1 BACKGROUND

The subject matter discussed in this book is variously referred to as *economic analysis, engineering economy,* and *economic decision analysis,* among other names. Traditionally, the application of economic analysis techniques in the comparison of engineering design alternatives has been referred to as *engineering economy.* However, the emergence of a widespread interest in economic analysis in public sector decision making has brought about greater use of the more general term, *economic analysis.*

Our approach to the subject will be a *cash flow approach.* A cash flow occurs when money actually changes hands from one individual to another or, for that matter, from one organization to another. Thus, money received and money dispersed (spent) constitute cash flows. We also will emphasize the fundamentals of economic analysis without dwelling on some of the philosophical issues associated with the subject; for those interested in pursuing the latter, references are provided.

## 1.2 THE PROBLEM-SOLVING PROCESS

Economic analyses are typically performed as a part of the overall problem-solving process. In designating a new or improved product, a manufacturing process, or a system to provide a desired service, the "problem solver" is involved in performing the following steps.

State A ⟶ [black box] ⟶ State B

FIGURE 1.1. Black box approach.

1. Formulation of the problem.
2. Analysis of the problem.
3. Search for alternative solutions to the problem.
4. Selection of the preferred solution.
5. Specification of the preferred solution.

The *formulation of the problem* involves the establishment of its boundaries, and it is aided by taking a black box approach. As Krick [7] describes the black box, an originating state of affairs (state A) and a desired state of affairs (state B) exist. Also, a transformation must take place in going from state A to state B, as depicted in Figure 1.1. More than one method of performing the transformation from state A to state B exists and there is unequal preferability of these methods. The solution to the problem is visualized as a black box of unknown, unspecified contents having input A and output B.

The *analysis of the problem* consists of a relatively detailed phrasing of the characteristics of the problem, including restrictions and the criteria to be used in evaluating alternatives. Considerable fact gathering is involved, including the real restrictions on the problem, which must be satisfied. Consequently, any budget, quality, safety, personnel, environmental, and service level constraints that may exist are identified.

The *search for alternative solutions to the problem* involves the use of the engineer's creativity in developing feasible solutions to the design problem. A number of aids to stimulate one's creativity are provided by Krick [7]. In particular, he suggests that one should:

1. Exert the necessary effort.
2. Not get bogged down in details too soon.
3. Make liberal use of the questioning attitude.
4. Seek many alternatives.
5. Avoid conservatism.
6. Avoid premature rejection.
7. Avoid premature acceptance.
8. Refer to analogous problems for ideas.
9. Consult others.
10. Attempt to divorce one's thoughts from the existing solution.
11. Try the group approach.

**12.** Remain conscious of the limitations of the mind in the process of idea generation.

Unfortunately, the natural tendency is to mix the *evaluation* of alternatives with the *generation* of alternative solutions. As a result, the search process usually terminates with the development of the first alternative that can be justified economically. By separating the search process from the selection process, we enhance the chance of generating a number of alternatives that can be justified economically.

The *selection of the preferred solution* consists of the measurement of the alternatives, using the *appropriate criteria*. The alternatives are compared with the constraints and infeasible alternatives are eliminated. The benefits produced by the feasible alternatives are then compared. Among the criteria considered for choosing among alternatives is the economic performance of each alternative.

The *specification of the preferred solution* consists of a detailed description of the solution to be implemented. Predictions of the performance characteristics of the solution to the problem are included in the specification.

---

Example 1.1 _____

As an illustration of the application of the problem-solving approach, consider the case of a locally owned electrical power generating company faced with an air pollution problem that had to be solved. The power plant was using low-grade coal with sulfur content of 1.9% by weight, and the air pollution control board for the state directed the power plant to comply with newly adopted standards for air quality. A consulting engineer was hired to study the problem and recommend the best course of action.

The engineer met with representatives of the air pollution control board, the power company, county and city officials, and suppliers of various fuel alternatives and associated power generating equipment. An analysis of the problem yielded an agreement to use a 20-year planning horizon, forecasts of power requirements over the planning horizon, and a specification of the criteria to be used in evaluating the alternatives. Specifically, a benefit-cost analysis was to be performed; a cost of $150/ton of sulfur pollutant released into the air was to be employed in the analysis.

A search for alternative solutions to the problem resulted in a consideration of bunker C fuel oil, low-sulfur furnace fuel oil, and natural gas. The methods described in Chapter Seven were employed to compare the alternatives using a benefit-cost analysis; bunker C fuel oil was the recommended energy source.

Based on the report provided by the consulting engineer, the city contracted for the installation of storage tanks for the fuel oil, piping, pumps, and the other equipment necessary for converting the furnaces from coal to fuel oil.[1]

---

[1] For additional discussion of the case study on which the above was based, see [4]

Example 1.2_____

As a second illustration of the problem-solving procedure, a leading manufacturer of automotive bearings was faced with the need to expand its distribution operations. A study team was formed to analyze the problem and develop a number of feasible alternative solutions. After analyzing the problem and projecting future distribution requirements, the following feasible alternatives were determined:

1. Consolidate all distribution activities and expand the existing distribution center, located in Michigan.
2. Consolidate all distribution activities and construct a new distribution center, location to be determined.
3. Decentralize the distribution function and build several new distribution centers geographically dispersed in the United States.

After considering the pros and cons of each alternative, the officers of the company directed the study team to pursue the second alternative.

An extensive plant location study was performed. The location study resulted in five candidate locations being selected for final consideration. The criteria used to make the final selection included:

1. Land cost and availability.
2. Labor availability and cost.
3. Proximity to supply and distribution points (present and future).
4. Taxes (property and income) and insurance rates.
5. Transportation (access to rail and interstate highways).
6. Community attitudes.
7. Building costs.

Based on site visits to each location, the officers of the company selected a site in Alabama and directed the study team to develop design alternatives for the material handling system to be used in the distribution center.

Applying the problem-solving procedure, four alternatives were obtained for evaluation. The first alternative involved the use of pallet racks, lift trucks, and flow racks; the second alternative included the use of an automated stacker crane system, lift trucks, conveyors, and flow racks; the third alternative suggested narrow-aisle, guided picking machines, high-rise shelving, conveyors, driverless tractor trains, and lift trucks; the fourth alternative consisted of an automated stacker crane system, a rail guided picking vehicle system, a sortation and accumulation conveyor system, and high-rise, narrow-aisle lift trucks.

A planning horizon of 10 years was used in performing an economic analysis of each design alternative. The fourth alternative was the most economical and was recommended to top management. Based on the detailed presentation and the economics involved, management approved a budget of $9 million to implement the recommendations of the study team.

---

As demonstrated by the preceding illustrations, engineers and systems analysts must be prepared to defend their solutions to problems. Economic performance is among the several criteria used to evaluate each alternative. Monetary considerations seldom can be ignored. If economics are not considered in the criteria used in the final evaluation of the alternatives, they are usually involved in an initial screening of the alternatives. In fact, money can be a constraint on the alternatives that can be considered, as well as the basis for the final selection.

This textbook treats in detail the step in the problem-solving process that involves the selection of the preferred solution. The process of measuring cash flows and benefits and the consideration of multiple objectives in selecting the preferred design are also treated.

In comparing alternatives, the *differences* in the alternatives will be emphasized. Consequently, the aspects of the alternatives that are the same normally will not be included in the analysis.

## 1.3 NONMONETARY CONSIDERATIONS

### 1.3.1 Multiple Criteria in the Decision Process

The principal emphasis of this textbook is on the use of logical methodology to choose an investment project from among several investment projects available to the decision maker, where the criterion for choice will be some single economic measure of effectiveness. The investment project chosen, however, may be different from the project recommended by the engineer, who based the recommendation solely on an economic criterion. Managers typically have multiple criteria to be considered in reaching a final decision about the alternative to be adopted. Among the factors to be considered are: quality, safety, environmental impact, community attitudes, labor-management relationships, cash flow position, risks, system reliability, system availability, system maintainability, system operability, system flexibility, impact on personnel levels, training requirements, comparisons with competitors, impact on the different units within the organization, ego, customers' preferences, capital requirements, and economic justification.

To illustrate a situation in which both economic factors and noneconomic, or nonmonetary, factors are involved, suppose that the management of a firm, because of a significant increase in market demand, has the choice between (1) going to a second shift of production using existing machinery, and (2) installing more highly mechanized machinery in order to meet the new production schedules. Clearly, a comparison of the costs for each of these alternatives over a planning horizon is in order, and one objective of management would no doubt be to meet production schedules at the lowest possible annual cost of production.

Although considerable investigation and study are required, many of the factors involved in a decision that may at first seem noneconomic can be expressed in (or reduced to) monetary values. For instance, in the above example, maintenance personnel may have to be trained in the installation and repair of the new machinery. Most likely, these training costs can be determined and charged to the "mechanization" alternative. However, other factors involved are not so easily reduced to monetary values and, indeed, some factors cannot be so reduced. For example, the installation of the mechanized machinery would probably reduce the number of personnel required with the present production system, whereas the second shift alternative would result in additional employees. If persons are laid off or transferred to other jobs within the firm, their salaries could be considered annual savings accruing to the mechanized machinery alternative and, conversely, the salaries of additional employees integral with the second shift alternative would be annual costs. However, layoffs or transfers could have a deleterious effect on the morale of the remaining employees. On the other hand, adding new personnel could have a beneficial effect on morale. Assessing the cost or gain of these expected changes in morale is virtually impossible. Nevertheless, the potential effect on production that morale changes may have must be considered by management.

Factors that affect a decision but cannot be expressed in monetary terms are often called *intangibles* or *irreducibles*. Practically all real-world business decisions involve both monetary and intangible factors. In the above production example, assume that the alternative of a second shift results in a total annual cost of $100,000 for both shifts over a 5-year planning horizon with no persons laid off or transferred or new employees hired. For the mechanized machinery alternative, there would be an estimated annual cost of $85,000 over the 5-year planning horizon with six persons laid off. Thus, this oversimplified decision is reduced to two measures: a monetary annual cost and the single intangible factor of employee morale. If management chooses the second shift alternative, this would imply that management places a higher subjective utility value on "good employee morale" than on the $15,000 difference in annual costs between the two alternatives. That is, the objective of good employee morale is more important than the objective of lowest annual cost (at least for this difference of $15,000). Instances of decision situations in which intangible considerations outweigh monetary ones are frequent, and the engineer should not become distraught over this fact. Within the hierarchy of management responsibilities for a firm, the higher the level of management, the more likely it is that intangible considerations will be given greater subjective weight. Engineering project proposals should therefore reflect a knowledge of intangibles that management will wish to consider.

Miller and Starr have made interesting observations in regard to the decision-making process.[2]

1. Being unable to satisfactorily describe goals in terms of one objective, *people customarily maintain various objectives.*

2. Multiple objectives are frequently in conflict with each other, and when they are, a *suboptimization* problem exists.

3. At best, we can only optimize as of *that time* when the decision is made. This will frequently produce a suboptimization when viewed in subsequent times.

4. Typically, decision problems are so complex that any attempt to discover *the* set of optimal actions is useless. Instead, people set their goals in terms of outcomes that are *good enough.*

5. Granted all the difficulties, human beings make every effort to be *rational* in resolving their decision problems.

Accepting the premise that multiple objectives are involved in a decision, it follows that different measures for these objectives may arise. For instance, two or more engineering project alternatives could each involve annual costs in dollars, weight in pounds, repairability values on an arbitrary index from 1 to 10, and so forth. The decision problem would be greatly simplified if the various measures could be transformed to a single measure, say, values on a "utility" scale. Then each measure (dollars, pounds, etc.) for a given alternative could be converted to a utility value, the new utility values could be weighted by their relative importance, and the weighted utility values could be aggregated by an appropriate functional form. A single utility value would thus result for each alternative, and a selection would be made by choosing the alternative having the maximum utility value. This area of study is the study of utility theory and value measurement—a very interesting area of study, but one with some controversy attached. The subject area is outside the scope of this book and selected references are given in the bibliography of this chapter. Some of the references stress mathematical theory [1, 3, 9, 10] and others offer an applied, but approximate, approach [5, 11].

There will be some additional discussion on multiple objectives in Chapter Nine. Otherwise it will be assumed throughout the book that only a single economic measure of effectiveness is relevant in comparing alternative projects.

## 1.3.2 Selling the Project

One of the more important ingredients in gaining approval from management for an economically and technically justified investment alternative is the sales

---

[2] David W. Miller and M. K. Starr, *Executive Decisions and Operations Research*, Second Edition, Prentice-Hall, 1969, pp. 52–53.

technique of the individual who presents the alternative to the manager. It is not unusual for a manager to adopt a weaker alternative because of the persuasive powers of the person who presented the alternative.

As Lutz [8] points out, *what is said* to support a project justification and *how it is said* influence the ultimate disposition of the funding request. As he puts it, "Communication content may outweigh the technical aspects of the proposal. The decision as to whether or not a project is funded often hinges on how successfully you communicate with the decision makers using their rules, words and decision criteria. Structuring your proposal for ease of acceptance is imperative and should be closely examined to try and increase the likelihood of obtaining capital budgeting dollars." Lutz argues that managers will often accept and implement a weak alternative because they:

1. Recognize the existence of a problem.
2. Understand the recommended solution.
3. Know how to implement the recommendation.
4. Anticipate immediate improvements with positive financial results.

A tendency exists for problem solvers to spend too much time trying to develop the recommended solution to the problem and too little time in determining the best way to sell the solution to management. Effective communication is essential; in particular, it is important to communicate with management in a way that can be understood and accepted [6].

It is important to write a proposal that includes an *executive summary*. The executive summary normally reduces the performance characteristics of the recommendation to a few pages. Technical aspects of the recommended solution and the technical details of the economic analysis are usually provided in the main report for the manager who wishes to obtain additional understanding of the information in the executive summary. A face-to-face presentation is also quite important in achieving effective communication.

Communication techniques such as the use of visual aids, voice control, dress, structure, and clarity of expression are necessary to sell the proposed solution. However, it is probably more important to know your audience well and to prepare your proposal using their language. Since the language of managers is often the language of finance, Chapter Two introduces some fundamental financial terminology used by managers.

## 1.4 PRESENT ECONOMY EXAMPLES

Most of the methodology presented in this book is based on the fundamental concept that money has a time value. Investment projects that have a life cycle of several years will almost invariably have associated revenues (or savings) and expenses that occur periodically over the life cycle.

The time value of money concept simply recognizes that a given dollar amount has different values at different discrete points on the continuous time scale. In order to adjust these different values to a common reference point on the time scale, an interest rate is used for the adjustment. This time value of money concept will be developed and illustrated in Chapter Three.

Projects that require the consideration of the time value of money, or an interest factor, typically involve an investment (or investments) of capital over a long period of time. There are many other situations in which time is not a significant economic factor. The choice between alternative production sequences and the choice between alternative materials for a particular product component are two general examples where a capital investment may not be required, and hence time is not an economic factor in the decision process. Degarmo [2], one of the early authors on the subject of engineering economy, has labeled these types of economic problems as "present economy studies."

Such types of economic decisions are not insignificant in the business and industrial world by any means, but only four examples are subsequently presented for illustrative purposes. The latter two examples do involve capital investments but the decision question posed can be answered without considering the time value of money.

Example 1.3_____

A particular metal component part can be machined on either an engine lathe or a turret lathe. In either case, the machines are not used exclusively for this particular part. Rather, parts are produced in batches according to a customer's order. If a customer's order is produced on the engine lathe, the following information is relevant:

Set-up time required per order               = negligible
Direct labor and machine time per part = 10 minutes
Direct labor rate                                    = $7.50/hour
Machine rate                                          = $12.00/hour
Cutting tool cost per part                      = $0.20/unit

If the part is produced on the turret lathe, a set-up of the machine is required each time a customer's order is processed. The set-up is required only once per order if the order size is less than or equal to 300 units. If the order size is greater than 300 units, then a set-up is required every 300 units. Relevant time and cost data is given below:

Set-up time required per order (both the operator and
   machine time is charged to the order)    = 2 hours
Direct labor and machine time per part          = 5 minutes
Direct labor rate                                           = $9.00/hour
Machine rate                                                 = $15/hour
Cutting tool cost per part                             = $0.10/unit

The material cost per unit is the same regardless of the machine used. Which is the economical machine to use for order sizes of 25, 100, and 500 units?

Let $x$ = the number of units produced per customer order. Then, it will be convenient to develop a cost per order formula for each type of machine. That is,

Engine lathe cost per order:

| | |
|---|---|
| Direct labor cost/minute | = $7.50/60 = $0.125/minute |
| Direct labor cost/unit | = ($0.125/minute)(10 minute/unit) |
| | = $1.25/unit |
| Machine cost/minute | = $12.00/60 = $0.20/minute |
| Machine cost/unit | = ($0.20/minute)(10 minute/unit) |
| | = $2.00 |
| Cutting tool cost/unit | = $0.20/unit |
| Thus, the cost per order is | |
| $(1.25 + 2.00 + 0.20)/ | |
| unit ($x$ units/order) | = $3.45$x$ |

Turret lathe cost per order:

| | |
|---|---|
| Set-up cost/order | = ($n$)(2 hour/set-up) |
| | × ($9.00/hour + $15.00/hour) |
| | = ($48/set-up)($n$) |
| where $n$ | = the number of set-ups |
| | required per order |
| Direct labor cost/minute | = $9.00/60 = $0.15/minute |
| Direct labor cost/unit | = ($0.15/minute)(5 minute/unit) |
| | = $0.75/unit |
| Machine cost/minute | = $15.00/60 = $0.25/minute |
| Machine cost/unit | = ($0.25/minute)(5 minute/unit) |
| | = $1.25/minute |
| Cutting tool cost/unit | = $0.10/unit |
| The cost per order is then | |
| $48 $n$ + $(0.75 + 1.25 + 0.10)/unit ($x$) | = $48 $n$ + $2.10 $x$ |

The comparative costs per order size of 25, 100, and 500 units are calculated as:

For an order size of 25 units

| | | |
|---|---|---|
| Engine lathe cost = $3.45(25) | | = $ 86.25 |
| Turret lathe cost = $48(1) + $2.10(25) | | = $100.50 |

For an order size of 100 units

| | | |
|---|---|---|
| Engine lathe cost = $3.45(100) | | = $345.00 |
| Turret lathe cost = $48(1) + $2.10(100) | | = $258.00 |

For an order size of 500 units

| | | |
|---|---|---|
| Engine lathe cost = $3.45(500) | | = $1725 |
| Turret lathe cost = $48(2) + $2.10(500) | | = $1146 |

Example 1.4_____

The SMA Corporation, a food processing firm, ships cartons of canned food to wholesale distributors via motor freight (tractor and trailer). The cartons are 8 inches wide × 12 inches long × 6 inches high and are stacked in layers on wooden pallets, with the largest surface area of the cartons (8 inches × 12 inches) resting on the pallet. The pallets are 40 inches long × 48 inches wide × 6 inches high and the cartons are arranged on the pallet in the pattern sketched in Figure 1.2. Three layers of cartons are stacked on each pallet and then "strapped" with metal bands to stabilize the load. Each pallet with 60 cartons is then termed a unit load. The total height of the unit load is 24 inches, as shown in Figure 1.3.

The inside dimensions of the trailer (closed van) are 34 feet long × 7 feet 6 inches wide × 6 feet 10 inches high. The freight cost to ship to a particular wholesale distributor is $1000/trailer load. The SMA Corporation therefore wishes to ship the maximum number of unit loads per trip. The company has a choice of two types of wooden pallets. The types are two-way entry and four-way entry, that is, a fork-lift truck (used to place and retrieve the unit loads from the trailer) can insert the forks into the pallet from two directions or from four directions, as show in Figure 1.4. The implication of this choice is that, for the two-way entry pallet, the fork-lift truck can only address the pallet perpendicular to the 48 inch dimension of the pallet. The two-way pallets cost $6 each and the four-way pallets cost $8 each. If the unit loads can be stacked only three tiers high inside the trailer (three unit loads per stack), which pallet style will result in the lowest freight cost per unit load? (Assume negligible clearance required between each adjacent stack and between a stack and the walls of the trailer).

The trailer is 7 feet 6 inches (or 90 inches) wide. Since the 40-inch × 48-inch, two-way entry pallet is 48 inches wide, two rows of pallets cannot be stacked in

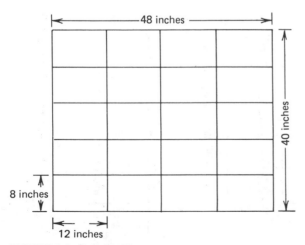

FIGURE 1.2. Pallet pattern.

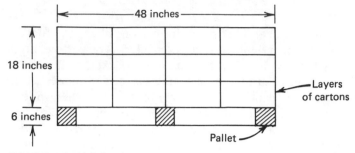

FIGURE 1.3. Unit load.

the trailer (in the width direction). The trailer is 34 feet (or 408 inches) long and the number of unit load stacks that can be placed in the trailer length direction is 408 inches/40 inches = 10.2, or 10 stacks. With 3 unit loads/stack, 10 stacks/row, and 1 row/trailer, 30 unit loads can be shipped per trailer. This results in a unit load cost per load of $1000/30 = $33.33/unit load. The unit cost of the pallets is $6, and the total cost per unit load shipped is $(33.33 + 6) = $39.33.

If the four-way entry pallet is used, two rows of pallets can be stacked in the trailer (in the width direction), with 10 inches of clearance between the rows. That is, 90-inch trailer width − 2(40 inch unit load width) = 10 inch clearance. With a trailer length of 408 inches, the number of load stacks per row is 408 inches/48 inch = 8.5, or 8 stacks/row. Thus, two rows/trailer × 8 stacks/row × 3 unit loads/stack = 48 unit loads/trailer. The freight cost per unit load shipped is then $1000/48 = $20.83. With a unit cost of $8/pallet, the total cost per unit load shipped is $28.33.

The four-way entry pallets should be purchased.

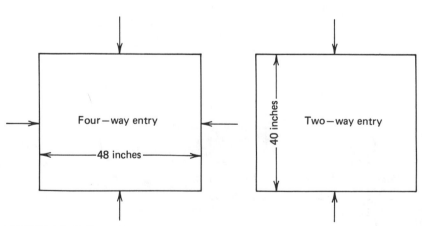

FIGURE 1.4. Pallet styles.

TABLE 1.1. Sales Volume Data for Example 1.5

| | Sales Volume | |
| --- | --- | --- |
| Style | Present Sales Volume (units/month) | Predicted Sales Volume (percent of present volume) |
| 1 | 860 | 120 |
| 2 | 5300 | 140 |
| 3 | 370 | 200 |
| 4 | 900 | 100 |
| 5 | 2200 | 80 |
| 6 | 230 | 150 |

Example 1.5

The Hoakie Hahn Company manufactures a single product but in six different styles. The market area for the product is intrastate only and the company also owns six retail outlet stores within the state. Because of customer preferences, only certain styles are inventoried at each outlet store. The decision at the present time concerns the warehouse space at each outlet store.

Based upon an analysis of past sales records and other current marketing information, an increase in future sales for each store is predicted. An increase in warehouse space at each store location is therefore required. If the company only has $100,000 available for warehouse expansion, how much capital must be borrowed to finance the construction of the needed space?

Relevant data are given in Tables 1.1 and 1.2. Although some of the warehouses have a small amount of unused space currently available, assume they are at capacity. Further, as a gross estimate, assume that the space required is proportional to the volume of product (style) manufactured per month.

TABLE 1.2. Space Requirement Data for Example 1.5

| | | Space Requirements | |
| --- | --- | --- | --- |
| Outlet Store | Styles Inventoried | Present Warehouse Space (square feet) | Cost of Additional Space (dollars/square foot) |
| A | 1,2,5 | 4,000 | 25 |
| B | 2,4 | 2,000 | 35 |
| C | 3,5,6 | 3,600 | 30 |
| D | 1,2,4 | 7,400 | 40 |
| E | 1,3,6 | 700 | 25 |
| F | 3,5,6 | 4,200 | 30 |

The predicted sales for each style are:

Style 1 = ( 860)(1.20) = 1032 units/month
Style 2 = (5300)(1.40) = 7420 units/month
Style 3 = ( 370)(2.00) = 740 units/month
Style 4 = ( 900)(1.00) = 900 units/month
Style 5 = (2200)(0.80) = 1760 units/month
Style 6 = ( 230)(1.50) = 345 units/month

The present sales indices for each outlet store are:

Store A = ( 860 + 5300 + 2200) = 8360 units/month
Store B = (5300 + 900) = 6200 units/month
Store C = ( 370 + 2200 + 230) = 2800 units/month
Store D = ( 860 + 5300 + 900) = 7060 units/month
Store E = ( 860 + 370 + 230) = 1460 units/month
Store F = ( 370 + 2200 + 230) = 2800 units/month

The indices for each outlet store, based on forecast sales, are:

Store A = (1032 + 7420 + 1760) = 10212 units/month
Store B = (7420 + 900) = 8320 units/month
Store C = ( 740 + 1760 + 345) = 2845 units/month
Store D = (1032 + 7420 + 900) = 9352 units/month
Store E = (1032 + 740 + 345) = 2117 units/month
Store F = ( 740 + 1760 + 345) = 2845 units/month

An estimate of the total warehouse space needed at each store is:

Store A = (10212/8360)(4000 square feet) = 4886 square feet
Store B = ( 8320/6200)(2000 square feet) = 2684 square feet
Store C = ( 2845/2800)(3600 square feet) = 3658 square feet
Store D = ( 9352/7060)(7400 square feet) = 9802 square feet
Store E = ( 2117/1460)( 700 square feet) = 1015 square feet
Store F = ( 2845/2800)(4200 square feet) = 4268 square feet

The cost of the additional warehouse space needed at each store is:

Store A = (4886 − 4000)($25) = $ 22,150
Store B = (2684 − 2000)($35) = 23,940
Store C = (3658 − 3600)($30) = 1,740
Store D = (9802 − 7400)($40) = 96,080
Store E = (1015 − 700)($25) = 7,875
Store F = (4268 − 4200)($30) = 2,040
Total                           $153,825

Since the $153,825 is greater than the $100,000, the company would have to borrow $53,825 to fiance all the estimated warehouse expansion required.

## Example 1.6

For one product that a company manufactures, the purchased raw material is 2-inch diameter × 10-feet long extruded aluminum bar stock. This material costs $1.05/pound. Since the company purchases an average of 100,000 pounds/month, it is considering manufacturing the raw material, which would require the purchase of an extrusion machine and a cutoff saw. Raw material for the extruder can be purchased in 4-inch × 4-inch × 57-inch long cast aluminum ingots for $0.50/pound. The company's process engineer estimates the extruding operation would cost $0.30/pound and the subsequent cutoff operation would cost $0.05/pound. The engineer also estimates there would be a 10% scrap loss of material at the extrusion stage and a 5.7% loss of material at the cutoff operations. (1) If the company manufactures the extruded rod, what are the estimated annual savings? (2) If the company manufactures the extruded rod, what is the production cost for each rod? (Aluminum density = 0.1 lb./in³).

(1) The present annual cost of the raw material is

(100,000 pounds/month)($1.05/pound)(12 months/year) = $1,260,000

If the company manufactures the extruded rod from cast ingots, the scrap losses must be considered to arrive at the average quantity of ingots to purchase each month. Thus,

let $x$ = the pounds of ingots purchased per month

Then, since there is a 10% loss at the extrusion stage, 90% or $0.90x$ will output from this stage. The quantity, $0.9x$, then enters the cutoff stage. Since there is a 5.7% loss at this stage, then 94.3% of the input quantity will output from the cutoff stage, or $(0.90x)(0.943) = .8487x$ is the quantity output from the cutoff operation. This quantity must equal 100,000 pounds/month. The monthly purchase quantity should therefore be

$$0.8487x = 100,000 \text{ pounds/month}$$

$$x = 117,827.3 \text{ pounds/month}$$

The monthly input and output quantities for each manufacturing stage are:

Extrusion input: 117,827.3 pounds
Extrusion output: (117,827.3)(.90) = 106,044.6 pounds
Cutoff input: 106,044.6 pounds
Cutoff output: (106,044.6)(.943) = 100,000 pounds

The monthly production costs are:

Raw material costs: (117,827.3 pounds/month)($0.50/pound)
   = $58,913.65/month
Extrusion costs: (117,827.3 pounds/month)($0.30)
   = $35,348.19/month
Cutoff costs: (106,044.6 pounds/month)($0.05)
   = $5302.23/month

The total costs are:

$$(\$58,913.65 + \$35,348.19 + \$5302.23)(12)$$
$$= \$1,194,768.80/\text{year}$$

Thus the annual savings from manufacturing the extruded rod are:

$$(\$1,260,000) - (\$1,194,768.80) = \$65,231.20$$

(2) The volume of each extruded rod is

$$(\pi)(10)(12) = 377.00 \text{ cubic inch}$$

and the weight of each rod is

$$(377.00 \text{ cubic inch})(0.1 \text{ pound/cubic inch}) = 37.70 \text{ pounds}$$

The number of rods extruded per month is

$$(100,000 \text{ pounds/month}) \div (37.70 \text{ pounds/unit}) = 2652.52 \text{ units/month}$$

The unit cost of an extruded rod is

$$\frac{(\$58,913.65 + \$35,348.19 + \$5,302.23)/\text{month}}{2652.52 \text{ units/month}}$$

or $37.54/unit.

## 1.5 OVERVIEW OF THE TEXT

The material in the book has been organized in a manner consistent with the logical sequence of steps followed in performing an economic evaluation of investment alternatives. Chapter Two provides a discussion of cost concepts and includes discussions of elements of *costs*: future costs, sunk costs, fixed costs, variable costs, average costs, and an introduction to project cost estimation.

After acquainting the reader with the types of data that are germane to investment projects, Chapter Three presents some important fundamental concepts involving the *time value of money*. In fact, Chapter Three provides the foundation for the remainder of the book; therefore the reader must understand the time value of money operations presented in this chapter.

Chapter Four presents various *economic measures for determining a project's worth*. Included are the measures of annual worth, future worth, present worth, rates of return, savings/investment ratios, and payback period. The impact of inflation and deflation on a project's economic feasibility is discussed.

The economic measures for determining project worth are used in Chapter Five for *comparing investment alternatives*. The subject of replacing capital assets is also emphasized in this chapter.

Chapter Six addresses the issues of *depreciation allowances* and *income taxes* and their incorporation in economic analyses; Chapter Seven presents *benefit-cost analysis* which is often used for the economic analysis of projects in the public sector.

In Chapter Eight, supplementary analysis techniques are presented, particularly the effects of *risk and uncertainty* on the analysis of economic investment alternatives. In modeling the effects of risk and uncertainty mathematically, two approaches are taken: a prescriptive (normative) approach and a descriptive approach. A prescriptive model prescribes the action that, in some sense, is optimal; a descriptive model describes the behavior of the situation modeled. Chapter Eight provides a descriptive treatment of the effects of risk and uncertainty by presenting the subjects of *break-even, sensitivity*, and *risk analysis*. Then, prescriptive models are presented in Chapter Nine and are classified as models for *decision making under risk and uncertainty*.

## BIBLIOGRAPHY

1. Chernoff, Herman, and Lincoln E. Moses, *Elementary Decision Theory*, Wiley, 1959.
2. Degarmo, E. Paul, *Engineering Economy*, Macmillan, 1960. (Also, later editions in 1967, 1973—with John R. Canada; 1979—with John R. Canada and William Sullivan).
3. Fishburn, Peter C., *Decisions and Value Theory*, Wiley, 1964.
4. Fleischcher, G. A., *Benefit-Cost Analysis Applied to Air Pollution Control: A Case Study Based on the Dixon, Tiller County/U.S.A. Reference Community*, National Communicable Disease Center, Bureau of Disease Prevention and Environmental Control, Public Health Service, U.S. Department of Health, Education, and Welfare, Atlanta, Ga., 1968.
5. Kepner, Charles H., and Benjamin B. Tregoe, *The Rational Manager*, McGraw-Hill, 1965.
6. Klausner, R. F., "Communicating Investment Proposals to Corporate Decision-Makers," *The Engineering Economist*, 17 (1), 1971, pp. 45–55.
7. Krick, E. V., *An Introduction to Engineering and Engineering Design*, Wiley, 1965.
8. Lutz, R. P., "How to Justify the Purchase of Material Handling Equipment to Management," *Proceedings of the 1976 MHI Material Handling Seminar*, The Material Handling Institute, Pittsburgh, Pa., 1976.
9. Raiffa, Howard, *Decision Analysis—Introductory Lectures on Choice Under Uncertainty*, Addison-Wesley, 1968.
10. Schlaifer, Robert, *Probability and Statistics for Business Decisions*, McGraw-Hill, 1959.
11. Starr, Martin Kenneth, *Product Design and Decision Theory*, Prentice-Hall, 1963, Chapters 1 and 6.

## PROBLEMS

1. For Example 1.3 in Chapter 1, determine the order size at which it will be economical to switch over from the engine lathe to the turret lathe. (1.4)

2. In Example 1.4 of Chapter 1, assume that the alternative pallets available to the company are 32 inch long × 32 inch wide × 6 inch high, and 40 inch long × 48 inch wide × 6 inch high, both of two-way entry type. The pallets cost $4 and $6, respectively. Find which pallet style will result in the lowest freight cost per carton. (1.4)

3. In Example 1.6 of Chapter 1, let there be a fixed overhead cost of $500/month connected with the extruder machine. (For simplicity, assume a zero overhead cost for the cut-off saw.) Find the volume of production at which it will be economical for the company to start manufacturing extruded rods. (1.4)

4. In Example 1.6 of Chapter 1, assume that the company has decided to manufacture extruded rods and has two alternative sources of aluminum ingots. One source provides 4 inch × 4 inch × 57 inch ingots for $0.50/pound and the other source provides 5 inch × 6 inch × 100 inch ingots for $0.40/pound. The process engineer estimates a 10% scrap loss for the 50 inch long ingot and a 15% scrap loss in the 100 inch long ingot at the extrusion stage. The cutoff loss is estimated at 5% for each of the ingot types. Assume the monthly consumption of the extruded bar stock is 100,000 pounds. Which size ingot should the company purchase? (1.4)

5. A small grocery store is considering buying a personal computer (PC) system to be able to keep complete inventory records. There is a choice of three alternative local suppliers. The complete specifications and cost details for each PC are given below:

| PC Type | A | B | C |
|---|---|---|---|
| Time required to perform all the desired activities (hours/day) | 4 | 6 | 3 |
| Hourly wage of the operator (dollars/hour) | $5 | $5 | $5 |
| Fixed overhead (dollars/day) | $8 | $5 | $10 |
| Cost of power (dollars/hour) | $0.30 | $0.30 | $0.60 |

Based on this economic data, which microcomputer would you recommend that the store management purchase? (Assume that the operator can do other productive work at the store when not using the microcomputer.) (1.4)

6. The relationship between cutting tool life ($T$) in minutes, and cutting speed ($V$), in feet/minute, is expressed by Taylor's equation:

$$VT^n = K$$

where $K$ is a constant. From the above equation, it is possible to mathematically derive an equation to find the tool life resulting in minimum unit cost. This equation is:

$$T_c = \left(t_c + \frac{c_c}{c_o}\right)\left(\frac{1}{n} - 1\right)$$

where $T_c$ = tool life for minimum unit cost, minutes
$t_c$ = tool change time, minutes
$c_c$ = cost per cutting edge, dollars/edge
$c_o$ = cost of labor and overhead, dollars/minute
$n$ = Taylors tool-life exponent

A metal machining company is evaluating three different types of cutting tool inserts for one of their high-volume NC turning operations. The specifications and cost information available for the alternate inserts are as follows:

| | Insert Type | | |
| --- | --- | --- | --- |
| | Tungsten Carbide | Coated Carbide | Ceramic |
| $n$ | 0.22 | 0.27 | 0.38 |
| $t_c$ (minutes) | 2 | 2 | 2 |
| $c_c$ (dollars/edge) | 0.20 | 0.90 | 1.50 |
| $c_o$ (dollars/minute) | 0.5 | 0.5 | 0.5 |
| $K$ | 150 | 250 | 550 |

Using $T_c$ for each type of insert, and the fact that the metal removal rate is directly proportional to the cutting speed, $V$, which insert type is preferred to maximize production? (1.4)

7. A group of 10 university students is planning a trip to Daytona Beach, Florida during the 7-day Spring break between quarters. This destination is 1000 miles away from campus. The group has decided to share automobile expenses equally and is considering three alternative car rentals. For simplicity, assume only the following cost data is relevant. The cost for Car A is $15/day plus $0.10/mile traveled, and the capacity of Car A is four people. The cost for Car B is $25/day plus $0.08/mile traveled, with a capacity of four people. The cost for Car C is $30/day plus $0.12 per mile traveled, with a capacity of six people. The group estimates that local travel in Daytona will be about 400 miles (1.4)
(a) What is the most economic alternative travel plan for the group of students?
(b) Given the answer to part (a), determine the per person cost for travel.

8. Mr. Joe Smith is considering the purchase of a new car for business purposes. He has a choice of three cars, the details of which are given below:

| Car Type | A | B | C |
| --- | --- | --- | --- |
| Fixed cost/month (dollars) | 800 | 700 | 1000 |
| Miles per gallon | 31 | 25 | 35 |
| Estimated cost of maintenance (dollars/mile) | 0.10 | 0.15 | 0.08 |

Mr. Smith estimates a travel of about 1000 miles/month. The cost of gas is $1.20/gallon. Which is the best car for Mr. Smith? What is the average cost per mile of travel? (1.4)

9. The process engineer in a medium-sized machine shop has the choice of machining a particular part on either of two machines. Orders for this part are received regularly, but on a recurring basis. The order size varies. When the order is processed on Machine A, four setup operations of 1 hour each are required. Once setup, the per unit machining time on Machine A is 0.5 hours.
When the order is processed on Machine B, eight setup operations of 0.75 hour each are required. Once setup, the per unit machining time on Machine B is 0.55 hours. The hourly wage rates for the machine operators are $5 and $4.50 for Machine A and Machine B, respectively. The hourly overhead rates (including setup time) are $10 and $9 for Machine A and Machine B, respectively. (1.4)
(a) Which machine is preferred if the order size is 100 units? 500 units?
(b) What is the break-even order size?

10. The American Grinding Company is faced with a decision to select one of the following two types of grinding wheels for a rough grinding operation. The criterion of selection is to minimize the unit cost per cubic inch of metal removed. The unit cost is given by the following expression:

$$\text{Unit cost} = \text{cost of machining} + \text{cost of wheel}$$

$$= \frac{c_o}{R} + \frac{c_w}{G_r}$$

where $c_o$ = Cost of labor and overhead, dollars/hour
  $R$ = Material removal rate, cubic inch/minute
  $c_w$ = Cost of wheel, dollars/cubic inch
  $G_r$ = Grinding ratio, cubic inch of metal/cubic inch of wheel

Also, the grinding ratio and the material removal rate are related by

$$G_r R^n = K$$

where $n$ and $K$ = constants.
The alternatives being considered are:

| | Grinding Wheel Type | |
|---|---|---|
| | A | B |
| $c_w$ | 0.1 | 0.09 |
| $c_o$ | 10 | 10 |
| $n$ | 1 | 1 |
| $K$ | 4400 | 2600 |

Determine a basis for choice between wheels. (1.4)

CHAPTER TWO
# COST CONCEPTS

## 2.1 INTRODUCTION

Engineering economic analysis is primarily concerned with comparing alternative projects on the basis of an economic measure of effectiveness. This comparison process utilizes a variety of cost terminologies and cost concepts, and it will be helpful to present them prior to the discussion in Chapter Four of the economic measures of effectiveness for comparing alternative projects. To implement the discussion of cost terminology, a typical production situation will now be described.

Let us assume that the business of a small manufacturing firm is job-shop machining. That is, the firm produces a variety of products and component parts according to customer order. Any given order may be for quantities of as few as five parts or as many as several hundred parts. The firm has periodically received orders to manufacture a part, which we will identify as Part No. 163H, for the Deetco Corporation. The part has been manufactured in a four-stage production sequence consisting of (1) sawing bar stock to length, (2) machining on an engine lathe, (3) machining on an upright drill press, and (4) packaging. The unit cost to produce Part No. 163H by this sequence has been $25, where the unit cost consists of the major cost elements of direct labor, direct materials, and overhead (prorated costs for insurance, taxes, electric power, marketing expenses, etc.). The firm is now in the process of negotiations with the Deetco Corporation to obtain a contract for producing 10,000 of these parts over a period of 4 years, or an average of 2500 units/year. A contract for this volume of parts is highly desirable but, in order to obtain the contract, the firm must lower the unit cost.

An engineer for the firm has been assigned to determine production methods to lower the unit cost. After study, the engineer recommends the purchase of a small turret lathe. With the turret lathe, the processing sequence of Part No. 163H would consist essentially of (1) machining bar stock on the turret lathe and (2) packaging. The estimated unit cost for Part No. 163H by this production method would be $15. Furthermore, the production rate by the new method would be increased over the old method because the turret lathe would replace the sawing, engine lathe, and drill press operations.

If the turret lathe is purchased, the saw, engine lathe, and drill press would not be sold but would be kept for other jobs for which the firm may receive orders. The turret lathe would be reserved for the production of Part No. 163H, but about 25% excess production capacity could be devoted to other jobs.

The incremental investment required to purchase the turret lathe and the new tooling required, as well as installing the machine, is $50,000. The physical life of the turret lathe is judged to be about 25 years, but present federal tax laws permit the investment capital to be recovered through annual depreciation charges in 5 years. At the end of 5 years, the firm estimates the market, or salvage, value of the turret lathe would be $25,000. If the maximum unit price that the Deetco Corporation will pay for Part No. 163H is $22, should the firm accept the contract for 10,000 parts and then purchase the turret lathe in order to execute the contract?

The question will not be answered in this chapter, but the example situation has been cited to illustrate one type of decision with which this text is concerned. In fact, additional information would be required if the question were to be answered using the methodology presented in later chapters. However, the reader can readily appreciate that certain cost figures in the above example must be determined or estimated before any rational decision can be reached as to the purchase of the turret lathe. The cost elements contributing to the installed first cost of $50,000, the unit production cost of $15, and the salvage value of $25,000 would be determined or estimated from information obtained from a variety of sources. These sources might typically be production records, accountant's records, manufacturer's catalogs, publications from the U.S. Government Printing Office, and so forth. The engineer should therefore be familiar with cost terminology, cost factors, and cost concepts as used by different specialists if effective economic comparisons and intelligent recommendations are to be made.

## 2.2 COST TERMINOLOGY

Because both cost definitions and cost concepts are included in this section, clarity will be achieved by the use of six categories: (1) life-cycle costs, (2) past and sunk costs, (3) future and opportunity costs, (4) direct, indirect, and over-head costs, (5) fixed and variable costs, and (6) average and marginal costs.

## 2.2.1 Life-Cycle Costs

The *life-cycle cost* for an item is the sum of all expenditures associated with the item during its entire service life. The term *item* should be interpreted in the general sense as a machine, a unit of equipment, a product line, a project, a building, a system, and so forth. Life-cycle costs may include engineering design and development costs, fabrication and testing costs, operating and maintenance costs, and disposal costs. Life-cycle costs may also be expressed as the summation of acquisition, operation, maintenance, and disposal costs. Thus, life-cycle cost terminology may vary from author to author, but the basic meaning of the term is clear.

This textbook is predominantly concerned with the economic justification of engineering projects, the replacement of existing projects or capital assets, and the economic comparison of alternative projects. For the purpose of these types of analyses, we will define life-cycle costs to consist of (1) first cost (or initial investment), (2) operating and maintenance costs, and (3) disposal costs. There is obviously a time period involved for the life cycle and, upon disposal, the item will have a salvage value.

The *first cost* of an item is considered the total initial investment required to get the item ready for service; such costs are usually nonrecurring during the life of the item. For the purchase of a numerically controlled machine tool, Steffy et al. [8] state that the first cost of the machine tool may consist of the following major elements: (1) the basic machine cost, (2) costs for training personnel, (3) shipping and installation costs, (4) initial tooling costs, and (5) supporting equipment costs. The installation costs may include, for example, the costs of preparing a foundation; vibration and noise insulation; providing heat, light, and power supply; and cost of testing. Supporting equipment costs may include computer hardware and software, and a spare parts inventory.

For some other item, a different set of first-cost elements may be appropriate. Some projects may include *working capital* for inventories, accounts receivable, and cash for wages, materials, and so forth. In any case, it is emphasized that the first cost of an item normally involves many more cost elements than just the basic purchase price. Whether the first-cost elements are aggregated or maintained separately depends on income tax considerations and whether or not a before-tax or after-tax economic analysis is desired. Certain income tax laws and depreciation methods are presented in Chapter Six, and further discussion on this particular point is therefore deferred.

*Operating* and *maintenance* costs are recurring costs that are necessary to operate and maintain an item during its useful life. Operating costs usually consist of labor, material, and overhead items. Depending upon the accounting system used by a firm, a wide range of cost factors may be included in the major cost classification of overhead. Typical overhead items are fuel or electric power, insurance premiums, inventory charges, indirect labor (as opposed to direct labor), administrative and management expenses, and so forth. It is usually assumed that *operating and maintenance costs* are annual costs, but maintenance costs may not be on a recurring, annual basis. That is, a regular, annual schedule of minor or preventive maintenance may be followed, or it could be policy that maintenance is performed only when necessary, such as when a major overhaul is required. In most cases, the maintenance policy would consist of both preventive maintenance and maintenance on an "as-needed" basis. In any case, repair and upkeep result in costs that must be recognized in the economic analysis of engineering projects.

When the life cycle of an item has ended, *disposal costs* usually result. Disposal costs may include labor and material costs for removal of the item, shipping costs, or special costs; an example of special costs being the cost of disposing of hazardous materials. Although disposal costs may be incurred at the end of the life cycle, most items have some monetary value at the time of disposal. This value is the *market* or *trade-in value* (i.e., the actual dollar worth for which the item may be sold at the time of disposal). Then, after deducting the cost of disposal, the net dollar worth at the time of disposal is termed the *salvage value*.

The market value, the disposal costs, and the salvage value are usually not known with certainty and therefore must be estimated. For an item that satisfies the United States Internal Revenue Service (IRS) definition of a capital asset and that decreases in value over time through physical deterioration, the IRS has approved various depreciation methods that can serve to estimate the rate of deterioration and consequent decrease in value of the asset. The value of the capital asset at the end of a given accounting period during the asset's life is termed the *book value*. The book value is thus an estimate of the market value; further discussion on depreciation accounting is deferred until Chapter Six. *Scrap value*, on the other hand, refers only to the value of the material of which the item is made. For example, a 4-year-old automobile may have a scrap value of $200 but a market value of $3500. A distinction between these terms is not particularly important for evaluating potential investment projects—and salvage value will be used as the general term to denote the end-of-life value. For example, a trade-in value of $3000 minus disposal costs of $500 equals a net salvage value of $2500 for some item.

The life cycle obviously involves a time horizon, and the end of an item's life may be judged from either a functional or an economic point of view. The economic life an an item is generally shorter than the functional life. For example, an engine lathe may remain functionally useful for 40 years or more, but, because of periodic advancements in machine design technology, newer engine lathes have higher production rates; the economically useful life of an engine lathe may only be 10 years. The economic life of an item is usually a matter of company policy that is greatly influenced by income tax considerations. In this textbook, the end of life for an item will mean the end of its economic life, rather than its functional life.

## 2.2.2 Past and Sunk Costs

*Past costs* are historical costs that have occurred for the item under consideration. *Sunk costs* are past costs that are unrecoverable. The distinction is perhaps best made through examples. Assume that an investor purchases 100 shares of common stock in the ABC Corporation through a broker at $25/share. In addition, the investor pays $85 in brokerage fees and other charges. Just 2 months later, and before receiving any dividend payments, the purchaser resells the 100 shares of common stock through the same broker at $35/share minus $105 for selling expenses. The purchaser realizes a net profit of $810 ($3500 − $2500 − $85 − $105) on these transactions. At the time of sale, the $2500 and $85 are past costs, but because these are recovered after the sales transaction, sunk costs are not incurred. If, on the other hand, the investor were to sell the 100 shares 2 months after purchase and the market price were $20/share, with a $70 charge for selling fees, the investor would incur a capital loss of $655 ($2000 − $2500 − $85 − $70). In this instance, some of the past costs would be recovered, but the $655 capital loss would be a sunk cost. If the investor reasons that the market price will decline further or if he or she simply needs the money, the $655 sunk cost should be ignored if the shares are to be sold for $20 each. However, sunk costs are not totally irrelevant to a present decision. They may qualify as capital losses and serve to offset capital gains or other taxable income and thus reduce income taxes paid. Examples in Chapter Six will illustrate this point. Also, past costs and sunk costs provide information that can improve the accuracy of estimating future costs for similar items.

Another example of sunk costs is the purchase and sale of an item of equipment. Assume the equipment is purchased for $10,000 and the salvage value at the end of 5 years of service is estimated to be $5000. For illustrative purposes, we will further assume that the annual decrease in value for the equipment through physical deterioration, or depreciation, is $1000. The $1000 annual cost of depreciation is a cost of production that, in theory, should be allocated to the output of the equipment. After allocating this and other manufacturing costs, general and administrative costs, and marketing costs to each unit of production, the total unit cost is determined. A profit is then added to each unit of production in order to arrive at the unit selling price. Thus, when a unit is sold, a portion of each sales dollar returns a portion of the depreciation expense. In this illustration, it is assumed that sales will return, or recover, the total estimated depreciation expense of $5000 (first cost minus estimated salvage value) for the 5-year period. However, if the equipment has a market value of only $2000 at the end of 5 years, there is a $3000 ($5000 − $2000) sunk cost. The $3000 capital loss represents an error in estimating the rate of depreciation, and the owner cannot insist that the equipment is worth $5000 when the market value for the 5-year-old equipment is, in fact, only $2000. If the equipment is kept, it is argued that the true value being kept is thus only $2000.

### 2.2.3 Future and Opportunity Costs

All costs that may occur in the future are termed *future costs*. These future costs may be operating costs for labor and materials, maintenance costs, over-haul costs, and disposal costs. In any case, by virtue of occurring in the future, these costs are rarely known with certainty and must therefore be estimated. This is, of course, also true for future revenues or savings if these are involved in a given project. Estimates of future costs or revenues are uncertain and subject to error. Thus, the economic analysis is simplified if certainty of future costs, revenues, or savings is assumed. This assumption is made until Chapter Eight, where concepts of probability are introduced.

The cost of forgoing the opportunity to earn interest, or a return, on invest-ment funds is termed an *opportunity cost*. This concept is best explained by means of illustrations. For example, if a person has $1000 and stores this cash in a home safe, the person is forgoing the opportunity to earn interest on the money by establishing a savings account in a local bank that pays, for example, 9% annual compound interest. (Of course, investments other than savings ac-counts are possible). For a 1-year period, the person is forgoing the opportunity to earn $(0.09)($1000) = $90$. The $90 amount is thus termed the opportunity cost associated with storing the $1000 in the home safe.

A similar illustration of an opportunity cost is to assume that a person has $5000 cash on hand. This amount is considered *equity capital* if the $5000 was not borrowed (i.e., there is no debt obligation involved). The person has avail-able secure investment opportunities such as establishing a personal savings account in a commercial bank or purchasing other financial instruments. From the available investment opportunities, suppose the optimum combination of risk (security level) and interest yield on the investment results in a 10% annual interest. Thus, the investment of $5000 would yield $(0.10)($5000) = $500$ each year. If the person, instead of investing the $5000, purchases an automobile for the same amount for personal use, the person will forgo the opportunity to earn $500 interest/year. The $500 amount is again termed an annual opportunity cost associated with purchasing the automobile.

The same logic applies in defining an annual opportunity cost for investments in business and engineering projects. The purchase of an item of production machinery with $20,000 of equity funds prevents this money from being in-vested elsewhere with greater security or higher profit potential. This concept of *opportunity cost* is fundamental to the study of engineering economy and is a cost element that is included in virtually all methodologies for comparing alter-native projects. In later chapters, the concept of opportunity costs will be discussed under the heading of *minimum attractive rate of return (MARR)*.

Some individuals define *MARR* as the *cost of capital*. As used in this text, the term *cost of capital* refers to the cost of obtaining funds for financing projects through debt obligations. These funds are usually obtained from external sources by (1) borrowing money from banks or other financial organizations (e.g., insurance companies and pension funds) and (2) issuing bonds. These debt obligations are normally long term, as opposed to short-term, obligations for the purchase of supplies and raw materials. The debt obligations result in interest payments on, say, a monthly, quarterly, semiannual, or annual basis. The interest payments are thus a cost of borrowed capital. Financing projects through issuing bonds is a method of obtaining capital funds that is probably less known to the reader than borrowing money from a bank at a stated interest rate. Some elaboration of bonds is therefore appropriate.

Bonds are issued by various organizational units—partnerships, corporations (profit or nonprofit), governmental units (municipal, state, and federal), or other legal entities. The sale of bonds represents a legal debt of the issuing organization; bonds are generally secured by its assets. Examples are mortgage bonds or collateral bonds. Debenture bonds, on the other hand, are promissory notes or just a promise to pay. In any case, the purchaser of a bond has legal claim to the assets of the issuing unit but has no ownership privileges in the issuing unit. Purchasers of the common or preferred stock of an organizational unit do hold ownership status but may or may not have voting privileges, depending on the stipulations of the particular stock issue. In the sense that bonds are debt obligations and not ownership shares, bonds are considered a more secure investment than either common or preferred stock. This statement should not be taken as a universal truth, however, since the security level for a bond or a stock depends on many factors, economic and otherwise; the principal factor is the financial soundness of the issuing unit. Further details on interest payments on bank loans and interest payments on issued bonds are covered in Chapters Three and Four, respectively.

Another method of financing engineering projects is through the use of *equity funds* (i.e., through ownership capital or cash that is debt free). The use of such funds incurs an opportunity cost, as mentioned previously, and will again be discussed in Chapter Four as the minimum attractive rate of return (*MARR*) for an investment project. However, the point is that the cost of capital is the cost of borrowed funds. Chapter Seven further discusses the cost of capital for financing public utility projects.

## 2.2.4 Direct, Indirect, and Overhead Costs

It will be helpful to provide definitions of direct, indirect, and overhead costs in the context of a manufacturing environment. A typical cost structure for manufacturing, adapted from Ostwald,[1] is provided in Figure 2.1.

The *cost of goods sold*, as shown in Figure 2.1, is the total cost of manufacturing a product. An amount of profit is then added to this total cost to arrive at

---

[1] Phillip F. Ostwald, *Cost Estimating for Engineering and Management*, Prentice-Hall, 1974, p. 55.

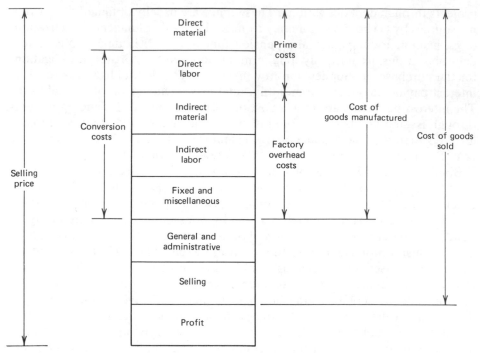

FIGURE 2.1. A cost structure for manufacturing.

a selling price. Such a cost structure is helpful in arriving at a unit cost, which is a primary objective of cost accounting. The term *cost of goods sold*, as used here, has a different meaning from the term *cost of goods sold* used in general accounting practice, particularly for retail businesses. As will be mentioned later in the section on accounting principles, general accounting defines the cost of goods sold to be beginning-of-the-year inventory plus purchases minus end-of-the-year inventory. Different meanings for the same terminology are unfortunate, but they do occur in the literature on accounting, and the reader is cautioned on this point. To simplify the treatment of the total cost of goods sold (as defined by Figure 2.1), the major cost elements can be defined as direct material, direct labor, and overhead costs.

*Direct material and labor costs* are the costs of material and labor that are easily measured and can be conveniently allocated to a specific operation, product, or project.

*Indirect costs for both labor and material*, on the other hand, are either very difficult or impossible to assign directly to a specific operation, product, or project. The expense of directly assigning such costs is prohibitive, and actual direct costs are therefore considered to be indirect for accounting purposes.

As an example of these different cost elements, suppose the raw material for a given part is a rectangular gray iron casting. The casting is milled on five sides, the unmachined surface is painted and air-dried, then four through holes are drilled and tapped. The finished parts are stacked in wooden boxes, 30/box, and are delivered to a customer.

In this example, the direct labor required per part to machine, paint, and package is probably readily determined. The labor required to receive the raw materials, handle parts between work stations, load boxes onto a truck, and deliver material to the customer is less easily identified and assigned to each part. This labor would be classified as indirect labor, especially if the labor in receiving, handling, shipping, and delivery is responsible for dealing with many different parts during the normal workday. The unit purchase price of the gray iron casting is an identifiable direct material cost. The cost of paint used per part may or may not be easily determined; if it is not, it is an example of indirect material cost. Also, any lubricating oils used during the machining processes would be an indirect material cost, not readily assigned on a cost per part basis.

*Overhead* costs consist of all costs of manufacturing other than direct material and direct labor. A given firm may identify different overhead categories such as factory overhead, general and administrative overhead, and marketing expenses. Furthermore, overhead amounts may be allocated to a total plant, departments within a plant, or even to a given item of equipment. Typical specific items of cost included in the general category of overhead are: indirect materials, indirect labor, taxes, insurance premiums, rent, maintenance and repairs, supervisory and administrative (technical, sales, and management) personnel, and utilities (water, electric power, etc.). Depreciation expenses are also usually included in the general overhead, but may occasionally be considered a part of direct costs. It is the task of cost accounting to assign a proportionate amount of these costs to various products manufactured or to services provided by a business organization.

## 2.2.5 Fixed and Variable Costs

*Fixed* costs do not vary in proportion to the quantity of output. General administrative expenses, taxes and insurance, rent, building and equipment depreciation, and utilities are examples of cost items that are usually invariant with production volume and hence are termed *fixed costs*. Such costs may be fixed only over a given range of production; they may then change and be fixed for another range of production. *Variable* costs vary in proportion to quantity of output. These costs are usually for direct material and direct labor.

Many cost items have both fixed and variable components. For example, a plant maintenance department may have a constant number of maintenance personnel at fixed salaries over a wide range of production output. However, the amount of maintenance work done and replacement parts required on equipment may vary in proportion to production output. Thus, total annual maintenance costs for a plant over several years would consist of both fixed and variable components. Indirect labor, equipment depreciation, and electrical power are other cost items that may consist of fixed and variable components. Determining the fixed and variable portion of such a cost item may not be possible; if it is possible, the expense of establishing detailed measurement techniques and accounting records may be prohibitive. A comprehensive discussion on this issue is outside the scope of this book, and the reader is referred to books on general cost accounting for further reading.

Certain total costs (TC), then, can be expressed as the sum of fixed (FC) and variable costs (VC). As an example, the total annual cost for operating a personal automobile for a given year might be expressed as

$$TC(x) = FC + VC(x)$$

where $x$ = miles per year. Costs for insurance, license tags, depreciation, certain maintenance, and interest on borrowed money if the automobile were financed are essentially fixed costs, independent of the miles traveled per year. Expenses for gasoline, oil, tire replacements, and certain maintenance are proportional to, or functional with, the mileage per year. One could argue, however, that depreciation expenses are comprised of both fixed and variable components. Arbitrarily assigning numerical values to the total cost function, assume that

$$TC(x) = \$950 + \$0.15x$$

is a valid relationship for a given year in question (the expression is restricted to a given year, since actual depreciation expenses, and hence the fixed expenses, may vary from year to year). This relationship is linear in terms of $x$, however, the variable cost component is often a nonlinear function. Figure 2.2 graphically illustrates the total cost function.

Now let us consider Figure 2.2 as a total cost function for a production line in a manufacturing firm where the output from the line is a single product. Furthermore, let it be assumed that each unit of production can be sold for $R and that the total revenue (TR) is a linear function of the production quantity:

$$TR(x) = \$Rx$$

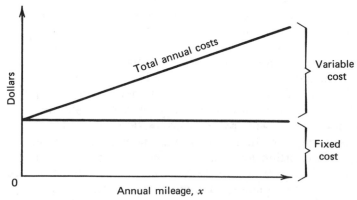

FIGURE 2.2. Total annual costs as a function of annual mileage.

Adding this functional relationship to Figure 2.2 and modifying the terminology for this example yields Figure 2.3.

It is noted from Figure 2.3 that the total annual revenue equals the total annual cost at point $A$ or at an annual production volume of $x^*$ units. Thus, at $x^*$,

$$TR(x^*) = TC(x^*)$$
$$= FC + VC(x^*)$$

and $x^*$ is termed the annual production volume required in order to *break even*. Certain important observations can now be made. If the production volume is less than $x^*$, an annual net loss will occur, the amount of which is equal to $TC(x) - TR(x)$, evaluated for a particular value of $x$. By the same token, if the production volume is greater than $x^*$, then an annual net revenue or profit will result (the shaded region of Figure 2.3). The amount of annual profit is equal to $TR(x) - TC(x)$, evaluated for a particular value of $x$.

It is generally desirable to have a "low" break-even value. For the general example of Figure 2.3, this can be accomplished in three independent ways: (1) increasing the slope of the total revenue line, (2) decreasing the slope of the variable cost line, and (3) decreasing the magnitude of the fixed cost line. Increasing the slope of the total revenue line means increasing the selling price of the product, which may be a poor marketing strategy in a competitive market environment where sales would be lost. Fixed costs, although not literally "fixed" in all cases, are difficult to reduce. Thus, reducing variable costs for direct material and labor usually offers the greatest opportunity to the engineer or analyst for profit improvement.

The concept of break-even analysis is general. Assuming that a break-even point exists, then, for two relationships $y = g(\cdot)$ and $y = h(\cdot)$ that are functions of a single variable $x$, the value of $x$ for break-even, say $x^*$, may be determined

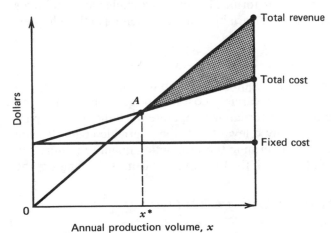

FIGURE 2.3. Relationship between total revenue and total cost as a function of annual production volume.

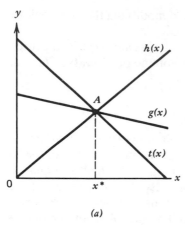

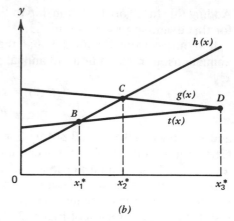

(a)                                              (b)

FIGURE 2.4. Graphical representation of break-even point. (a) A single break-even value. (b) Three break-even values.

from equating $g(x) = h(x)$ and solving for $x^*$. The concept can be extended to more than two functions of a single variable, say $y = h(x)$, $y = g(x)$, and $y = t(x)$. If these are all linear functions, then Figure 2.4 depicts two of the possible results.

In Figure 2.4b, there is no unique break-even value of $x$ involving all three functional relationships. The linear equations $y = h(x)$ and $y = t(x)$ intersect at point $B$ or $x = x_1^*$, which is then the break-even value for these two relationships. Point $C$, or $x = x_2^*$, is the break-even value for $y = h(x)$ and $y = g(x)$. Point $D$, or $x = x_3^*$, is the break-even value for $y = g(x)$ and $y = t(x)$.

The concept of break-even analyis also extends to nonlinear functions, with one or more break-even values, and functions of more than a single variable, which may be of linear or nonlinear form. However, examples and problems dealing only with functions of a single variable will be presented in this chapter.

### Example 2.1

The cost of tooling and direct labor required to set up for a machining job on a turret lathe is $300. Once set up, the variable cost to produce one finished unit consists of $2.50 for material and $1 for labor to operate the lathe. For simplicity, it is assumed these are the only relevant fixed and variable costs. If each finished unit can be sold for $5, determine the production quantity required to break even and the net profit (or loss) if the lot size is 1000 units. Letting $x =$ the production volume in units, then

$$TR(x) = TC(x) = FC + VC(x)$$

and

$$\$5x = \$300 + (\$2.50 + \$1.00)x$$

Solving yields $x = x^*$ (the break-even value), or

$$x^* = \$300/\$1.50 = 200 \text{ units}$$

For a production output of 1000 units, the net profit, $P$, is calculated to be

$$P = \underbrace{\$5(1000 \text{ units})}_{TR(x)} - \underbrace{(\$2.50 + \$1.00)(1000 \text{ units})}_{TC(x)} + \$300$$

$$= \$1200$$

---

## Example 2.2[2]

This example concerns the decision of selecting between two alternative methods of processing crude oil in a producing oil field, where the basis for the decision is the number of barrels of crude oil processed per year. The two methods of processing the crude oil are (1) a manually operated tank battery or (2) an automated tank battery. The tank batteries consist of heaters, treaters, storage tanks, and so forth, that remove salt water and sediment from crude oil prior to its entrance to pipelines for transport to an oil refinery.

For each alternative, fixed costs and variable costs are involved. Fixed costs include items such as pumper labor, maintenance (fixed over the production quantity of interest), taxes, certain energy costs (power to operate control panels and motors in continuous operation), and, for manual tank batteries, a cost for oil "shrinkage." Variable costs for chemical additives, heating, and noncontinuous operating motors are proportionate with the volume of oil being processed. This relationship is assumed to be linear over the production quantity of interest, since the data given in Tables 2.1 and 2.2 are based on a production quantity of 500 barrels/day.

In addition to the fixed and variable costs given in Table 2.1 for the automatic tank battery operation, other annual fixed costs are:

$D_1$ = annual cost of depreciation and interest

= \$3082

$M_1$ = annual cost of maintenance, taxes, and labor

= \$5485

Letting $x$ = the number of barrels of oil processed per year, the total annual cost, $TC_1(x)$, for the automatic tank battery operations is given by

$$TC_1(x) = FC_1 + VC_1(x)$$

$$= (\$982 + \$3082 + \$5485)$$

$$+ \$0.01136x$$

$$= \$9549 + \$0.01136x$$

[2] The example is taken, with slight modification, from Ferguson and Shamblin [3] by permission of the publisher.

TABLE 2.1. Cost Data for Automatic Tank
Battery Operations

| | |
|---|---|
| Fixed cost | |
|   Control panel power | $0.15/day |
|   Circulating pump power (3 horsepower) | 0.82/day |
|   Maintenance | 1.00/day |
|   Meter calibration | 0.40/day |
|   Chemical pump power (1/4 horsepower) | 0.32/day |
|     Total | $2.69/day |
|     or $982/year ($2.69/day × 365 days) | |
| Variable cost | |
|   Pipeline pump (5 horsepower @ 50% utilization) | $0.63/day |
|   Chemical additives (7.5 quarts/day) | 3.75/day |
|   Inhibitor (2 quarts/day) | 1.00/day |
|   Gas (10.8 MCF/day × $0.0275/MCF) | 0.30/day |
|     Total | $5.68/day |
|     or $0.01136/barrel based on 500 barrels/day | |

In addition to the fixed and variable costs given in Table 2.2 for the manual tank battery operations, other annual fixed costs are:

$$D_2 = \text{annual cost of depreciation and interest}$$
$$= \$2017$$

$$M_2 = \text{annual cost of maintenance, taxes, and labor}$$
$$= \$7921$$

Then, the total annual cost, $TC_2(x)$, for the manual tank battery operation is given by

$$TC_2(x) = FC_2 + VC_2(x)$$
$$= (\$358 + \$2017 + \$7921)$$
$$+ \$0.00810x$$
$$= \$10{,}296 + \$0.00810x$$

By equating the two total cost functions, the break-even production volume can be determined as

$$TC_1(x) = TC_2(x)$$

$$\$9549 + \$0.01136x = \$10{,}296 + \$0.00810x$$

$$x^* = 229{,}141 \text{ barrels/year}$$

The interpretation of the break-even point in this example is that $x^* = 229{,}141$ barrels/year is the *point of indifference* between the choice of the two alternatives. If production volume is less than $x^*$, then the first alternative, or the automatic tank battery operation, would be preferred. For instance, if $x =$

TABLE 2.2. Cost Data for Manual Tank Battery Operation

| | |
|---|---|
| Fixed cost | |
| Chemical pump power | $0.16/day |
| Circulating pump power | 0.82/day |
| Total | $0.98/day |
| or $358/year ($0.98/day × 365 days) | |
| Variable cost | |
| Chemical additives (7.5 quarts/day) | $375/day |
| Gas | 0.30/day |
| Total | $4.05/day |
| or $0.00810/barrel based on 500 barrels/day | |

200,000 barrels/year, then $TC_1(x) = \$11{,}821$ and $TC_2(x) = \$11{,}916$. Similarly, if production volume is greater than $x^*$, the manual tank battery operation would be preferred.

## 2.2.6 Average and Marginal Cost

The *average cost* of one unit of output (unit cost) is the ratio of total cost and quantity of output (miles traveled, production volume, etc.). That is,

$$AC(x) = \frac{TC(x)}{x}$$

where $AC(x)$ = average cost per unit of $x$
$\quad TC(x)$ = total cost for $x$ units of output
$\quad\quad x$ = output quantity

The average cost is usually a variable function of the output quantity, and normally decreases with an increasing quantity of output. Using the automobile example from Section 2.2.5, which had a total cost function of $\$(950 + 0.15x)$, the average cost, in dollars per mile, is given by

$$AC(x) = \frac{950 + 0.15x}{x}$$

$$= \frac{950}{x} + 0.15$$

If the automobile travels 10,000 miles/year, then the average operating costs is $\$(950/10000 + 0.15) = \$0.245$/mile. For a total annual travel distance of 20,000 miles, the unit operating cost decreases to $0.1975/mile.

It should be noted from the average cost function of this example, $\$(950/x + 0.15)$, that as the output quantity $x$ increases, the proportion of the fixed cost allocated to each unit of output decreases. This relationship is a fundamental economic principle often referred to as the "economies of scale," a principle underlying the economic benefits of mass production. Such a relationship assumes, however, that the variable cost coefficient remains constant over the range of the output variable $x$. In a production environment, it is very likely that the variable cost coefficient will increase as the production volume increases (due to increased maintenance expenses, defective product, etc.).

For a total cost function that is continuous in the output (or independent) variable $x$, marginal cost is defined as the derivative of the total cost function (dependent variable) with respect to $x$, or $dTC(x)/dx$. This is true for continuous functions that are linear or nonlinear in the output variable $x$. In the special case of a continuous total cost function that is linear in $x$, such as $TC(x) = \$(950 + 0.15x)$, then $dTC(x)/dx = \$0.15$. In this case, marginal cost is the constant value $\$0.15$, and is the cost required to increase the output quantity $x$ by one unit.

If the total cost function is discontinuous and defined only for discrete values of $x$ (for example, $x = 1, 2, 3, . . .$), then difference equations must be used to determine marginal costs. For example, $TC(6) - TC(5)$ is the marginal cost of increasing the output quantity from $x = 5$ to $x = 6$. Thus, in the discrete case, marginal cost is always the cost required to increase the output quantity $x$ by one unit at a specified level of output. This is true for a discrete total cost function which has either a linear or nonlinear trend.

The concept of marginalism is general and applies to other mathematical functions as well. For example, marginal revenues can be determined from total revenue functions, marginal profit values can be determined from total profit functions, and so forth. If these functions are defined for discrete values of $x$ or are continuous functions that are linear in $x$, then marginal revenue (profit) is the additional revenue (profit) received from selling one more unit of the output quantity $x$, at a specified level of output.

Marginal and average values corresponding to a specified output quantity are generally different. If the marginal cost is smaller than the average cost, an increase in output will result in a reduction of unit cost. This can be seen by recalling the familiar automobile problem, when $TC(x) = \$950 + \$0.15x$. The average cost is $AC(x) = \$950/x + \$0.15$ and the marginal cost is $MC(x) = \$0.15$. Thus, for all nonnegative finite values of $x$, marginal cost is always smaller than the average cost and the unit cost will continue to decrease as $x$ is increased. Such a relationship is not true in general for nonlinear total cost functions. The following example will illustrate some of the cost, revenue, and profit relationships.

## Example 2.3

A small firm blends and bags chemicals, primarily for home gardening purposes. The market area for the firm is local and all sales are to wholesale distributors. For one pesticide dust product, sales and production cost records over the past 10 seasons have been reviewed and analyzed. The following equations *approximate* the relationships among selling price, sales volume, production costs, and profit before income taxes. (The functional form for selling price implicitly reflects a fundamental relationship between price and demand. Namely, as the selling price is decreased, demand for the item increases. Alternatively, in order to increase the demand, the selling price must be reduced).

Let $t$ = number of tons per season

$SP(t)$ = selling price in order to sell $t$ tons

$\quad = \$(800 - 0.8t)$

$TR(t)$ = total revenue when $t$ tons are sold at a particular selling price

$\quad = $ Selling price $\times$ Demand

$\quad = \$(800 - 0.8t)t$

$\quad = \$(800t - 0.8t^2)$

$MR(t)$ = the marginal revenue at a sales volume of $t$ tons

$\quad = \dfrac{dTR(t)}{dt} = \$(800 - 1.6t)$

$TC(t)$ = the total production cost for $t$ tons

$\quad = \$(10{,}000 + 400t)$

$TP(t)$ = total profit when $t$ tons are sold

$\quad = TR(t) - TC(t)$

$\quad = \$(800t - 0.8t^2) - \$(10{,}000 + 400t)$

$\quad = \$(-0.8t^2 + 400t - 10{,}000)$

$AP(t)$ = average profit per ton when $t$ tons are sold

$\quad = TP(t)/t$

$\quad = \$(-0.8t + 400 - 10{,}000/t)$

The equations apply for the range, $0 \le t < 1000$.

We first note that the total revenue will be maximized when 500 tons are produced and sold per season. That is, by calculus,

$$\frac{dTR(t)}{dt} = 800 - 2(0.8)t = 0 \quad \text{and} \quad t = 500 \text{ tons}$$

The total revenues with a sales volume of 500 tons is

$$TR(500) = \$800(500) - \$0.8(500)^2$$
$$= \$200{,}000$$

The marginal revenue at the output level of 500 tons is

$$MR(500) = \$800 - \$1.6(500) = 0$$

Thus, the rate of change in the $TR(t)$ function with respect to $t$ is zero when $TR(t)$ is evaluated at $t = 500$. The $TR(t)$ function is a strictly concave function with a unique maximum value at $TR(500)$. For sales from $t = 1$ to 500, the total revenue function is increasing at a decreasing rate. For sales volumes from $t = 500$ to 1000, total revenues are decreasing at an increasing rate.

Maximizing total revenues is not the issue in this example however. We wish to maximize profits. Again, by calculus,

$$\frac{dTP(t)}{dt} = 2(-0.8)t + 400 = 0 \quad \text{and} \quad t = 250 \text{ tons}$$

Thus, total profit will be maximized for a sales volume of 250 tons, and the maximum profit per season would be

$$TP(250) = -\$0.8(250)^2 + \$400(250) - \$10,000$$
$$= \$40,000$$

The average profit per ton when 250 tons are sold is

$$AP(250) = -0.8(250) + 400 - 10,000/250$$
$$= \$160/\text{ton}$$

Finally, we note that there are two break-even points in this example. By equating $TR(t) = TC(t)$

$$800t - 0.8t^2 = 10,000 + 400t$$

which yields

$$-0.8t^2 + 400t - 10,000 = 0$$

Solving for the positive roots of this quadratic equation yields

$$t = 26.39, \ 473.61$$

For a sales volume in the range, $26.39 \le t \le 473.61$, the firm will make a profit. Sale volumes outside this range will result in total costs exceeding total revenues and a net loss to the firm.

## 2.3 ESTIMATION

The estimation of future events or the outcomes of present actions taken is obviously a fact of life for every individual, group, and organization. Family budgeting, weather forecasts, market forecasts of demand for consumer products, and predicting the annual national revenues from income taxation are only a few examples of the almost limitless number and variety of estimates that are made in personal and business lives. In this textbook, we are concerned with estimation in the specific context of factors relevant to comparing alternative engineering/investment projects and making a selection from these projects. The annual revenues or savings, the initial and annual recurring costs, the life of a project, and the future salvage value of capital assets such as buildings and equipment that may be associated with a given project are rarely, if ever, known with certainty.

Many different terms pertain to the general subject of estimation. No attempt will be made in this text to enumerate and explain all the terms exhaustively; selected terminology will be given throughout the book as needed to explain the topics contained in a given chapter or section. Furthermore, an in-depth study of estimation procedures and the accuracy of estimated values is the study of mathematical statistics and probability theory, about which a vast literature exists.

Chapters Three to Seven are concerned with comparing alternative projects when the estimated values for relevant factors are single-valued or point estimates. In such cases, the single-valued estimates are considered certain to occur. Moving a step toward realism, an interval estimate could be made for the value of a given factor such as annual costs. That is, a high and low value or range might be estimated for the factor. Chapter Eight discusses factors that contribute to uncertainty in estimated values and considers deviations from single-valued or point estimates. Chapter Nine continues the discussion on decision making with the topics of risk and uncertainty, defining these terms and posing methodology for selecting a feasible alternative.

Although it is difficult to state precisely in quantitative terms, there is a relationship between the accuracy of an estimate and the cost of making the estimate. Intuitively, as more detailed information is obtained to provide the basis for an estimate and as more mathematical preciseness is exercised in calculating the estimate, the more accurate the estimate should be. However, as the level of detail increases, the greater the cost involved in making the estimate. Ostwald[3] has conceptualized this notion by the function

$$C_T = C(M) + C(E)$$

where $C_T$ = the total cost of making the estimate, dollars
$C(M)$ = the functional cost of making the estimate, dollars
$C(E)$ = the functional cost of errors in the estimate, dollars

As depicted in Figure 2.5, the total cost of making an estimate reaches a minimum value when the amount of detail reaches a value $D_1$. Quantitatively defining the amount of detail is at best difficult and may be a practical impossibility. However, this concept of the total cost of an estimate varying with the amount of detail involved in making the estimate is realistic and is important to the general subject of estimation. In the abbreviated discussion on cost estimation techniques that follows, it will be noted that the individual techniques are based on varying amounts of detail with implied differences in the cost of making the estimate.

The text by Ostwald [7] is a primary reference on the subject of cost estimating. He defines four categories of estimated items: operations, products, projects, and systems. The categories are based on the scope of activity involved, but the distinction among them is not sharp and unequivocal. An *operation* is

[3] Op. cit., p. 455.

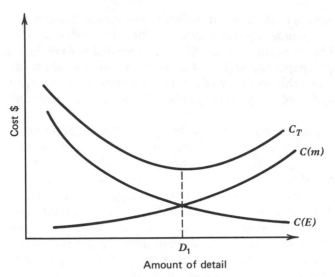

FIGURE 2.5. Cost of increasing detail.

considered a basic or lowest-level activity. Machining, assembly, and painting are examples of operations for cost estimation purposes. Therefore, operations are subelements of manufacturing a product or providing a service. Producing a given product or providing a given service may be a subelement of a project, and a system may consist of several projects. A system may, for example, be defined as a total manufacturing plant (or several plants), a regional health care program of facilities and personnel, or a total program of national defense.

Generally, the larger the scope of the categorical item being estimated, the more difficult it is to make accurate estimates of the costs and revenues involved. Ostwald presents a comprehensive discussion of estimating techniques for all categories, and Vinton [9] provides procedural detail for estimating the cost of certain manufacturing operations and manufactured products.

## 2.3.1 Project Estimation

For all categories of estimated items previously mentioned, three principal classes of estimates, based on accuracy and degree of detail, may be defined: (1) order-of-magnitude estimates, (2) preliminary estimates, and (3) detailed estimates. *Order-of-magnitude estimates* are usually gross estimates based on experience and judgment and made without formal examination of the details involved. Accuracies of ±50%, relative to actual cost, can be expected.

*Preliminary estimates* are also gross estimates, but more consideration is given to detail in making the estimate than for order-of-magnitude estimates. Certain subelements of the overall task are individually estimated, engineering specifications are considered, and so forth. Estimating the cost for manufacturing new components or products before designs and production plans are complete is an example of preliminary cost estimation. This type of estimate enables process engineering and product engineering groups to compare alternative designs or manufacturing methods and assess the economic impact of them. An accuracy level of about ±20% of actual cost can be expected with preliminary estimates.

*Detailed estimates* are expected to result in an accuracy level of ±5% of actual cost. In preparing the estimate, each subelement of the overall task is considered, and an attempt is made to assign a realistic cost to the subelement. Pricing a product or contract bidding usually involves detailed estimates of the costs involved.

Estimating the cost of a project may involve all of the above types of estimates. The time available to make the estimate, the information available, the experience and knowledge of the persons making the estimate, and the type and scope of the project are all factors that affect the desired accuracy of the project's cost estimate and the cost of making the estimate. When a comparison of alternative projects is based on future cash flows, the cost and revenue figures are necessarily estimates. These estimates will normally vary from actual cost and revenue figures.

In subsequent chapters, examples illustrating the principles of engineering economic analysis will primarily involve the following estimated items: first cost, operating costs, maintenance costs, recurring revenues or savings, useful life, salvage value, income taxes and cost of capital. These cost items have been discussed previously in this chapter under the section on cost terminology.

Estimates of the functionally useful physical life of an item of equipment may be obtained from manufacturers and suppliers. Alternatively, if a company repeatedly buys a particular item of equipment and records its life, these records may be used for obtaining an estimate of the functionally useful life of the item. For example, records may reveal that 10% of the items survive for 4 years, 20% survive for 5 years, 50% survive for 6 years, 15% survive for 7 years, and the remaining 5% survive for 8 years. The mean, or expected, life for any given item of equipment is thus equal to [(0.10)(4) + (0.20)(5) + (0.50)(6) + (0.15)(7) + (0.05)(8)] or 6.15 years.

In making an economic comparison of alternative projects, the analyst may only be interested in a particular period of time or planning horizon. This planning horizon may be (and often is) different from the functionally useful life of one or more of the projects under consideration. For example, a job contract may cover a 5-year period when the functionally useful lives of the projects under consideration are 10 years. The planning horizon adopted might then be 5 years, and only the relevant cost and revenue figures for the 5-year period are used in the comparison. Further discussion on this point is deferred until Chapter Five.

## 2.3.2 General Sources of Data

There are many sources for providing data to make the various estimates required in comparing alternative investment projects. Morris [5] devotes a chapter to the general subject of gathering data for economic analysis purposes and cites both internal and external sources of information. Examples of sources within a firm are sales records, production control records, inventory records, quality control records, purchasing department records, work measurement and other industrial engineering studies, maintenance records, and personnel records. Some, all, or additional records are input into the accounting function of the firm, which then compiles various financial reports for management. A principal objective of the accounting system is to determine per-unit costs for direct materials, direct labor, and overhead involved in manufacturing a product or providing a service. Since overhead costs, by definition, cannot be allocated as direct charges to a given product or service, they are prorated among the various products or services by somewhat arbitrary methods, some of which are presented later in the chapter. Thus, a caution is raised concerning the use of cost accounting data for estimating the cost of overhead items associated with a project. Nevertheless, the accounting system can and usually does serve as an important, if not primary, internal source of detailed estimates on operating costs, maintenance costs, and material costs, among others.

Sources of data external to the firm may be grouped into two general classes: published information that is generally available, and information (published or otherwise) available on request. Available published information includes the vast literature of trade journals, professional society journals, U.S. government publications, reference handbooks, other books, and technical directories. Information not generally available except by request includes many sources listed in the previous category. For instance, many professional societies and trade associations publish handbooks, other books, special reports, and research bulletins that are available on request. Manufacturers of equipment and distributors of equipment are excellent sources of technical data, and most will readily supply this information without charge. Additionally, various government agencies, commercial banks (particularly holding companies involved in leasing buildings and equipment), and research organizations (commercial, governmental, industrial, and educational) may be sources of data to aid the estimating process.

## 2.4 GENERAL ACCOUNTING PRINCIPLES

As already mentioned, the engineer should have some understanding of basic accounting practice and cost accounting techniques in order to obtain data from the firm's accounting system. If accounting is classified into general accounting and cost accounting, then cost accounting is judged more important to the engineer as a source of data for making cost estimates pertinent to engineering projects. Cost accounting will therefore receive the greater emphasis in this text; in either case, the treatment of accounting is cursory and is directed toward fundamental accounting concepts instead of comprehensive accounting detail.

According to Niven and Ohman [6], an accounting system is concerned with the four primary functions of *recording, classifying, summarizing,* and *interpreting* the financial data of an organization, whether nonprofit or a business organization. The discussion to follow assumes a business organization, and, in virtually all such organizations, general accounting information is summarized in at least two basic financial reports:

1. A *balance sheet,* or statement of financial conditions, provides a summary listing of the assets, liabilities, and net worth accounts of the firm as of a particular date.

2. A *profit and loss statement* (or *income statement*) shows the revenues and expenses incurred by the firm during a stated period of time—a month, quarter, or year.

## 2.4.1  Balance Sheet

Records of financial transactions and a variety of internal reports serve as inputs to the accounting system, providing information on sales and other revenues and the expenses incurred in obtaining the revenues. Revenues and expenses for a specified period are then summarized on the profit and loss statement. The net profit or loss resulting for this period is then transferred to the net worth section of the balance sheet. Thus, although the two basic financial reports provide different financial pictures of the firm, they are directly related. Before illustrating this fact, discussion of certain terminology is necessary.

The items listed on a balance sheet are usually classified into three main groups: assets, liabilities, and net worth items. Subgroups may also be identified, such as current and fixed assets or liabilities. *Assets* are properties owned by the firm, and *liabilities* are debts owed by the firm against these assets. The dollar difference between assets and liabilities is the *net worth* of the business, which measures the investment made by the owners or stockholders of the business plus any accumulated profits left in the business by the owners or stockholders. Another term for net worth is *owners' equity.* A fundamental accounting equation is thus defined:

$$\text{Assets} - \text{Liabilities} = \text{Net Worth}$$

Rewriting, we have

$$\text{Assets} = \text{Liabilities} + \text{Net Worth}$$

and the usual format of a balance sheet follows the equation in this form.

*Current assets* include cash and other assets that can be readily converted into cash; an arbitrary period of one year is usually assumed as a criterion for conversion. Similarly, *current liabilities* are the debts that are due and payable within 1 year from the date of the balance sheet in question. *Fixed assets,* then, are the properties owned by the firm that are not readily converted into cash within a 1-year period, and *fixed liabilities* are long-term debts due and payable after 1 year from the date of the balance sheet. Typical current-asset items are cash, accounts receivable, notes receivable, raw material inventory, work in process, finished goods inventory, and prepaid expenses. Fixed-asset items are land, buildings, equipment, furniture, and fixtures. Items that are typically listed under current liabilities are accounts payable, notes payable, interest payable, taxes payable, prepaid income, and dividends payable. Fixed liabilities may also be notes payable, bonds payable, mortgages payable, and so forth. Net worth items appearing on a balance sheet are less standard and, to a degree, depend on whether the business is a sole proprietorship, a partnership, or a corporation. The size of the corporation is also an influencing factor on item designation. However, items such as capital stock, retained earnings, capital surplus, or earned surplus appear under the net worth group. An example of a balance sheet for a hypothesized firm is exhibited in Table 2.3.

The sample balance sheet in Table 2.3 is "balanced" since assets = liabilities + net worth. It gives a statement of the financial condition of the organization as of a specific date—the close of an accounting period. Although depreciation will be covered in Chapter Six, note the fixed asset portion of the balance sheet. For example, the building originally cost $200,000, and depreciation expenses have been charged annually, so that the total depreciation charges (as of the date of the balance sheet) have been $50,000, the amount entered as *depreciation reserve* for the building. In theory, the depreciation reserve is an accumulated amount of funds held to repurchase the asset when its functionally useful life terminates. The first cost of the depreciable asset (building, in this case) minus the amount in the depreciation reserve equals the *book value*. If the book value were a true estimate of the salvage, or market, value, then the sum of the amount in the depreciation reserve plus the book value provides the firm with an amount of funds equal to the original purchase price. This sum can be applied toward the purchase of a replacement asset. It is the usual case, however, that the market value of the asset when sold differs from the book value. The difference results in either a capital gain or loss and affects the firm's income taxes. (This point will be covered more fully in Chapter Six.)

A similar explanation applies to the fixed asset of equipment. In this particular balance sheet, the equipment account is an aggregate for all the equipment owned by the company instead of an individual listing of each equipment item. There could be separate equipment accounts, grouped according to equipment class. In any case, a company normally keeps individual records on equipment items, which are then summarized on the balance sheet.

## 2.4.2 Income Statement

The second basic financial report compiled by the accounting system is the *income statement* or *profit and loss* statement. For the current accounting

TABLE 2.3. Sample Balance Sheet

Jax Tool and Engineering Company, Inc.
Balance Sheet
December 31, 19—

*Assets*

| | | |
|---|---|---|
| Current assets | | |
| Cash | $ 25,000 | |
| Accounts receivable | 115,000 | |
| Raw materials | 8,500 | |
| Work in process | 7,000 | |
| Finished goods inventory | 3,000 | |
| Small tool inventory | 12,500 | |
| Total current assets | | $171,000 |
| | | |
| Fixed assets | | |
| Land | | $ 30,000 |
| Building | $200,000 | |
| Less: Depreciation Reserve | 50,000 | 150,000 |
| | | |
| Equipment | $750,000 | |
| Less: Depreciation Reserve | 150,000 | 600,000 |
| | | |
| Office equipment | 10,000 | |
| Total fixed assets | | $790,000 |
| Total assets | | $961,000 |

*Liabilities and Net Worth*

| | | |
|---|---|---|
| Current liabilities | | |
| Accounts payable | $ 32,000 | |
| Taxes payable | 15,000 | |
| Total current liabilities | | $ 47,000 |
| | | |
| Fixed liabilities | | |
| Mortgage loan payable | $130,000 | |
| Equipment loan payable | 350,000 | |
| Total fixed liabilities | | $480,000 |
| Total liabilities | | $527,000 |
| | | |
| Common stock | $325,000 | |
| Retained earnings | 80,000 | |
| Earned surplus (for current year) | 29,000 | |
| | | |
| Total equity | | $434,000 |
| Total liabilities and equity | | $961,000 |

period, the income statement provides management with (1) a summary of the revenues received, (2) a summary of the expenses incurred to obtain the revenues, and (3) the profit or loss resulting from business operations. The format of the income statement varies, and the revenue and expense items depend on the type of business involved. Thus, the income statement given in Table 2.4 is illustrative only and still concerns our hypothesized Jax Tool and Engineering Company. Let us further assume that the period of time covered is 1 year, which has ended as of the date of the balance sheet given in Table 2.3.

The format for the income statement in Table 2.4 is an oversimplification of such a statement for even a small manufacturing firm. For example, the *cost of goods sold* entry may be considerably more detailed to reflect multiple product lines, the depreciation items may be detailed to a greater extent to reflect multiple classes of assets, and several common expense items—such as employee benefits contributions, insurance premiums, and advertising—are not included.

Another version of the income statement is shown in Table 2.5. This format is typical for a retail business, and the major difference in form concerns the method of determining the cost of goods sold item. Incidentally, this format basically follows that of Schedule C of IRS Form 1040 for small businesses.

---

TABLE 2.4. Sample Income Statement

Jax Tool and Engineering Co., Inc.
Income Statement
Year Ended December 31, 19___

| | | |
|---|---|---|
| Sales | | $1,200,000 |
| Less cost of goods manufactured: | | |
| Direct labor | $420,000 | |
| Direct materials | 302,000 | |
| Indirect labor | 112,000 | |
| Depreciation | 98,000 | |
| Repairs and maintenance | 41,500 | |
| Utilities | 11,500 | 985,000 |
| Gross profit | | $ 215,000 |
| | | |
| Less other expenses: | | |
| Administration | $ 76,000 | |
| Marketing | 49,000 | |
| Interest payments | 35,000 | 160,000 |
| Net profit before taxes | | $ 55,000 |
| | | |
| Less income taxes | | 26,000 |
| Net profit | | $ 29,000 |

---

TABLE 2.5. Sample Income Statement

Jax Tool and Engineering Co., Inc.
Income Statement
Year Ended December 31, 19__

| | | |
|---|---:|---:|
| Sales | | $1,200,000 |
| Less cost of goods sold: | | |
| Inventory, January 1, 19__ | $ 26,000 | |
| Plus purchases | 432,000 | |
| | $458,000 | |
| Less inventory, December 31, 19__ | 44,000 | 414,000 |
| Gross profit | | $ 786,000 |
| | | |
| Less expenses: | | |
| Direct labor | $420,000 | |
| Depreciation—building | 10,000 | |
| Depreciation—equipment | 30,000 | |
| Repairs and maintenance | 41,500 | |
| Indirect labor | 218,000 | |
| Utilities | 9,800 | |
| Supplies, tooling | 1,700 | $ 731,000 |
| Net profit before income taxes | | $ 55,000 |
| Less income taxes | | 26,000 |
| Net profit (posted to earned surplus) | | $ 29,000 |

## 2.5 COST ACCOUNTING PRINCIPLES

### 2.5.1 Cost Allocation Methods

The balance sheet and the income statement in Section 2.4 are considerably removed both in time and in detail from decisions at the usual engineering project level. More important to the engineer as a source of cost information is the cost accounting system within a particular firm. The firm may be involved in manufacturing or providing services, and if it is involved in manufacturing, production may be on a job-shop or process basis. There are some fundamental differences in cost accounting procedures for determining manufacturing costs versus determining the cost of providing a service; also, there are differences in accounting procedures if manufacturing is on a job-shop or process basis. In order to concentrate on basic principles instead of details, the cost accounting system assumed will be that of a job-shop manufacturing firm. Thus, the emphasis will be on determining the per-order costs for a job order.

The total cost of producing any job order consists of direct material, direct labor, and overhead costs. An additional item of cost could be special tooling or equipment purchases strictly for the job order in question. This definition of total cost does not detail the overhead cost into factory overhead, general overhead, and marketing expenses in order to simplify the presentation. Materials for a given job order may include purchased parts and in-house fabricated parts, and the cost for direct materials is determined primarily from purchase invoices. Questions of scrap allowances and averaging material costs, which fluctuate over time, present problems in obtaining accurate direct material costs, but determining such costs is reasonably straightforward. Direct labor time spent on a job order is normally recorded by operators on labor time cards, and the direct labor cost is determined by applying the appropriate labor cost rates. The labor rates, as determined by the accounting system, will normally include the cost of employee fringe benefits in addition to the basic hourly rate. Although accurately determining the direct labor cost for a given job is a major accounting problem, it is more readily determined than the overhead cost.

Overhead costs cannot be allocated as direct charges to any single job order and must, therefore, be prorated among all the job orders on some arbitrary basis. Common methods for distribution are:

**1.** The rate per direct labor hour.

**2.** A percentage of direct labor cost.

**3.** A percentage of prime cost (direct material plus direct labor cost).

If a single overhead rate for the entire manufacturing firm is to be used, this would, of course, be an average overhead rate for the entire factory, assuming the expense of providing and using an hour of factory facilities is about the same throughout the factory. To illustrate the three methods listed, it is assumed that a company experienced the following costs in the previous year 19__.

| | |
|---|---|
| Total direct labor hours | 48,000 |
| Total direct labor cost | $480,000 |
| Total direct material cost | $600,000 |
| Total Overhead cost | $360,000 |

Then, the *overhead rate per direct labor hour* would be

$$\text{Rate} = \frac{\text{Overhead cost}}{\text{Direct labor hours}}$$

$$= \frac{\$360,000}{48,000}$$

$$= \$7.50/\text{direct labor hour}$$

If a particular job order requires 40 hours of direct labor with an average rate of $12.50/hour and $850 of direct materials, then the total cost of the job would be computed as

| | | |
|---|---|---|
| Direct material cost | | = $ 850 |
| Direct labor cost = 40 hours × $12.50/hour | = | 500 |
| Overhead cost = 40 hours × $ 7.50/hour | = | 300 |
| Total cost | | $1650 |

Determining the overhead rate as a *percentage of direct labor cost* for this company will yield

$$\% = \frac{\text{Overhead cost}}{\text{Direct labor cost}} (100\%)$$

$$= \frac{\$360,000}{\$480,000} (100\%)$$

$$= 75\%$$

For the same job above, the total cost would be computed as

| | | |
|---|---|---|
| Direct material cost | = | $ 850 |
| Direct labor cost | = | 500 |
| Overhead cost = (0.75)($500) | = | 375 |
| Total cost | | $1725 |

Determining the overhead rate as a *percentage of prime cost* for this company will yield

$$\% = \frac{\text{Overhead cost}}{\text{Direct labor cost} + \text{Direct material cost}} (100\%)$$

$$= \frac{\$360,000}{\$1,080,000} (100\%)$$

$$= 33\tfrac{1}{3}\%$$

For the same job above, the total cost would be computed as

| | | |
|---|---|---|
| Direct material cost | = | $ 850 |
| Direct labor cost | = | 500 |
| Overhead cost = (0.333)($1350) | = | 450 |
| Total cost | | $1800 |

Determining the overhead cost for a job order by the *rate per direct labor hour* method will yield the same result as the *percentage of direct labor cost* method, provided that the rate per direct labor hour used on the job in question is equal to the average factory labor rate. The "percentage of prime cost" method will necessarily yield a different assignment of overhead to a job order than the other two methods. The choice among these three methods is arbitrary; indeed, other methods are used by cost accountants in distributing overhead costs to a given job order. The *rate per direct labor hour* method is perhaps most commonly used.

Whatever method is chosen from the above for distributing overhead costs to job orders in a current year, the rates or percentages are based on the previous year's cost figures. Thus, overhead rates may change from year to year within a particular firm.

Since an average overhead rate for the entire factory may very well be too gross an estimate when actual overhead costs differ among departments within the factory, cost accounting may determine individual overhead rates for departments or cost centers. Furthermore, the hourly rates for direct labor may vary among these cost centers. A further refinement is to determine overhead rates for individual machines within cost centers. Then, as particular job orders progress through cost centers (departments and/or machines), the direct labor time (or machine time) spent on the job order in the various cost centers is recorded, the appropriate labor or machine rates and overhead rates are applied, and the total cost for the job is calculated.

The following example illustrates the variety of methods used to distribute overhead to a given cost center; the total overhead for the cost center will then be distributed by yet another method.

Example 2.4_____

The following information has been accumulated for the Deetco Company's two departments during the past year. (See Table 2.6.)

The Deetco Company distributes depreciation overhead based on (1) the first cost of equipment in each department, (2) a zero salvage value of the equipment in 10 years, and (3) a constant annual (or straight-line) rate of depreciation. All overhead other than depreciation is first distributed to each department according to the number of employees in each department, and an overhead rate per direct labor hour is computed.

What selling price should the company quote on Job Order D if raw material costs are estimated as $900, estimated direct labor hours required in Departments A and B are 30 hours and 100 hours, respectively, and profit is to be calculated as 25% of selling price?

For Department A, the total overhead allocation is determined as follows:

$$
\begin{aligned}
\text{Annual depreciation} && &= \$\ 25,000 \\
\text{Other factory overhead} &= (14/23)(\$150,000) &= &\quad 91,304 \\
\text{General overhead} &= (14/23)(\$350,000) &= &\quad \underline{213,043} \\
\text{Total overhead costs} && &\quad \$329,347
\end{aligned}
$$

TABLE 2.6. Data for Example 2.4

|  | Department A | Department B | Total |
|---|---|---|---|
| Direct materials cost | $720,000 | $240,000 | $960,000 |
| Direct labor cost | $260,000 | $140,000 | $400,000 |
| Direct labor hours | 25,200 | 16,200 | 41,400 |
| Number of employees | 14 | 9 | 23 |
| First cost of equipment | $250,000 | $200,000 | $450,000 |
| Annual depreciation | $ 25,000 | $ 20,000 | $ 45,000 |
| Other factory overhead |  |  | $150,000 |
| General overhead |  |  | $350,000 |

Thus, the overhead rate for Department A per direct labor hour is

$$\text{Rate} = \frac{\$329,347}{25,200 \text{ hours}}$$

$$= \$13.07/\text{direct labor hour}$$

For Department B, the total overhead allocation is calculated as follows:

| | | |
|---|---|---|
| Annual depreciation | = | $ 20,000 |
| Other factory overhead = (9/23)($150,000) | = | 58,696 |
| General overhead = (9/23)($350,000) | = | 136,957 |
| Total overhead costs | | $215,653 |

Thus, the overhead rate for Department B per direct labor hour is

$$\text{Rate} = \frac{\$215,653}{16,200 \text{ hours}}$$

$$= \$13.31/\text{direct labor hour}$$

The estimated total cost for Job Order D is then computed as

| | | |
|---|---|---|
| Direct material cost | = | $ 900.00 |
| Direct labor cost for Dept. A = ($260000/25200)(30 hours) | = | 309.52 |
| Overhead cost for Dept. A = ($13.07)(30 hours) | = | 392.10 |
| Direct labor cost for Dept. B = ($140000/16200)(100 hours) | = | 864.20 |
| Overhead cost for Dept. B = ($13.31)(100 hours) | = | 1331.00 |
| Total cost | | $3796.82 |

If $x$ = the selling price of Job Order D, then

$$x = \text{Total cost} + \text{Profit}$$

$$= \$3796.82 + (0.25)x$$

and

$$x = \frac{3796.82}{0.75} = \$5062.43$$

## 2.5.2 Standard Costs

Although the first task of cost accounting is to determine per-item or per-order costs, another major purpose of cost accounting is to interpret financial data so that management can (1) measure changes in production efficiency and (2) judge the adequacy of production performance. Establishing cost standards can be of great assistance in achieving these objectives. A standard-cost system involves in advance of manufacture (1) the preparation of standard rates for material, labor, and overhead, and (2) the application of these rates to the standard quantities of material and labor required for a job order, or for each production operation required to complete the job order.[4]

Since a process-type manufacturing firm such as an oil refinery outputs the same product (or a few products) over a long time period, cost standards are more readily determined for process firms than for job-shop firms where the variety of output is large and varies with customer order. However, the number and type of production operations required to complete various job orders are finite for a given manufacturing firm. Each job order is, of course, made up of single units. Thus, a standard amount of material can be determined for each unit, and standard labor times and machine times can be determined for each unit. It is usually the responsibility of the work measurement function within the firm to determine these standard quantities. Then, by applying standard unit material costs and standard labor rates, standard unit costs for material, labor, and overhead can be determined. The standard costs then serve as a basis for measuring production efficiency and performance over time. Deviations from standard costs may be caused by several factors, especially (1) raw material price variations and (2) actual quantities of material and labor used versus the standard amounts of these items. This latter factor is the one of primary concern in determining production efficiency and performance, measures of which provide information to management to aid in cost control.

## 2.6 SUMMARY

In this chapter we provided an introduction to the language of accountants, financial analysts, and managers. To be successful in selling engineering designs, one must learn to communicate; to communicate effectively, all parties must speak the same language. Additionally, the data sources for economic analyses are often to be found in the accounting systems; hence, it is essential that the engineer be familiar with accounting principles. To complete the coverage of cost concepts, a brief introduction to cost estimation was presented.

## BIBLIOGRAPHY

1. Apple, James M., *Material Handling Systems Design,* Ronald Press, 1972, Chapters 13 to 15.
2. Blank, Leland T., and Tarquin, Anthony J., *Engineering Economy,* Second Edition, McGraw-Hill, 1983.
3. Ferguson, Earl J., and Shamblin, James E., "Break-Even Analysis," *The Journal of Industrial Engineering,* XVIII (8), August 1967.
4. Kepner, Charles H., and Tregoe, Benjamin B., *The Rational Manager,* McGraw-Hill, 1965.

---

[4] Frank W. Wilson and Philip D. Harvey (Editors), *Tool Engineers Handbook,* Second Edition, McGraw-Hill, 1959, pp. 2–23.

5. Morris, William T., *The Analysis of Management Decisions*, Revised Edition, Irwin, 1964, Chapter 5.

6. Niven, William, and Ohman, Anka, *Basic Accounting Procedures*, Prentice-Hall, 1964, p. 2.

7. Ostwald, Phillip F., *Cost Estimating for Engineering and Management*, Prentice-Hall, 1974.

8. Steffy, Wilbert, Smith, Donald H., and Souter, Donald, *Economic Guidelines for Justifying Capital Purchases*, Industrial Development Division, University of Michigan, Ann Arbor, Mich., 1973, p. 89.

9. Vinton, Ivan R. (Editor), *Realistic Cost Estimating For Manufacturing*, Society of Manufacturing Engineers, Dearborn, Mich., 1968.

## PROBLEMS

1. A small manufacturing firm is converting from a manual inventory system to a microcomputer system. The incremental investment required would be $5000. Due to the fast increase in technology, it is assumed that the salvage value will be zero any time after the system is purchased. It is estimated that a net annual savings of $1000 in operating costs will result when the new computer is purchased. Assume that the time value of money is zero. (2.2.5)
   (a) How many years of savings are required to recover the incremental investment?
   (b) What annual savings are required to recover the incremental investment in 4 years?

2. An engine lathe or a turret lathe can be used to produce a job order of Part 173. If the job order is produced on the engine lathe, Machinist B performs all required operations (i.e., the tooling setup and the operation of the lathe during the machining cycle). On the engine lathe, a tooling setup time of 15 minutes and machining time of 30 minutes is required to produce each part. Machinist B earns $9/hour and the overhead rate for each machining hour of engine lathe operation is $14.00. The tooling costs per lot for production on the engine lathe are estimated to be $500.
   If the job order is produced on the turret lathe, Machinist A must do the initial tooling setup for the job, and 4 hours are required for this. Once the setup is done, no further tooling setup per part is required, and Machinist B will operate the turret lathe for the machining cycles. Machinist A earns $12/hour. The machining time for Part No. 173 is 10 minutes on the turret lathe, and the overhead rate for each machining hour is $25. The tooling costs per lot for production on the turret lathe is $750. For what job-order size would the turret lathe be economically preferred to the engine lathe? (2.2.5)

3. In a stable economic environment, the A. B. Jax Specialty Foundry Co. can produce a maximum of 1500 gray iron railroad car wheels per month and sell these for $300 each. Now, after prolonged labor problems in the coal industry, the foundry can only sell an average of 500 wheels/month at $250 each. The coal industry strike is temporary, but the future over the next 6 months to a year appears very uncertain.

The foundry has a depreciable investment in buildings and equipment of $300,000, and the value decreases at a rate of 6⅔% each year.

At a production rate of 500 wheels/month, direct labor costs will increase from $55/wheel at maximum production rate to $75/wheel. Direct material costs per wheel will remain at $50. Other annual overhead costs theoretically vary in linear fashion from $60,000 at zero output to $105,000 at maximum production capacity. Would you recommend that the A. B. Jax Specialty Foundry Co. shut down temporarily or continue to operate at this reduced capacity? (2.2.5)

4. A commercial machine shop regularly produces a stainless steel component for a major electronics manufacturer. The machine shop purchases the component from a nearby specialty steel company in semifinished condition, performs drilling and milling operations on the part, and ships it to the electronics firm.

The machining operations are those that can readily be performed on a tape-controlled drill press with a turret head. Thus, the management of the machine shop feels the purchase of such a machine to produce only this part is economically justified. An engineer is then assigned the task of determining the production quantity for break-even, assuming a time value of money (minimum attractive rate of return) equal to zero.

The engineer compiles the following information and cost estimates. The tape-controlled machine will have an installed first cost of $60,000, which includes the necessary electronic software, cutting tools and holders, and work-holding devices. Training of the operator for the machine is included in the purchase price. The economic life of the machine is assumed to be 10 years with a salvage value of $20,000 at that time. The decrease in asset value is estimated at $4000/year and judged to be an annual fixed cost. Other fixed costs are $1500/year. The steel parts are sold to the electronics firm for $15.30/unit. The variable unit costs are estimated as $1.70 for direct labor, $7.00 for direct material, and $3.50 for overhead (excluding depreciation of the machine—the $4000/year fixed cost mentioned previously). What annual sales volume (number of parts) is required in order to break even on the machine purchase if linearity is assumed? (2.2.5)

5. A subsidiary plant of a major furniture company manufactures wooden pallets primarily for the parent company but also sells pallets to other industrial customers. The pallet produced is essentially of standard size and design. Minor modifications in pallet design occur but, relative to the standard design, the quantity sold is negligible and the subsidiary plant can be considered a single-product plant. The subsidiary plant has the capacity to produce 450,000 pallets/year. Presently, the plant is operating at 70% of capacity. The average selling price of a pallet is $14.75 with a variable cost per pallet of $9.50 (unit revenue and cost rates are linearly related to production quantity). At zero output, the subsidiary plant's annual fixed costs are about $550,000 and are approximately constant up to the maximum production quantity per year. (2.2.5)

(a) With the present 70% of capacity production, what is the expected annual profit or loss for the subsidiary plant?

(b) What annual volume of sales is required in order for the plant to break even?

(c) What would be the annual profit or loss if the plant were operating at 90% of capacity?

**6.** A first production run of a new product is made with the following data compiled:

Direct material cost = $ 5.00/unit
Direct labor cost   = $ 2.90/unit
Overhead cost       = $ 4.10/unit
Selling price       = $20.00/unit

The above costs are based on all good units produced and no rejects or scrap. Parts are inspected only once, after all the manufacturing operations are performed but prior to shipping. During the production run, a scrap rate of 30% has occurred. What is the maximum scrap rate permissible in order to break even (i.e., total costs equal to total revenues)? (2.2.5)

**7.** A typical gasoline-powered farm tractor with PTO rating of 35 horsepower has an operating cost of about $2.50/hour of use (fuel, lubricants, oil filters, repairs, and maintenance), excluding labor cost, and an annual fixed cost of about $950 (for depreciation, insurances and taxes, housing, and opportunity costs).
Similar data for a diesel-powered tractor of the same size are $2.15/hour of use for operating cost and $1180 annual fixed cost.[5]
**(a)** What is the number of operating hours per year for break-even between the two tractors? (2.2.5)
**(b)** If the estimated number of operating hours per year is 1000, what annual savings are estimated if the diesel-powered tractor is purchased instead of the gasoline-powered tractor? (2.2.5)

**8.** K. Z. Moley purchased a small retail restaurant business and opened on July 1, 19—. At the date of opening, he had invested $20,000 of equity funds with the following breakdown: $10,000 in equipment, $7000 in inventory items, and $3000 in operating cash.
**(a)** Prepare a balance sheet for the business as of July 1, 19—. (2.4.1)
The data below summarize the gross sales and expenses for the restaurant during the first three-month period.

| | |
|---|---|
| Gross sales | $32,000 |
| Purchases | 12,500 |
| Salaries | 6,000 |
| Advertising expense | 1,000 |
| Rent expense | 1,800 |
| Expense for utilities | 900 |
| Expense for misc. supplies | 600 |

For the purchases, $10,000 was paid with cash and $2500 is still owed. At the close of the 3-month period, the end-of-period inventory is worth $5500.
**(b)** Prepare an income statement for the 3-month period covered. (2.4.2)
**(c)** If, at the end of the 3-month period on October 1, 19—, the net worth (or ownership) account is $22,000, determine the amount of the cash account in order to balance the accounting equation as of October 1, 19—. (2.4.1)

---

[5] Example problem based on Publication 510 (revised February 1974) entitled "Farm Machinery Performance and Costs," Extension Division, Virginia Polytechnic Institute and State University, Blacksburg, Virginia.

9. A successful building contractor purchased a 150-acre farm to operate on a part-time basis and raise beef cattle. The purchase price of the farm was $120,000. The contractor paid $40,000 cash and financed the remainder over 10 years at a 12% annual simple interest rate with a mortgage, payable to a local bank. Soon after the purchase of the farm, the contractor purchased cattle for $25,000, paid $10,000 in cash, and gave a promissory note to the seller for the balance. The note carried a 10% annual simple interest rate and was to be paid off within 5 years. During the first full year, the farm operation resulted in the following revenues and expenses.

| | |
|---|---:|
| Calves sold | $11,250 |
| Labor expenses | 1,000 |
| Expenses for machinery hired | 2,500 |
| Veterinarian fees | 350 |
| Fertilizer purchased | 1,800 |
| Property taxes | 950 |
| Expenses for repairs | 750 |
| Interest expenses | 11,100 |
| Expenses for miscellaneous supplies | 250 |

Prepare an income statement for the farming operation and determine the net profit (loss) before income taxes. (2.4.2)

10. An income statement for the WAC Company covering a calendar year ending December 31 is as follows.

## INCOME STATEMENT

| | | |
|---|---:|---:|
| Gross income from sales | | $247,000 |
| Less: cost of goods sold | | 138,800 |
| Net income from sales | | $108,200 |
| Operating expenses | | |
| Rent | $ 9,700 | |
| Salaries | 30,200 | |
| Depreciation | 5,800 | |
| Advertising | 4,300 | |
| Insurance | 1,500 | $ 51,500 |
| Net profit before income taxes | | 56,700 |
| Less: income taxes | | 23,973 |
| Net profit after income taxes | | $ 32,727 |

The following Balance Sheet Accounts, as of December 31, are:

| | |
|---|---:|
| Cash | $94,227 |
| Accounts receivable | 8,000 |
| Notes payable | 25,000 |
| Raw material inventory | 10,000 |
| Work-in-process inventory | 15,000 |
| Accounts payable | 6,000 |
| Declared dividends | 20,000 |
| Finished goods inventory | 18,500 |
| Land | 30,000 |

|                                        |        |
|----------------------------------------|--------|
| Original cost of building              | 80,000 |
| Building—Reserve for depreciation      | 8,000  |
| Original cost of equipment             | 40,000 |
| Equipment—Reserve for depreciation     | 4,000  |

**(a)** Determine the Earned Surplus for the year in question.
**(b)** Prepare a Balance Sheet as of December 31. (2.4.1, 2.4.2)

11. An oil refinery produces one base type of crude oil for the southeastern U.S. market. The two equations below give the relationships which approximate the total cost and the total profit per day in dollars. (*Note:* This problem is for illustration only, the total cost function coefficients are not necessarily realistic.)

$$\text{Total cost, TC}_t = 50,000 + 20.2t + 0.0001t^2$$

$$\text{Total profit, TP}_t = (SP)(t) - \text{TC}_t$$

where $t$ = amount of crude oil produced, barrels/day
  $SP$ = sales price of crude oil, dollars/barrel
    = \$30/barrel

**(a)** At what level of production is the cost/barrel minimum? What is the minimum cost/barrel? (2.2.6)
**(b)** What is the maximum daily profit that the company can make? At what volume of production is the maximum daily profit attainable? (2.2.6)
**(c)** Between what range of daily production is profit possible? (2.2.6)

12. Production of a particular type of annual crop is a function of fertilizer used and is given by the following relationship:

$$P(t) = \text{crop production, barrels/acre}$$

$$= 0.417t - 0.00125t^2$$

where $t$ = amount of fertilizer used, pounds/acre.
The crop can be sold at a price of \$15/barrel. Given that one barrel of crop weighs about 120 pounds, the total cost of crop production is

$$\text{TC}(t) = \$(220 + 2t) \text{ per acre}$$

**(a)** How much fertilizer should the farmer use for this crop per acre of land to maximize his annual profit? (2.2.6)
**(b)** What is the maximum annual profit per acre of land for this crop? (2.2.6)
**(c)** Between what range of use of fertilizer is a profit possible? (2.2.6)

13. Assume the total annual inventory cost for a particular item in inventory is

$$\text{TC}(Q) = \text{annual cost of ordering} + \text{annual cost of carrying inventory}$$

$$= P_c \left(\frac{A}{Q}\right) + (15 + 0.75Q)$$

where $P_c$ = cost of preparing a purchase order (constant, regardless of the quantity ordered)
    = \$5 per purchase order
  $Q$ = the number of items ordered each time a purchase order is placed
  $A$ = total annual demand for this item = 1000

Note that the total annual inventory cost $TC(Q)$ varies with the variable $Q$. (2.2.6)

**(a)** What is the most economic order quantity for this item to minimize annual inventory cost?

**(b)** What will be the corresponding total annual inventory cost?

**(c)** If a personal computer is used to prepare the purchase order and the purchase order preparation cost ($P_c$) reduces to $4/order, what impact will it have on the economic order quantity?

**14.** A worker in a company is on an incentive system. The worker earns $1.50 for each part produced. In *addition,* the worker earns a guaranteed wage of $5/hour regardless of production output. The product is sold at a price of $(5 - 0.4t)$, dollars/unit, where $t$ is the number of units produced per hour. (2.2.6)

**(a)** At what rate of production will the company earn the most profit from this worker's output? (Assume all units produced can be sold.)

**(b)** What is the amount of maximum profit per hour?

**(c)** At what rate of production will the revenue be maximized? What is the corresponding profit per hour?

**(d)** Between what range of production can the company make profit?

**15.** The flow of current (amperes) through a conductor is inversely proportional to its resistance (ohms). The resistance of the wire is also inversely proportional to its nominal diameter in inches. These relations are given as follows:

$$I = \frac{V}{R} \quad \text{and} \quad R = \frac{11}{D}$$

where $I$ = amount of flow of current, amperes
$\quad V$ = voltage = 110 volts
$\quad R$ = resistance, ohms
$\quad D$ = diameter of the conductor, inches
The selling price of current is $0.06/ampere. The cost of transmission of the current is given as follows:

$$\text{Cost} = 0.1D^2 + 0.5D, \text{ dollars/ampere}$$

**(a)** What is the optimum diameter of the conductor in order to maximize profit per ampere? (2.2.6)

**(b)** What is the maximum profit per ampere? (2.2.6)

**(c)** Up to what maximum diameter of the conductor will any profit be possible? (2.2.6)

**16.** An order for 5000 units of Part D-142 is received by the J. T. Kling Engineering Company, a small machine shop. The finished dimensions of the rectangular part are $1\frac{7}{8}$ inch $\times$ $1\frac{13}{16}$ inch $\times$ 4 inch (neglecting tolerances). The raw material for this part is purchased 2 inch $\times$ 2 inch $\times$ 8 foot SAE 1020 steel rectangular bar stock, with each unit costing $22 and yielding 20 part blanks. The basic manufacturing sequence, with standard machining times per part, and machine overhead rates (per machining hour) is given in the table.

| Operating | Standard Time per Part | Machine Overhead Rate |
|---|---|---|
| Cutoff on power hacksaw | 1 minute | $0.60/hour |
| Mill two sides; deburr | 4 minutes | 1.00/hour |
| Drill three 3/8-inch diameter holes | 2 minutes | 0.80/hour |
| Surface grind one side; deburr | 2 minutes | 0.70/hour |
| Package | 0.25 minutes | — |

The direct labor time per part is the same as the machine (or operation) time per part. The tooling cost for this job order is estimated to be $500. Excluding tooling costs and machine overhead, other factory overhead costs (for indirect labor, utilities, indirect materials, etc.) are $12/direct labor hour. The average direct labor hour rate (including fringe benefits) is $8/hour. (2.5.1)

(a) Determine the total estimated costs for the job order of 5000 units.

(b) Determine the unit selling price if profit is to be 50% of the total cost.

17. The welding department of a mining equipment manufacturing plant consists of four cost centers: manual arc welding (A), semiautomatic welding (B), furnace brazing and heat treating (C), and finishing (D). Some oxyacetylene cutting is also done in Center B. Assume that it is possible to allocate departmental overhead expenses directly to each cost center and that the following data for the welding department were compiled last year by the accounting system.

| Cost Center | Departmental Expenses | Direct Labor | | Direct Material |
|---|---|---|---|---|
| | | Hours | Cost | |
| A | $21,000 | 10,000 | $75,000 | $12,000 |
| B | 10,000 | 4,000 | 18,000 | 12,000 |
| C | 5,400 | 1,500 | 9,000 | 4,500 |
| D | 4,800 | 2,800 | 11,200 | 3,000 |

Compute the overhead rate (or rates) applicable by the following methods. (2.5.2)

(a) Blanket (departmental) percentage of direct labor cost.

(b) Blanket percentage of prime cost.

(c) Blanket hourly rate per direct labor hour.

(d) Percentage of direct labor cost for each cost center.

(e) Percentage of prime cost for each cost center.

(f) Rate per direct labor hour for each cost center.

18. Given the welding department and four cost centers found in Problem 17, the direct labor hours and costs for each cost center remain the same, but new and additional data are given below (assume cost data are for the previous year).

| Cost Center | Square Feet Occupied | Cost of Machinery | Number of Direct Labor Employees |
|---|---|---|---|
| A | 900 | $10,000 | 5 |
| B | 400 | 9,000 | 2 |
| C | 600 | 11,700 | 1 |
| D | 500 | 4,800 | 1 |
| Total | 2,400 | $35,500 | 9 |

Expenses other than for direct labor and materials chargeable to the welding department last year were

| | |
|---|---|
| Maintenance | $ 6,400 |
| Gas and electricity | 20,000 |
| Supervision and other indirect labor | 30,000 |
| Miscellaneous supplies | 7,500 |
| Equipment depreciation | 4,600 |
| Building depreciation | 8,000 |

Determine an overhead rate per direct labor hour for each cost center if the welding department expenses above are first allocated to each cost center as follows. (2.5.1)

(a) Maintenance expenses and equipment depreciation expenses are allocated according to the value of equipment (percent of total) in each cost center.

(b) Building depreciation expenses are allocated according to the floor space occupied by each cost center.

(c) Supervision and other indirect labor expenses are allocated according to the number of direct labor employees of each cost center.

(d) Supplies and gas and electricity expenses are allocated according to the number of direct labor hours for each cost center.

CHAPTER THREE

# TIME VALUE OF MONEY OPERATIONS

## 3.1 INTRODUCTION

Design alternatives are normally compared by using a host of different criteria, including system performance and economic performance. Among the system performance characteristics that are of concern, quality, safety, and customer service considerations are of primary importance. Among the economic performance characteristics normally considered are initial investment requirements, return on investment, and the cash flow (CF) profile. Since the cash flow profiles are usually quite different among the several design alternatives, in order to compare the economic performances of the alternatives, one must compensate for the differences in the timing of cash flows. A fundamental concept underlies much of the material covered in the text: *money has a time value*. By this, we mean that the *value* of a given sum of money depends on *when* the money is received.

Example 3.1 _____

To illustrate the concept of the time value of money, suppose a wealthy individual approaches you and says, "Because of your outstanding ability to manage money, I am prepared to present you with a tax-free gift of $100. However, if you prefer, I will postpone the presentation for 1 year, at which time I will guarantee that you will receive a tax-free gift of $X." Would you choose to receive the $100 now or the $X 1 year from now if $X equaled (1) $100, (2) $110, (3) $200, (4) $1000?

In presenting this situation to numerous students, no students preferred receiving $X in case (1), very few students preferred $X in case (2), most students preferred receiving $200 a year from now, and all students indicated a preference for $1000 a year from now. The point is that the value of $100 1 year from now was perceived to be less than the value of $100 at present. For most students, the value of $110 1 year from now was believed to be less than the value of $100 at present. Only a few students felt that $200 a year from now was less valuable than $100 at present. All students believed $1000 a year from now was more valuable than $100 at present. Thus, for each individual student, some value (or range of values) of $X exists for which one would be indifferent between receiving $100 now versus receiving $X a year from now. If, for example, one is indifferent for $X equal to $125, then we would conclude that $125 occurring one year from now has a *present value of* $100 *for that particular individual.*

---

Some would argue that the reason the students prefer receiving $X today rather than receiving $X 1 year later is due to the effects of *inflation.* They would contend that the purchasing power of $X decreases over time during periods of inflation. In fact, most people have experienced the impact of inflation and cost of living increases.

While it is true that the *value, economic worth,* or *purchasing power of money* changes over time during periods of inflation and deflation, it is also true that people will prefer to receive $X today rather than 1 year later even when there is no inflation or deflation. One reason for this preference is the *opportunity cost of money.*

As noted in Chapter Two, "the cost of forgoing the opportunity to earn interest, or a return, on investment funds is termed an *opportunity cost.*" The $X received today can be invested at some interest rate so that 1 year later the investment will be worth more than $X.

Because people favor current consumption to postponed consumption, it is necessary that postponed consumption be compensated or rewarded with "interest" or a "return on investment." Without such compensation, consumption would not be postponed and capital would not be available for investment.

The concept of inflation and its consideration in evaluating investment alternatives is extremely important. However, before addressing the subject of inflation it is essential that a number of fundamental concepts be addressed related to the time value of money in the comparison of the economic performances of design alternatives. In this chapter we examine a number of mathematical operations that are based on the time value of money, with an emphasis on modeling cash flow profiles. A consideration of inflation is reserved for Chapter Four.

The following examples are intended to illustrate the time value of money in the absence of inflation or deflation.

TABLE 3.1.  Cash Flow Profiles for Two
Investment Alternatives

| End of Year (EOY) | CF | | |
|---|---|---|---|
| | A | B | (A − B) Difference |
| 0 | −$10,000 | −$10,000 | $0 |
| 1 | + 7,000 | + 1,000 | +6,000 |
| 2 | + 5,000 | + 3,000 | +2,000 |
| 3 | + 3,000 | + 5,000 | −2,000 |
| 4 | + 1,000 | + 7,000 | −6,000 |

Example 3.2_____

To continue our consideration of different cash flow situations, examine closely the two cash flow profiles given in Table 3.1. Both alternatives involve an investment of $10,000 in ventures that last for 4 years. Alternative A involves an investment in a minicomputer by a consulting engineer who is planning on providing computerized design capability for clients. Since the engineer antici- pates that competition will develop very quickly if the plan proves to be suc- cessful, a declining revenue profile is anticipated.

Alternative B involves an investment in a land development venture by a group of individuals. Different parcels of land are to be sold over a 4-year period. The land is anticipated to increase in value. There will be differences in the sizes of parcels to be sold. Consequently, an increasing revenue profile is anticipated.

The consulting engineer has available funds sufficient to undertake either investment, but not both. The cash flows shown are cash flows after taxes and other expenses have been deducted. Both investments result in $16,000 being received over the 4-year period; hence, a net cash flow of $6,000 occurs in both cases.

Which would you prefer? If you prefer Alternative B, then you are not acting in a manner consistent with the concept that money has a time value. The $6000 difference at the end of the first year is worth more than the $6000 difference at the end of the fourth year. Likewise, the $2000 difference at the end of the second year is worth more than the $2000 difference at the end of the third year.

Example 3.3_____

As another illustration of the impact of the time value of money on the prefer- ence between investment alternatives, consider investment alternatives C and D, having the cash flow profiles depicted in Figure 3.1. The cash flow diagrams indicate that the positive cash flows for Alternative C are identical to those for

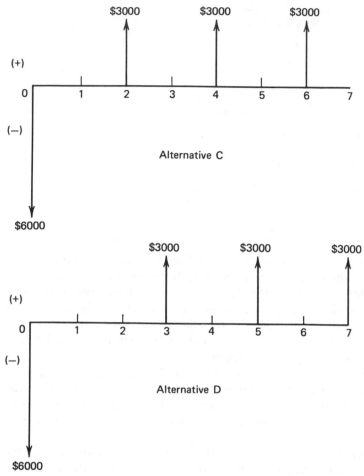

FIGURE 3.1. Cash flow diagrams for Alternatives C and D.

Alternative D, except that the former occur 1 year sooner; both alternatives require an investment of $6000. *If exactly one of the alternatives must be selected,* then Alternative C would be preferred to Alternative D, based on the time value of money.

---

Example 3.4_____

A third illustration of the effect of the time value of money on the selection of the preferred investment alternative is presented in Figure 3.2. Either Alternative E or Alternative F must be selected; the only differences in the performance characteristics of the two alternatives are economic differences. As shown in Figure 3.2, the economic differences reduce to a situation in which the receipt of $100 is delayed in order to receive $200 a year later. For this

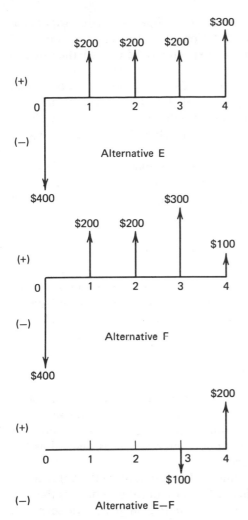

FIGURE 3.2. Cash flow diagrams for Alternatives E and F.

illustration, we would conclude that most of the students polled earlier would prefer Alternative E to Alternative F, since most of the students preferred $X when it equaled $200.

In order to provide motivation and to provide a familiar scenario, the treatment of time value of money operations will be presented primarily in the context of personal finance. However, each of the concepts examined in this chapter can and does occur in the business world.

In this chapter we emphasize end-of-period cash flows and end-of-period compounding.[1] Depending on the financial institution involved, in personal

---

[1] The exception to this will be the treatment of continuous cash flows.

finance transactions, savings accounts might not pay interest on deposits made in "the middle of a compounding period." Consequently, answers obtained using the methods we describe should not be expected to be exactly the same as those provided by the financial institution.

The beginning-of-period cash flows can be handled very easily by noting that the end of period $t$ is the beginning of period $t + 1$. To illustrate, rental payments might be made at the beginning of each month. However, one can think of the payment made at the beginning of, say, March as having been made at the end of February.

In Chapter Four, end-of-year cash flows are assumed unless otherwise noted. There, it is realized that monetary transactions take place during a calendar year, but it is convenient to ignore any compounding effects within a year and deal directly with end-of-year cash flows.

## 3.2 INTEREST CALCULATIONS

In considering the time value of money, it is convenient to represent mathematically the relationship between the current or *present* value of a single sum of money and its *future* value. Letting time be measured in years, if a single sum of money has a current or *present* value of $P$, its value in $n$ years would be equal to

$$F_n = P + I_n$$

where $F_n$ is the accumulated value of $P$ over $n$ years, or the *future* value of $P$, and $I_n$ is the increase in the value of $P$ over $n$ years. $I_n$ is referred to as the accumulated *interest* in borrowing and lending transactions and is a function of $P$, $n$, and the *annual interest rate*, $i$. The annual interest rate is defined as the change in value for \$1 over a 1-year period.

Over the years, two approaches have emerged for computing the value of $I_n$. The first approach considers $I_n$ to be a linear function of time. Since $i$ is the rate of change over a 1-year period, it is argued that $P$ changes in value by an amount $Pi$ each year. Hence, it is concluded that $I_n$ is the product of $P$, $i$, and $n$, or,

$$I_n = Pin$$

and

$$F_n = P(1 + in)$$

This is called the *simple interest* approach.

The second approach used to compute the value of $I_n$ is to interpret $i$ as the rate of change in the accumulated value of money. Hence, it is argued that the following relation holds,

$$I_n = iF_{n-1}$$

and

$$F_n = F_{n-1}(1 + i)$$

This approach is referred to as the *compound interest* approach.

The approach to be used in any particular situation depends on how the interest rate is defined. Since practically all monetary transactions are based currently on compound interest rates instead of on simple interest rates, we will assume compounding occurs unless otherwise stated.

A convenient method of representing the time value of money is to visualize positive and negative cash flows as though they were generated by a borrower and a lender. In particular, suppose you loaned $1000 to an individual who agreed to pay you interest at a rate of 10%/year. The $1000 is referred to as the *principal* amount. At the end of 1 year, you would receive $1100 from the borrower. Thus, we might say that $1100 1 year from now is worth $1000 today based on a 10% interest rate or, conversely, we might say that $1000 today has a value of $1100 1 year from now based on a 10% interest rate.

If the individual borrowed the $1100 for an additional year, then you would be owed $1210, since the interest on $1100 for 1 year equals $(0.10) \times (\$1100)$, or $110. Equivalently, borrowing $1000 for 2 years at 10% interest yields $1210 owed if interest is *compounded* annually. Compound interest involves the computation of interest charges during a time period based on the unpaid principal amount plus any accumulated interest charges up to the beginning of the time period. The interest amount due for a given interest period converts to principal for the purpose of calculating the interest amount due in the subsequent interest period. The relationship between the principal amount and the compounding of interest is given in Table 3.2.

Recall that whenever the interest charge for any time period is based only on the unpaid principal amount and not on any accumulated interest charges,

TABLE 3.2. Illustrating the Effect of Compound Interest

| End of Period | (A) Amount Owed | (B) Interest for Next Period | (C) = (A) + (B) Amount Owed for Next Period* | |
|---|---|---|---|---|
| 0 | $P$ | $Pi$ | $P + Pi$ | $= P(1 + i)$ |
| 1 | $P(1 + i)$ | $P(1 + i)i$ | $P(1 + i) + P(1 + i)i$ | $= P(1 + i)^2$ |
| 2 | $P(1 + i)^2$ | $P(1 + i)^2 i$ | $P(1 + i)^2 + P(1 + i)^2 i$ | $= P(1 + i)^3$ |
| 3 | $P(1 + i)^3$ | $P(1 + i)^3 i$ | $P(1 + i)^3 + P(1 + i)^3 i$ | $= P(1 + i)^4$ |
| ⋮ | ⋮ | ⋮ | ⋮ | |
| $n - 1$ | $P(1 + i)^{n-1}$ | $P(1 + i)^{n-1}i$ | $P(1 + i)^{n-1} + P(1 + i)^{n-1}i = P(1 + i)^n$ | |
| $n$ | $P(1 + i)^n$ | | | |

* Notice, the value in column (C) for the end of period $(n - 1)$ provides the value in column (A) for the end of period $n$.

simple interest calculations apply. The interest due for a given interest period does not convert to principal for the purpose of calculating the interest amount due in the subsequent interest period. In the preceding illustration, the amount owed at the end of 2 years would be the $1000 principal plus the interest charges of $(0.10)(\$1000) = \$100$ each year or $1200 total. Thus, in this example, the $1210 for the compounding case compares with the $1200 for the simple interest case. (In some cases, the difference is much more dramatic!)

### Example 3.5

Person A borrows $4000 from Person B and agrees to pay $1000 plus accrued interest at the end of the first year and $3000 plus the accrued interest at the end of the fourth year. What are the amounts for the two payments if 8% annual simple interest applies? For the first year, the payment is $1000 + (0.08)($4000) = $1320. For the fourth year, the payment is $3000 + (0.08)($4000 − $1000)(3) = $3720.

## 3.3 SINGLE SUMS OF MONEY

To illustrate the mathematical operations involved in modeling cash flow profiles using compound interest, first consider the investment of a single sum of money, $P$, in a savings account for $n$ interest periods. Let the interest rate per interest period be denoted by $i$ and let the accumulated total in the fund $n$ periods in the future be denoted by $F$. As shown in Table 3.2, assuming no monies are withdrawn during the interim, the amount in the fund after $n$ periods equals $P(1 + i)^n$. As a convenience in computing values of $F$ (the future worth) when given values of $P$ (the present worth), the quantity $(1 + i)^n$ is tabulated in Appendix A for various values of $i$ and $n$. The quantity $(1 + i)^n$ is referred to as the *single sum, future worth factor* and is denoted $(F|P\ i,n)$. The expression $(F|P\ i,n)$ is read as the $F$, given $P$ factor at $i\%$ for $n$ periods. The above discussion is summarized as follows.

Let $P$ = the equivalent value of an amount of money at time zero, or present worth
$F$ = the equivalent value of an amount of money at time $n$, or future worth
$i$ = the interest rate per interest period
$n$ = the number of interest periods

Thus, the future worth is related to the present worth as follows:

$$\boxed{F = P(1 + i)^n} \tag{3.1}$$

or equivalently,

$$F = P(F|P\ i,n) \tag{3.2}$$

A cash flow diagram depicting the relationship between $F$ and $P$ is given in Figure 3.3. Remember that $F$ occurs $n$ periods after $P$.

Example 3.6_____

An individual borrows $1000 at 12% compounded annually. The loan is paid back after 5 years. How much should be repaid?

Using the compound interest tables in Appendix A for 12% and five periods, the value of the $(F|P\ 12,5)$ factor is found to be 1.7623. Thus,

$$F = P(F|P\ 6,5)$$
$$= \$1000(1 + 0.12)^5$$
$$= \$1000(1.7623)$$
$$= \$1762.30$$

The amount to be repaid equals $1762.30.

_____

Since we are able to determine conveniently values of $F$ when given values of $P$, $i$, and $n$, it is a simple matter to determine values of $P$ when given values of $F$, $i$, and $n$. In particular, since

$$F = P(1 + i)^n$$

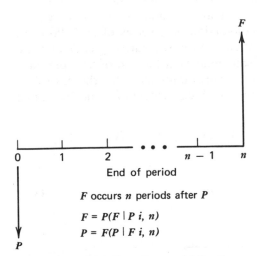

FIGURE 3.3. Cash flow diagram of the time relationship between $P$ and $F$.

on dividing both sides by $(1 + i)^n$, we find that the present worth and future worth have the relation

$$P = F(1 + i)^{-n} \qquad (3.3)$$

or

$$P = F(P|F\ i,n) \qquad (3.4)$$

where $(1 + i)^{-n}$ and $(P|F\ i,n)$ are referred to as the *single sum, present worth factor*.

Example 3.7 ————————————————————————

To illustrate the computation of $P$ given $F$, $i$, and $n$, suppose you wish to accumulate \$10,000 in a savings account 4 years from now and the account pays interest at a rate of 9% compounded annually. How much must be deposited today?

$$P = F(P|F\ 9,4)$$
$$= \$10,000(0.7084)$$
$$= \$7084$$

## 3.4 SERIES OF CASH FLOWS

Having considered the transformation of a single sum of money to a future worth equivalent when given a present worth amount and vice versa, we generalize that discussion to consider the conversion of a series of cash flows to present worth and future worth equivalents. In particular, let $A_t$ denote the magnitude of a cash flow (receipt or disbursement) at the end of time period $t$. Using discrete compounding, the present worth equivalent for the cash flow series is equal to the sum of the present worth equivalents for the individual cash flows. Consequently,

$$P = A_1(1 + i)^{-1} + A_2(1 + i)^{-2} + \cdots + A_{n-1}(1 + i)^{-(n-1)} + A_n(1 + i)^{-n} \qquad (3.5)$$

or, using the summation notation,

$$P = \sum_{t=1}^{n} A_t(1 + i)^{-t} \qquad (3.6)$$

or, equivalently,

$$P = \sum_{t=1}^{n} A_t(P|F\ i,t) \qquad (3.7)$$

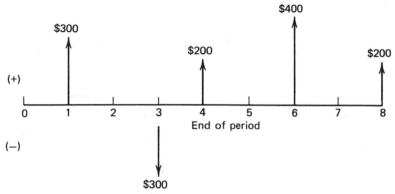

FIGURE 3.4. Series of cash flows.

## Example 3.8

Consider the series of cash flows depicted by the cash flow diagram given in Figure 3.4. Using an interest rate of 6%/interest period, the present worth equivalent is given by

$$P = \$300(P|F\ 6,1) - \$300(P|F\ 6,3) + \$200(P|F\ 6,4)$$
$$+ \$400(P|F\ 6,6) + \$200(P|F\ 6,8)$$
$$= \$300(0.9434) - \$300(0.8396) + \$200(0.7921)$$
$$+ \$400(0.7050) + \$200(0.6274)$$
$$= \$597.04$$

The future worth equivalent is equal to the sum of the future worth equivalents for the individual cash flows. Thus,

$$F = A_1(1 + i)^{n-1} + A_2(1 + i)^{n-2} + \cdots + A_{n-1}(1 + i) + A_n \qquad (3.8)$$

or, using the summation notation,

$$F = \sum_{t=1}^{n} A_t(1 + i)^{n-t} \qquad (3.9)$$

$$\boxed{F = \sum_{t=1}^{n} A_t(F|P\ i, n - t)} \qquad (3.10)$$

Alternately, since we know the value of future worth is given by

$$F = P(1 + i)^n \qquad (3.11)$$

substituting Equation 3.6 into Equation 3.11 yields

$$F = (1 + i)^n \sum_{t=1}^{n} A_t(1 + i)^{-t}$$

Hence,

$$F = \sum_{t=1}^{n} A_t (1 + i)^{n-t} \qquad (3.12)$$

Example 3.9 _____

Given the series of cash flows in Figure 3.4, determine the future worth at the end of the eighth period using an interest rate of 6%/interest period.

$$F = \$300(F|P\ 6,7) - \$300(F|P\ 6,5) + \$200(F|P\ 6,4)$$
$$\quad + \$400(F|P\ 6,2) + \$200$$
$$= \$300(1.5036) - \$300(1.3382) + \$200(1.2625)$$
$$\quad + \$400(1.1236) + \$200$$
$$= \$951.56$$

Alternately, since we know the present worth is equal to \$597.04

$$F = P(F|P\ 6,8)$$
$$\quad = \$597.04(1.5938)$$
$$\quad = \$951.56$$

Note that in computing the future worth of a series of cash flows, *the future worth amount obtained occurs at the end of time period n*. Thus, if a cash flow occurs at the end of time period $n$, it earns no interest in computing the future worth amount at the end of the $n$th period.

Obtaining the present worth and future worth equivalents of cash flow series by summing the individual present worths and future worths, respectively, can be quite time consuming if many cash flows are included in the series. However, with the development and widespread use of pocket calculators (some of which have the capability of performing compound interest calculations automatically), microcomputers, minicomputers, and time-sharing systems, it is not uncommon to treat all series in the manner described above.

When such hardware is not available, it is possible to use more efficient solution procedures if the cash flow series have one of the following forms.

Uniform Series of Cash Flows

$$A_t = A \qquad\qquad t = 1, \ldots, n$$

Gradient Series of Cash Flows

$$A_t = \begin{cases} 0 & t = 1 \\ A_{t-1} + G & t = 2, \ldots, n \end{cases}$$

## Geometric Series of Cash Flows

$$A_t = \begin{cases} A & t = 1 \\ A_{t-1}(1 + j) & t = 2, \ldots, n \end{cases}$$

### 3.4.1 Uniform Series of Cash Flows

A uniform series of cash flows exists when all of the cash flows in a series are equal. In the case of a uniform series the present worth equivalent is given by

$$P = \sum_{t=1}^{n} A(1 + i)^{-t} \tag{3.13}$$

where $A$ is the magnitude of an individual cash flow in the series.

Letting $X = (1 + i)^{-1}$ and bringing $A$ outside the summation yields

$$P = A \sum_{t=1}^{n} X^t$$

$$= AX \sum_{t=1}^{n} X^{t-1}$$

Letting $h = t - 1$ gives the geometric series

$$P = AX \sum_{h=0}^{n-1} X^h \tag{3.14}$$

Since the summation in Equation 3.14 represents the first $n$ terms of a geometric series, the closed form value for the summation is given by

$$\sum_{h=0}^{n-1} X^h = \frac{1 - X^n}{1 - X} \tag{3.15}$$

Hence, on substituting Equation 3.15 into Equation 3.14, we obtain

$$P = AX \left( \frac{1 - X^n}{1 - X} \right)$$

Replacing $X$ with $(1 + i)^{-1}$ yields the following relationship between $P$ and $A$.

$$P = A \left[ \frac{(1 + i)^n - 1}{i(1 + i)^n} \right] \tag{3.16}$$

more commonly expressed as

$$P = A(P|A\ i,n) \tag{3.17}$$

where $(P|A\ i,n)$ is referred to as the *uniform series, present worth factor* and is tabulated in Appendix A for various values of $i$ and $n$.

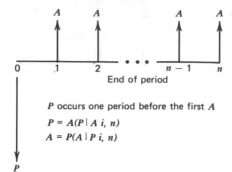

P occurs one period before the first A

$P = A(P|A\ i,\ n)$

$A = P(A|P\ i,\ n)$

P

FIGURE 3.5. Cash flow diagram of the relationship between P and A.

## Example 3.10

An individual wishes to deposit a single sum of money in a savings account so that five equal annual withdrawals of $2000 can be made before depleting the fund. If the first withdrawal is to occur 1 year after the deposit and the fund pays interest at a rate of 12% compounded annually, how much should be deposited?

Because of the relationship of P and A, as depicted in Figure 3.5, in which P occurs one period before the first A, we see that

$$P = A(P|A\ 12,5)$$
$$= \$2000(3.6048)$$
$$= \$7209.60$$

Thus, if $7209.60 is deposited in a fund paying 12% compounded annually, then five equal annual withdrawals of $2000 can be made.

## Example 3.11

In Example 3.10, suppose that the first withdrawal will not occur until 3 years after the deposit.

As depicted in Figure 3.6, the value of P to be determined occurs at $t = 0$, whereas a straightforward application of the $(P|A\ 12,5)$ factor will yield a single sum equivalent at $t = 2$. Consequently, the value obtained at $t = 2$ must be moved backward in time to $t = 0$. The latter operation is easily performed using the $(P|F\ 12,2)$ factor. Therefore,

$$P = A(P|A\ 12,5)(P|F\ 12,2)$$
$$= \$2000(3.6048)(0.7972)$$
$$= \$5747.49$$

Deferring the first withdrawal for 2 years reduces the amount of the deposit by $7209.60 − $5747.49 = $1462.11.

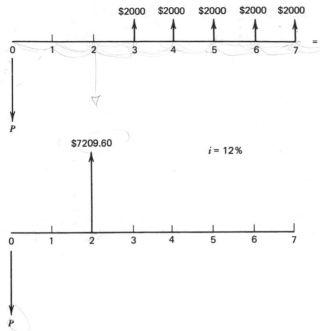

FIGURE 3.6. Equivalent cash flow diagrams.

The reciprocal relationship between $P$ and $A$ can be expressed as

$$A = P \left[ \frac{i(1 + i)^n}{(1 + i)^n - 1} \right] \tag{3.18}$$

or as

$$A = P(A|P\ i,n) \tag{3.19}$$

The expression $(A|P\ i,n)$ is called the *capital recovery factor* for reasons that will become clear in Chapter Four. The $(A|P\ i,n)$ factor is used frequently in both personal financing and in comparing economic investment alternatives.

Example 3.12

Suppose $10,000 is deposited into an account that pays interest at a rate of 15% compounded annually. If 10 equal, annual withdrawals are made from the account, with the first withdrawal occurring 1 year after the deposit, how much can be withdrawn each year in order to deplete the fund with the last withdrawal?

Since we know that $A$ and $P$ are related by

$$A = P(A|P\ i,n)$$

then

$$A = P(A|P\ 15,10)$$
$$= \$10,000(0.1993)$$
$$= \$1993$$

Example 3.13

Suppose that in Example 3.12 the first withdrawal is delayed for 2 years, as depicted in Figure 3.7. How much can be withdrawn each of the 10 years?

The amount in the fund at $t = 2$ equals

$$V_2 = P(F|P\ 15,2)$$
$$= \$10,000(1.3225)$$
$$= \$13,225$$

Therefore, the size of the equal annual withdrawals will be

$$A = V_2(A|P\ 15,10)$$
$$= \$13,225(0.1993)$$
$$= \$2635.74$$

Thus, delaying the first withdrawal for 2 years increases the size of each withdrawal by \$642.74.

The future worth of a uniform series is obtained by recalling that

$$F = P(1 + i)^n \tag{3.20}$$

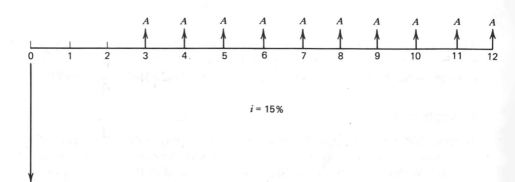

FIGURE 3.7. Cash flow diagram of deferred payment example.

Substituting Equation 3.16 into Equation 3.20 for $P$ and reducing yields

$$F = A \left[ \frac{(1 + i)^n - 1}{i} \right]$$ (3.21)

or, equivalently,

$$F = A(F|A\ i,n)$$ (3.22)

where $(F|A\ i,n)$ is referred to as the *uniform series, future worth factor.*

Example 3.14_____

If annual deposits of \$1000 are made into a savings account for 30 years, how much will be in the fund immediately after the last deposit if the fund pays interest at a rate of 8% compounded annually?

$$F = A(F|A\ 8,30)$$
$$= \$1000(113.2831)$$
$$= \$113,283.10$$

The reciprocal relationship between $A$ and $F$ is easily obtained from Equation 3.21. Specifically, we find that

$$A = F \left[ \frac{i}{(1 + i)^n - 1} \right]$$ (3.23)

or, equivalently,

$$A = F(A|F\ i,n)$$ (3.24)

The expression $(A|F\ i,n)$ is referred to as the *sinking fund factor,* since the factor is used to determine the size of a deposit one should place (sink) in a fund

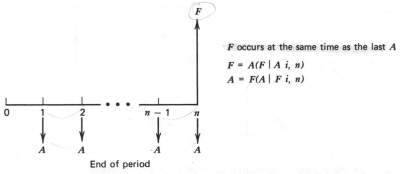

$F$ occurs at the same time as the last $A$
$F = A(F\ |\ A\ i, n)$
$A = F(A\ |\ F\ i, n)$

End of period

FIGURE 3.8. Cash flow diagram of the relationship between $A$ and $F$.

in order to accumulate a desired future amount. As depicted in Figure 3.8, $F$ *occurs at the same time as the last $A$*. Thus, the last $A$ or deposit earns no interest.

## Example 3.15

If \$150,000 is to be accumulated in 35 years, how much must be deposited annually in a fund paying 8% compounded annually in order to accumulate the desired amount immediately after the last deposit?

$$A = F(A|F\ 8,35)$$
$$= \$150,000(0.0058)$$
$$= \$870$$

## 3.4.2 Gradient Series of Cash Flows

A gradient series of cash flows occurs when the value of a given cash flow is greater than the value of the previous cash flow by a constant amount, $G$. Consider the series of cash flows depicted in Figure 3.9. The series can be represented by the sum of a uniform series and a gradient series. By conven-

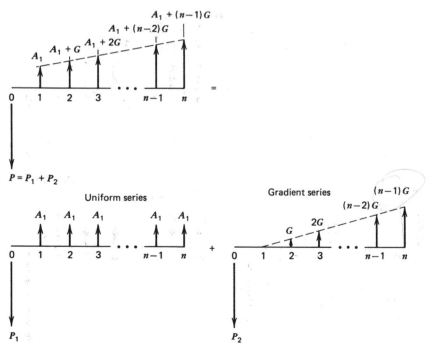

FIGURE 3.9. Cash flow diagram of a combination of uniform and gradient series.

tion, the gradient series is defined to have the first positive cash flow occur at the end of the second time period. The size of the cash flow in the gradient series occurring at the end of period $t$ is given by

$$A_t = (t - 1)G \quad t = 1, \ldots, n \tag{3.25}$$

The gradient series arises when the value of an individual cash flow differs by a constant, $G$, from the preceding cash flow. As an illustration, if an individual receives an annual bonus and the size of the bonus increases by \$100 each year, then the series is a gradient series. Also, operating and maintenance costs tend to increase over time because of both inflation effects and a gradual deterioration of equipment; such costs are often approximated by a gradient series.

The present worth equivalent of a gradient series is obtained by recalling

$$P = \sum_{t=1}^{n} A_t(1 + i)^{-t} \tag{3.26}$$

Substituting Equation 3.25 into Equation 3.26 gives

$$P = \sum_{t=1}^{n} (t - 1)G(1 + i)^{-t} \tag{3.27}$$

or, equivalently,

$$P = G \sum_{t=1}^{n} (t - 1)(1 + i)^{-t} \tag{3.28}$$

As an exercise you may wish to show that the summation reduces to

$$P = G \left[ \frac{1 - (1 + ni)(1 + i)^{-n})}{i^2} \right] \tag{3.29}$$

Expressing the term in brackets in terms of interest factors already treated yields

$$P = G \left[ \frac{(P|A\ i,n) - n(P|F\ i,n)}{i} \right] \tag{3.30}$$

or, equivalently,

$$P = G(P|G\ i,n) \tag{3.31}$$

where $(P|G\ i,n)$ is the *gradient series, present worth factor* and is tabulated in Appendix A.

A uniform series equivalent to the gradient series is obtained by multiplying the value of the gradient series present worth factor by the value of the $(A|P\ i,n)$ factor to obtain

$$A = G \left[ \frac{1}{i} - \frac{n}{i} (A|F\ i,n) \right]$$

or, equivalently,

$$A = G(A|G\ i,n) \tag{3.32}$$

where the factor $(A|G\ i,n)$ is referred to as the *gradient-to-uniform series conversion factor* and is tabulated in Appendix A. To obtain the future worth equivalent of a gradient series at time $n$, multiply the value of the $(A|G\ i,n)$ factor by the value of the $(F|A\ i,n)$ factor.

It is not uncommon to encounter a cash flow series that is the sum or difference of a uniform series and a gradient series. To determine present worth and future worth equivalents of such a composite, one can deal with each special type of series separately.

## Example 3.16

An individual deposits an annual bonus into a savings account that pays 8% compounded annually. The size of the bonus increases by \$100/year; the initial bonus was \$300. Determine how much will be in the fund immediately after the fifth deposit.

A cash flow diagram for this example is given in Figure 3.10. Note that the

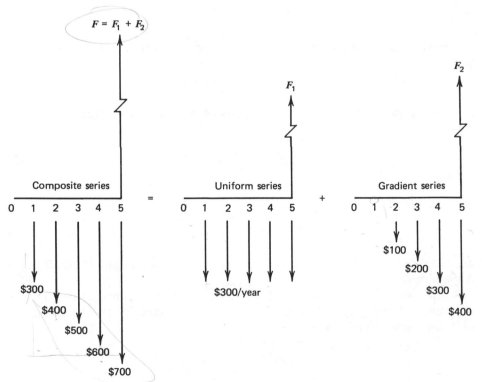

FIGURE 3.10. Cash flow diagram for a gradient series example.

cash flow series consists of the sum of a uniform series of $300 and a gradient series with $G$ equal to $100. Converting the gradient series to a uniform series gives

$$A = G(A|G\ 8,5)$$
$$= \$100(1.8465)$$
$$= \$184.65$$

*(Notice that n equals 5 even though only four positive cash flows are present in the gradient series.)*

The cash flow series given in Figure 3.10 is equivalent to a uniform series having cash flows equal to $300 + $184.65 or $484.65. Converting the uniform series to a future worth equivalent:

$$F = A(F|A\ 8,5)$$
$$. = \$484.65(5.8666)$$
$$= \$2843.25$$

Thus, $2843.25 will be in the fund immediately after the fifth deposit.

---

Example 3.17_____

Five annual deposits are made into a fund that pays interest at a rate of 8% compounded annually. The first deposit equals $800; the second deposit equals $700; the third deposit equals $600; the fourth deposit equals $500; and the fifth deposit equals $400. Determine the amount in the fund immediately after the fifth deposit.

As depicted by the cash flow diagrams in Figure 3.11, the cash flow series can be represented by the difference in a uniform series of $800 and a gradient series of $100. The uniform series equivalent of the gradient series is given by

$$A = G(A|G\ 8,5)$$
$$= \$100(1.8465)$$
$$= \$184.65$$

Therefore, a uniform series having cash flows equal to $800 − $184.65, or $615.35, is equivalent to the original cash flow series. The future worth equiva-

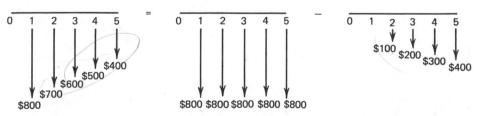

FIGURE 3.11. Cash flow diagrams for the decreasing gradient series example.

lent is found to be

$$F = A(F|A\ 8,5)$$
$$= \$615.35(5.8666)$$
$$= \$3610.01$$

### 3.4.3 Geometric Series of Cash Flows

The geometric cash flow series, as depicted in Figure 3.12, occurs when the size of a cash flow increases (decreases) by a fixed percent from one time period to the next. If $j$ denotes the percent change in the size of a cash flow from one period to the next, the size of the $t$th cash flow can be given by

$$A_t = A_{t-1}(1 + j) \qquad t = 2, \ldots, n$$

or, more conveniently,

$$A_t = A_1(1 + j)^{t-1} \qquad t = 1, \ldots, n \tag{3.33}$$

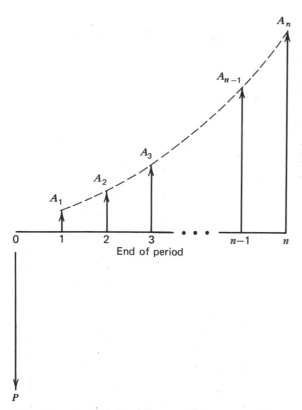

FIGURE 3.12. Cash flow diagram of the geometric series.

The geometric series is used to represent the growth (positive $j$) or decay (negative $j$) of costs and revenues undergoing annual percentage changes. As an illustration, if labor costs increase by 10% a year, then the resulting series representation of labor costs will be a geometric series.

The present worth equivalent of the cash flow series is obtained by substituting Equation 3.33 into Equation 3.6 to obtain

$$P = \sum_{t=1}^{n} A_1(1 + j)^{t-1}(1 + i)^{-t} \tag{3.34}$$

or

$$P = A_1(1 + j)^{-1} \sum_{t=1}^{n} \left(\frac{1 + j}{1 + i}\right)^t \tag{3.35}$$

As an exercise, the reader may wish to show that the following relationship results:

$$P = \begin{cases} A_1 \left[\dfrac{1 - (1 + j)^n(1 + i)^{-n}}{i - j}\right] & i \neq j \\[3mm] \dfrac{nA_1}{1 + i} & i = j \end{cases} \tag{3.36}$$

or

$$P = A_1(P|A_1\ i,j,n) \tag{3.37}$$

where $(P|A_1\ i,j,n)$ is the *geometric series, present worth factor* and is tabulated in Appendix A for various values of $i$, $j$, and $n$.

For the case of $j \geq 0$ and $i \neq j$, the relationship between $P$ and $A$ can be conveniently expressed in terms of compound interest factors previously considered

$$P = A_1 \left[\frac{1 - (F|P\ j,n)(P|F\ i,n)}{i - j}\right] \qquad i \neq j, \qquad j \geq 0 \tag{3.38}$$

Example 3.18_____

Labor costs have been increasing at an annual rate of 8%. A firm wishes to set aside funds to cover labor costs for the next 5 years. Determine how much must be set aside today if the money will be invested and will earn interest at a rate of 10%. The labor cost next year will be $50,000.

For this example, $A_1 = \$50{,}000$, $j = 8\%$, $i = 10\%$, and $n = 5$. The present worth equivalent will be

$$P = A_1(P|A_1\ 10,8,5)$$
$$= \$50{,}000(4.3831)$$
$$= \$219{,}155$$

If labor costs increase at an annual rate of 10%,

$$P = \frac{nA_1}{1 + i}$$

$$= \frac{5(\$50,000)}{1.1}$$

$$= \$227,272.73$$

The future worth equivalent of the geometric series is obtained by multiplying the value of the geometric series present worth factor and the $(F|P\ i,n)$ factor to obtain

$$F = \begin{cases} A_1 \left[ \dfrac{(1 + i)^n - (1 + j)^n}{i - j} \right] & i \neq j \\ nA_1(1 + i)^{n-1} & i = j \end{cases} \tag{3.39}$$

or

$$\boxed{F = A_1(F|A_1\ i,j,n)} \tag{3.40}$$

where $(F|A_1\ i,n)$ is the *geometric series, future worth factor* and is tabulated in Appendix A. From Equation 3.39, notice that $(F|A_1\ i,j,n) = (F|A_1\ j,i,n)$.

Example 3.19_____

An individual receives an annual bonus and deposits it in a savings account that pays 8% compounded annually. The size of the bonus increases by 10% each year; the initial deposit was $500. Determine how much will be in the fund immediately after the tenth deposit.

In this case, $A_1 = \$500$, $i = 8\%$, $j = 10\%$, and $n = 10$. Thus, the value of $F$ is given by

$$F = A_1(F|A_1\ 8,10,10) = A_1(F|A_1\ 10,8,10)$$
$$= \$500(21.7409)$$
$$= \$10,870.45$$

The interest factors developed to this point are summarized in Table 3.3. Values of the factors given in Table 3.3 are provided in Appendix A.

## 3.5 MULTIPLE COMPOUNDING PERIODS IN A YEAR

Not all interest rates are stipulated as annual compounding rates. For example, if the interest rate is stipulated as 12% compounded quarterly, then the interest period is a 3-month period with 3% interest rate/interest period. Thus, if $1000

TABLE 3.3. Summary of Discrete Compounding
Interest Factors

| To Find | Given | Factor | Symbol |
|---------|-------|--------|--------|
| $P$ | $F$ | $(1 + i)^{-n}$ | $(P\|F\ i,n)$ |
| $F$ | $P$ | $(1 + i)^n$ | $(F\|P\ i,n)$ |
| $P$ | $A$ | $\dfrac{(1 + i)^n - 1}{i(1 + i)^n}$ | $(P\|A\ i,n)$ |
| $A$ | $P$ | $\dfrac{i(1 + i)^n}{(1 + i)^n - 1}$ | $(A\|P\ i,n)$ |
| $F$ | $A$ | $\dfrac{(1 + i)^n - 1}{i}$ | $(F\|A\ i,n)$ |
| $A$ | $F$ | $\dfrac{i}{(1 + i)^n - 1}$ | $(A\|F\ i,n)$ |
| $P$ | $G$ | $\dfrac{1 - (1 + ni)(1 + i)^{-n}}{i^2}$ | $(P\|G\ i,n)$ |
| $A$ | $G$ | $\dfrac{(1 + i)^n - (1 + ni)}{i[(1 + i)^n - 1]}$ | $(A\|G\ i,n)$ |
| $P$ | $A_1, j$ | $\dfrac{1 - (1 + j)^n(1 + i)^{-n}}{i - j}$ | $(P\|A_1\ i,j,n)^*$ |
| $F$ | $A_1, j$ | $\dfrac{(1 + i)^n - (1 + j)^n}{i - j}$ | $(F\|A_1\ i,j,n)^*$ |

$^* i \neq j.$

is borrowed at an interest rate of 12% compounded quarterly, then the amount owed at the end of 5 years or 20 interest periods is obtained as follows:

$$F = P(F|P\ 3,20)$$
$$= \$1000(1 + 0.03)^{20}$$
$$= \$1000(1.8061)$$
$$= \$1806.10$$

With quarterly compounding, $1806.10 is to be repaid.

## Example 3.20

As another illustration of multiple compounding periods within a year, suppose interest is stated as 8% compounded quarterly. If you borrow $1000 for 1 year, how much must be repaid?

Using an interest rate of 2%/3-month period, after four interest periods the amount owed is given by:

$$F = P(F|P\ 2,4)$$
$$= \$1000(1.0824)$$
$$= \$1082.40$$

Thus, $1082.40 would be owed. Note that if interest were stated as 8.24% compounded annually,

$$F = P(F|P\ 8.24,1)$$
$$= \$1000(1.0824)$$
$$= \$1082.40$$

and the same amount would be owed.

---

From the results of Example 3.20 it can be concluded that 8% compounded quarterly is equivalent to 8.24% compounded annually. The rate of 8% is referred to as the *nominal annual interest rate*; the rate of 8.24% is referred to as the *effective annual interest rate*. The effective interest rate is defined as the annual compounding rate that is equivalent to the stated interest rate.

Letting $r$ denote the nominal annual interest rate, $m$ denote the number of compounding periods per year, $i$ denote the interest rate per interest period, and $i_{\text{eff}}$ denote the effective interest rate per year,

$$i_{\text{eff}} = \left(1 + \frac{r}{m}\right)^m - 1$$
$$= (1 + i)^m - 1$$

$$\boxed{i_{\text{eff}} = \left(F|P\ \frac{r}{m},\ m\right) - 1} \qquad (3.41)$$

To illustrate with 8% compounded semiannually, $r$ equals 0.08, $m$ equals 2, and

$$i_{\text{eff}} = (1 + 0.04)^2 - 1$$
$$= (F|P\ 4,2) - 1$$
$$= 0.0816$$

Thus, 8% compounded semiannually is equivalent to 8.16% compounded annually (i.e., 8.16% is the effective annual interest rate when interest is 8% compounded semiannually).

Suppose interest is stated as 18% compounded monthly. The effective annual interest rate is given by

$$i_{\text{eff}} = (1 + 0.015)^{12} - 1$$
$$= (F|P\ 1.5,12) - 1$$
$$= 0.1956 \qquad \text{or} \qquad 19.56\%$$

Example 3.21 _____

An individual borrowed $1000 and paid off the loan with interest after 4.5 years. The amount paid was $1500. What was the effective annual interest rate for this transaction? Letting the interest period be a 6-month period, it is seen that the payment of $1500 and the debt of $1000 are related by the expression

$$F = P(F|P\ i,n)$$

Thus,

$$\$1500 = \$1000(F|P\ i,9)$$

or

$$\$1500 = \$1000(1 + i)^9$$

Dividing both sides by $1000 gives

$$1.50 = (1 + i)^9$$

Taking the logarithm of both sides yields

$$\log 1.50 = 9 \log (1 + i)$$

*Handwritten annotations:*
$$\ln 1.50 = 9 \ln(1+i)$$
$$\tfrac{1}{9} \ln 1.50 = \ln(1+i)$$
$$e^{-1/9 \ln 1.50} = e^{\ln(1+i)}$$
$$e^{-0.0450} = 1+i \qquad i = .044$$

Dividing both sides by 9 and taking the antilog of the result provides the relation

$$(1 + i) = 1.046$$
$$i = 0.046 \qquad 4.4\%$$

Thus, the 6-month interest rate is approximately 4.6%. Computing the effective annual interest rate yields

$$i_{\text{eff}} = (1 + i)^2 - 1$$
$$= (1 + 0.046)^2 - 1$$
$$= 0.0943$$

The effective annual interest rate for the loan transaction was approximately 9.43%.

Instead of using logarithms, we could have searched the interest tables for a value of $i$ that yielded a value of 1.50 for the $(F|P\ i,9)$ factor. With interpolation $i$ would be found to equal approximately 4.6%. The computation of the effective annual interest rate would follow the same procedure used to determine the effective annual interest rate for 9.2% compounded semiannually.

---

With the passage of the Truth-in-Lending Law, anyone borrowing money must be informed of the total amount of interest paid and the *annual percentage rate* associated with the transaction. Depending on how the lender computes the annual percentage rate, the percentage quoted can be either the nominal annual interest rate, the effective annual interest rate, or something else entirely. To illustrate, many revolving charge accounts charge an interest rate of 1.5%/month, which is equivalent to 18% compounded monthly or 19.56% compounded annually. In many cases, the annual percentage rate is stated to be 18%.

Since, in personal financing, the annual percentage rate might be computed differently among the alternative sources of funds, the effective annual interest rate is a very good basis for comparing alternative financing plans. The effective annual interest rate is also useful when the timing of cash flows does not coincide with the end of interest periods.

A number of situations involve cash flows occurring at intervals that are not the same as the compounding intervals. Consider the following two possibilities: money is compounded *more* frequently than the occurrence of cash flows and money is compounded *less* frequently than the occurrence of cash flows.

### Example 3.22

Suppose an individual makes monthly deposits into a fund that pays interest at a rate of 8% compounded quarterly. Depending on how the financial institution interprets the rate of "8% compounded quarterly," money deposited during a quarter might not earn any interest. We interpret 8% compounded quarterly to mean that deposits earn interest equivalent to 8% compounded quarterly. Hence, the monthly interest rate to be applied to the monthly deposits should be equivalent to 8% compounded quarterly. Letting $i$ be the monthly rate, we note that the effective interest rate for $i\%$/month should be the same as for 8% compounded quarterly. Therefore

$$i_{\text{eff}} = (1 + i)^{12} - 1$$
$$= \left(1 + \frac{0.08}{4}\right)^4 - 1$$

or

$$(1 + i)^{12} = (1.02)^4$$

or

$$(1 + i)^3 = 1.02$$

Thus,

$$1 + i = (1.02)^{1/3}$$
$$i = (1.02)^{1/3} - 1$$
$$i = 0.0066227$$

Consequently, a monthly interest rate of 0.66227% can be used to determine the balance in the fund at the end of any month.

### Example 3.23

Suppose quarterly deposits are made into a fund that pays interest at a rate of 6% compounded monthly. By the same arguments used in the previous example, we know that

$$i_{eff} = (1 + i)^4 - 1$$

$$= \left(1 + \frac{0.06}{12}\right)^{12} - 1$$

or

$$(1 + i)^4 = (1.005)^{12}$$
$$i = (1.005)^3 - 1$$
$$= 0.015075$$

Thus, a quarterly interest rate of 1.5075% can be used to determine the balance in the fund at the end of any quarter.

## 3.6 CONTINUOUS COMPOUNDING

In the discussion of the effective interest rate in the previous section it was noted that as the frequency of compounding in a year increases the effective interest rate increases. Since monetary transactions occur daily or hourly in most businesses and money is normally "put to work" for the business as soon as it is received, compounding is occurring quite frequently. If one wishes to account explicitly for such rapid compounding, then *continuous compounding relations should be used*. Continuous compounding means that each year is divided into an infinite number of interest periods. Mathematically, the single payment compound amount factor under continuous compounding is given by

$$\lim_{m \to \infty} \left(1 + \frac{r}{m}\right)^{mn} = e^{rn}$$

where $n$ is the number of years, $m$ is the number of interest periods per year, and $r$ is the nominal annual interest rate. Given $P$, $r$, and $n$, the value of $F$ can be computed using continuous compounding as follows.

$$F = P\, e^{rn} \qquad\qquad (3.42)$$

or

$$F = P(F|P\ r,n)_{\infty} \qquad\qquad (3.43)$$

where $(F|P\ r,n)_{\infty}$ denotes the *continuous compounding, single sum, future worth factor.* The subscript ∞ is provided to denote that continuous compounding is being used. The interest tables for continuous compounding are given in Appendix B.

Example 3.24 _____

If $2000 is invested in a fund that pays interest at a rate of 12% compounded continuously, after 5 years the cumulative amount in the fund will total

$$F = P(F|P\ 12\%,5)_\infty$$
$$= \$2000(1.8221)$$
$$= \$3644.20$$

Thus, a withdrawal of $3644.20 will deplete the fund after 5 years.

_____

The effective interest rate under continuous compounding is easily obtained using the relation

$$\boxed{i_{eff} = e^r - 1} \tag{3.44}$$

or

$$\boxed{i_{eff} = (F|P\ r,1)_\infty - 1} \tag{3.45}$$

To illustrate, if interest is 12% compounded continuously, then the effective interest rate is given by

$$i_{eff} = (F|P\ 12,1)_\infty - 1$$
$$= 0.1275$$

Thus, 12.75% compounded annually is equivalent to 12% compounded continuously.

The inverse relationship between $F$ and $P$ indicates that

$$\boxed{P = F\,e^{-rn}} \tag{3.46}$$

or

$$\boxed{P = F(P|F\ r,n)_\infty} \tag{3.47}$$

where $(P|F\ r,n)_\infty$ is called the *continuous compounding, single sum, present worth factor.*

## 3.6.1 Discrete Flows

If it is assumed that cash flows are discretely spaced over time, then the continuous compound relations for the uniform, gradient, and geometric series can be obtained. Substituting $e^{-rn}$ for $(1 + i)^{-n}$, $e^r - 1$ for $i$, and $e^{rn}$ for $(1 + i)^n$ in the remaining discrete compounding formulas yields the continuous compound interest factors summarized in Table 3.4. Values for these factors are provided in Appendix B.

### Example 3.25

To illustrate the use of the continuous compound interest factors, suppose $1000 is deposited each year into an account that pays interest at a rate of 12% compounded continuously. Determine both the amount in the account immediately after the tenth deposit and the present worth equivalent for 10 deposits.

The amount in the fund immediately after the tenth deposit is given by the relation

$$F = \$1000(F|A\ 12,10)_\infty$$
$$= \$1000(18.1974)$$
$$= \$18,197.40$$

The present worth equivalent for 10 deposits is obtained using the relation

$$P = \$1000(P|A\ 12,10)_\infty$$
$$= \$1000(5.4810)$$
$$= \$5481$$

---

In the case of the geometric series the size of the $t$th cash flow will be assumed to be given by

$$A_t = A_{t-1}e^c \qquad t = 2, \ldots, n \tag{3.48}$$

or, equivalently,

$$A_t = A_1 e^{(t-1)c} \qquad t = 1, \ldots, n \tag{3.49}$$

where $c$ is the nominal compound rate of increase in the size of the cash flow. The resulting expressions for the *continuous compounding, geometric series present worth factor* and the *continuous compounding, geometric series future worth factor* are given, respectively, by $(P|A_1\ r,c,n)_\infty$ and $(F|A_1\ r,c,n)_\infty$, as given in Table 3.4 for the case of $r \neq c$. As an exercise you may wish to derive the appropriate expressions when $r = c$.

### Example 3.26

An individual receives an annual bonus and deposits it in a savings account that pays 8% compounded continuously. The size of the bonus increases each year

TABLE 3.4. Summary of Continuous Compounding Interest Factors for Discrete Flows

| To Find | Given | Factor | Symbol |
|---------|-------|--------|--------|
| $P$ | $F$ | $e^{-rn}$ | $(P\|F\ r,n)_\infty$ |
| $F$ | $P$ | $e^{rn}$ | $(F\|P\ r,n)_\infty$ |
| $F$ | $A$ | $\dfrac{e^{rn} - 1}{e^r - 1}$ | $(F\|A\ r,n)_\infty$ |
| $A$ | $F$ | $\dfrac{e^r - 1}{e^{rn} - 1}$ | $(A\|F\ r,n)_\infty$ |
| $P$ | $A$ | $\dfrac{e^{rn} - 1}{e^{rn}(e^r - 1)}$ | $(P\|A\ r,n)_\infty$ |
| $A$ | $P$ | $\dfrac{e^{rn}(e^r - 1)}{e^{rn} - 1}$ | $(A\|P\ r,n)_\infty$ |
| $P$ | $G$ | $\dfrac{e^{rn} - 1 - n(e^r - 1)}{e^{rn}(e^r - 1)^2}$ | $(P\|G\ r,n)_\infty$ |
| $A$ | $G$ | $\dfrac{1}{e^r - 1} - \dfrac{n}{e^{rn} - 1}$ | $(A\|G\ r,n)_\infty$ |
| $P$ | $A_1,c$ | $\dfrac{1 - e^{(c-r)n}}{e^r - e^c}$ | $(P\|A_1\ r,c,n)_\infty{}^*$ |
| $F$ | $A_1,c$ | $\dfrac{e^{rn} - e^{cn}}{e^r - e^c}$ | $(F\|A_1\ r,c,n)_\infty{}^*$ |

$*\ r \neq c.$

at a rate of 10% compounded continuously; the initial deposit was $500. Determine how much will be in the fund immediately after the tenth deposit.

In this case, $A_1 = \$500$, $r = 8\%$, $c = 10\%$, and $n = 10$. Thus, the value of $F$ is given by

$$F = A_1(F|A_1\ 8,10,10)_\infty$$
$$= \$500(22.5162)$$
$$= \$11,258.10$$

$\dfrac{e^{.08(10)} - e^{.10(10)}}{e^{.08} - e^{.10}} = 22.5162$

$24.596$ —In book,

Comparing the result with that obtained in Example 3.19, the effect of continuous compounding increases the amount in the fund by $\$11,258.10 - \$10,870.45 = \$387.65$.

If discrete flows occur during a year, it is necessary to define $r$ consistent with the spacing of cash flows. To illustrate, suppose semiannual deposits are made into an account paying 12% compounded continuously. In this case the nominal semiannual rate should be determined such that its effective interest rate is 12.75%, or $r = 6.18\%$.

Since the differences in discrete and continuous compounding are not great in most cases, it is not uncommon to see discrete compounding used when

2.22554

continuous compounding is more appropriate. The arguments given for this are that errors in estimating the cash flows will probably offset any attempts to be very precise by using continuous compounding and that the interest rate used in discrete compounding is actually the effective interest rate resulting from continuous compounding.

### 3.6.2 Continuous Flow

Thus far only discrete cash flows have been assumed. It was assumed that cash flows occurred at, say, the end of the year. In some cases money is expended throughout the year on a somewhat uniform basis. Costs for labor, carrying inventory, and operating and maintaining equipment are typical examples. Others include capital improvement projects that conserve energy, water, or process steam.

Consequently, as a mathematical convenience, instead of assuming that money flows in discrete increments at the end of monthly, weekly, daily, or hourly time periods, it is assumed that money flows continuously during the time period at a uniform rate. Instead of having a uniform series of discrete cash flows of magnitude A, it assumed that a *total* of $\bar{A}$ dollars flows uniformly and continuously throughout a given time period. Such an approach to modeling cash flows is referred to as the *continuous flow* approach.

To illustrate the continuous flow concept, suppose you are to divide \$1000 into $k$ equal amounts to be deposited at equally spaced points in time during a year. The interest rate per period is defined to be $r/k$, where $r$ is the nominal rate. Thus, the present worth of the series of $k$ equal amounts is

$$P = \frac{\$1000}{k} \left(P|A\ \frac{r}{k},\ k\right)$$

or

$$P = \frac{\$1000}{k} \left[\frac{(1 + (r/k))^k - 1}{(r/k)(1 + (r/k))^k}\right]$$

which reduces to

$$P = \$1000 \left[\frac{1}{r} - \frac{1}{r(1 + (r/k))^k}\right]$$

Taking the limit of $P$ as $k$ approaches infinity gives

$$\lim_{k\to\infty} P = \$1000 \left(\frac{1}{r} - \frac{1}{re^r}\right)$$

or

$$P = \$1000 \left(\frac{e^r - 1}{re^r}\right)$$

In general, for $n$ years,

$$P = \bar{A}\left(\frac{e^{rn} - 1}{re^{rn}}\right)$$

(3.50)

or

$$P = \bar{A}(P|\bar{A}\ r,n)$$

(3.51)

where $(P|\bar{A}\ r,n)$ is referred to as the *continuous flow, continuous compounding uniform series present worth factor* and is tabulated in Appendix B.

The remaining continuous flow, continuous compound interest factors are summarized in Table 3.5. Values of the factors are given in Appendix B for various values of $r$ and $n$. With continuous flow and continuous compounding the continuous annual cash flow, $\bar{A}$, is equivalent to $\bar{A}r/(e^r - 1)$, when discrete flow and continuous compounding is used. Hence, the discrete flow equivalent of $\bar{A}$ is given by

$$A = \frac{\bar{A}(e^r - 1)}{r}$$

or

$$A = \bar{A}(F|\bar{A}\ r,1)$$

## Example 3.27

What are the present worth and future worth equivalents of a uniform series of continuous cash flows totalling $10,000/year for 10 years when interest is compounded continuously at a rate of 20%/year?

TABLE 3.5. Summary of Continuous Compounding Interest Factors for Continuous Flows

| To Find | Given | Factor | Symbol |
|---------|-------|--------|--------|
| $P$ | $\bar{A}$ | $\dfrac{e^{rn} - 1}{re^{rn}}$ | $(P|\bar{A}\ r,n)$ |
| $\bar{A}$ | $P$ | $\dfrac{re^{rn}}{e^{rn} - 1}$ | $(\bar{A}|P\ r,n)$ |
| $F$ | $\bar{A}$ | $\dfrac{e^{rn} - 1}{r}$ | $(F|\bar{A}\ r,n)$ |
| $\bar{A}$ | $F$ | $\dfrac{r}{e^{rn} - 1}$ | $(\bar{A}|F\ r,n)$ |

The present worth equivalent is given by

$$P = \$10,000(P|\bar{A}\ 20,10)$$
$$= \$10,000(4.3233)$$
$$= \$43,233$$

The future worth equivalent is given by

$$F = \$10,000(F|\bar{A}\ 20,10)$$
$$= \$10,000(31.9453)$$
$$= \$319,453$$

Continuous flow and continuous compounding are concepts that are used to represent more closely the realities of business transactions. In fact, several of the computer programs developed by industries and governmental agencies and used to perform economic analyses incorporate both concepts. We have presented both concepts since each has experienced increased usage during the past decade.

## 3.7 EQUIVALENCE

Throughout the preceding discussion we have used the term *equivalence* without defining what was meant by the term. This was done intentionally in order to introduce subtly the notion that two cash flow series or profiles are *equivalent* at some specified interest rate, $k\%$, if their present worths are *equal* using an interest rate of $k\%$.

Example 3.28 _____

The two cash flow profiles shown in Figure 3.13 are equivalent at 15%, since each has a present worth of $424.80. You may wish to verify this claim.

Of course, if the present worths of two cash flow profiles are equal at $k\%$, their "worths" at any given point in time will be equal when using a discount (interest) rate of $k\%$. Likewise, if two cash flow profiles are equivalent at $k\%$, their respective uniform series equivalents will be equal when expressed over the same time period. Hence we could state that the two cash flow profiles shown in Figure 3.13 are equivalent at 15%, since each has a worth of $1299.46 at the end of period eight; alternately, we could assert their equivalence at 15% on the basis that each can be represented by a uniform series of $126.72/period over the interval [1, 8].

Example 3.29 _____

What single sum of money at $t = 6$ is equivalent to the cash flow profile shown in Figure 3.14 if $i = 10\%$?

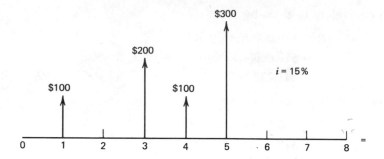

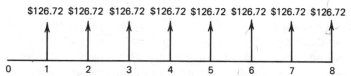

FIGURE 3.13. Cash flow diagrams for the equivalence example.

The present worth of the cash flow profile is given by

$$P = -\$400(P|F\ 10,1) + \$100(P|A\ 10,3)(P|F\ 10,1)$$
$$+\$100(P|A\ 10,3)(P|F\ 10,5)$$
$$= -\$400(0.9091) + \$100(2.4869)(0.9091) + \$100(2.4869)(0.6209)$$
$$= \$16.85$$

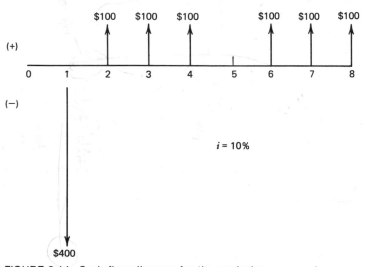

FIGURE 3.14. Cash flow diagram for the equivalence example.

96 • TIME VALUE OF MONEY OPERATIONS

Moving $16.85 forward in time to $t = 6$ gives

$$F = \$16.85(F|P\ 10,6)$$
$$= \$16.85(1.7716)$$
$$= \$29.86$$

Thus, at 10% a positive cash flow of $29.86 at $t = 6$ is equivalent to the cash flow profile shown in Figure 3.14.

Example 3.30

Using an 8% discount rate, what uniform series over five periods, [1, 5], is equivalent to the cash flow profile given in Figure 3.15?

The cash flow profile in Figure 3.15a consists of the difference in a uniform series of $500 and a gradient series, with $G = \$100$. A uniform series equivalent

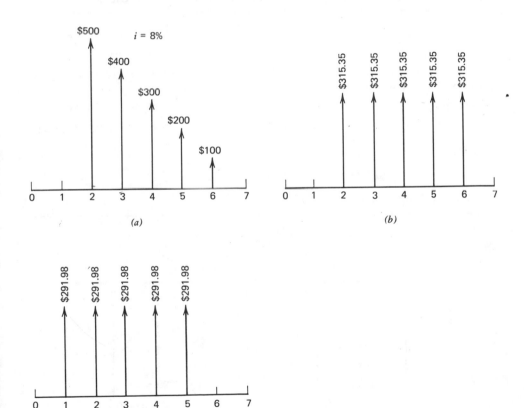

FIGURE 3.15. Cash flow diagrams for the equivalence example.

of the cash flow profile can be obtained for the interval [2, 6] as follows:

$$A = \$500 - \$100(A|G\ 8,5)$$
$$= \$500 - \$100(1.8465)$$
$$= \$315.35$$

The uniform series of $315.35 over the interval [2, 6] must be converted to a uniform series over the interval [1, 5]. Thus, *each* of the five cash flows must be moved back in time one period. The discounted value of $315.35 over one time period using an 8% discount rate is

$$P = \$315.35(P|F\ 8,1)$$
$$= \$315.35(0.9259)$$
$$= \$291.98$$

Consequently, a uniform series of $291.98 over the interval [1, 5] is equivalent to the cash flow profile given in Figure 3.15a. If you have doubts concerning the equivalence, compare their present worths using an 8% interest rate.

---

Example 3.31 _____

Determine the value of $X$ that makes the two cash flows, as given in Figure 3.16, equivalent when a discount rate of 15% is used.

Equating the future worths of the two cash flow profiles at $t = 4$ gives

$$\$200(F|A\ 15,4) + \$100(F|A\ 15,3) + \$100 = [\$200 + X(A|G\ 15,4)](F|A\ 15,4)$$

Canceling $200(F|A 15,4) on both sides yields

$$\$100(3.4725) + \$100 = X(1.3263)(4.9934)$$

Solving for $X$ gives a value of $67.53.

---

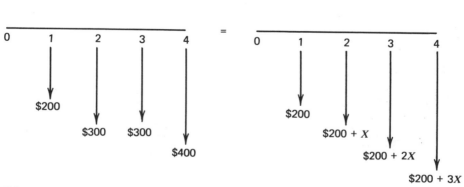

FIGURE 3.16. Cash flow diagrams for the equivalence example.

Example 3.32_____

For what interest (discount) rate are the two cash flow profiles shown in Figure 3.17 equivalent?

Converting each cash flow profile to a uniform series over the interval [1, 5], gives

$$-\$4000(A|P\ i,5) + \$1500 = -\$7000(A|P\ i,5) + \$1500 + \$500(A|G\ i,5)$$

or

$$\$3000(A|P\ i,5) = \$500(A|G\ i,5)$$

which reduces to

$$(A|G\ i,5) = 6(A|P\ i,5)$$

On searching through the interest tables at $n = 5$, it is found that the $(A|G\ i,5)$ factor is six times the value of the $(A|P\ i,5)$ factor for an interest rate between 12% and 15%. Specifically, with a 12% interest rate,

$$(A|G\ 12,5) - 6(A|P\ 12,5) = 1.7746 - 6(0.2774) = 0.1102$$

and, using a 15% interest rate,

$$(A|G\ 15,5) - 6(A|P\ 15,5) = 1.7228 - 6(0.2983) = -0.0670$$

Interpolating for $i$ gives

$$i = 0.12 + \frac{(0.15 - 0.12)(0.1102)}{(0.1102 + 0.0670)}$$

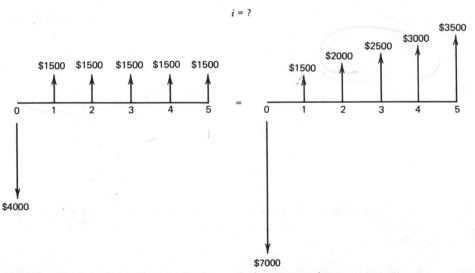

FIGURE 3.17. Cash flow diagrams for the equivalence example.

or

$$i = 0.1386$$

Therefore, using a discount rate of approximately 13.86% will establish an equivalence relationship between the cash flow profiles given in Figure 3.17.

---

Example 3.33

A firm purchases a machine for $30,000, keeps it for 5 years, and sells it for $6000. During the time the machine was owned by the company, operating and maintenance costs totaled $8000 the first year, $9000 the second year, $10,000 the third year, $11,000 the fourth year, and $12,000 the fifth year. The firm uses a 15% interest rate in performing economic analyses. Determine the single sum of money occurring at (1) $t = 0$ and (2) $t = 5$, which is equivalent to the cash flow history for the machine. Also, determine the uniform series occurring over the interval [1, 5] that is equivalent to the cash flow profile for the machine.

In the economic analysis literature, the single sum equivalent at time zero for a cash flow profile is called the *present worth* or *present value* for the cash flow profile. We will use the term present worth and denote it as $PW$. For the example,

$$PW = -\$30,000 - \$8000(P|A\ 15,5) - \$1000(P|G\ 15,5)$$
$$+ \$6000(P|F\ 15,5)$$
$$= -\$59,609.57$$

Hence, a single expenditure of $59,609.57 at time zero would have been equivalent to the cash flows experienced during the ownership of the machine.

The single sum equivalent at the end of the life of a project is termed the *future worth* or *future value* for the project. We denote the future worth by $FW$ and compute its value as follows:

$$FW = -\$30,000(F|P\ 15,5) - [\$8000 + 1000(A|G\ 15,5)](F|A\ 15,5)$$
$$+ \$6000$$
$$= -\$119,897$$

Alternately, the future worth can be obtained from the present worth:

$$FW = PW(F|P\ i,n)$$
$$= -\$59,609.57(F|P\ 15,5)$$
$$= -\$119,898.69$$

(The difference of $1.69 is due to round-off error in the tables.) Hence, a single expenditure of $119,897 at time 5 would have been equivalent to the cash flows associated with the machine.

A uniform series equivalent for a series of yearly cash flows is referred to as the *annual worth* or *equivalent uniform annual cost* for the project. The latter

EUAC

expression is most appropriate for the type example under consideration, since the resulting uniform series is a cost, not an income, series. The annual worth designation is $AW$; $EUAC$ denotes the equivalent uniform annual cost. We will use both designations throughout the text.

For the example problem, the $EUAC$ determination is performed as follows:

$$EUAC = \$30,000(A|P\ 15,5) + [\$8000 + \$1000(A|G\ 15,5)]$$
$$-\$6000(A|F\ 15,5)$$
$$= \$17,782$$

Hence an annual expenditure of \$17,782/year for 5 years is equivalent to the cash flow profile associated with the machine investment. Alternately, the equivalent uniform annual cost can be obtained from either the present worth or the future worth.

$$EUAC = -PW(A|P\ i,n)$$
$$EUAC = -FW(A|F\ i,n)$$

Present worth, future worth, and annual worth computations are used often in comparing economic investment alternatives having different cash flow profiles. Consequently, we will have need for $PW$, $FW$, and $AW$ calculations for the discussion of alternative comparisons in Chapter Four.

## 3.8 PRINCIPAL AMOUNT AND INTEREST AMOUNT IN LOAN PAYMENTS

With both personal and corporate investments that are financed from borrowed funds, income taxes are affected by the amount of interest paid. Hence, it is quite important to know how much of each payment is interest and how much is reducing the principal amount borrowed initially. To illustrate this situation, suppose you borrowed \$10,000 and paid it back using four equal annual installments, with interest computed at 10% compounded annually. The payment size is computed to be

$$A = P(A|P\ 10,4)$$
$$= \$10,000(0.3155)$$
$$= \$3155$$

The interest accumulation the first year is 0.10(10,000), or \$1000. Therefore, the first payment consists of a \$1000 interest payment and a \$2155 principal payment. The unpaid balance at the beginning of the second year is \$10,000 − \$2155, or \$7845; consequently, the interest charge the second year is 0.10(\$7845), or \$784.50. Thus, the second payment consists of a \$784.50 interest payment and a principal payment of \$3155 − \$784.50, or \$2370.50. The unpaid balance at the beginning of the third year is \$7845 − \$2370.50, or \$5474.50, so the interest charge the third year is 0.10(\$5474.50), or \$547.45.

Thus, the third payment consists of a $547.45 interest payment and a principal payment of $3155 − $547.45, or $2607.55. The unpaid balance at the beginning of the fourth year is $5474.50 − $2607.55, or $2866.95; the interest charge the fourth year is 0.10($2866.95), or $286.70. Thus, the fourth payment consists of a $286.70 interest payment and a principal payment of $3155 − $286.70, or $2868.30 (the principal payment in the fourth payment should equal the unpaid balance of $2866.95 at the beginning of the fourth year. The difference of $1.35 is due to round-off errors in computing the payment size.)

The amount of principal remaining to be repaid immediately after making payment $(t - 1)$ can be found by stripping off the interest on the remaining $n - t + 1$ payments; letting $U_{t-1}$ denote the unpaid principal after making payment $t - 1$, we know that

$$U_{t-1} = A(P|A\ i,n - t + 1) \tag{3.52}$$

or

$$U_{t-1} = A \sum_{j=1}^{n-t+1} (1 + i)^{-j}$$

where $A$ denotes the size of the individual payments and $i$ represents the interest rate used to compute $A$.

The amount by which payment $t$ reduces the unpaid principal will be designated $E_t$ and is given by the relation, $E_t = U_{t-1} - U_t$. Hence,

$$E_t = A \sum_{j=1}^{n-t+1} (1 + i)^{-j} - A \sum_{j=1}^{n-t} (1 + i)^{-j} \tag{3.53}$$

or

$$E_t = A \left[ \sum_{j=1}^{n-t} (1 + i)^{-j} + (1 + i)^{-(n-t+1)} \right] - A \sum_{j=1}^{n-t} (1 + i)^{-j}$$

Thus,

$$E_t = A(1 + i)^{-(n-t+1)} \tag{3.54}$$

or

$$\boxed{E_t = A(P|F\ i,n - t + 1)} \tag{3.55}$$

Recalling how the payment size is determined, we express $E_t$ as

$$\boxed{E_t = P(A|P\ i,n)(P|F\ i,n - t + 1)} \tag{3.56}$$

where $P$ is the original principal amount borrowed, and $E_t$ is the amount of payment $t$, which is an *equity* payment (i.e., payment against principal). Letting

$I_t$ be the amount of payment $t$, which is an interest payment, it is seen that

$$I_t = A - E_t$$

$$\boxed{I_t = A[1 - (P|F\ i, n - t + 1)]} \qquad (3.57)$$

Example 3.34——————————————————————————————

To illustrate the use of the formulas for computing the values of $E_t$ and $I_t$, suppose \$10,000 is borrowed at 12% annual interest and repaid with five equal annual payments. The payment size is found to be

$$A = \$10,000(A|P\ 12,5)$$
$$A = \$10,000(0.2774)$$
$$= \$2774$$

Thus

$$E_1 = \$2774(P|F\ 12,5) = \$1574, \qquad I_1 = \$1200$$
$$E_2 = \$2774(P|F\ 12,4) = \$1763, \qquad I_2 = \$1011$$
$$E_3 = \$2774(P|F\ 12,3) = \$1975, \qquad I_3 = \$799$$
$$E_4 = \$2774(P|F\ 12,2) = \$2211, \qquad I_4 = \$563$$
$$E_5 = \$2774(P|F\ 12,1) = \$2477, \qquad I_5 = \$297$$

Example 3.35——————————————————————————————

An individual purchases a \$130,000 house and makes a down payment of \$30,000. The remaining \$100,000 is financed over a 25-year period at 12% compounded monthly. The monthly house payment is computed to be

$$A = \$100,000(A|P\ 1,300)$$
$$= \$1053$$

The individual keeps the house for 5 years and decides to sell it. How much equity is there in the house?

The amount of principal remaining to be paid can be determined by computing the present worth of the remaining 240 monthly payments using a 1% interest rate, or

$$P = \$1053(P|A\ 1,240)$$
$$= \$95,633$$

Thus, the individual's equity equals (\$100,000 − \$95,633) or \$4367, plus the value of the down payment, or $34,367. It is interesting (or perhaps depressing if you are a homeowner) to determine that over the 5-year period the individual made payments totaling $63,180, of which $58,813 was for interest, that is, $58,813 = $63,180 − $4367.

## 3.9 CHANGING INTEREST RATES

The preceding discussion assumed that the interest rate did not change during the time period of concern. Recent experience indicates that such a situation is not likely if the time period of interest extends over several years (i.e., more than one interest rate may be applicable). Considering a single sum of money and discrete compounding, if $i_t$ denotes the interest rate appropriate during time period $t$, the future worth equivalent for a single sum of money can be expressed as

$$F = P(1 + i_1)(1 + i_2) \ldots (1 + i_{n-1})(1 + i_n) \tag{3.58}$$

and the inverse relation

$$P = F(1 + i_n)^{-1}(1 + i_{n-1})^{-1} \ldots (1 + i_2)^{-1}(1 + i_1)^{-1} \tag{3.59}$$

Example 3.36 _____

Consider the situation depicted in Figure 3.18 in which an individual deposited $1000 in a savings account that paid interest at an annual compounding rate of 8% for the first 3 years, 10% for the next 4 years, and 12% for the next 2 years. How much was in the fund at the end of the ninth year?

Letting $V_t$ denote the value of the account at the end of time period $t$, we see

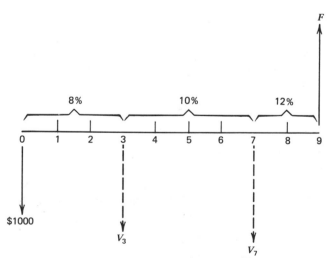

FIGURE 3.18. Cash flow diagram for a changing interest rate example.

that

$$V_3 = \$1000(F|P\ 8,3)$$
$$= \$1000(1.2597)$$
$$= \$1259.70$$

Likewise,

$$V_7 = \$1259.7(F|P\ 10,4)$$
$$= \$1259.7(1.4641)$$
$$= \$1844.33$$

Similarly,

$$F = V_9 = \$1844.33(F|P\ 12,2)$$
$$= \$1844.33(1.2544)$$
$$= \$2313.53$$

Alternately, the amount in the account at the end of nine years is given by

$$F = \$1000(1.08)(1.08)(1.08)(1.10)(1.10)(1.10)(1.10)(1.12)(1.12)$$
$$= \$2313.55$$

(The difference in answers is due to cumulative round-off errors in the interest tables.)

---

Extending the consideration of changing interest rates to series of cash flows, the present worth of a series of cash flows can be represented as

$$P = A_1(1 + i_1)^{-1} + A_2(1 + i_1)^{-1}(1 + i_2)^{-1} + \cdots$$
$$+ A_n(1 + i_1)^{-1}(1 + i_2)^{-1} \cdots (1 + i_n)^{-1} \tag{3.60}$$

The future worth of a series of cash flows can be given by

$$F = A_n + A_{n-1}(1 + i_n) + A_{n-2}(1 + i_{n-1})(1 + i_n) + \cdots$$
$$+ A_1(1 + i_2)(1 + i_3) \cdots (1 + i_{n-1})(1 + i_n) \tag{3.61}$$

Example 3.37_____

Consider the cash flow diagram given in Figure 3.19 with the appropriate interest rates indicated. Determine the present worth, future worth, and uniform series equivalents for the cash flow series.

Computing the present worth gives

$$P = \$200(P|F\ 10,1) - \$200(P|F\ 10,1)(P|F\ 10,1)$$
$$+ \$300(P|F\ 8,1)(P|F\ 10,1)(P|F\ 10,1)$$
$$+ \$200(P|F\ 12,1)(P|F\ 8,1)(P|F\ 8,1)(P|F\ 10,1)(P|F\ 10,1)$$
$$= \$200(P|F\ 10,1) - \$200(P|F\ 10,2)$$
$$+ \$300(P|F\ 10,2)(P|F\ 8,1) + \$200(P|F\ 10,2)(P|F\ 8,2)(P|F\ 12,1)$$
$$= \$372.62$$

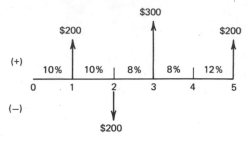

FIGURE 3.19. Cash flow diagram for a
changing interest rate example.

The future worth is given by

$$F = \$200 + \$300(F|P\ 8,1)(F|P\ 12,1) - \$200(F|P\ 8,2)(F|P\ 12,1)$$
$$+ \$200(F|P\ 10,1)(F|P\ 8,2)(F|P\ 12,1)$$
$$= \$589.01$$

The uniform series equivalent is obtained as follows.

$$P = A(P|F\ 10,1) + A(P|F\ 10,2) + A(P|F\ 8,1)(P|F\ 10,2)$$
$$+ A(P|F\ 8,2)(P|F\ 10,2) + A(P|F\ 12,1)(P|F\ 8,2)(P|F\ 10,2)$$
$$\$372.62 = A[(0.9091) + (0.8264) + (0.9259)(0.8264) + (0.8573)(0.8264)$$
$$+ (0.8929)(0.8573)(0.8264)]$$
$$\$372.62 = 3.842\ A$$
$$A = \$96.99$$

Thus, $96.99/time period for five time periods is equivalent to the original cash flow series.

## 3.10 SUMMARY

In this chapter we developed the time value of money concept and defined a number of mathematical operations consistent with that concept. To motivate the discussion, we emphasized the subject from the viewpoint of personal financing. In subsequent chapters we apply the concepts developed in this chapter to the study of investment alternatives from the viewpoint of an ongoing enterprise.

## PROBLEMS

1. Person A sells Person B a used automobile for $2000. Person B pays $500 cash down and gives Person A three personal notes for the remainder due. The principal

of each note is therefore $500. One note is due at the end of the first year; the second note due at the end of the second year; and the third note is due at the end of the third year. The annual simple interest rate agreed on is 15%. How much total interest will Person B pay Person A? (3.2)

2. A debt of $2000 is incurred at $t = 0$. An annual simple interest rate of 16% on the unpaid balance is agreed on. Three equal payments of $890.50 each at $t = 1, 2, 3$ will pay off this debt and the relevant interest due. For *each* payment, what is (a) the payment on the principal, and (b) the interest amount paid? (3.2)

3. If $5000 is deposited at $t = 0$ into a fund paying 9% compounded per period, what sum will be accumulated at the end of ten periods, or $t = 10$? What would be the sum accumulated if the fund paid 12% compounded per period? (3.3)

4. How long does it take a deposit in a 12% fund to triple in value? (3.3)

5. If a deposit of $5000 at $t = 0$ amounts to $14,930 at the end of the sixth compounding period, what value of $i$ is involved? (3.3)

6. How much money today is equivalent to $2000 in 6 years, with interest at 9% compounded annually? (3.3)

7. If a fund pays 10% compounded annually, what single deposit is required at $t = 0$ in order to accumulate $12,000 in the fund at the end of the tenth year $(t - 10)$? (3.3)

8. Susan deposits $2000, $1200, and $500 at $t = 1, 2, 3$, respectively. If the fund pays 8% compounded per period, what sum will be accumulated in the fund at (a) $t = 3$ and (b) $t = 4$? (3.4)

9. How much should be deposited at $t = 0$, into a fund paying 10% compounded per period, in order to withdraw $2000 at $t = 1$, $1500 at $t = 3$, and $700 at $t = 7$ and the fund be depleted? (3.4)

10. Barbara annually deposits $A_t$ in an account at time $t$, $t = 1, \ldots , 20$, where

$$A_t = 25t(1.10)^{t-1}$$

If the fund pays 10%, how much is in the fund immediately after the seventeenth deposit?

*Note.*

$$\sum_{t=1}^{n} t = n(n + 1)/2$$

$$\sum_{t=1}^{n} tx^t = \frac{(x - 1)(n + 1)x^{n+1} - x^{n+2} + x}{(x - 1)^2}$$

(3.4)

11. Bill deposits $2000 in a savings account that pays 10% compounded annually. Exactly 3 years later he deposits $3000. Two years after the $3000 deposit $4000 is deposited. Then, 4 years after the $4000 deposit, half of the accumulated funds is transferred to a fund that pays 12% compounded annually. How much money will be in each fund, 4 years after the transfer? (3.4)

**12.** A debt of $1000 is incurred at $t = 0$. What is the amount of three equal payments at $t = 1, 2, 3$ that will repay the debt if "money is worth" 8% compounded per period? (3.4.1)

**13.** Five deposits of $500 each are made at $t = 1, 2, 3, 4, 5$ into a fund paying 8% compounded per period. How much will be accumulated in the fund at (a) $t = 5$, and (b) $t = 9$? (3.4.1)

**14.** What equal, annual deposits must be made at $t = 2, 3, 4, 5$ in order to accumulate $10,000 at $t = 6$ if "money is worth" 10% compounded annually? (3.4.1)

**15.** A deposit of $X is placed into a fund paying 10% compounded annually at $t = 0$. If withdrawals of $9464 can be made at $t = 1, 2, 3$, and 4 such that the fund is depleted with the last withdrawal, show that the value of $X$ is $30,000 by (a) use of the interest factor $(P|A\ 10,4)$, and (b) use of the interest factors: $(P|F\ 10,1)$, $(P|F\ 10,2)$, $(P|F\ 10,3)$, and $(P|F\ 10,4)$. (3.4.1)

**16.** If you know the values of the $(F|P, i,n)$ factor for $n = 1, 2, \ldots, n$ show how you would determine the value for the $(A|P\ 6,n)$ factor. (3.4.1)

**17.** With 12% interest compounded annually, how much money will be accumulated in a fund at the time of the fifth deposit, if equal deposits of $800 are made annually? (3.4.1)

**18.** Kim deposits $1000 in a savings account; 6 years after the deposit, half of the account balance is withdrawn; $2000 is deposited annually for 6 more years, with the first deposit occuring 1 year after the withdrawal; the total balance is withdrawn at the end of the fifteenth year. If the account earns interest at a rate of 9%, how much is withdrawn (a) at the end of 6 years, and (b) at the end of 15 years? (3.4.1)

**19.** John borrows $10,000 at 15% compounded annually and wishes to pay the loan back over a 5-year period with annual payments. However, the second payment is to be $500 greater than the first payment; the third payment is to be $500 greater than the second payment; the fourth payment is to be $500 greater than the third payment; and the fifth payment is to be $300 greater than the fourth payment. Determine the size of the first payment. (3.4.2)

**20.** By the use of a uniform gradient series formula and any other appropriate formulas, determine the future worth, $F$, at $t = 15$ of the following deposits: $1000 at $t = 8$, $900 at $t = 9$, $800 at $t = 10$, and $700 at $t = 11$. Assume the fund pays 15% compounded per period. (3.4.2)

**21.** A debt of $X is incurred at $t = 0$ (purchase of land). It is agreed that payments of $5000, $4000, $3000, and $2000 at $t = 4, 5, 6$, and 7, respectively, will satisfy the debt if 12% compounded/period is the appropriate interest rate. By use of the *uniform gradient series factor* and any other relevant interest factor(s), determine the amount of the debt, $X. (3.4.2)

**22.** Mary works for a company that pays an annual bonus, the size of which is based on experience with the company. After 1 year with the company the bonus equals $500. The size of the bonus thereafter compounds at an annual rate of 5%. Mary decides to place half the bonus in a fund that pays 10% compounded annually.

(a) How much money will be in the fund immediately prior to the sixth deposit?

(b) What is the answer to (a) if the fund compounds at 5% annually? (3.4.3)

23. Carol places $1000 in a fund at the end of 1985. She places end-of-year deposits in the fund until the end of 2002, when her last deposit is made. The fund pays 10% compounded annually. If the size of a deposit at the end of year $t$, $A_t$, equals $0.90A_{t-1}$, how much will be in the fund immediately after her last deposit? (3.4.3)

24. Terri wishes to make a single deposit $P$ at $t = 0$ into a fund paying 12% compounded quarterly such that $1000 payments are received at $t = 1, 2, 3$, and 4 (periods are three-month intervals) and a single payment of $7500 is received at $t = 12$. What single deposit is required? (3.5)

25. Jim deposits $3000 in a money market fund. The fund pays interest at a rate of 12% compounded annually. Just 3 years after making the single deposit he withdraws half the accumulated money in his account. Then, 5 years after the initial deposit, he withdraws all of the accumulated money remaining in the account. How much does he withdraw 5 years after his initial deposit? (3.5)

26. A woman borrows $5000 at 18%/year compounded monthly. She wishes to repay the loan with 12 end-of-month payments. She wishes to make her first payment 3 months after receiving the $5000. She also wishes that, after the first payment, the size of each payment be 10% greater than the previous payment. What is the size of her sixth payment? (3.5)

27. A man borrows $25,000 at 12% compounded quarterly. He wishes to repay the money with 10 equal semiannual installments. What must be the size of the payment if the first payment is made 1 year after obtaining the $25,000? (3.5)

28. Monthly deposits of $150 are made into an account paying 8% compounded quarterly. Ten monthly deposits are made. Determine how much will be accumulated in the account 6 months after the last deposit. (3.5)

29. A person makes four consecutive annual deposits of $1000 in a savings account that pays interest at a rate of 10% compounded semiannually. How much money will be in the account 2 years after the last deposit? (3.5)

30. Jaime makes monthly deposits of $300 in a savings account that pays interest at a rate equivalent to 8% compounded quarterly. How much money should be in the account immediately after the sixtieth deposit? If no interest is earned on money deposited during a quarter and the first deposit coincides with the beginning of a quarter, what will be the account balance immediately after the sixtieth deposit? (3.5)

31. An individual makes semiannual deposits of $1000 into an account that pays interest equivalent to 12% compounded quarterly. Determine the account balance immediately *before* the tenth deposit. (3.5)

32. An individual borrows $7500 at an interest rate of 16%/year compounded semiannually and desires to repay the money with five equal end-of-year payments, with the first payment made 2 years after receiving the $7500. What should be the size of the annual payment? (3.5)

33. What is the effective interest rate for 12% compounded monthly? (3.5)

34. What monthly payments are required in order to pay off a $20,000 debt in 5 years if the *nominal* rate of interest charged on the unpaid balance is 9% compounded annually. (3.8.3)

35. What monthly deposits must be made in order to accumulate $10,000 in 5 years if a fund pays 15% compounded monthly? (Assume the first deposit is made at the end of the first month and the final deposit is made at the end of the fifth year.) (3.5)

36. Dick borrows $20,000 at 15% compounded semiannually. He pays back the loan with four equal semiannual payments, with the first payments made 1 year after receiving the $20,000. What should be the size of each of the four payments? (3.5)

37. Lynne borrows $5000 at 1.5%/month. She desires to repay the money using equal monthly payments for 10 months. Lynne makes four such payments and decides to pay off the remaining debt with one lump sum payment at the time for the fifth payment. What should the size of this payment be if interest is truly compounded at a rate of 1.5%/month? (3.5)

38. Leon borrows $5000 and pays the loan off, with the interest, after 2 years. He pays back $6000. What is the effective interest rate for this transaction? (3.5)

39. Approximately how long will it take a deposit to triple in value if money is worth 10% compounded semiannually? (3.5)

40. Mr. Wright borrows $8000 from a bank that charges interest at 16% compounded semiannually. Mr. Wright is to pay the money back with six equal payments. However, the first payment is to be made immediately on receipt of the $8000. Successive payments are spaced 1 year apart.
    (a) Determine the size of the equal annual payment.
    (b) At the time of the fourth payment, suppose Mr. Wright decides to pay off the loan with one lump sum payment. How much should be paid? Include the fourth payment. (3.5)

41. Bob borrows $1000 from the Shady Deal Finance Company. He is told the interest rate is merely 2%/month, and his payment is computed as follows.

$$\text{Payback period} = 30 \text{ months}$$
$$\text{Interest} = 30(0.02)(\$1000) = \$600$$
$$\text{Credit investigation and insurance} = \$50$$
$$\text{Total amount owed} = \$1650$$
$$\text{Payment size} = \$1650/30 = \$55 \text{ per month}$$

What is the approximate effective interest rate for this transaction? (3.5)

42. Assume a person deposits $1000 now, $2000 2 years from now, and $5000 5 years from now into a fund paying 12% compounded semiannually.
    (a) What sum of money will have accumulated in the fund at the end of the sixth year?
    (b) What equal deposits of size $A$, made every 6 months (with the first deposit at $t = 0$ and the last deposit at the end of the sixth year), are equivalent to the three deposits stated above? (3.5)

**43.** Semiannual deposits of $500 are made into a fund paying 8% compounded continuously. What is the accumulated value in the fund after 10 such deposits? (3.6.1)

**44.** Four equal, quarterly deposits of $1000 each are made at $t = 0, 1, 2$, and 3 (time periods are 3-month intervals) into a fund that pays 8% compounded *continuously*. Then, at $t = 7$ and $t = 10$, withdrawals of size $A$ are made so that the fund is depleted at $t = 10$. What is the size of the withdrawals? (3.6.1)

**45.** Mike receives an annual bonus from his employer. He wishes to deposit the bonus in a fund that pays interest at a rate of 6% compounded continuously. His first bonus is $1000. The size of his bonus is expected to increase at a rate of 4%/year. How much money will be in the fund immediately *before* the tenth deposit? (3.6)

**46.** Operating and maintenance costs for year $k$ are given as $C_k = (e^{0.10})C_{k-1}$, $k = 2, 3,$ ... , 15, with $C_1 = \$1000$. Determine the equivalent uniform annual operating and maintenance cost based on continuous compounding with a nominal interest rate of (a) 10% and (b) 20%. (3.6.1)

**47.** A person borrows $10,000 and wishes to pay it back with 10 equal annual payments. What will the size of the payments be if the interest charge is 10% compounded (a) annually, (b) semiannually, and (c) continuously? (3.6.1)

**48.** For the cash flow series given below, determine (a) the present worth equivalent, (b) the future worth equivalent, and (c) the uniform annual series equivalent using a 10% continuous compound interest rate. (3.6.1)

| EOY | Cash Flow | EOY | Cash Flow |
|-----|-----------|-----|-----------|
| 0 | −$10,000 | 6 | $5,000e^{0.10}$ |
| 1 | 1,000 | 7 | $5,000e^{0.20}$ |
| 2 | 2,000 | 8 | $5,000e^{0.30}$ |
| 3 | 3,000 | 9 | $5,000e^{0.40}$ |
| 4 | 4,000 | 10 | $5,000e^{0.50}$ |
| 5 | 5,000 | | |

**49.** Labor costs occur continuously during the year. At the end of each year a new labor contract becomes effective for the following year. Let $\bar{A}_j$ denote the cumulative labor cost occurring uniformly during year $j$, where $\bar{A}_j = 1.05\bar{A}_{j-1}$. If money is worth 15% compounded continuously and $\bar{A}_1 = \$25,000$, determine the present worth equivalent for 5 years of labor costs. (3.6.2)

**50.** Operating and maintenance costs for a production machine occur continuously during the year. If the total annual operating and maintenance cost is $24,000 and money is worth 25% compounded continuously, what single sum of money at the present is equivalent to 5 years of operating and maintenance costs? (3.6.2)

**51.** Consumption of fuel oil occurs continuously in a manufacturing plant. The total monthly cost in January is $25,000. The demand for fuel oil increases at a monthly continuous compound rate of 1.5%. Determine the equivalent end-of-year (future worth) cost for one year's consumption using a monthly nominal continuous compound rate of 2.0%. (3.6.2)

**52.** For the continuous cash flows shown below, determine (a) the present worth equivalent and (b) the future worth equivalent using a continuous compound interest rate of 20%. (3.6.2)

| Year | Continuous Cash Flow, $\bar{A}$ |
|------|--------------------------------|
| 1 | $10,000 |
| 2 | 12,000 |
| 3 | 14,000 |
| 4 | 16,000 |
| 5 | 18,000 |

**53.** Solve Problem 52 for the following cash flow series:

| Year | Continuous Cash Flow, $\bar{A}$ |
|------|--------------------------------|
| 1 | $10,000 |
| 2 | $10,000e^{0.20}$ |
| 3 | $10,000e^{0.40}$ |
| 4 | $10,000e^{0.60}$ |
| 5 | $10,000e^{0.80}$ |

(3.6.2)

**54.** A firm buys a new computer that costs $100,000. It may either pay cash now or pay $20,000 down and $30,000/year for 3 years. If the firm can earn 15% on investments, which would you suggest? (3.7)

**55.** Hugh invests $10,000 in a venture that returns him $500(0.80)^{t-1}$ at the end of year $t$ for $t = 1, \ldots , 20$. With an interest rate of 10%, what is the equivalent uniform annual profit (cost) for the venture? (3.4.3)

**56.** Hugh is considering two investment alternatives. Alternative A requires an initial investment of $10,000; it will yield incomes of $3000, $3500, $4000, and $4500 over its 4-year life. Alternative B requires an initial investment of $12,000; it is anticipated that the revenue received will increase at a compound rate of 10%/year. Based on an interest rate of 10% compounded annually, what must be the revenue the first year for B in order for A and B to be equivalent? (3.7)

**57.** It is desired to determine the size of the uniform series over the time period [2,5] that is equivalent to the cash flow profile shown below using an interest rate of 10%. (3.7)

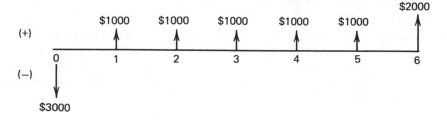

**58.** Given the following cash flows, what single sum at $t = 4$ is equivalent to the given data. Assume $i = 15\%$. (3.7)

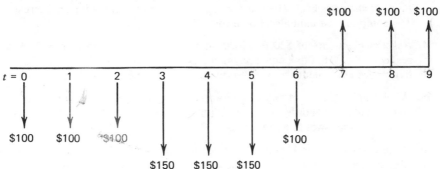

**59.** A machine is purchased at $t = 0$ for $40,000 (including installation costs). Net annual revenues resulting from operating the machine are $15,000. The machine is sold at the end of 8 years, $t = 8$, for $5000. The cash flow series for the machine is equivalent to what single sum, $X, at (a) $t = 6$ and (b) $t = 8$ if money is worth 15% compounded continuously? (3.7)

**60.** Assume the following two investment plans.
   **(a)** Purchase for $4000 (a negative cash flow) and receive (1) $400 at the end of each 6 months for 4 years ($t = 1, 2, \ldots, 8$), and (2) a single payment of $2000 at the end of the fourth year.
   **(b)** Purchase for $3850 and receive (1) $900 at the *beginning* of each year for the 4-year period, and (2) a single payment of $X at the end of the fourth year.
   If money is worth 8% compounded semiannually, what is the value of $X$ so that the two investments are equivalent? (3.7)

**61.** Don invests $5000 and receives $600 each year for 10 years, at which time he sells out for $1000. With a 15% interest rate, what equal annual cost (profit) is equivalent to the venture? (3.7)

**62.** Consider the following cash flow series.

| EOY | CF |
|---|---|
| 0 | −$10,000 |
| 1 | 6,500 |
| 2 | 6,000 |
| 3 | 5,500 |
| 4 | 5,000 |
| 5 | 4,500 |
| 6 | 4,000 |
| 7 | 3,500 |
| 8 | 3,000 |

At 10% annual compound interest, what uniform annual cash flow is equivalent to the above cash flow series? (3.7)

**63.** Susan borrows $1000 from a bank at $t = 0$ at 10% simple interest for 2 years. She pays the total interest due for the 2-year period at $t = 0$ and thus receives $800 at $t = 0$. If she pays back $1000 at $t = 2$, Susan is, in effect, paying an interest rate of $X\%$ compounded annually. Solve for $X$. (3.7)

**64.** Assume payments of $2000, $3000, and $5000 are received at $t = 3$, 4, and 5, respectively. What five equal payments occurring at $t = 1, 2, 3, 4$, and 5, respectively, are equivalent if $i = 10\%$ compounded per period? (3.7)

**65.** What single deposit of size $X into a fund paying 10% compounded annually is required at $t = 0$ in order to make withdrawals of $500 each at $t = 4, 5, 6$, and 7 and a single withdrawal of $1000 at $t = 10$? (3.7)

**66.** If the withdrawals in Problem 65 are immediately placed into another fund paying 12% compounded annually, what amount will be accumulated in this fund at $t = 10$? (3.7)

**67.** A college student borrows $6000 and repays the loan with four quarterly payments of $600 during the first year and four quarterly payments of $1500 during the second year after receiving the $6000 loan. Determine the effective interest rate for the loan transaction. (3.7)

**68.** Quarterly deposits of $500 are made at $t = 1, 2, 3, 4, 5, 6, 7$. Then withdrawals of size $A$ are made at $t = 12, 13, 14, 15$ and the fund is depleted with the last withdrawal. If the fund pays 8% compounded quarterly, what is the value of $A$? (3.7)

**69.** Given the cash flow diagram shown below and a nominal interest rate of 10% compounded continuously, solve for the value of an equivalent amount at (a) $t = 5$, (b) $t = 10$, and (c) $t = 15$. (*Note.* Upward arrows represent cash inflows and downward arrows represent cash outflows.)

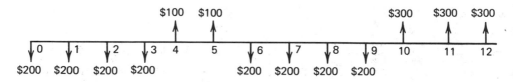

**70.** What single sum of money at $t = 4$ is equivalent to the cash flow profile shown below? Use a 12% interest rate in your analysis. (3.7)

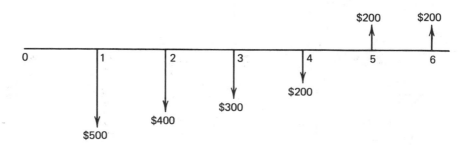

**71.** Determine the value of $X$ so that the following cash flow series are equivalent at 12% interest. (3.7)

| EOY | CF(A) | CF(B) |
|---|---|---|
| 0 | −$8,000 | −$15,000 |
| 1 | 6,000 | 4,000 |
| 2 | 5,000 | 3,000 + X |
| 3 | 4,000 | 2,000 + 2X |
| 4 | 5,000 | 3,000 + 3X |
| 5 | 6,000 | 4,000 + 4X |
| 6 | 5,000 | 3,000 + 5X |

**72.** Given the cash flow profiles shown below, determine the value of $X$ so that the two cash flow profiles are equivalent at a 10% interest rate (3.7)

| EOY | CF(A) | CF(B) |
|---|---|---|
| 1 | −$12,000 | −$10,000 |
| 2 | 1,000 | 7,000 |
| 3 | 3,000 | 6,000 + 0.5X |
| 4 | 5,000 | 5,000 + 1.0X |
| 5 | 7,000 | 4,000 + 1.5X |
| 6 | 9,000 | 3,000 + 2.0X |

**73.** Given the two cash flow profiles shown below, for what value of $X$ are the two series equivalent using an interest rate of 15% (3.7)

| EOY | CF(A) | CF(B) |
|---|---|---|
| 0 | −$200,000 | −$140,000 |
| 1 | 24,000 | 16,000 |
| 2 | 32,000 | 16,000 |
| 3 | 40,000 | 16,000 + X |
| 4 | 48,000 | 16,000 + 2X |
| 5 | 56,000 | 16,000 + 3X |
| 6 | 64,000 | 16,000 + 4X |
| 7 | 72,000 | 16,000 + 5X |
| 8 | 80,000 | 16,000 + 6X |

**74.** Given the cash flow profiles shown below, determine the value of $X$ such that the two cash flow profiles are equivalent at 20% compounded annually. (3.7)

| EOY | CF(A) | CF(B) |
|-----|-------|-------|
| 1 | −$12,000 | $ − X |
| 2 | 1,000 | 7,000 |
| 3 | 4,000 | 9,000 |
| 4 | 6,000 | 10,000 |
| 5 | 7,000 | 10,000 |
| 6 | 5,000 | 7,000 |

**75.** Given the two cash flow profiles shown below, for what value of $X$ are the two series equivalent using an interest rate of 25%? (3.7)

| EOY | CF(A) | CF(B) |
|-----|-------|-------|
| 0 | −$100,000 | −$70,000 |
| 1 | 12,000 | 8,000 |
| 2 | 16,000 | 8,000 |
| 3 | 20,000 | 8,000 |
| 4 | 24,000 | 8,000 + X |
| 5 | 28,000 | 8,000 + 2X |
| 6 | 32,000 | 8,000 + 3X |
| 7 | 36,000 | 8,000 + 4X |
| 8 | 40,000 | 8,000 + 5X |

**76.** Given the three cash flow profiles shown below, determine the values of $X$ and $Y$ so that all three cash flow profiles are equivalent at an annual interest rate of 20%. (3.7)

| EOY | CF(A) | CF(B) | CF(C) |
|-----|-------|-------|-------|
| 0 | −$1,000 | −$2,500 | $ Y |
| 1 | X | 3,000 | Y |
| 2 | 1.5X | 2,500 | Y |
| 3 | 2.0X | 2,000 | 2Y |
| 4 | 2.5X | 1,500 | 2Y |
| 5 | 3.0X | 1,000 | 2Y |

**77.** Given the cash flow profiles shown below, determine the value of $x$ such that the two cash flow profiles are equivalent at 25% *compounded continuously.* (3.7)

| EOY | CF(A) | CF(B) |
|---|---|---|
| 1984 | -$75,000.00 | $ - X |
| 1985 | 10,000.00 | 5,000 |
| 1986 | 12,500.00 | 10,000 |
| 1987 | 15,625.00 | 15,000 |
| 1988 | 19,531.25 | 20,000 |
| 1989 | 24,414.06 | 25,000 |
| 1990 | 30,517.58 | 30,000 |

**78.** Given the two cash flow profiles shown below, for what value of $x$ are the two series equivalent using an interest rate of 15%? (3.7)

| EOY | CF(A) | CF(B) |
|---|---|---|
| 0 | $60,000 | -$80,000 |
| 1 | 15,000 | 15,000 |
| 2 | $15,000(1 + x)$ | 20,000 |
| 3 | $15,000(1 + x)^2$ | 25,000 |
| 4 | $15,000(1 + x)^3$ | 30,000 |
| 5 | $15,000(1 + x)^4$ | 35,000 |

**79.** The Ajax Chemical Company sells a particular chemical to the A. C. Works. A 5-year contract has been negotiated with the following terms. A constant volume, $V_t$, measured in cubic feet will be supplied uniformly and continuously during year $t$, with $V_t = 1.10V_{t-1}$. The price per cubic foot supplied during year $t$ will be $p_t$, where $p_t = 1.08p_{t-1}$. The Ajax Chemical Company wishes to determine the equivalent *present value* of the contract using a continuous compounding rate of 20%/year. Express the equivalent present value in terms of $p_1$ and $V_1$. (3.7)

**80.** Given the following profile of continuous flows, determine the equivalent series of annual discrete cash flows using a continuous compounding rate of 15%/year. (3.7)

| EOY | $\overline{A}$ |
|---|---|
| 1 | $10,000 |
| 2 | 12,000 |
| 3 | 14,000 |
| 4 | 16,000 |
| 5 | 18,000 |

**81.** If a machine costs $10,000 and lasts for 10 years, at which time it is sold for $2000, what equal annual cost over its life is equivalent to these two cash flows with a 30% annual interest rate? (3.7)

**82.** At what interest rate are the following two cash flow profiles equivalent? (3.7)

| EOY | CF(A) | CF(B) |
|-----|-------|-------|
| 0 | $10,000 | — |
| 1 | — | $1,000 |
| 2 | — | 2,000 |
| 3 | — | 3,000 |
| 4 | — | 4,000 |
| 5 | — | 5,000 |

**83.** An investment opportunity offered by commercial banks is certificates of deposit. For example, a certificate of deposit (C.D.) having a 4-year maturity date may have a stated interest rate of "10%, payable quarterly." This statement means that every 3 months, the issuing bank would pay the C.D. holder an amount of 0.025 times the face value (purchase price) of the C.D. Assuming a $5000, 4-year, 10% C.D. were purchased at $t = 0$, quarterly payments of $125 would be received at $t = 1, 2, 3, \ldots , 16$ and, at $t = 16$, the C.D. would be redeemed by the bank for the original purchase price, or $5000. (This payment series is an application of simple interest on a quarterly basis).

It is also possible for the purchaser of the C.D. to arrange with the bank for each interest payment to be deposited into a regular savings account that pays, say 8% compounded quarterly. Let it be assumed that this is done; the first deposit of $125 occurs at $t = 1$, and the last deposit is made at $t = 16$.

  **(a)** For the initial investment of $5000, what total sum of money will be received by the purchaser at the end of the fourth year (C.D. redemption value plus savings account balance)?

  **(b)** Now consider only the initial investment, $P = $5000$, and the $F$ value calculated from (a). What effective annual compound interest rate relates the two single sums of money?

  **(c)** For $P$ and $F$ from (b) above, what annual simple interest rate relates the two single sums of money? (3.7)

**84.** Martin borrows $5000 at 12% compounded annually. Five equal annual payments are used to repay the loan, with the first payment occurring 1 year after receiving the $5000. Determine the interest amount included in each payment. (3.8)

**85.** Martin borrows $8000 at 10% compounded annually. Six equal annual payments are used to repay the loan, with the first payment occurring 2 years after receiving the $8000. Determine the payment on principal included in each payment. (3.8)

**86.** Joe borrows $10,000 at 12% compounded monthly. He is to pay off the loan with 60 monthly payments. One month after making the thirtieth payment, he elects to pay off the unpaid balance on the note. How much should he repay? (3.8)

**87.** Mr. and Mrs. Kerr purchase a boat for $50,000: they make a down payment of $15,000 and finance the balance over a 5-year period. Monthly payments are made. Determine the amount of interest paid each year if the nominal annual interest rate is 15%. (3.8)

88. Patricia makes five annual deposits of $1000 in a money market account that pays interest at a rate of 12% compounded annually. One year after making the last deposit, the interest rate changes to 10% compounded annually. Five years after the last deposit the accumulated money is withdrawn from the account. How much is withdrawn? (3.9)

89. Jerry deposits $10,000 in a money market account paying 8% compounded annually for the first 2 years and 10% compounded annually for the next 2 years. Four annual withdrawals are made from the savings account. The size of the withdrawal increases by $1000/year, with the first withdrawal occurring 1 year after the deposit. Determine the size of the last withdrawal, which depletes the balance of the account. (3.9)

90. Assume that six equal deposits of $500 are made at $t = 1, 2, 3, 4, 5$, and 6 (3-month periods) into a fund paying 10% *compounded quarterly*. This interest applies until $t = 16$. The accumulated sum is withdrawn at this time and immediately deposited into a fund paying 15% *compounded continuously*. Beginning with $t = 17$, determine the amount of three equal payments, $A$, that may be withdrawn such that the fund is depleted at $t = 19$. (3.9)

91. Sam borrows $10,000 and repays the loan with four equal annual payments. The interest rate for the first 2 years of the loan is 8% compounded continuously, and for the third and fourth years of the loan it is 10% compounded continuously. Determine the size of the annual payment. (3.9)

92. Linda deposits $5000 in a savings account. One year after the initial deposit, $1000 is withdrawn. Two years after the first withdrawal, $4000 is deposited in the account. Three years after the second deposit, $2000 is withdrawn from the account. Four years after the second withdrawal, all funds are withdrawn from the account. During the period of time the savings account was in use, the bank paid 8% compounded continuously for the first 2 years, 10% compounded annually for the next 4 years, and 12% compounded quarterly for the remainder of the time the account was in use. Determine the amount of the final withdrawal. (3.9)

93. John borrows $5000 and repays the loan with three equal annual payments. The interest rate for the first year of the loan is 12% compounded annually, for the second year of the loan is 15% compounded annually, and for the third year of the loan is 18% compounded annually. Determine the size of the equal annual payment. (3.8.2)

94. Ken deposits $1000 in a fund that pays interest at a continuous compound rate of $r_t$ for the $t$th year after the initial deposit, where

$$r_t = \begin{cases} 0.08 + 0.005t, & t = 1, 2, 3, 4, 5, 6 \\ 0.11 & t = 7, 8, 9, 10, \ldots \end{cases}$$

Thus a continuous compound rate of $8\frac{1}{2}$% is earned during the first year, 9% is earned during the second year, $9\frac{1}{2}$% is earned during the third year, and so on. A maximum of 11% compounded continuously is earned during the sixth and each successive year. If the person withdraws $500 2 years after the initial deposit, how much money will be in the fund 6 years after the initial deposit? (3.9)

**95.** Robert deposits $2500 in an account that pays interest at a rate of 7% compounded annually. Two years after the deposit, the savings account begins paying interest at a rate of 8% compounded continuously. Five years after the deposit, the savings account begins paying interest at a rate of 9% compounded semiannually.

    **(a)** How much money should be in the savings account 10 years after the initial deposit?

    **(b)** What annual compound interest rate is equivalent to the interest pattern of the savings account over the 10-year period? (3.9)

**96.** Deborah deposits $1000 in a fund each year for a 10-year period. The fund initially pays 8% compounded annually. Immediately after she makes the sixth deposit, the fund begins paying 10% compounded annually. Deborah removes the money from the fund 3 years after her last deposit. How much should she be able to withdraw at that time? (3.9)

**97.** Based on discrete cash flows and discrete compounding, compute the present worth and annual worth for the following situation: (3.9)

| EOY | CF | Interest Rate During Period |
|---|---|---|
| 0 | −$10,000 | 0 |
| 1 | 2,000 | 0.08 |
| 2 | 4,000 | 0.09 |
| 3 | 6,000 | 0.10 |
| 4 | 8,000 | 0.11 |
| 5 | 10,000 | 0.12 |

**98.** Solve problem 97 for the case of continuous compounding with the interest rates shown interpreted as nominal rates. (3.9)

**99.** Maria deposits $2000 in a savings account that pays 7% compounded annually; 2 years after the deposit, the interest rate increases to 8% compounded annually. A second deposit of $2000 is made immediately after the interest rate changes to 8%. How much will be in the fund 5 years after the second deposit? (3.9)

**100.** Based on continuous cash flows and continuous compounding, compute the equivalent discrete uniform annual series for the following situation: (3.9)

| Year | $\bar{A}$ | Nominal Interest Rate During Year |
|---|---|---|
| 1 | $2,000e^{0.05}$ | 0.25 |
| 2 | $3,000e^{0.10}$ | 0.20 |
| 3 | $4,000e^{0.15}$ | 0.15 |
| 4 | $5,000e^{0.20}$ | 0.10 |
| 5 | $6,000e^{0.25}$ | 0.05 |

## CHAPTER FOUR

# MEASURING THE WORTH OF INVESTMENTS

## 4.1 INTRODUCTION

Chapter One recommended that engineers and systems analysts solve problems by formulating and analyzing the problem, generating a number of feasible solutions (alternatives) to the problem, comparing the investment alternatives, selecting the preferred solution, and implementing the solution. The process of evaluating the investment alternatives and selecting the preferred solution is the subject of the remainder of this book.

In this chapter we develop a number of methods and criteria for measuring the worth of investments. This material forms the basis for techniques used in Chapter Five to compare alternative investments and select those preferred from the economic point of view. The methods and criteria of this chapter must be understood and properly applied if valid economic analyses are to be assured.

Before we can measure the worth of an investment, we must specify the *time value of money* to be used. This interest rate is influenced by many factors, and represents the minimum return we must have on our investment capital. Then, we must select from among several *measures of merit* for establishing the worth of an investment. Most of these are *equivalent* methods. That is, they will give the same decision regarding the desirability of an investment. Examples are presented covering each method; however, an important special class of investments called bonds is also examined to illustrate and integrate some of the methods. *Capital recovery cost* must often be calculated to determine the annual cost of an asset. Various frequently used formulas for calculating the capital recovery cost are presented, all of which provide the same answers. Finally, explicit treatment of *inflation* and *deflation* may be necessary in an economic analysis. In recent years, the effects of inflation have become much more important. All of the above topics are treated in this chapter.

## 4.2 SPECIFYING THE TIME VALUE OF MONEY

An important step in measuring the worth of an investment involves the specification of the interest or discount rate to be used. Except where other intangible benefits are involved, the discount rate should be greater than the cost of securing additional capital. Indeed, it should be greater than the cost of capital by an amount that will cover unprofitable investments that a firm must make for nonmonetary reasons. Examples of the latter would include investments in antipollution equipment, safety devices, and recreational facilities for employees. The discount rate that is specified establishes the firm's minimum attractive rate of return (*MARR*) in order for an investment to be justified.

### 4.2.1 The Cost of Capital

It costs money to obtain money for investment. The cost of capital was first discussed in Chapter Two. There, it was noted that funds for financing projects may consist of *debt capital* and *equity* (or ownership) *capital*. Most often, a major project is financed by a mix of both. In trying to establish the time value of money, it is important that we understand why money has a cost, even when it is our own money that we use.

Debt capital such as a loan results in a specific obligation to pay back both interest and principal. A short-term loan may be obtained from a bank or an insurance company. The interest paid for the use of the money is a cost of capital. Fortunately, not all of the interest is lost. Interest is deductible from income for tax purposes.[1] For example, even though we pay $I$ in interest during a year, that $I$ may be subtracted from taxable income, resulting in a tax savings of $I \times T$ where $T$ is our tax rate (46% for large companies). That is, the effective interest payment is $I(1 - T)$. Therefore, for a simple loan, the effective after-tax interest rate ($k_l$) is the effective interest rate from Equation 3.41 times $1 - T$. It is given by

$$k_l = \left[ \left( 1 + \frac{r}{m} \right)^m - 1 \right] (1 - T) \tag{4.1}$$

---

[1] Tax effects will be only briefly considered here. Income taxes are treated in considerable detail in Chapter Six.

where $r$ is the loan's nominal annual interest rate and $m$ is the number of payment periods per year.

Example 4.1 _____

A company has taken out a short-term loan with the bank at a rate of 18%/year with quarterly payments. Assuming the tax rate is .46, what is the cost of capital for this debt funding?

 Answer:

$$k_l = \left[\left(1 + \frac{r}{m}\right)^m - 1\right](1 - T)$$

$$= \left[\left(1 + \frac{.18}{4}\right)^4 - 1\right](1 - .46) = .104$$

or 10.4%.

_____

Debt capital can also be raised through the sale of bonds. Common debenture bonds have a stated face value, annual rate of interest, payment interval, and maturation period (e.g., $1000, 12%, semiannual, and 15 years). When a bond is sold, not necessarily for its face value, the firm receives money for immediate investment. In return, it gives a promise to pay interest on the face value $m$ times per year at an annual rate $r$. In addition, the firm promises to repay the face value of the bond at the end of a specified period of time. Much like a loan, bond interest paid by a firm is deductible from income for tax purposes. If the amount paid for the bond equals the face value, and if any costs of preparing and selling the bond are neglected, the effective after-tax interest rate ($k_b$) is

$$k_b = \left[\left(1 + \frac{r}{m}\right)^m - 1\right](1 - T) \tag{4.2}$$

Equity capital is sometimes called ownership capital. One source of equity capital is through the sale of stock. Money raised through a stock issue is available to the company for investment. Surprisingly, there is no guarantee of a return to an investor in stock. Of course, the investor expects to receive periodic dividends and also hopes that the trading price of the stock will increase, resulting in a gain when the stock is sold to someone else in the future. Company management usually decides upon the amount of dividends to declare for stockholders after the earnings of the firm are known for the year. Management desires to seek a suitable balance between the amount of dividends paid to keep stockholders happy and the amount of earnings retained in the company for investment in the future. Perhaps the most simplistic estimate of the cost of equity capital involving stock ($k_s$) is

$$k_s = \frac{D}{P_s} \tag{4.3}$$

where $D$ is the dividend per share of stock and $P_s$ is the present trading price or current value of the stock. Since dividends are paid out of after-tax earnings, $k_s$ is already an after-tax rate. This estimate assumes that dividends will remain constant into the future.

## Example 4.2

A company issues dividends of $6/share on stock selling at $50 per share on the financial market. What is the cost of capital for this equity funding?

Answer:

$$k_s = \frac{D}{P_s}$$

$$= \frac{\$6}{\$50} = .12$$

or 12.0%.

Another source of equity capital is the earnings retained within the company and not declared as dividends. Even though this money appears free, there is a cost of capital. Remember, stockholders are owners of the company and therefore the retained earnings are theirs. The cost of using that money for investment carries an opportunity cost, and that is the cost of dividends forgone by the stockholders from their own earnings. For this reason, one estimate of the cost of capital for retained earnings ($k_r$) is

$$k_r = \frac{D}{P_s} = k_s \tag{4.4}$$

The overall weighted average cost of capital ($k_a$) is simply the individual source costs of capital weighted by the fraction of total funding from that source. That is,

$$k_a = k_l p_l + k_b p_b + k_s p_s + k_r p_r \tag{4.5}$$

where $p_l$, $p_b$, $p_s$, and $p_r$ are the fractions of funding from loans, bonds, stocks, and retained earnings, respectively.

## Example 4.3

The capital structure of our firm is made up as follows:

|  | Funding | Fraction |
|---|---|---|
| Loans | $ 1,000,000 | 1/13 |
| Bonds | 1,800,000 | 1.8/13 |
| Stock | 8,000,000 | 8/13 |
| Retained earnings | 2,200,000 | 2.2/13 |
|  | $13,000,000 | 13/13 |

The costs of capital for these various sources are 10.4, 7.2, 12.0, and 12.0%, respectively. What is the overall weighted average cost of capital for the company?

Answer:

$$k_a = .104(1/13) + .072(1.8/13) + .120(8/13) + .120(2.2/13)$$
$$= .1121$$

or 11.21%.

---

Taxes were briefly introduced in order to describe more realistically the cost of capital. Further discussion of taxes will be deferred until Chapter Six where much more complete coverage is given. Also, the discussion on cost of capital was maintained at a very elementary level. An excellent and in-depth treatment of the cost of capital, with reference to many other specific sources, is contained in [6].

## 4.2.2 The Minimum Attractive Rate of Return

Some firms establish a standard discount rate or minimum attractive rate of return to be used in all economy studies; others maintain a flexible posture. As one firm described it,

The XYZ Company return on investment (ratio of earnings to gross investment) has been, on the average over the past five years, approximately equal to a rate-of-return of 15% per year. Accordingly, 15% per year is being established as a tentative *minimum requirement for investment alternatives* whose results are primarily measurable in quantitative dollar terms. The principle being applied is that such alternatives should be expected to maintain or to improve overall return-on-investment performance for the XYZ Company. The minimum requirement is based on overall XYZ financial results (as opposed to results for various parts of the company) in order to avoid situations in which investment alternatives of a given level of attractiveness are unknowingly undertaken in one part of the company and rejected in another.

The 15% minimum attractive rate-of-return standard is intended as a guide, rather than as a hard-and-fast decision rule. Furthermore, it is intended to apply to alternatives having risks of the kind usually associated with investments which are primarily in plant and facilities. For alternatives involving expenditures with substantially lower risks, such as those solely for inventories or those in buy-or-lease alternatives, a lower minimum requirement may apply.

Other approaches that are used to establish the MARR include:

1. Add a fixed percentage to the firm's cost of capital.
2. Average rate of return over the past 5 years is used as this year's *MARR*.
3. Use different *MARR* for different planning horizons.
4. Use different *MARR* for different magnitudes of initial investment.

5. Use different *MARR* for new ventures than for cost improvement projects.

6. Use as a management tool to stimulate or discourage capital investments, depending on the overall economic condition of the firm.

7. Use the average stockholder's return on investment for all companies in the same industry group.

There are a number of different approaches used by companies in establishing the discount rate to be used in performing economic analyses. The issue is not a simple one.

The proper determination of the discount rate has been the subject of considerable controversy in the economic analysis literature for many years. In some ways we are only slightly closer to agreement today than we were 20 years ago. In many cases, a particular alternative will be preferred over a range of possible discount rates; in other cases, the alternative preferred will be quite sensitive to the discount rate used. Hence, depending upon the situation under study, it might not be necessary to specify a particular value for the discount rate—a range of possible values might suffice. We examine this situation in more depth in Chapter Eight.

Subsequently, it will be convenient to refer to the interest rate or discount rate used as the *MARR* and to interpret its value using the opportunity cost concept. The argument will be made that money should not be invested in an alternative if it cannot earn a return at least as great as the *MARR*, since it is reasoned that other opportunities for investment exist that will yield returns equal to the *MARR*.

In the case of the public sector, a different interpretation is required in determining the discount rate to use. Since this chapter emphasizes economic analysis in the private sector, Chapter Seven discusses establishing the discount rate for the public sector.

## 4.3 METHODS OF MEASURING INVESTMENT WORTH

As noted in Chapter Three in discussing equivalence, the worth of an investment can be measured in a number of ways. Present worth (*PW*) and annual worth (*AW*) methods are two commonly used approaches. Among the several methods of measuring investment worth are:

1. Present worth method (*PW*).
2. Annual worth method (*AW*).
3. Future worth method (*FW*).
4. Internal rate of return method (*IRR*).
5. External rate of return method (*ERR*).
6. Savings/investment ratio method (*SIR*).

7. Payback period method (*PBP*).

8. Capitalized worth method (*CW*).

Each of the above measures of merit or measures of effectiveness has been used numerous times in evaluating real-world investments. They may be described briefly as follows:

1. Present worth method converts all cash flows to a single sum equivalent at time zero using $i = MARR$.

2. Annual worth method converts all cash flows to an equivalent uniform annual series of cash flows over the planning horizon using $i = MARR$.

3. Future worth method converts all cash flows to a single sum equivalent at the end of the planning horizon using $i = MARR$.

4. Internal rate of return method determines the interest rate that yields a future worth (or present worth or annual worth) of zero, implicitly assuming reinvestment of recovered funds at the *IRR*.

5. External rate of return method determines the interest rate that yields a future worth of zero, explicitly assuming reinvestment of recovered funds at the *MARR*.

6. Savings/investment ratio method determines the ratio of the present worth of savings to the present worth of the investment.

7. Payback period method determines how long at a zero interest rate it will take to recover the initial investment.

8. Capitalized worth method determines the single sum at time zero that is equivalent at $i = MARR$ to a cash flow pattern that continues indefinitely.

With the exception of the payback period and capitalized worth methods, all of the measures listed are equivalent methods of measuring investment worth, Hence, applying each of the first six measures of merit to the same investment alternative will yield the same recommendation.

Since the present worth, annual worth, future worth, internal rate of return, external rate of return, and savings/investment ratio methods are equivalent, why does more than one of the methods exist? The primary reason for having different, but equivalent, measures of effectiveness for economic alternatives appears to be the differences in preferences among managers. Some individuals (and firms) prefer to express the net economic worth of an investment alternative as a single sum amount; hence, either the present worth method or the future worth method is used. Other individuals prefer to see the net economic worth spread out uniformly over the planning horizon, so the annual worth method is used by them. Yet another group of individuals wishes to express the net economic worth as a rate or percentage; consequently, one of the rate-of-return methods would be preferred. Finally, some individuals prefer to see the net economic worth expressed as a percentage of the investment required; the savings/investment ratio is one method of providing such information.

Since many organizations have established procedures for performing economic analyses, it seems worthwhile to consider in this chapter the more popular measures of merit that are used. Among those listed, it appears that the present worth, rate-of-return, and payback period methods are currently the most popular. However, the U.S. Postal Service and a number of governmental agencies have adopted some version of the savings/investment (or benefit/cost) ratio method for purposes of comparing investment alternatives; hence, it is popular in some sectors.

### 4.3.1 Present Worth Method

The present worth of Investment Alternative $j$ can be represented as

$$PW_j(i) = \sum_{t=0}^{n} A_{jt}(1 + i)^{-t}$$ (4.6)

with $PW_j(i)$ = present worth of Alternative $j$ using $MARR$ of $i\%$
  $n$ = planning horizon
  $A_{jt}$ = net cash flow for Alternative $j$ at the end of period $t$
  $i = MARR$

The *present worth method* is the most popular measure of merit available. When it is used in Chapter Five to compare alternatives, the one having the greatest present worth is the alternative recommended.

### Example 4.4

A pressure vessel was purchased for $16,000, kept for 5 years, and sold for $3000. Annual operating and maintenance costs were $4000. Using a 12% minimum attractive rate of return, what was the present worth for the investment?

$PW_1(12\%) = -\$16,000 - \$4000(P|A\ 12,5) + \$3000(P|F\ 12,5)$
$= -\$16,000 - \$4000(3.6048) + \$3000(.5674)$
$= -\$28,717$

We can see that a single expenditure of $28,717 at time zero is equivalent to the cash flow profile for the pressure vessel.

Example 4.4 deals with an investment where no returns other than the salvage value are shown, resulting in a negative present worth. Example 4.5 considers an investment having a series of positive net cash flows.

### Example 4.5

Improved tooling for numerical control machinery will cost $10,000, last 6 years, and have no salvage value at that time. Due to this investment, net income will increase by $2525 during each of the first 3 years and by $3840 during each of the remaining 3 years. Using a 15% *MARR*, what is the present worth for the investment?

$$PW_1(15\%) = -\$10,000 + \$2525(P|A\ 15,3) + \$3840(P|A\ 15,3)(P|F\ 15,3)$$
$$= -\$10,000 + \$2525(2.2832) + \$3840(2.2832)(.6575)$$
$$= \$1529.70$$

A single sum of $1529.70 at time zero is equivalent to the cash flow profile for this investment.

---

### 4.3.2 Annual Worth Method

The annual worth of Investment Alternative $j$ can be computed as

$$AW_j(i) = \left[\sum_{t=0}^{n} A_{jt}(P|F\ i,t)\right](A|P\ i,n) \qquad (4.7)$$

or

$$AW_j(i) = PW_j(i)(A|P\ i,n) \qquad (4.8)$$

where $AW_j(i)$ denotes the annual worth of Alternative $j$ using $i = MARR$. The *annual worth method* and the *present worth method* are related by a constant multiplier, $(A|P\ i,n)$, and are therefore equivalent measures of investment worth. The investment alternative having the greatest annual worth will be preferred when the annual worth method is used to compare alternatives in Chapter Five.

---

Example 4.6_____

Let us determine the annual worth for the pressure vessel of Example 4.4 in two different ways. Recall that $P = \$16,000$, $n = 5$, $F = \$3000$, annual operating and maintenance was $4000, and $i = 12\%$.

$$AW_1(12\%) = -\$16,000(A|P\ 12,5) - \$4000 + \$3000(A|F\ 12,5)$$
$$= -\$16,000(.2774) - \$4000 + \$3000(.1574)$$
$$= -7966.20/\text{year}$$

Or, using Equation 4.8, and the answer to Example 4.3,

$$AW_1(12\%) = (-\$28,717)(A|P\ 12,5)$$
$$= (-\$28,717)(.2774)$$
$$= -\$7966.10/\text{year}$$

The minor difference in answers is due to round-off error. An expenditure of

$7966.20 at the end of each of 5 years is equivalent to the cash flow profile for the pressure vessel.

---

Example 4.7————————————————————————————————

We want to determine the annual worth for the improved numerical control tooling of Example 4.5. Recall that the tooling lasts 6 years and costs $10,000, but net incomes increase by $2525/year during the first 3 years and by $3840 thereafter. *MARR* is 15%.

$$AW_1(15\%) = -\$10,000(A|P\ 15,6) + \$2525 + \$1315(F|A\ 15,3)(A|F\ 15,6)$$
$$= -\$10,000(.2642) + \$2525 + \$1315(3.4725)(.1142)$$
$$= \$404.48/\text{year}$$

or

$$AW_1(15\%) = PW_1(15\%)(A|P\ 15,6)$$
$$= \$1529.70(.2642)$$
$$= \$404.15/\text{year}$$

Note that the difference between $404.48/year and $404.15/year is due to round-off errors.

---

### 4.3.3 Future Worth Method

The future worth of Investment Alternative $j$ can be determined using the relationship

$$FW_j(i) = \sum_{t=0}^{n} A_{jt}(1 + i)^{n-t} \qquad (4.9)$$

or

$$FW_j(i) = PW_j(i)(F|P\ i,n) \qquad (4.10)$$

or

$$FW_j(i) = AW_j(i)(F|A\ i,n) \qquad (4.11)$$

where $FW_j(i)$ is defined as the future worth of Investment $j$ using a *MARR* of $i\%$. The *future worth method* is equivalent to the *present worth method* and the *annual worth method,* since the ratio of $FW_j(i)$ and $PW_j(i)$ equals a constant, $(F|P\ i,n)$, and the ratio of $FW_j(i)$ and $AW_j(i)$ equals a constant, $(F|A\ i,n)$. The alternative having the greatest future worth will be preferred when the future worth method is used in Chapter Five to compare alternatives.

Example 4.8_____

For the previous two examples, the future worth is given by

$$FW_1(15\%) = -\$10,000(F|P\ 15,6) + \$2525(F|A\ 15,6) + \$1315(F|A\ 15,3)$$
$$= -\$10,000(2.3131) + \$2525(8.7537) + \$1315(3.4725)$$
$$= \$3538.43$$

or

$$FW_1(15\%) = PW_1(15\%)(F|P\ 15,6)$$
$$= \$1529.70(2.3131)$$
$$= \$3538.35$$

or

$$FW_1(15\%) = AW_1(15\%)(F|A\ 15,6)$$
$$= \$404.48(8.7537)$$
$$= \$3540.70$$

Again, differences in answers are due to round-off errors.

_____

### 4.3.4 Internal Rate of Return Method

The rate of return for an alternative can be defined as the interest rate that equates the future worth to zero. Letting $i_j^*$ denote the rate of return for Alternative $j$, Equation 4.9 becomes

$$0 = \sum_{t=0}^{n} A_{jt}(1 + i_j^*)^{n-t} \qquad (4.12)$$

This method of defining the rate of return is referred to in the economic analysis literature as the discounted *cash flow rate of return, internal rate of return,* and the *true rate of return.* We prefer the term *internal rate of return (IRR).*

Note that the present worth and the annual worth can be obtained by multiplying both sides of Equation 4.9 by the appropriate interest factor [i.e., $(P|F\ i_j^*,n)$ and $(A|F\ i_j^*,n)$]. Hence, the internal rate of return can also be defined to be the interest rate that yields either a present worth or an annual worth of zero. Depending on the form of a particular cash flow profile, it might be more convenient to use a present worth or an annual worth formulation to determine the internal rate of return for an alternative.

It is important to understand the definition of rate of return inherent in the use of the *IRR* method. In particular, the internal rate of return on an investment can be defined as *the rate of interest earned on the unrecovered balance of an investment.* This concept was demonstrated in Chapter Three in discussing the amount of a loan payment that was principal. It is illustrated again in Table 4.1, where \$10,000 is invested to obtain the receipts shown over a 6-year period. $A_t$ denotes the cash flow at the *end* of period $t$, $B_t$ represents the

TABLE 4.1. Data Illustrating the Meaning of
the Internal Rate of Return

| $t$ | $A_t$ | $B_t$ | $I_t^a$ | $E_t$ |
|---|---|---|---|---|
| 0 | −10,000 | | | −10,000 |
| 1 | 2,525 | −10,000 | −2,000 | − 9,475 |
| 2 | 2,525 | − 9,475 | −1,895 | − 8,845 |
| 3 | 2,525 | − 8,845 | −1,769 | − 8,089 |
| 4 | 3,840 | − 8,089 | −1,618 | − 5,867 |
| 5 | 3,840 | − 5,867 | −1,173 | − 3,200 |
| 6 | 3,840 | − 3,200 | − 640 | 0 |

* Based on $i = 0.20$.

unrecovered balance at the *beginning* of period $t$, $E_t$ is the unrecovered balance
at the *end* of period $t$, and $I_t$ is defined as the interest on the unrecovered
balance *during* period $t$. The following relationships exist.

$$E_0 = A_0$$
$$B_t = E_{t-1} \qquad\qquad t = 1, \ldots, n$$
$$I_t = B_t i \qquad\qquad t = 1, \ldots, n$$
$$E_t = A_t + B_t + I_t \qquad t = 1, \ldots, n$$

If $i$ is the internal rate of return, then $E_n$ will equal zero. As indicated in Table
4.1, if $i$ is 20%, $E_n$ is approximately zero. Consequently, $i^*$ is approximately
20%. The equivalence of $E_n$ being zero and the future worth being zero is easily
understood by recognizing that $E_n$ is actually the future worth of the cash flow
profile. To see why this is true, notice that

$$E_n = A_n + B_n + I_n$$

Employing the definition of $I_n$,

$$E_n = A_n + B_n(1 + i)$$

By the relationship between $B_n$ and $E_{n-1}$, it is seen that

$$E_n = A_n + E_{n-1}(1 + i)$$

Since a similar relationship exists between $E_{n-1}$ and $E_{n-2}$, we note that

$$E_n = A_n + A_{n-1}(1 + i) + E_{n-2}(1 + i)^2$$

Generalizing, the recursive relationship between $E_t$ and $E_{t-1}$ gives

$$E_n = A_n + A_{n-1}(1 + i) + A_{n-2}(1 + i)^2 + \cdots + A_0(1 + i)^n$$

Hence, we see that

$$E_n = FW(i\%)$$

as anticipated.

The above example illustrates that the time value of money operations involved in the *IRR* method are equivalent to assuming that all monies received are reinvested and earn interest at a rate equal to the internal rate of return. This can also be seen mathematically by letting

$$A_{jt} = \text{net cash flow for Investment } j \text{ in period } t$$

$$R_{jt} = \begin{cases} A_{jt}, & \text{if } A_{jt} \geq 0 \\ 0, & \text{otherwise} \end{cases}$$

$$C_{jt} = \begin{cases} A_{jt}, & \text{if } A_{jt} < 0 \\ 0, & \text{otherwise} \end{cases}$$

$r_t$ = reinvestment rate for positive cash flows occurring in period $t$

$i'$ = the rate of return for negative cash flows

Then the following relationship can be defined.

$$\sum_{t=0}^{n} R_{jt}(1 + r_t)^{n-t} = \sum_{t=0}^{n} C_{jt}(1 + i')^{n-t} \qquad (4.13)$$

Note that the future worth of reinvested monies received must equal the future worth of investments.

If $r_t$ equals $i'$, Equation 4.13 becomes

$$0 = \sum_{t=0}^{n} (R_{jt} - C_{jt})(1 + i')^{n-t} \qquad (4.14)$$

Letting $A_{jt}$ equal $R_{jt} - C_{jt}$ defines the *IRR* method given by Equation 4.12. Hence, we see that the rate of return obtained using the *IRR* method can be interpreted as the reinvestment rate for all recovered funds.

Since the determination of the rate of return involves solving Equation 4.12 for $i_j^*$, it is seen that (for a given investment $j$) it is necessary to determine the values of $x$ that satisfy the $n$-degree polynomial $0 = A_0 x^n + A_1 x^{n-1} + \cdots + A_{n-1}x + A_n$ where $x = (1 + i_n^*)$. In general, there can exist $n$ distinct roots (values of $x$) for an $n$-degree polynomial; however, most cash flow profiles encountered in practice will have a unique root (internal rate of return).

The number of real positive roots of an $n$-degree polynomial with real coefficients is less than or equal to the number of changes of sign in the sequence of cash flows, $A_0, A_1, \ldots, A_{n-1}, A_n$. Since the typical cash flow pattern begins with a negative cash flow, followed by positive cash flows, a unique root will normally exist.

| TABLE 4.2. Cash Flow Profile | |
|---|---|
| EOY | CF |
| 0 | −$1,000 |
| 1 | 4,100 |
| 2 | − 5,580 |
| 3 | 2,520 |

Example 4.9 _____

As an illustration of a cash flow profile having multiple roots, consider the data given in Table 4.2. The future worth of the cash flow series will be zero using a 20, 40, or 50% interest rate.

$$FW_1(20\%) = -\$1000(1.2)^3 + \$4100(1.2)^2 - \$5580(1.2) + \$2520 = 0$$

$$FW_1(40\%) = -\$1000(1.4)^3 + \$4100(1.4)^2 - \$5580(1.4) + \$2520 = 0$$

$$FW_1(50\%) = -\$1000(1.5)^3 + \$4100(1.5)^2 - \$5580(1.5) + \$2520 = 0$$

A plot of the future worth for this example is given in Figure 4.1. The future worth polynomial is a third-degree polynomial and there are three changes of sign in the ordered sequence of cash flows $(-, +, -, +)$. In this case there are three unique positive real roots, corresponding to $i = .20$, $i = .40$, and $i = .50$. This is seen by factoring the future worth polynomial which can be written as

$$FW_1(i) = \$1000(1.2 - x)(1.4 - x)(1.5 - x)$$

where $x = (1 + i)$.

_____

Multiple rates of return as seen in Example 4.9 are difficult to interpret properly without other information such as the future worth, present worth, or annual worth. For example, the pattern of cash flows shown in Table 4.2 is preferred to investing at the *MARR* if $0\% \leq MARR < 20\%$ or $40\% < MARR < 50\%$. This is easily seen by looking at Figure 4.1, but is not immediately obvious from the three roots to the future worth equation.

The difficulty of interpreting the polynomial roots in complex cash flow profiles is a severe drawback to the use of the *IRR* approach. For additional discussion of the subject of multiple roots in internal rate of return calculations, as well as their interpretation, see Bernhard [3, 4, 5] and Bussey [6].

### 4.3.5 External Rate-of-Return Method

The possibility of multiple roots occurring in the internal rate of return calculation, coupled with the reinvestment assumptions concerning recovered funds,

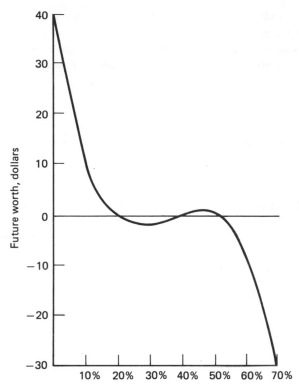

FIGURE 4.1. Plot of future worth for Example 4.9.

has led to the development of an alternative rate-of-return method, called the *external rate of return method*. The external rate of return (*ERR*) method consists of the determination of the value of $i'$ that satisfies Equation 4.13 which is repeated here for convenience.

$$\sum_{t=0}^{n} R_{jt}(1 + r_t)^{n-t} = \sum_{t=0}^{n} C_{jt}(1 + i')^{n-t}$$  (4.15)

In this equation, $i'$ is now known as the external rate of return. As defined previously, $R_{jt}$ and $C_{jt}$ are positive and negative net cash flows for Investment $j$ during period $t$, respectively, and $r_t$ is the reinvestment rate for positive cash flows occurring in period $t$. Normally, $r_t$ equals the minimum attractive rate of return, since the *MARR* reflects the opportunity cost for money available for investment. That is, we assume that positive net cash flows can always be reinvested at the *MARR*.

The external rate of return method is one of several rate of return approaches that consider the explicit reinvestment of positive cash flows, positive net cash flows, accumulated cash flows, and so forth. The *ERR* method offers a distinct advantage over the *IRR* approach, in that there is a unique value of $i'$ that satisfies Equation 4.15. That is, there is only one external rate of return for a given cash flow profile. If $i'$ exceeds the *MARR*, the investment is preferred over investing at the *MARR*.

Example 4.10_____

Let us reconsider the cash flows of Example 4.9 in Table 4.2. The value of the *ERR* may be determined by solving the equation

$$4100(1 + r_t)^2 + 2520 = 1000(1 + i')^3 + 5580(1 + i')$$

for $i'$, where $r_t = MARR$. This results in the following series of values for a wide range of the *MARR*.

| MARR Value $(r_t)$ | External Rate of Return $(i')$ |
|---|---|
| .10 | .101302 |
| .15 | .150458 |
| .20 | .200000 |
| .25 | .249817 |
| .30 | .299812 |
| .35 | .349898 |
| .40 | .400000 |
| .45 | .450053 |
| .50 | .500000 |
| .55 | .549795 |
| .70 | .697893 |

There is not much difference between the value of $r_t$ and its corresponding value $i'$. This is simply a function of the particular cash flow profile and is not important (note that the future worths in Figure 4.1 did not deviate far from zero either). What *is* important is that the *ERR* exceeds the *MARR* in the range $0 \le MARR < 20\%$ and $40\% < MARR < 50\%$. These are precisely the same ranges for which the future worths were positive in Figure 4.1.

This example illustrates the consistency of the *ERR* method with the future worth, present worth, and annual worth methods when evaluating a single investment, even when that investment has a complex cash flow pattern.

Example 4.11_____

For the data provided in Table 4.1, suppose money received from the initial investment is reinvested and earns 15% interest. At the end of the sixth year the reinvested funds total

$$\$2525(F|A\ 15,6) + \$1315(F|A\ 15,3)$$
$$= \$2525(8.7537) + \$1315(3.4725)$$
$$= \$26,669.43$$

Consequently, the *external rate of return* is defined as the interest rate such that the future worth of the $10,000 investment equals $26,669.43. Thus,

$$\$10,000(1 + i')^6 = \$26,669.43$$

Taking the logarithm and solving for $i$ yields a value of 17.76% as the external rate of return.

---

Example 4.12

Now, rework Example 4.11 assuming recovered funds are reinvested at 20%. The future worth of the reinvested funds will be

$$\$2525(F|A\ 20,6) + \$1315(F|A\ 20,3)$$
$$= \$2525(9.9299) + \$1315(3.6400)$$
$$= \$29,859.60$$

Setting the future worth of the $10,000 investment equal to $29,859.60 yields

$$\$10,000(1 + i')^6 = \$29,859.60$$

Solving for $i'$ gives a value of 20% as the external rate of return, as anticipated.

---

### 4.3.6 Savings/Investment Ratio Method

The *savings/investment ratio method* can be defined in many ways; two typical approaches will be described. The *savings/investment ratio* to be used for analyzing and comparing alternatives in Chapter Five is given as

$$SIR_j(i) = \frac{\sum_{t=0}^{n} R_{jt}(1 + i)^{-t}}{\sum_{t=0}^{n} C_{jt}(1 + i)^{-t}} \tag{4.16}$$

where $SIR_j(i)$ is the savings/investment ratio for Investment Alternative $j$ based on a *MARR* of $i\%$. We again define $A_{jt}$, $R_{jt}$, and $C_{jt}$ as follows:

$A_{jt}$ = net cash flow for Investment $j$ in period $t$

$$R_{jt} = \begin{cases} A_{jt}, & \text{if } A_{jt} \geq 0 \\ 0, & \text{otherwise} \end{cases}$$

$$C_{jt} = \begin{cases} A_{jt}, & \text{if } A_{jt} < 0 \\ 0, & \text{otherwise} \end{cases}$$

Note that this *SIR* formulation is the present worth of net positive cash flows divided by the present worth of net negative cash flows. Therefore, for a project to be preferred over investing at the *MARR*, the ratio must be greater than one.

Another formulation of the savings/investment ratio is defined as follows:

$$SIR_j(i) = \frac{\sum_{t=0}^{n} A_{jt}(1 + i)^{-t}}{\sum_{t=0}^{n} C_{jt}(1 + i)^{-t}}$$

(4.17)

where $A_{jt}$ and $C_{jt}$ are as defined previously. Note that this formulation is simply the present worth of all cash flows divided by the present worth of all negative cash flows. A project having a ratio greater than zero is preferable to investing at the *MARR*.

Example 4.13 _____

Consider the cash flows given in Table 4.1 and let the *MARR* be 15%. The present worth of the net positive cash flows is

$$\sum_{t=0}^{n} R_{1t}(1 + i)^{-t} = \$2525(P|A\ 15,3) + \$3840(P|A\ 15,3)(P|F\ 15,3)$$

$$= \$2525(2.2832) + \$3840(2.2832)(.6575)$$

$$= \$11,529.70$$

The present worth of the net negative cash flows is

$$\sum_{t=0}^{n} C_{1t}(1 + i)^{-t} = -\$10,000$$

The present worth of all net cash flows is

$$\sum_{t=0}^{n} A_{1t}(1 + i)^{-t} = -\$10,000 + \$11,529.70 = \$1529.70$$

The savings/investment ratio using Equation 4.16 is given by

$$SIR_1(15\%) = \frac{\$11,529.70}{\$10,000} = 1.15297$$

Using the savings/investment ratio given by Equation 4.17 results in

$$SIR_1(15\%) = \frac{\$1529.70}{\$10,000} = .15297$$

Note that these ratios are greater than one and zero, respectively, both of them indicating that an investment resulting in the cash flow profile of Table 4.1 is preferable to investing at the *MARR*.

_____

An alternative label for the savings/investment ratio is the *benefit-cost ratio*. The savings are interpreted as benefits derived from a venture and the investment corresponds to the cost of providing the benefits. The benefit-cost ratio is used extensively in government and public sector economic analyses. Like the savings/investment ratio, the benefit-cost ratio has many possible formulations. We explore benefit-cost analysis more fully in Chapter Seven.

### 4.3.7 Payback Period Method

Each of the preceding methods for measuring project worth is equivalent or consistent. That is, each will provide the same decision regarding the desirability of investing in a project versus investing at the *MARR*. The *payback period method,* unfortunately, is not an equivalent method. Yet, it is used frequently in economic analyses; therefore, we should know about its good points as well as its pitfalls.

The payback period method involves the determination of the length of time required to recover the initial investment based on a zero interest rate. Letting $C_{j0}$ denote the initial cost of Investment Alternative $j$, and $R_{jt}$ denote the net revenue received from Investment $j$ during period $t$, if we assume no other negative net cash flows occur, then the smallest value of $m_j$ such that

$$\sum_{t=1}^{m_j} R_{jt} \geq C_{j0} \tag{4.18}$$

defines the payback period for Investment $j$. The investment alternative having the smallest payback period is the preferred alternative using the payback period method.

Example 4.14 _____

Based on the data given in Table 4.1,

$$\sum_{t=1}^{3} R_{1t} = \$2525 + \$2525 + \$2525 = \$7575 < C_{10} = \$10{,}000$$

and

$$\sum_{t=1}^{4} R_{1t} = \$2525 + \$2525 + \$2525 + \$3840 = \$11{,}415 > C_{10} = \$10{,}000$$

Consequently, $m$ equals 4, indicating that four years are required to pay back the original investment.

A number of variations of the payback or payout method have been used by different organizations. For example, if negative net cash flows occur after year zero, the payback period might be defined as the smallest value of $m_j$ in which the sum of the net cash flows $\sum_{t=0}^{m_j} A_{jt}$ first exceeds zero. Also, the payback period may be noninteger, assuming cash flows are uniformly distributed throughout the year. However, the basic deficiencies of the payback period method are present in all variations with which we are familiar; the *timing* of cash flows and the *duration* of the project are ignored.

To illustrate the deficiencies of the payback period, consider the cash flow profiles depicted in Figure 4.2a. Using the payback period, profile B would be preferred, even though we would most likely choose profile A. Likewise, in

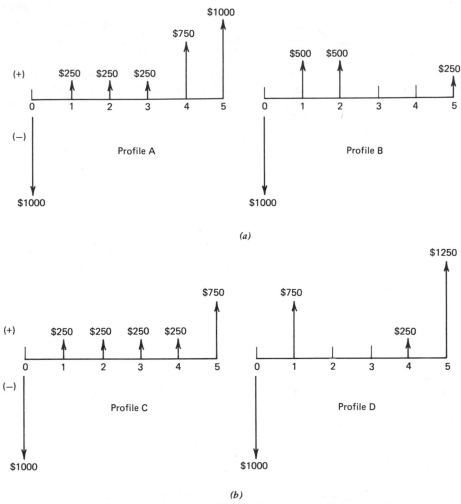

FIGURE 4.2. Four different cash flow profiles for illustrating deficiencies of payback period.

Figure 4.2b, profiles C and D are equally preferred using the payback period, even though profile D is clearly better.

Despite its obvious deficiencies, the payback period method continues to be one of the most popular methods of judging the desirability of investing in a project. The reasons for its popularity include the following:

1. Does not require interest rate calculations.

2. Does not require a decision concerning the discount rate (*MARR*) to use.

3. Is explained and understood easily.

4. Reflects a manager's attitudes when investment capital is limited.

5. Hedges against uncertainty of future cash flows.

6. Provides a rough measure of the liquidity of an investment.

The payback period method is recommended as a supplementary method of measuring investment worth. In particular, it is suggested that the payback period method be used in addition to a "time value of money"-based method.

### 4.3.8 Perpetuities and Capitalized Worth

A specialized type of cash flow series is a perpetuity, a uniform series of cash flows which continues indefinitely. This is a special case since an infinite series of cash flows would rarely be encountered in the business world; rather, a finite series of cash flows is the general rule. However, for such very long-term investment projects as bridges, highways, forest harvesting, or the establishment of endowment funds where the estimated life is 50 years or more, an infinite cash flow series may be appropriate.

If a present value $P$ is deposited into a fund at interest rate $i$ per period so that a payment of size $A$ may be withdrawn each and every period forever, then the following relation holds between $P$, $A$, and $i$.

$$Pi = A \qquad (4.19)$$

Thus, as depicted in Figure 4.3, $P$ is a present value that will pay out equal payments of size $A$ indefinitely if the interest rate per period is $i$. The present value $P$ is termed the *capitalized worth* of $A$, the size of each of the perpetual payments.

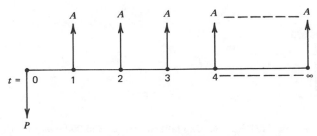

FIGURE 4.3. An infinite cash flow series.

## Example 4.15

What deposit at $t = 0$ into a fund paying 9½% annually is required in order to pay out $5000 each year forever? The solution is straightforward from

$$P = \frac{A}{i} = \$5000/.095 = \$52,631.58$$

By means of the subsequent example, let us now broaden the meaning of capitalized worth to be that present value that would pay for the first cost of some project and provide for its perpetual maintenance at an interest rate $i$.

## Example 4.16

Project ABC consists of the following requirements:

1. A $50,000 first cost at $t = 0$.
2. A $5000 expense every year.
3. A $25,000 expense every third year forever, with the first expense occurring at $t = 3$.

What is the capitalized cost of Project ABC if $i = 15\%$ annually?

It will be instructive to determine the capitalized cost of each requirement separately and then sum the results. First, the capitalized cost of "$25,000 every third year forever" may be determined from any of three points of view. One view is that a value $P$ is required at the beginning of a 3-year period such that, with interest compounded at 15% annually, a sum of $P$ + $25,000 will accrue at the end of the 3-year period. Thus, $25,000 would be withdrawn; thereby leaving the value $P$ to repeat the cycle indefinitely each three-year period. This logic is illustrated in Figure 4.4.

Thus,

$$P(F|P\ 15,3) = P + \$25,000$$

or

$$1.5209P = P + \$25,000$$

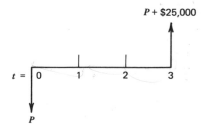

FIGURE 4.4. Capitalized cost of $25,000 every third year—first view.

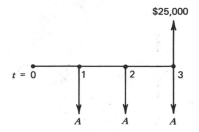

$25,000

$$t = 0$$

1  2  3

A  A  A

FIGURE 4.5. Capitalized cost of $25,000 every third year—second view.

and

$$P = \$47,993.86$$

which is the capitalized cost of the requirement.

A second view is to reason that, for each 3-year period, three equal deposits of size $A$ are required that will amount to $25,000 at the end of the third year. Then, if these payments of size $A$ occur every year, $25,000 will be available every third year forever. The present value $P$ that yields the required payment $A$ is the solution, as shown in Figure 4.5.

The required payment $A$ is given by

$$A = F(A|F\ 15,3) = \$25,000(0.2880)$$
$$= \$7200$$

Then,

$$P = \frac{A}{i} = \frac{\$7200}{0.15} = \$48,000.00$$

which is approximately the same answer as from the first approach. The difference of $6.14 is due to rounding in the interest factor calculations.

A third point of view for the $25,000 requirement is to consider the infinite series depicted in Figure 4.6, such that $P = A/i$ would yield the desired result. However, the value of $i$ required is the effective interest rate for a 3-year period. Thus,

$$i = [(1 + 0.15)^3 - 1] = 0.5209$$

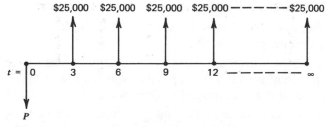

$25,000  $25,000  $25,000  $25,000 ――――――$25,000

$$t = 0$$

3  6  9  12 ―――――― ∞

P

FIGURE 4.6. Capitalized cost of $25,000 every third year— third view.

and

$$P = \frac{\$25,000}{0.5209} = \$47,996.16$$

The capitalized cost of Project ABC is computed as:

1. Capitalized cost of $50,000 first cost            = $ 50,000
2. Capitalized cost of $5000 every year = $5000/0.15 = $ 33,333
3. Capitalized cost of $25,000 every third year       = $ 48,000

                          Total capitalized cost    $131,333

The capitalized worth of $131,333 would provide approximately $A = Pi = \$131,333(0.15) = \$19,700$ every year forever.

---

The capitalized worth may be looked upon as a present worth of some cash flow pattern that repeats indefinitely.

Example 4.17_____

Recall that the annual worth of the cash flows in Example 4.7 was calculated to be $404.48/year (these were the same cash flows later shown in Table 4.1). If this cash flow pattern were to repeat every 6 years forever, the capitalized value would be

$$P = \frac{A}{i} = \frac{\$404.88}{.15} = \$2699.20$$

Note that this is not drastically higher than the present worth of $1529.70 calculated in Example 4.5 for the first 6 years. This is because the distant cash flows are substantially discounted at $i = 15\%$ in the capitalized worth calculation.

---

## 4.4 BOND PROBLEMS

Although bonds are important financial instruments in the business world as investment opportunities, there are additional reasons for considering bond problems in an economic analysis textbook.

1. The issuance and sale of bonds is a mechanism by which capital may be raised to finance engineering projects.
2. Bond problems illustrate the notion of equivalence. That is, the purchase price of a bond is equivalent (has the same present value) to the returns from the bond at an appropriate compound interest rate.
3. Bond problems are convenient investment opportunities, which may be used to illustrate the various methods of measuring investment worth, especially the present worth and internal rate of return methods.

The latter two reasons are the primary ones for treating bond problems in this text. Subsequent discussion presents the appropriate terminology and illustrates the three types of problems that are possible.

An organizational unit desiring to raise capital may issue bonds totaling, say, $1 million, $5 million, $25 million, or more. A financial brokerage firm usually handles the issue on a commission basis and sells smaller amounts to other organizational units or individual investors. Individual bonds are normally issued in even denominations such as $500, $1000, or $5000. The stated value on the individual bond is termed the *face* or *par value*. The par value is to be repaid by the issuing organization at the end of a specified period of time, say 5, 10, 15, 20, or even 50 years. Thus, the issuing unit is obligated to *redeem* the bond at par value at *maturity*. Furthermore, the issuing unit is obligated to pay a stipulated *bond rate* on the face value during the interim between date of issuance and date of redemption. This might be 10% payable quarterly, 9½% payable semiannually, 11% payable annually, and so forth. For the purpose of the problems to follow, it is emphasized that the bond rate applies to the par value of the bond.

### Example 4.18

A person purchases a $5000, 5-year bond on the date of issuance for $5000. The stated bond rate is "10% semiannually," and the interest payments are received on schedule until the bond is redeemed at maturity for $5000. The bond rate per interest period is 10%/2 = 5%. Thus, the bond holder receives (0.10/2)($5000) = $250 payments every 6 months. A cash flow diagram for the duration of the investment is given in Figure 4.7, where time periods are 6-month intervals.

It is noted from Figure 4.7 that the $5000 expenditure at $t = 0$ yields $250 each interest period for 10 periods and a $5000 redemption value at $t = 0$. Thus,

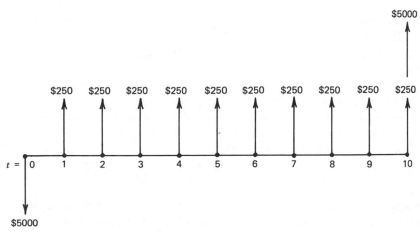

FIGURE 4.7. Cash flow for a $5000 bond.

the $5000 investment at $t = 0$ yielded the revenues from $t = 1$ through $t = 10$. What annual interest rate or internal rate of return did the $5000 investment yield?

One might intuitively answer the above question by stating that the $5000 investment was exactly returned (no loss or gain in capital) at redemption and, since during the interim an interest rate of 10% semiannual was received, then the yield *must* be 10% semiannual. Accepting this argument for the moment, one might further pose certain hypotheses concerning the transaction. For instance, it is hypothesized that the $5000 at $t = 0$ is equivalent (has the same present worth) to the revenue cash flows if the time value of money is 10% compounded semiannually. That this hypothesis is true is shown by the relation

$$P = A(P|A\ 5,10) + F(P|F\ 5,10)$$

or

$$\$5000 = (0.05)(\$5000)(7.7217) + \$5000(0.6139),$$

and

$$\$5000 = \$5000$$

A second hypothesis concerning the transaction is that if more than the par value is paid for the bond at $t = 0$ and all revenue figures remain the same, then an internal rate of return less than the bond rate of 10% semiannual will be received on the investment. For example, if $5500 were paid for the bond at $t = 0$, the present worth of the $5500 outgo is not equal (equivalent) to the present worth of the revenues at a 10% semiannual yield. This is shown by

$$\$5500 \neq (0.05)(\$5000)(P|A\ 5,10) + \$5000(P|F\ 5,10)$$
$$\$5500 \neq \$250(7.7217) + \$5000(0.6139)$$
$$\$5500 \neq \$5000$$

This result now raises the pertinent question, what *is* the rate of return on the investment if $5500 is paid for the bond and the revenues remain the same? Intuition may suggest that the yield will be less than 10% semiannual because the purchase value of $5500 decreases (a loss of investment capital) to $5000 on redemption. Furthermore, the semiannual payments of $250 are not equal to 5% of the $5500 purchase price. In order to answer the question more precisely, let us answer the alternate question of "what interest rate per period (or yield per period) will make the future worth of all cash flows equal to zero?" That is, what interest rate satisfies the following equation?

$$-\$5500(F|P\ i,10) + (.05)(\$5000)(F|A\ i,10) + \$5000 = \$0$$

The solution to the above equation gives the answer to the original question, and the task is to solve for the positive roots of the polynomial. However, an approximate solution to avoid such tedium is by trial and error. An iterative procedure follows.

For $i = 3\%$ (6% semiannual)

$$-\$5500(F|P \ 3,10) + (.05)(\$5000)(F|A \ 3,10) + \$5000 \neq \$0$$

$$-\$5500(1.3439) + (\$250)(11.4639) + \$5000 \neq \$0$$

$$\$474.53 \neq \$0$$

For $i = 4\%$ (8% semiannual)

$$-\$5500(F|P \ 4,10) + (.05)(\$5000)(F|A \ 4,10) + \$5000 \neq \$0$$

$$-\$5500(1.4802) + (\$250)(12.0061) + \$5000 \neq \$0$$

$$-\$139.58 \neq \$0$$

From these two trials (for $i = 3\%$ and $i = 4\%$), the future worth of \$0 is bracketed, as shown in Table 4.3. Using the data of Table 4.3, we can solve for $X$ by linear interpolation, or

$$\frac{0.04 - 0.03}{-\$139.58 - \$474.53} = \frac{0.04 - X}{-\$139.58 - \$0.00}$$

or

$$X = 0.3772 \quad \text{or} \quad 3.772\%$$

Thus, the equivalent yield on the \$5500 investment is approximately 3.772%/period, or $(2)(3.772) = 7.544\%$ semiannually, or an effective annual yield of $[(1 + 0.03772)^2 - 1](100\%) = 7.686\%$. The figures support the *a priori* intuition of an internal rate of return less than 10% semiannual, and the second hypothesis is accepted.

---

Bond problems arise in economic analysis because many bonds trade daily through financial markets such as the New York Stock Exchange. Thus, bonds may be purchased for less than, greater than, or equal to par value, depending on the economic environment. They may also be sold for less than, greater than, or equal to par value. Furthermore, once purchased, bonds may be kept for a variable number of interest periods before being sold. A variety of situations can occur, but only three basic types of bond problems can occur. These

TABLE 4.3. Bond Yield Interpolation

| For $i =$ | 0.03 | X | 0.04 |
|---|---|---|---|
| FW of revenues = | \$474.53 | \$0.00 | -\$139.58 |

will be presented after formalizing the discussion thus far. We now employ the following notation.

$P$ = the purchase price of a bond
$F$ = the sales price (or redemption value) of a bond
$V$ = the par or face value of a bond
$r$ = the bond rate per interest period
$i$ = the yield rate per interest period
$n$ = the number of interest payments received by the bondholder
$A = Vr$ = the interest payment received

The general expression relating these is

$$P = Vr(P|A\ i,n) + F(P|F\ i,n) \tag{4.20}$$

Now, the three types of bond problems follow:

**1.** Given $P,r,n,V$, and a desired $i$, find the sales price $F$.

**2.** Given $F,r,n,V$, and a desired $i$, find the purchase price $P$.

**3.** Given $P,F,r,n$, and $V$, find the yield $i$ that has been earned on the investment.

Each of these cases is illustrated in the following examples.

### Example 4.19

A $1000, 12% semiannual bond is purchased for $1050 at an arbitrary $t = 0$. If the bond is sold at the end of 3 years and six interest payments, what must be the selling price if this bond is to be a good investment compared to investing at *MARR* of 10% nominal. The cash flow for this example is given by Figure 4.8.

The present worth of the cash flow depicted in Figure 4.8 follows.

$$P = Vr(P|A\ 5,6) + F(P|F\ 5,6)$$

or

$$\$1050 = (\$1000)(0.06)(5.0757) + F(0.7462).$$

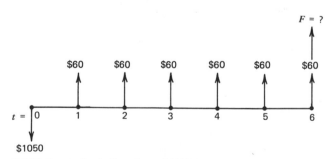

FIGURE 4.8. Cash flow for a $1000 bond—determine sales price.

Solving the above equation for $F$ yields a value of $999.01. As long as the selling price at the end of three years is at least $999.01 the bond is preferred over investing at the *MARR*.

Example 4.20

If a $1000, 12% semiannual bond is purchased at an arbitrary $t = 0$, held for 3 years and six interest payments, and redeemed at par value, what must the purchase price have been in order for the bond to be preferred over investing at a *MARR* of 14% nominal?

From the cash flow given in Figure 4.9, the present worth of future cash flows is

$$P = Vr(P|A\ 7{,}6) + F(P|F\ 7{,}6)$$

or

$$P = (\$1000)(0.06)(4.7665) + \$1000(0.6663)$$

and

$$P = \$952.29.$$

As long as the purchase price did not exceed $952.29, the bond was better than investing at the *MARR*.

Example 4.21

If a $1000, 12% quarterly bond is purchased at $t = 0$ for $1020 and sold 3 years later for $950, (1) what was the quarterly yield on the investment, and (2) what was the effective rate of return? From the cash flow diagram in Figure 4.10, and setting the present worth equal to zero, we have

$$-\$1020 + (\$1000)(0.03)(P|A\ i{,}12) + \$950(P|F\ i{,}12) = \$0$$

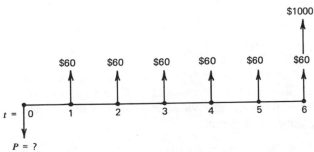

FIGURE 4.9. Cash flow for a $1000 bond—determine purchase price.

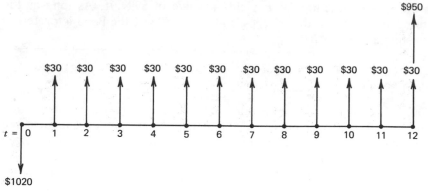

FIGURE 4.10. Cash flow for a $1000 bond—determine yield.

It is then necessary to solve for the unknown $i$ by trial and error as follows:

For $i = 2\%$,

$$-\$1020 + \$30(10.5753) + \$950(0.7885) \neq \$0$$

$$-\$46.33 \neq \$0$$

For $i = 2\frac{1}{2}\%$,

$$-\$1020 + \$30(10.2578) + \$950(0.7436) \neq \$0$$

$$\$5.85 \neq \$0$$

Then, interpolating gives

$$\frac{0.025 - 0.020}{\$5.85 - (-\$46.33)} = \frac{0.025 - X}{\$5.85 - \$0}$$

$$X = 0.02444 \quad \text{or} \quad 2.444\% \text{ per quarter}$$

The effective internal rate of return is

$$[(1 + 0.02444)^4 - 1]100\% = 10.140\%$$

## 4.5 CAPITAL RECOVERY FORMULA

In engineering economic analyses, it is common to refer to the *capital recovery cost* of an asset. Figure 4.11 illustrates an investment of $\$P$ in an asset having a life of $n$ years and disposed of for a salvage value of $\$F$. The capital recovery cost is a uniform annual amount defined as

$$CR = P(A|P\ i,n) - F(A|F\ i,n) \tag{4.21}$$

The salvage value at year $n$ is an income, and therefore a negative cost.

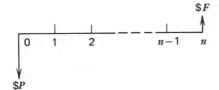

FIGURE 4.11. Cash flow diagram of investment in an asset.

## Example 4.22

A stand-alone minicomputer is purchased for $P = \$82,000$, has a service life of $n = 7$ years, and a salvage value of $F = \$5000$. At an interest rate of 15%, the capital recovery cost is

$$CR = \$82,000(A|P\ 15,7) - \$5000(A|F\ 15,7)$$
$$= \$82,000(.2404) - \$5000(.0904)$$
$$= \$19,260.80/\text{year}$$

From Chapter Three and the interest tables we know that

$$(A|P\ i,n) = (A|F\ i,n) + i \qquad (4.22)$$

and similarly

$$(A|F\ i,n) = (A|P\ i,n) - i \qquad (4.23)$$

Substituting Equations 4.22 and 4.23 into Equation 4.21 leads to two more capital recovery cost formulas. They are, respectively,

$$CR = (P\text{-}F)(A|F\ i,n) + Pi \qquad (4.24)$$

and

$$CR = (P\text{-}F)(A|P\ i,n) + Fi \qquad (4.25)$$

Among the three methods of computing capital recovery cost, Equation 4.25 appears to be the most popular. We tend to use Equation 4.21 since it is a direct application of the interest factors to the actual cash flows. The relationship between capital recovery cost and depreciation or cost recovery is explained in Chapter Six.

## Example 4.23

We can apply Equations 4.24 and 4.25 to the previous example problem having $P = \$82,000$, $n = 7$, $F = \$5000$, and $i = 15\%$. Using Equation 4.24, we have

$$CR = (\$82,000 - \$5,000)(A|F\ 15,7) + \$82,000(.15)$$
$$= \$77,000(.0904) + \$82,000(.15)$$
$$= \$19,260.80/\text{year}$$

Using Equation 4.25 we have

$$CR = (\$82,000 - \$5000)(A|P\ 15,7) + \$5000(.15)$$
$$= \$77,000(.2404) + \$5000(.15)$$
$$= \$19,260.80/\text{year}$$

## 4.6 INFLATIONARY EFFECTS

We are all aware of inflation. Inflation can be defined as an increase in the amount of currency in circulation, resulting in a relatively sharp and sudden fall in its value and, therefore, a rise in prices. Although we have become accustomed to living in inflationary times, we are often very uncertain about how inflation should be treated, if at all, in an economic analysis. This section will present some background material on inflation, followed by two equivalent approaches for performing economic analyses under inflation.

### 4.6.1 Background on Inflation

One of the most popular indicators of the *general inflation rate* ($j$) is the Consumer Price Index (CPI). Table 4.4 presents the CPI during recent years. The general inflation rate ($j$) is an overall measure of the annual decrease in purchasing power of a unit of currency such as the dollar. It is determined by the price changes of a wide range of goods and services. Prices of some of these goods and services may vary considerably, with some rising quickly and others rising not so quickly, or even falling. In general, the inflation rate is a weighted average of the effective annual escalation rates ($e_k$) of many goods and services. A negative value of the inflation rate $j$ is known as deflation.

The *effective annual escalation rate* ($e_k$) is the annual rate of increase in price for a particular ($k$) good or service. The value of $e_k$ includes the effects of general inflation $j$ and may be lower or higher than the value of $j$. For example, technological breakthroughs in electronics have actually driven down the price of some items, causing $e_k$ to be negative. Other values of $e_k$ may far exceed the value of $j$. Table 4.5 compares the CPI and effective annual escalation rates for

TABLE 4.4. An Indicator of the General Inflation Rate

| Year | Consumer Price Index*<br>(Base Is 100 in 1967) | CPI Rate of<br>Inflation in Percent |
|------|------------------------------|---------------------------|
| 1978 | 200.9 | — |
| 1979 | 225.4 | 12.20 |
| 1980 | 253.9 | 12.64 |
| 1981 | 279.9 | 10.24 |
| 1982 | 294.1 | 5.07 |

*CPI Detailed Report,* U.S. Department of Labor, Bureau of Labor Statistics, October issue of each year.

TABLE 4.5. Comparison of CPI and Annual Escalation Rates

| Year | Fuel Oil, Coal, Bottled Gas | | Televisions | | Poultry | | General Inflation Rate |
|---|---|---|---|---|---|---|---|
| | CPI | Escalation Rate | CPI | Escalation Rate | CPI | Escalation Rate | |
| 1978 | 300.1 | — | 102.1 | — | 177.3 | — | — |
| 1979 | 470.8 | 56.9% | 103.4 | 1.3% | 170.3 | -3.9% | 12.20% |
| 1980 | 558.7 | 18.7% | 104.7 | 1.3% | 209.1 | 22.8% | 12.64% |
| 1981 | 672.7 | 20.4% | 105.0 | .3% | 196.6 | -6.0% | 10.24% |
| 1982 | 677.2 | .7% | 103.5 | -1.4% | 195.4 | -.6% | 5.07% |
| Compound Annual Rate (annual rate needed to get from 1978 CPI to 1982 CPI) | $e_1 = 22.6\%$ $\left(\dfrac{677.2}{300.1} = 1.226^4\right)$ | | $e_2 = .3\%$ $\left(\dfrac{103.5}{102.1} = 1.003^4\right)$ | | $e_3 = 2.5\%$ $\left(\dfrac{195.4}{177.3} = 1.025^4\right)$ | | $j = 10.0\%$ |

three commodities. Table 4.5 shows that the value of $e_k$ may vary considerably from year to year, but it can be expressed as a compound annual rate over a longer period of time.

Knowledge of the inflation and effective escalation rates permits us to distinguish between *"then-current"* and *"constant worth"* dollars. The *"then-current"* value of an amount at the end of period $t(T_t)$ is the actual number of dollars that will change hands, be allowed as depreciation or cost recovery, and so forth. For example, we tend to think of our future salary in terms of then-current dollars. If the effective escalation rate is $e_k$, the then-current dollars can be expressed as

$$T_t = T_0(1 + e_k)^t \tag{4.26}$$

Thus, the set of then-current cash flows constitutes a *geometric series* (i.e., $T_t = A_1(1 + e_k)^{t-1}$) where $A_1 = T_0(1 + e_k)$.

*"Constant worth"* dollars $(C_t)$ represent the purchasing power at time $t$ of an amount relative to some base point in time, usually year zero. This can be thought of as the then-current dollar amount with the general inflation content removed. That is,

$$C_t = \frac{T_t}{(1 + j)^t} = \frac{T_0(1 + e_k)^t}{(1 + j)^t} \tag{4.27}$$

Note that if the effective escalation rate $(e_k)$ equals the general inflation rate $(j)$, the constant worth dollar amount is fixed over time, equal to $T_0$. Also note that when the set of constant worth cash flows constitutes a *uniform series* (that is, $C_t = A$, $t = 1, 2, \ldots, n$), the set of then-current cash flows constitutes another geometric series [that is, $T_t = A_1(1 + j)^{t-1}$] where $A_1 = A(1 + j)$.

## Example 4.24

Raw materials for use on a 3-year project are expected to have an annual escalation rate of 10%. The general inflation rate is also expected to be 10%/year. A unit of raw material costs $500 today. The then-current and constant worth costs per unit during each of the next 3 years are given in Table 4.6. Values of $T_0 = \$500$, $e_k = .10$, and $j = .10$ are used in Equations 4.26 and 4.27 to generate the table.

TABLE 4.6. Then-Current and Constant Worth Cost of Raw Material

| Year | Then-Current Dollars | Constant Worth Dollars |
|------|---------------------|------------------------|
| 1 | 550 | 500 |
| 2 | 605 | 500 |
| 3 | 665.50 | 500 |

TABLE 4.7. Then-Current and Constant Worth Cost of
Energy (per 1000 kilowatt-hours)

| Year | Then-Current Dollars | Constant Worth Dollars |
|------|---------------------|------------------------|
| 1 | 67.10 | 61.00 |
| 2 | 81.86 | 67.65 |
| 3 | 99.87 | 75.04 |

Example 4.25 _____

Energy on the 3-year project of Example 4.24 currently costs $55.00/thousand
kilowatt-hours. Energy costs will escalate at a rate of 22% and inflation will
remain constant at 10%/year. The then-current and constant worth energy rates
are presented in Table 4.7. Now, $T_0$ = $55/1000 kilowatt-hours, $e_k$ = .22, and
$j$ = .10.

## 4.6.2 Economic Analysis Under Inflation

In order to properly account for the time value of money, an appropriate
discount or interest rate must be determined. Actually, two discount rates are
useful to explain economic analysis under inflation.

The *real discount rate* ($i_r$) does not include an adjustment for the expected
general inflation rate. It represents the earning power of money, or the *MARR*,
in the absence of inflation, and is used to move constant worth amounts of
money backward and forward in time. Many agree that the real discount rate is
approximately constant over time. The *combined discount rate* ($i_c$) includes an
adjustment for the general inflation rate. It takes into account the earning
power of money as well as inflation and is expressed as

$$1 + i_c = (1 + i_r)(1 + j) \tag{4.28}$$

or

$$i_c = i_r + j + i_r j \tag{4.29}$$

It is used to move then-current amounts of money backward and forward in
time.

Economic analysis under inflation is quite simple. Either one of two procedures should be followed:

1. Express all cash flows in terms of their *"then-current"* amounts and use
an interest rate that combines the real discount rate and the inflation rate,
$i_c = i_r + j + i_r j$.
2. Express all cash flows in terms of their *"constant worth"* amounts and
use the real discount rate alone without an inflation rate component.

Once either of these approaches is selected, the analysis can proceed as usual.

Using the first approach to account for inflationary effects, the present worth equivalent of a series $T_t$ of cash flows is computed as follows:

$$PW(i_c) = \sum_{t=0}^{n} T_t(1 + i_r)^{-t}(1 + j)^{-t}$$

$$= \sum_{t=0}^{n} T_t(1 + i_r + j + i_r j)^{-t}$$

$$= \sum_{t=0}^{n} T_t(1 + i_c)^{-t} \tag{4.30}$$

The second approach computes the present worth equivalent using constant worth dollar amounts as follows:

$$PW(i_r) = \sum_{t=0}^{n} C_t(1 + i_r)^{-t} \tag{4.31}$$

In other sections of the book we implicitly assume that either the discount rate used with then-current dollars includes a component for inflation, or the cash flows are in the form of constant worth dollars. The former is more common in industry today.

## Example 4.26

To illustrate these approaches for dealing with inflation, consider two cash flow series for a project over a 3-year period. Expenses are $10,000 to start, with operating and maintenance costs of $2000 in the first year, increasing at an escalation rate of 15%/year. Incomes begin in year 1 at $8000 and deescalate at a rate of 20%/year. If the general inflation rate is 10% and the real discount rate is 7%, Tables 4.8 and 4.9 calculate the present worth of the project under approaches 1 and 2, respectively. Note that $e_1 = .15$, $e_2 = -.20$, $i_r = .07$, and $j = .10$. Had the then-current net cash flows been discounted using only the real discount rate $i_r = 7\%$, the present worth would have been $1208.91 instead of the $-$424.79 obtained considering inflation. Consequently, what would have appeared to be a profitable investment without considering inflation actually has a negative present worth when the effects of inflation are included in the analysis.

Other aspects of inflation that tend to complicate the analysis are the differences that may exist in inflation rates for not only various types of cash flows, but also for different regions of the country and world. The firm that has numerous plants scattered throughout not only the same country, but also the world, must cope with the economic differences that exist among the locations, as well as the differences in inflation rates for labor, equipment, materials, utilities, and supplies.

TABLE 4.8. Then-Current Project Analysis

| End of Year, $t$ | Then-Current Expenses, $e_1 = .15$ | Then-Current Incomes, $e_2 = -.20$ | Net Cash Flow, $T_t$ | Discount Factor, $(1 + i_r + j + i_r j)^{-t}$ $= (1 + .177)^{-t}$ | Discounted Net Cash Flow, $T_t(1 + .177)^{-t}$ |
|---|---|---|---|---|---|
| 0 | -$10,000 | — | -$10,000 | 1.000 | -$10,000.00 |
| 1 | -$ 2,000 | $8,000 | $ 6,000 | .850 | $ 5,097.71 |
| 2 | -$ 2,300 | $6,400 | $ 4,100 | .722 | $ 2,959.59 |
| 3 | -$ 2,645 | $5,120 | $ 2,475 | .613 | $ 1,517.91 |
| | | | | Total | -$ 424.79 |

TABLE 4.9. Constant Worth Project Analysis

| End of Year, $t$ | Constant Worth Expenses | Constant Worth Incomes | Net Cash Flow, $C_t$ | Discount Factor $(1 + i_r)^{-t}$ $= (1 + .07)^{-t}$ | Discounted Net Cash Flow, $C_t(1 + .07)^{-t}$ |
|---|---|---|---|---|---|
| 0 | −$10,000.00 | — | −$10,000.00 | 1.000 | −$10,000.00 |
| 1 | −$ 1,818.18 | $7,272.73 | $ 5,454.55 | .935 | $ 5,097.71 |
| 2 | −$ 1,900.83 | $5,289.26 | $ 3,388.43 | .873 | $ 2,959.59 |
| 3 | −$ 1,987.23 | $3,846.73 | $ 1,859.50 | .816 | $ 1,517.92 |
| | | | | Total | −$   424.79 |

As a further complication in dealing with inflation, the inflation rate tends to change from one time period to the next. If $j_t$ denotes the inflation rate for period $t$ and $i_{rt}$ denotes the real discount rate for period $t$, the present worth can be expressed in then-current dollars as

$$PW(i_{rt}, j_t) = T_0 + \sum_{k=1}^{n} T_k \prod_{t=1}^{k} (1 + j_t)^{-1}(1 + i_{rt})^{-1} \qquad (4.32)$$

Inflation is a much discussed subject in the area of economic investment analysis. Some argue that inflation effects can be ignored, since inflation will affect all investments in roughly the same way. Thus, it is argued, the relative differences in the alternatives will be approximately the same with or without inflation considered. Others argue that the inflation rate during the past decade has been so dynamic that an accurate prediction of the true inflation rate and its impact on future cash flows is not possible. Another argument for ignoring explicitly the effects of inflation is that it is accounted for implicitly, since cash flow estimates for the future are made by individuals conditioned by an inflationary economy. Thus, it is argued, any estimates of future cash flows probably incorporate implicitly inflationary effects. A final argument for ignoring inflation in comparing investment alternatives involving only negative cash flows is that an alternative that is preferred by ignoring inflation effects will be even more attractive when effects of inflation are incorporated in the analysis.

The above arguments are certainly valid in some instances. However, counterarguments can be given for each. Actually, it is not difficult to explicitly account for inflation. Use of either of the two approaches illustrated is recommended.

## 4.7 SUMMARY

In this chapter we developed the bases for estimating the worth of an investment. We considered selection of the minimum attractive rate of return, numerous measures of investment worth, bond problems, capital recovery, and the effects of inflation. Now, we are prepared to use these concepts in analyzing and selecting from among several alternative investments. This will be done in Chapter Five.

## BIBLIOGRAPHY

1. Adler, M., "The True Rate of Return and the Reinvestment Rate," *The Engineering Economist, 15* (3), 1970, pp. 185–187.

2. Beenhakker, H. L., "Discounting Indices Proposed for Capital Investment Evaluation: A Further Examination," *The Engineering Economist, 18* (3), 1973, pp. 149–168.

3. Bernhard, R. H., "On the Inconsistency of the Soper and Sturm-Kaplan Conditions for Uniqueness of the Internal Rate of Return," *The Journal of Industrial Engineering, 18* (8), 1967, pp. 498–500.

4. Bernhard, R. H., "A Comprehensive Comparison and Critique of Discounting Indices Proposed for Capital Investment Evaluation," *The Engineering Economist, 16* (3), 1971, pp. 157–186.

5. Bernhard, R. H., " 'Modified' Rates of Return for Investment Project Evaluation—A Comparison and Critique," *The Engineering Economist, 24* (3), 1979, pp. 161–168.

6. Bussey, L. E., *The Economic Analysis of Industrial Projects,* Prentice-Hall, 1978.

7. Canada, J. R., "Rate of Return: A Comparison Between the Discounted Cash Flow Model and a Model Which Assumes an Explicit Reinvestment Rate for the Uniform Income Flow Case," *The Engineering Economist, 9* (3), 1964, pp. 1–15.

8. de Faro, C., "On the Internal Rate of Return Criterion," *The Engineering Economist, 19* (3), 1974, pp. 165–194.

9. Freidenfelds, J. and Kennedy, M., "Price Inflation and Long-Term Present-Worth Studies," *The Engineering Economist, 24* (3), 1979, pp. 143–160.

10. Grant, E. L., "Reinvestment of Cash Flow Controversy—A Review of: FINANCIAL ANALYSIS IN CAPITAL BUDGETING by Pearson Hunt," *The Engineering Economist, 11* (3), 1966, pp. 23–29.

11. Jeynes, P. H., "Minimum Acceptable Return," *The Engineering Economist, 9* (4), 1964, pp. 9–25.

12. Jeynes, P. H., "Comment on Ralph O. Swalm's Letter," *The Engineering Economist, 10* (3), 1965, pp. 38–42.

13. Jeynes, P. H., "The Significance of Reinvestment Rate," *The Engineering Economist, 11* (1), 1965, pp. 1–9.

14. Kirshenbaum, P. S., "A Resolution of the Multiple Rate-of-Return Paradox," *The Engineering Economist, 10* (1), 1964, pp. 11–16.

15. Lin, S. A., "The Modified Internal Rate of Return and Investment Criterion," *The Engineering Economist, 21* (4), pp. 237–248.

16. Oakford, R. V., Bhimjee, S. A., and Jucker, J. V., "The Internal Rate of Return, the Pseudo Internal Rate of Return, and NPV and Their Use in Financial Decision Making," *The Engineering Economist, 22* (3), pp. 187–202.

17. Sharp, G. P. and Guzman-Garza, A., "Borrowing Interest Rate as a Function of the Debt-Equity Ratio in Capital Budgeting Models," *The Engineering Economist, 26* (4), 1981, pp. 293–315.

18. Sullivan, W. G., and Bontadelli, J. A., "The Industrial Engineer and Inflation," *Industrial Engineering, 12* (3), March 1980, pp. 24–33.

19. Swalm, R. O., "On Calculating the Rate of Return on an Investment," *The Journal of Industrial engineering, 9* (2), 1958, pp. 99–103.

20. Swalm, R. O., "Comment on 'Minimum Acceptable Return' by Paul H. Jeynes," *The Engineering Economist, 10* (3) 1965, pp. 35–38.

21. Teichroew, D., Robicheck, A. A., and Montalbano, M., "Mathematical Analysis of Rates of Return Under Certainty," *Management Science, 11* (3), 1965, pp. 395–403.

22. Waters, R. C. and Bullock, R. L., "Inflation and Replacement Decisions," *The Engineering Economist, 21* (4), pp. 249–258.

23. Weaver, J. B., "False and Multiple Solutions by the Discounted Cash Flow Method for Determining Interest Rate of Return," *The Engineering Economist, 3* (4), 1958, pp. 1–31.

24. White, J. A., Case, K. E., and Agee, M. H., "Rate of Return: An Explicit Reinvestment Rate Approach," *Proceedings of the 1976 AIIE Conference,* American Institute of Industrial Engineers, Norcross, Ga., 1976.

## PROBLEMS

1. Rex Electric Company has borrowed money from First National Bank on a short-term loan at 18%/year with monthly payments. The company's tax rate is 46%. What is the after-tax cost of capital for this money? (4.2.1)

2. Ajax Machinery borrows money at 16%/year to be paid back semiannually. Assume a tax rate of 46%.
   (a) Determine the effective before-tax cost of capital in percent. (4.2.1)
   (b) Determine the effective after-tax cost of capital in percent. (4.2.1)

3. Precision Components, Inc. issues bonds promising semiannual interest payments at an annual interest rate of 11%. The bonds are all sold for face value, and the costs of preparing and selling the bonds are relatively negligible. What is the after-tax cost of this debt funding to Precision Components if their tax rate is 46%? (4.2.1)

4. Jericho Steel calculates that they must pay interest of $30 each quarter to holders of bonds having $1000 face value. Clearly and explicitly state any necessary assumptions and
   (a) Calculate the before-tax interest rate of this debt funding. (4.2.1)
   (b) Calculate the after-tax interest rate of this debt funding. (4.2.1)

5. Witherspoon, Inc. is a publicly held firm having a present trading price of $35 per share of common stock. The Board of Directors has just voted a dividend of $3 per share to be paid to all shareholders. What is the cost of this equity capital in percent? Is this a before-tax cost or an after-tax cost of capital? (4.2.1)

6. Dunbar Chemicals plans to invest its retained earnings in capital improvements and expansion ventures. The company's stock trades for $76 per share on the financial market, and dividends recently paid were $8 per share. Is there a cost to Dunbar of using its own retained earnings? If so, what is the cost of capital in percent? (4.2.1)

7. The capital makeup of JWS Corporation is as follows:

|  | Amount | Cost of Capital, Percent |
| --- | --- | --- |
| Loans | $ 800,000 | 9.6 |
| Bonds | $1,400,000 | 8.2 |
| Stock | $2,100,000 | 11.9 |
| Retained earnings | $ 700,000 | 11.9 |

What is the overall average cost of capital for JWS? (4.2.1)

8. Systems Analysis, Inc. has the following capital structure:

|  | Amount | Other Information |
| --- | --- | --- |
| Loans | $2,600,000 | 14% before-tax rate, semiannual compounding |
| Bonds | $4,100,000 | 12% bond interest rate, quarterly payment |
| Stock | $5,000,000 | $6.50 dividends on selling price of $65 per share |
| Retained earnings | $3,300,000 |  |

Systems' tax rate is 46%. Determine the firm's overall average after-tax cost of capital. (4.2.1)

9. Brock Associates invested $40,000 in a business venture with the following cash flow results:

| EOY | CF | EOY | CF | EOY | CF |
| --- | --- | --- | --- | --- | --- |
| 0 | −$40,000 | 3 | $11,000 | 6 | $11,000 |
| 1 | 5,000 | 4 | 14,000 | 7 | 8,000 |
| 2 | 8,000 | 5 | 14,000 | 8 | 5,000 |

If *MARR* is 15%, determine the following.
**(a)** Present worth. (4.3.1)
**(b)** Annual worth. (4.3.2)
**(c)** Future worth. (4.3.3)
**(d)** Internal rate of return. (4.3.4)
**(e)** External rate of return. (4.3.5)
**(f)** Savings/investment ratio. (4.3.6)
**(g)** Payback period. (4.3.7)

10. Shrewd Endeavors, Inc. invested $70,000 in a business venture with the following cash flow results:

| EOY | CF | EOY | CF | EOY | CF |
|---|---|---|---|---|---|
| 0 | -$70,000 | 7 | $14,000 | 14 | $7,000 |
| 1 | 20,000 | 8 | 13,000 | 15 | 6,000 |
| 2 | 19,000 | 9 | 12,000 | 16 | 5,000 |
| 3 | 18,000 | 10 | 11,000 | 17 | 4,000 |
| 4 | 17,000 | 11 | 10,000 | 18 | 3,000 |
| 5 | 16,000 | 12 | 9,000 | 19 | 2,000 |
| 6 | 15,000 | 13 | 8,000 | 20 | 1,000 |

Assuming *MARR* to be 20%, determine the following.
**(a)** Present worth. (4.3.1)
**(b)** Annual worth. (4.3.2)
**(c)** Future worth. (4.3.3)
**(d)** Internal rate of return. (4.3.4)
**(e)** External rate of return. (4.3.5)
**(f)** Savings/investment ratio. (4.3.6)
**(g)** Payback period. (4.3.7)

11. An investment of $10,000 is to be made on an analog computer that will last for 6 years and have a zero salvage value at that time. Operating and maintenance costs are projected to be $1000 the first 3 years and $1500 the last 3 years. The minimum attractive rate of return is specified to be 12%. Determine for this investment the following.
**(a)** Present worth. (4.3.1)
**(b)** Annual worth. (4.3.2)
**(c)** Future worth. (4.3.3)

12. Tulsa Precision, Inc. borrows $100,000 to purchase a new numerically controlled milling machine and pays the loan back over a 4-year period with equal payments. Interest on the loan is 15% compounded annually. The machine is estimated to have annual operating and maintenance costs of $24,000/year and have a life of 10 years. Salvage value is estimated to be $25,000. The firm has a *MARR* of 12%. Determine for this investment the following.
**(a)** Present worth. (4.3.1)
**(b)** Annual worth. (4.3.2)
**(c)** Future worth. (4.3.3)

13. A utility vehicle is purchased for $13,000, kept for 4 years, and sold for $2000. Annual operating and maintenance costs were $2500. Using a 15% minimum attractive rate of return, determine the following.
   (a) Present worth. (4.3.1)
   (b) Annual worth. (4.3.2)
   (c) Future worth. (4.3.3)

14. An investment of $20,000 for a new condenser is being considered. Estimated salvage value of the condenser is $5000 at the end of an estimated life of 8 years. Annual gross income each year for the 8 years is $8500. Annual operating expenses are $2300. Assume money is worth 25% compounded annually. Determine the following measures of investment worth; for each, state whether or not your results indicate the new condenser should be purchased.
   (a) Present worth. (4.3.1)
   (b) Annual worth. (4.3.2)
   (c) Future worth. (4.3.3)
   (d) Internal rate of return. (4.3.4)
   (e) External rate of return. (4.3.5)
   (f) Savings/investment ratio. (4.3.6)

15. Smith Investors places $10,000 in an investment fund. One year after making the investment, Smith receives $1500 and continues to receive $1500 annually until 10 such amounts are received. Smith receives nothing further until 15 years after the initial investment, at which time $15,000 is received. Over the 15-year period, what are each of the following if $MARR = 12\%$?
   (a) Present worth (4.3.1)
   (b) Annual worth. (4.3.2)
   (c) Future worth. (4.3.3)
   (d) Internal rate of return. (4.3.4)
   (e) External rate of return. (4.3.5)
   (f) Savings/investment ratio. (4.3.6)
   (g) Payback period. (4.3.7)

16. Many Chemicals, Unlimited purchases a computer-controlled filter for $100,000. Half of the purchase price is borrowed from a bank at 15% compounded annually. The loan is to be paid back with equal annual payments over a 3-year period. The filter is expected to last 10 years, at which time it will have a salvage value of $4000. Over the 10-year period the operating and maintenance costs are anticipated to equal $20,000/year; however, $45,000 in fines for pollution will be avoided. The firm expects to earn 12% on its investments. Determine each of the following measures of investment worth and state whether or not the filter purchase was economically sound.
   (a) Present worth. (4.3.1)
   (b) Annual worth. (4.3.2)
   (c) Future worth. (4.3.3)
   (d) Internal rate of return. (4.3.4)
   (e) External rate of return. (4.3.5)
   (f) Savings/investment ratio. (4.3.6)

17. Owners of an economy motel chain are considering building a new 200-unit motel in Stillwater, Oklahoma. The present worth cost of building the motel is $6,000,000;

the firm estimates furnishings for the motel will cost an additional $700,000 and will require replacement every 5 years. Annual operating and maintenance costs for the facility are estimated to be $800,000. The average rate for a unit is anticipated to be $40/day. A 15-year planning horizon is used by the firm in evaluating new ventures of this type; a terminal salvage value of 15% of the original building cost is anticipated; furnishings are estimated to have no salvage value at the end of each 5-year replacement interval. Assuming an average daily occupancy percentage of 70%, 80%, and 90%, a *MARR* of 20%, 365 operating days/year, and ignoring the cost of the land, should the motel be built? Base your decision upon the following values.

(a) Present worth. (4.3.1)
(b) Annual worth. (4.3.2)
(c) Future worth. (4.3.3)
(d) Internal rate of return. (4.3.4)
(e) External rate of return. (4.3.5)
(f) Savings/investment ratio. (4.3.6)

18. Growth Fertilizer Company purchases a gravity settling tank by borrowing the $30,000 purchase price. The loan is to be repaid with four equal annual payments at an annual compound rate of 12%. It is anticipated that the tank will be used for 9 years and then be sold for $2000. Annual operating and maintenance expenses are estimated to be $9000/year. A savings of $15,000/year is realized over the present filtration system. The firm uses a *MARR* of 15% for its economic analyses. Determine the following.

(a) Present worth. (4.3.1)
(b) Annual worth. (4.3.2)
(c) Future worth. (4.3.3)
(d) Internal rate of return. (4.3.4)
(e) External rate of return. (4.3.5)
(f) Savings/investment ratio. (4.3.6)
(g) Payback period. (4.3.7)

19. A firm purchases an analog computer for $15,000; the computer is used for 4 years and is then sold for $2000. Annual disbursement for operating and maintenance costs equals $3000/year over the 4-year period. The computer replaced an older manual system that cost $8000/year. On the basis of a 20% *MARR,* determine if the decision to buy the analog computer was economically sound. Use the following measures of investment worth.

(a) Present worth. (4.3.1)
(b) Annual worth. (4.3.2)
(c) Future worth. (4.3.3)
(d) Internal rate of return. (4.3.4)
(e) External rate of return. (4.3.5)
(f) Savings/investment ratio. (4.3.6)

20. If a fund pays 12% compounded annually, what deposit is required today such that $1000 can be withdrawn every year forever? (4.3.8)

21. If a fund pays 12% compounded annually, what deposit is required today such that $5000 can be withdrawn every 5 years forever? (4.3.8)

22. A scholarship fund pays $2000, $1000, $2000, and $1000 at the end of quarters 1, 2,

3, and 4 each year. If the scholarship fund earns 12% compounded quarterly, how much money must be in the fund for this series of payments to continue forever? (4.3.8)

23. A scholarship fund pays $2000, $1000, $2000, and $1000 at the *beginning* of quarters 1, 2, 3, and 4 each year. If the scholarship fund earns 12% compounded quarterly, how much money must be in the fund for this series of payments to continue forever? (4.3.8)

24. Maintenance on a reservoir is cyclic with the following *costs* occurring over a 5-year period: $5000, $2000, $7000, $0, and $2000. It is anticipated that the sequence will repeat itself every 5 years forever. Determine the capitalized cost of the maintenance costs based on a time value of money of 12%. (4.3.8)

25. A flood control project has a construction cost at $t = 0$ of $200,000, an annual maintenance cost of $8000, and a major repair at 5-year intervals projected to cost $50,000 with the first such repair occurring at $t = 5$. If interest is 10% annually, determine the amount of money needed at $t = 0$ to provide for construction and perpetual upkeep. (4.3.8)

26. A firm can invest in a venture which costs $200,000 and returns $70,000/year at the end of each of 4 years. This investment can be renewed perpetually every 4 years. If the firm's *MARR* is 20%, determine the capitalized worth of an infinite series of these investments. (4.3.8)

27. A bond is purchased for $900 and kept for 10 years, at which time it matures at a face value of $1000. During the 10-year period $50 is received every 6 months (i.e., 20 receipts of $50 each). What is the rate of return for the investment? (4.4)

28. Dr. Quack purchases a $1000 face value bond for $1050. It matures in 15 years after paying $120 at the end of each year. What rate of return was earned on the investment? (4.4)

29. Jerry buys a $2000 bond for $1800. The bond has a bond rate of 12% with bond premiums paid annually. If the bond is kept for 8 years and sold for par value, determine the equivalent annual return (rate of return) for the bond investment. (4.4)

30. Susan is considering purchasing a bond having a face value of $2500 and a bond rate of 12% payable semiannually. The bond has a remaining life of 12 years. How much should be paid for the bond in order to earn a rate of return of 14% compounded semiannually? Assume the bond will be redeemed for face value. (4.4)

31. Mr. Rich wishes to purchase a bond having a face value of $10,000 and a bond rate of 12% payable annually. The bond has a remaining life of 8 years. In order to earn an 18% rate of return on the investment, what amount should be paid for the bond? If he can buy the bond cheaper, does his rate of return increase or decrease? (4.4)

32. A person wishes to sell a bond that has a face value of $1000. The bond has a bond rate of 9% with bond premiums paid annually. Four years ago, $900 was paid for the bond. At least a 12% return on investment is desired. What must be the minimum selling price for the bond in order to make the desired return on investment? (4.4)

**33.** A $5000, 10-year, 12% semiannual bond is purchased at $t = 0$ by Mr. Rich for par value. After receiving the twelfth dividend, Mr. Rich sells the bond to Mrs. Richer at a price to yield a 10% annual nominal rate of return on his original purchase price.
**(a)** What was the selling price Mr. Rich received? (4.4)
**(b)** If Mrs. Richer keeps the bond until maturity and redeems it for $5500, what approximate annual nominal rate of return on the bond investment will she have earned? (4.4)

**34.** Determine the capital recovery cost of a ditching machine having a first cost of $35,000, a salvage value of $3000, and a life of 6 years. Use an interest rate of 20%. Solve this using each of the three different capital recovery formulas. (4.5)

**35.** Determine the capital recovery cost of a bulldozer costing $200,000 having a life of 7 years, and a salvage value of $50,000. Use an interest rate of 15%. Solve this using each of the three different capital recovery formulas. (4.5)

**36.** Determine the capital recovery cost for chemical processing equipment costing $100,000, having a life of 6 years, and a salvage value of $-$25,000. Use an interest rate of 15%. (4.5)

**37.** John A. Smooth purchases an automobile costing $21,000. It will be kept for 3 years and then sold. How much must he receive for the car at that time in order for the capital recovery cost to be a mere $5000/year if $i = 12\%$? (4.5)

**38.** Crush Autosmashers can purchase a new electromagnet for moving cars at a cost of $12,000. When scrapped out, the electromagnet will be worth $500. If Crush's *MARR* is 15%, how many years must the unit last so that its capital recovery cost is $2500/year or less? (4.5)

**39.** The Acme Manufacturing company purchased an automatic transfer machine for $25,000 installed. At the end of 10 years the company paid $4000 to have the machine removed and junked. The operating and maintenance cost history was as follows.

| EOY | O&M Cost |
| --- | --- |
| 1 | $2,000 |
| 2 | 2,000 |
| 3 | 2,000 |
| 4 | 2,000 |
| 5 | 3,500 |
| 6 | 3,500 |
| 7 | 3,500 |
| 8 | 3,500 |
| 9 | 5,000 |
| 10 | 5,000 |

The company has a 15% *MARR*. Supply the values of $U$, $V$, $W$, $X$, and $Y$ in the following equation for computing the equivalent uniform annual cost for the machine. (4.5)

$$(\$25,000 - U)(A/F\,15\%, 10) + V(0.15) + W + \$1500\,[(F/A\,15\%, 6)$$
$$+ (F/A\,15\%, X](Y, 15\%, 10)$$

**40.** A robot is installed in the painting area of an assembly line at a cost of $50,000. After 8 years, it is donated to a high school, costing the company $2000 for removal and transportation. Operating and maintenance costs began in year 1 at $3000/year and increased by $300 each year thereafter. The company has an 18% *MARR*. Supply the values of *U, V, W, X, Y,* and *Z* in the following equation for computing the equivalent uniform annual cost of the robot. (4.5)

$$(U - V)(A/P18, W) + (-2000)X + 3000 + Y(Z, 18,8)$$

**41.** A distillation column is purchased for $300,000. Operating and maintenance costs for the first year are $30,000. Thereafter, operating and maintenance costs increase by 10%/year over the previous year's costs. At the end of 8 years the column is sold for $50,00. During the life of the investment, revenue was produced that could be related directly to the investment in the column. The revenue the first year was $75,000. Thereafter, revenue increased by $10,000 over the previous year's revenue. Using a *MARR* of 20%, determine the equivalent annual worth for the investment. (4.5)

**42.** Suppose the overall Consumer Price Index (CPI) takes on the following values:

| Year | CPI (Base Is 100 in 1967) |
|------|---------------------------|
| 1982 | 294.1 |
| 1983 | 311.7 |
| 1984 | 339.8 |
| 1985 | 380.6 |
| 1986 | 418.6 |
| 1987 | 431.2 |

**(a)** Determine the general inflation rate for each of the last 5 years. (4.6.1)
**(b)** Determine a compound annual inflation rate from 1982 to 1987. (4.6.1)
**(c)** Determine a compound annual inflation rate from 1967 to 1987. (4.6.1)

**43.** Assume the television CPI behaves as follows over the following years:

| Year | CPI (Base Is 100 in 1967) |
|------|---------------------------|
| 1982 | 103.5 |
| 1983 | 105.5 |
| 1984 | 108.1 |
| 1985 | 101.3 |
| 1986 | 106.9 |
| 1987 | 112.0 |

**(a)** Determine the television cost escalation rate for each of the last 5 years. (4.6.1)
**(b)** Determine a compound annual inflation rate from 1982 to 1987. (4.6.1)
**(c)** Determine a compound annual inflation rate from 1967 to 1987. (4.6.1)

**44.** Labor costs over a 4-year period have been forecast in then-current dollars as follows: $10,000, $12,000, $15,000, and $17,500. The general inflation rate for the 4 years is forecast to be 9%. Determine the constant worth dollar amounts of labor over each of the 4 years. (4.6.1)

**45.** Yearly labor costs of a highway maintenance group are currently $420,000/year. If labor rates escalate at a 13% rate and general inflation increases at 9%, determine for each of the next 4 years the labor cost in then-current and constant worth dollars. (4.6.1)

**46.** Big Screen, Inc. televisions are selling today for $2000. If the yearly escalation rate is −3% and the general inflation rate is 10%, what will a Big Screen television set cost 5 years from now in then-current and constant worth dollars? (4.6.1)

**47.** If you desire a real return of 4% on your money, excluding inflation, and inflation is running at 8%, what combined discount rate should you be seeking? (4.6.2)

**48.** Mellin Transformers Co. uses a *MARR* of 15% in all alternative evaluations. Inflation is running at 10%. What real discount rate, exclusive of inflation, are they implicitly using? (4.6.2)

**49.** The following material costs are anticipated over a 5-year period: $9000, $11,000, $14,000, $18,000, and $23,000. It is estimated that a 9% inflation rate will apply over the time period in question. The material costs given above are expressed in then-current dollars. The time value of money, excluding inflation, is estimated to be 4%. Determine the present worth equivalent for material cost using the following.
**(a)** Then-current costs (4.6.2)
**(b)** Constant worth costs (4.6.2)

**50.** Zero Defects Electronics has current labor and materials costs of $200,000 and $300,000 per year, respectively. Over the next 3 years, labor costs will escalate at a rate of 13%; material will escalate at a rate of −6% (deescalation). Inflation will average 9% and a real return of 5% is desired. Determine the following.
**(a)** Then-current labor and material costs for each year and their discounted present worth (4.6.2)
**(b)** Constant-worth labor and material costs for each year and their discounted present worth (4.6.2)

**51.** A landfill has a first cost of $270,000. Annual operating and maintenance costs for the first year will be $40,000. These costs will increase at 11%/year. Income for dumping rights at the landfill will be held fixed at $120,000/year. The landfill will be in operation for 10 years. Inflation will average 8% and a real return of 3.6% is desired.
**(a)** Determine the present worth of this project using then-current dollars. (4.6.2)
**(b)** Determine the present worth of this project using constant worth dollars. (4.6.2)

**52.** The inflation rates for 4 years are forecast to be 9%, 7%, 10%, and 10%. The interest rate exclusive of inflation is anticipated to be 6%, 5%, 4%, and 5% over this same period. If labor is projected to be $1000, $1500, $2000, and $1000 in then-current dollars, during those years, determine the present worth equivalent for labor cost. (4.6.2)

# CHAPTER FIVE
# COMPARISON OF ALTERNATIVES

## 5.1 INTRODUCTION

In this chapter we apply time value of money concepts and measures of investment worth to the comparison of economic investment alternatives. Although multiple objectives are often involved in performing a comparison of alternatives, for now we concentrate on the comparison of *mutually exclusive* alternatives on the basis of monetary considerations alone. Mutually exclusive alternatives mean that no more than one alternative can be chosen. The adage of "not being able to have one's cake and eat it too" illustrates the notion of mutually exclusive alternatives.

A systematic approach that can be used in comparing economic investment alternatives is summarized as follows:

1. Define the set of feasible, mutually exclusive economic investment alternatives to be compared.
2. Define the planning horizon to be used in the economic study.
3. Develop the cash flow profiles for each alternative.
4. Compare the alternatives using a specified time value of money and measure(s) of merit.
5. Perform supplementary analyses.
6. Select the preferred alternative.

The procedures for comparing investment alternatives outlined in this chapter are intended to aid in making better measurements of the *quantitative aspects* of capital investment alternatives. Following development of the six-step approach, more detail is provided on comparing alternatives having unequal lives. Also, two special classes of alternatives are examined further. These include alternatives having no positive cash flows, and those involving the possible replacement of existing assets. Regardless of the types of alternatives being considered, it *cannot be too strongly emphasized that no economic evaluation can replace the sound judgment of experienced managers concerned with both the quantitative and nonquantitative aspects of investment alternatives.* Typical of the aspects of alternatives not considered in this chapter are safety, personnel considerations, product quality, environmental effects, and engineering and construction capability. Such factors are relevant and often control decisions on capital expenditures. However, our concern is with the monetary aspects of the alternatives. Approaches that can be used to assimilate multiple objectives are treated in Chapter Nine.

## 5.2 DEFINING MUTUALLY EXCLUSIVE ALTERNATIVES

An individual alternative selected from a set of mutually exclusive alternatives can be made up of several *investment proposals*. Investment proposals are distinguished from investment alternatives by Thuesen et al. [32] by noting that investment alternatives are decision options; investment proposals are single projects or undertakings that are being considered as investment possibilities.

Example 5.1 _____

As an illustration of the distinction between investment proposals and investment alternatives, consider a distribution center that receives pallet loads of product, stores the product, and ships pallet loads of product to various customer locations. A new distribution center is to be constructed, and the following proposals have been made:

1. Proposed methods of moving materials from receiving to storage and from storage to shipping include:
   a. Conventional lift trucks for operating in 12-foot aisles.
   b. Narrow aisle lift trucks for operating in 5-foot aisles.
   c. Driverless tractor system.
   d. Towline conveyor system.
   e. Pallet conveyor system.
2. Proposed methods of placing materials in and removing materials from storage include:
   a. Conventional lift trucks for operating in 12-foot aisles.
   b. Narrow aisle lift trucks for operating in 5-foot aisles.
   c. Narrow aisle, operator driven, rail guided storage/retrieval vehicle.
   d. Narrow aisle, automated, rail guided storage/retrieval vehicle.

**3.** Proposed methods of storing materials include:
  **a.** Stacking pallet loads of material (8 feet high, 12-foot aisles).
  **b.** Conventional pallet rack (20 feet high, 12-foot aisles).
  **c.** Flow rack (20 feet high, 12-foot aisles).
  **d.** Narrow aisle, pallet rack (20 feet high, 5-foot aisles).
  **e.** Flow rack (20 feet high, 5-foot aisles).
  **f.** Medium height, pallet rack (35 feet high, 5-foot aisles).
  **g.** High rise, pallet rack (70 feet high, 5-foot aisles).

Given the set of proposals, alternative designs for the material handling system can be obtained by combining a proposed method of moving materials from receiving to storage, a proposed method of placing materials in storage, a proposed method of storage, a proposed method of removing materials from storage, and a proposed method of transporting materials from storage to shipping. Some of the combinations of proposals will be eliminated because of their incompatibility. For example, lift trucks requiring 12-foot aisles cannot be used to place materials in and remove materials from storage when 5-foot aisles are used. Other combinations might be eliminated because of budget limitations; a desire to minimize the variation in types of equipment due to maintainability, availability, reliability, flexibility, and operability considerations; ceiling height limitations; physical characteristics of the product (crushable product might require the use of storage racks); and a host of other considerations. Characteristically, experience and judgment are used to trim the list of possible combinations to a manageable number.

---

Example 5.2_____

To illustrate the formation of mutually exclusive investment alternatives from a set of investment proposals, consider a situation involving $m$ investment proposals. Let $x_j$ be defined to be 0 if proposal $j$ is not included in an alternative and let $x_j$ be defined to be 1 if proposal $j$ is included in an alternative. Using the binary variable $x_j$ we can form $2^m$ mutually exclusive alternatives. Thus, if there are three investment proposals, we can form eight mutually exclusive investment alternatives, as depicted in Table 5.1

---

Among the alternatives formed, some might not be feasible, depending on the restrictions or constraints placed on the problem. To illustrate, there might be a budget limitation that precludes the possibility of combining all three proposals; thus, Alternative 8 would be eliminated. Additionally, some of the proposals might be *mutually exclusive proposals*. For example, Proposals A and B might be alternative computer designs and only one is to be selected; in this case Alternative 7 would be eliminated from consideration. Other proposals might

TABLE 5.1. Developing Mutually Exclusive Investment Alternatives from Investment Proposals

| Alternative | Proposals | | | Explanation |
|---|---|---|---|---|
| | $x_A$ | $x_B$ | $x_C$ | |
| 1 | 0 | 0 | 0 | Do nothing (proposals A, B, and C not included) |
| 2 | 0 | 0 | 1 | Accept proposal C only |
| 3 | 0 | 1 | 0 | Accept proposal B only |
| 4 | 0 | 1 | 1 | Accept proposals B and C only |
| 5 | 1 | 0 | 0 | Accept proposal A only |
| 6 | 1 | 0 | 1 | Accept proposals A and C only |
| 7 | 1 | 1 | 0 | Accept proposals A and B only |
| 8 | 1 | 1 | 1 | Accept all three proposals |

be *contingent proposals* so that one proposal cannot be selected unless another proposal is also selected. As an illustration of a contingent proposal, Proposal C might involve the procurement of computer terminals, which require the selection of the computer design associated with Proposal B. In such a situation, Alternatives 2 and 6 would be infeasible. Thus, depending on the restrictions present, the number of feasible mutually exclusive alternatives that result can be considerably less than $2^m$.

In many organizations there is a rather formalized hierarchy for determining how the organization will invest its funds. Typically, the entry point in this hierarchy involves an individual analyst or engineer who is given an assignment to solve a problem; the problem may be one requiring the design of a new product, the improvement of an existing manufacturing process, or the development of an improved system for performing a service. The individual performs the steps involved in the problem-solving procedure and recommends the preferred solution to the problem. In arriving at the preferred solution, a number of alternative solutions are normally compared.

The preferred solution is usually forwarded to the next level of the hierarchy for approval. In fact, one would expect many preferred solutions to various problems to be forwarded to the second level of the hierarchy for approval. Each preferred solution becomes an investment *proposal,* the resulting set of mutually exclusive investment *alternatives* are formed, and the process of comparing economic investment alternatives is repeated. This sequence of operations is usually performed in various forms at each level of the hierarchy until, ultimately, the preferred solution by the individual analyst or engineer is accepted or rejected. In this textbook we concentrate on the process of comparing investment alternatives at the first level of the hierarchy; however, the need for such comparisons at many levels of the organization should be kept in mind.

## 5.3 DEFINING THE PLANNING HORIZON

In comparing investment alternatives, it is important to compare them over a common period of time. We define that period of time to be the *planning horizon*. In the case of investments in, say, equipment to perform a required service, the period of time over which the service is required might be used as the planning horizon. Likewise, in one-shot investment alternatives the period of time over which receipts continue to occur might define the planning horizon.

In a sense, the planning horizon defines the width of a "window" that is used to view the cash flows generated by an alternative. In order to make an objective evaluation, *the same window must be used in viewing each alternative.*

In some cases the planning horizon is easily determined; in other cases the duration of one or more projects is sufficiently uncertain to cause concern over the time period to use. Some commonly used methods for determining the planning horizon to use in economy studies include:

1. Least common multiple of lives for the set of feasible, mutually exclusive alternatives, denoted $\hat{T}$.
2. Shortest life among alternatives, denoted $T_s$.
3. Longest life among alternatives, denoted $T_l$.
4. Some other period of time less than $T_s$.
5. Some other period of time greater than $T_s$ but less than $T_l$.
6. Some other period of time greater than $T_l$.

The *least common multiple of lives* is the most common method of selecting the planning horizon. In this method, each alternative's cash flow profile is assumed to repeat in the future until a time is reached when all alternatives under consideration conclude at the same time. Using such a procedure when three alternatives are being considered and the individual lives are 3 years, 6 years, and 5 years yields a planning horizon of $\hat{T} = 30$ years. If the lives had been 3 years, 6 years, and 6 years, $\hat{T} = 6$ years. Clearly, strict reliance on $\hat{T}$ as the planning horizon is not advisable. Unfortunately, selection of this method is usually made implicitly, not explicitly, when one simply calculates and compares the annual worths of unequal lived alternatives. This practice is covered in more detail in Section 5.9.

In using the *shortest life among alternatives* to define the planning horizon, estimates are required for the values of the unused portions of the lives of the remaining alternatives. Thus, for the situation considered above, with $T_s = 3$ years, the salvage or residual values at the end of 3 years' use must be assessed for the other two alternatives.

If the *longest life among alternatives*, $T_l$, is used in determining the planning horizon, some difficult decisions must be made concerning the period of time between $T_s$ and $T_l$. If the alternative selected is to provide a necessary service, that service must continue throughout the planning horizon, regardless of the alternative selected. Consequently, when the shortest life alternative reaches the end of its project life, it must be replaced with some other asset capable of performing the required service. However, since technological developments will probably take place during the period of time $T_s$, new and improved candidates will be available for selection at time $T_s$. Thus, the specification of the cash flows for the shortest life alternative during the period of time from $T_s$ to $T_l$ is a difficult undertaking. As a result, $T_l$ is seldom used as the planning horizon.

Many organizations have adopted a *standard planning horizon* for all economic alternatives in order to ensure some consistency in economic analyses. Letting $T$ denote the planning horizon specified by the organization, different approaches are recommended, depending on whether $T < T_s$, $T_s < T < T_l$, or $T > T_l$. If $T < T_s$, each alternative's cash flow profile must be truncated with salvage or residual values estimated at time $T$. If $T_s < T < T_l$, alternatives having a life less than $T$ require that plans and cash flows be specified up through time $T$; those having lives longer than $T$ must be truncated. If $T > T_l$, all alternatives require that plans and cash flows be specified through the lesser of $\hat{T}$ and $T$. That is, when the planning horizon is greater than or equal to the least common multiple of lives, it is recommended that the economic analysis be based on a period of time equal to the least common multiple of lives. The reason is that at time $\hat{T}$ a new economic analysis can be performed based on the alternatives available at that time. After $\hat{T}$ years new alternatives might be available; furthermore, we can more accurately estimate the values of cash flows occurring after $\hat{T}$ if we wait until nearer time $\hat{T}$ to make the estimates.

## Example 5.3

To illustrate the difficulties associated with the selection of the planning horizon, consider the three mutually exclusive alternatives whose cash flow profiles are given in Table 5.2. Alternatives 1, 2, and 3 have anticipated lives of 3, 6, and 5 years, respectively. Each has revenues and costs which are used to determine the net cash flows. Alternative 1 involves used equipment having a useful life of 3 years. Much of this older equipment is readily available at virtually no cost other than for operating and maintenance. Alternatives 2 and 3 involve the purchase and operation of new equipment which will last for up to 6 and 5 years, respectively; estimated salvage values for the equipment are shown as a function of time.

Using a least common multiple of lives approach, a planning horizon of $\hat{T} = 30$ years would be used. Using a 30-year planning horizon requires answers to the following questions. What cash flows are anticipated for years 4 to 30 if Alternative 1 is selected? What will be the cash flows for Alternative 2 for years 7 to 30? What about Alternative 3 for years 6 to 30?

As shown in Table 5.3, the traditional approach is to assume that Alternative 1 will be repeated ten times, Alternative 2 will be repeated five times, Alterna-

TABLE 5.2. Cash Flow Profiles for Three Mutually Exclusive Investment
Alternatives Having Unequal Lives

| End of Year, $t$ | Revenues, $R_t$ | Costs, $C_t$ | Net Cash Flows, $R_t - C_t$ | Salvage Value If Sold at Time $t$ |
|---|---|---|---|---|
| Alternative 1 | | | | |
| 0 | — | — | — | $0 |
| 1–3 | $27,500 | $23,000 | $4,500 | 0 |
| Alternative 2 | | | | |
| 0 | — | $50,000 | −$50,000 | $50,000 |
| 1 | $30,000 | 10,000 | 20,000 | 35,000 |
| 2 | 30,000 | 10,000 | 20,000 | 25,000 |
| 3 | 30,000 | 10,000 | 20,000 | 15,000 |
| 4 | 30,000 | 10,000 | 20,000 | 5,000 |
| 5 | 30,000 | 10,000 | 20,000 | 0 |
| 6 | 30,000 | 10,000 | 20,000 | 0 |
| Alternative 3 | | | | |
| 0 | — | $75,000 | −$75,000 | $75,000 |
| 1 | $27,500 | 7,500 | 20,000 | 55,000 |
| 2 | 32,500 | 7,500 | 25,000 | 40,000 |
| 3 | 37,500 | 7,500 | 30,000 | 25,000 |
| 4 | 42,500 | 7,500 | 35,000 | 10,000 |
| 5 | 47,500 | 7,500 | 40,000 | 0 |

tive 3 will be repeated six times, and that identical cash flows occur during
these repeating life cycles. Inflation effects, as well as technological improvements and consumer demand changes tend to invalidate such assumptions.

If the shortest life approach is used, a planning horizon of 3 years would be
used. In such a case, the estimated salvage values of Alternatives 2 and 3
should be indicated at the end of year 3, as denoted in Table 5.3 by an asterisk.

Using the longest life approach yields a 6-year planning horizon. In this
instance a decision must be made concerning the cash flows of Alternative 1 in
years 4, 5, and 6 and Alternative 3 in year 6. In the case of Alternative 1, we
assume that another piece of used equipment will be employed, resulting in
resumption of the $4500 net cash flows during years 4, 5, and 6. For Alternative
3, we *could* assume that another new unit is purchased for $75,000 at the end of
year 5. This would provide service and net income of $20,000 during year 6, and
then have a salvage value of $55,000. This 1-year purchase is obviously not
economically sound. Therefore, we assume that a piece of readily available
used equipment is employed for year 6, earning $4500.

Suppose a standard planning horizon of 2, 5, or 8 years is used. The same
questions that arise under planning horizons of $\hat{T}$, $T_s$, or $T_l$ apply when one of
these other horizons is used. As shown in Table 5.3, Alternatives 1, 2, and 3 are

either truncated at, or repeated until, the end of the planning horizon at which time estimates of terminal salvage values are provided.

Although the use of a standard planning horizon has the benefit of a consistent approach in comparing investment alternatives, there are also some dangers that should be recognized. In some cases the major benefits associated with an alternative might occur in the later stages of its project life. If the planning horizon is less than the project life, such alternatives would seldom be accepted. Just such a practice caused one major textile firm to lose its strong position in the industry. A major modernization of the processing departments had been proposed, but its benefits would not be realized until the bugs had been worked out of the new system, all personnel were trained under the new system, and the marketing people had regained the lost customers. Unfortunately, the planning horizon specified by the firm was too short in duration and the modernization plan was not approved.

TABLE 5.3. Cash Flow Profiles for Various Planning Horizons

| End of Year, $t$ | Net Cash Flows for Alternatives | | |
|---|---|---|---|
| | $A_{1t}$ | $A_{2t}$ | $A_{3t}$ |
| $T = \hat{T} = 30$ years | | | |
| 0 | $4,500 | −$50,000 | −$75,000 |
| 1 | 4,500 | 20,000 | 20,000 |
| 2 | 4,500 | 20,000 | 25,000 |
| 3 | 4,500 | 20,000 | 30,000 |
| 4 | 4,500 | 20,000 | 35,000 |
| 5 | 4,500 | 20,000 | −75,000 + 40,000 |
| 6 | 4,500 | −50,000 + 20,000 | 20,000 |
| 7 | 4,500 | 20,000 | 25,000 |
| 8 | 4,500 | 20,000 | 30,000 |
| . | . | . | . |
| . | . | . | . |
| . | . | . | . |
| 29 | 4,500 | 20,000 | 35,000 |
| 30 | 4,500 | 20,000 | 40,000 |
| $T = T_s = 3$ years | | | |
| 0 | — | −$50,000 | −$75,000 |
| 1 | $4,500 | 20,000 | 20,000 |
| 2 | 4,500 | 20,000 | 25,000 |
| 3 | 4,500 | 20,000 + 15,000* | 30,000 + 25,000* |
| $T = T_l = 6$ years | | | |
| 0 | — | −$50,000 | −$75,000 |
| 1 | $4,500 | 20,000 | 20,000 |
| 2 | 4,500 | 20,000 | 25,000 |
| 3 | 4,500 | 20,000 | 30,000 |
| 4 | 4,500 | 20,000 | 35,000 |
| 5 | 4,500 | 20,000 | 40,000 |
| 6 | 4,500 | 20,000 | 4,500 (continued) |

TABLE 5.3. (*Continued*)

| End of Year, t | Net Cash Flows for Alternatives | | |
|---|---|---|---|
| | $A_{1t}$ | $A_{2t}$ | $A_{3t}$ |
| $T = 2$ years $< T_s$ | | | |
| 0 | — | −$50,000 | −$75,000 |
| 1 | $4,500 | 20,000 | 20,000 |
| 2 | 4,500 | 20,000 + 25,000* | 25,000 + 40,000* |
| $T_s < T = 5$ years $< T_l$ | | | |
| 0 | — | −$50,000 | −$75,000 |
| 1 | $4,500 | 20,000 | 20,000 |
| 2 | 4,500 | 20,000 | 25,000 |
| 3 | 4,500 | 20,000 | 30,000 |
| 4 | 4,500 | 20,000 | 35,000 |
| 5 | 4,500 | 20,000 + 0* | 40,000 |
| $T = 8$ years $> T_l$ | | | |
| 0 | — | −$50,000 | −$75,000 |
| 1 | $4,500 | 20,000 | 20,000 |
| 2 | 4,500 | 20,000 | 25,000 |
| 3 | 4,500 | 20,000 | 30,000 |
| 4 | 4,500 | 20,000 | 35,000 |
| 5 | 4,500 | 20,000 | 40,000 − 75,000 |
| 6 | 4,500 | 20,000 − 50,000 | 20,000 |
| 7 | 4,500 | 20,000 | 25,000 |
| 8 | 4,500 | 20,000 + 25,000* | 30,000 + 25,000* |

* Denotes salvage value.

## Example 5.4

As a second illustration of the planning horizon selection process, consider the two cash flow diagrams given in Figure 5.1. The two alternatives are mutually exclusive, *one-shot* investments. We are unable to predict what investment

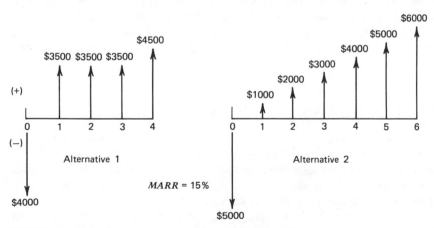

FIGURE 5.1. Cash flow diagrams for the example problem.

alternatives will be available in the future, but we do anticipate that recovered capital can be reinvested and earn a 15% return.

For this type situation, a 6-year planning horizon is suggested, with zero cash flows occurring in years 5 and 6 with Alternative 1. At the end of 6 years the net future worths for the two alternatives will be

$$FW_1(15\%) = \$4500(F|P\ 15,2) + \$3500(P|A\ 15,3)(F|P\ 15,6)$$
$$-\$4000(F|P\ 15,6)$$
$$= \$4500(1.3225) + \$3500(2.2832)(2.3131)$$
$$- \$4000(2.3131)$$
$$= \$15,183.29$$
$$FW_2(15\%) = \$1000(F|A\ 15,6) + \$1000(A|G\ 15,6)(F|A\ 15,6)$$
$$- \$5000(F|P\ 15,6)$$
$$= \$1000(8.7537) + \$1000(2.0972)(8.7537)$$
$$- \$5000(2.3131)$$
$$= \$15,546.46$$

Thus, we would recommend Alternative 2.

---

If we did not give careful thought to the situation involved and simply calculated the annual worth over 4- and 6-year horizons, respectively, values of $3000.15/year and $1776.20/year would result in favor of Alternative 1. This would be comparable to blindly assuming a least common multiple of lives planning horizon of 12 years, with identical cash flows in repeating life cycles. Hence, it is important to consider the particular situation involved and specify the planning horizon instead of employing a rule of thumb for establishing planning horizons that does not consider the nature of the investments.

It appears that the preferred approach would be to have a "flexible" standard planning horizon. All of the "routine" economic analyses would be based on the standard planning horizon of, say, 5 to 10 years; nonroutine economic analyses would be based on a planning horizon that was appropriate for the situation.

## 5.4 DEVELOPING CASH FLOW PROFILES

Once the set of mutually exclusive alternatives has been specified and the planning horizon decision has been made, the cash flow profiles can be developed for the alternatives. As has been emphasized, the cash flow profiles should be developed by giving careful consideration to *future* conditions instead of relying completely on *past* cash flows. The cash flows for an investment alternative are obtained by aggregating the cash flows for all investment proposals included in the investment alternative.

TABLE 5.4. Cash Flow Profiles for Three
Investment Proposals

| End of Year, $t$ | Net Cash Flows for Proposals | | |
|---|---|---|---|
| | $CF_{At}$ | $CF_{Bt}$ | $CF_{Ct}$ |
| 0 | −$20,000 | −$30,000 | −$50,000 |
| 1 | − 4,000 | 4,000 | − 5,000 |
| 2 | 2,000 | 6,000 | 10,000 |
| 3 | 8,000 | 8,000 | 25,000 |
| 4 | 14,000 | 10,000 | 40,000 |
| 5 | 25,000 | 20,000 | 10,000 |

Example 5.5

To illustrate the approach to be taken, suppose a planning horizon of 5 years is used and there are three investment proposals. Cash flow profiles for the proposals are given in Table 5.4. A budget limitation of $50,000 is available for investment among the proposals. Proposal B is contingent on Proposal A, and Proposals A and C are mutually exclusive. Based on the restrictions associated with the combinations of proposals, only four investment alternatives are to be considered. These are developed in Table 5.5. Alternative 0 is the "do nothing" alternative; Alternative 1 involves Proposal C alone; Alternative 2 involves Proposal A alone; and Alternative 3 involves a combination of Proposals A and B. The cash flow profiles for the four alternatives are given in Table 5.6.

The "do nothing" alternative is the status quo condition and serves as the base against which other alternatives are considered. In some cases the "do nothing" alternative is not feasible (e.g., failing to comply with pollution standards). Moreover, the do nothing alternative does not necessarily have zero

TABLE 5.5. Developing Mutually Exclusive
Investment Alternatives

| Feasible Alternatives | Proposals | | | Investment Required |
|---|---|---|---|---|
| | $X_A$ | $X_B$ | $X_C$ | |
| 0 | 0 | 0 | 0 | $ 0 |
| 1 | 0 | 0 | 1 | 50,000 |
| | 0 | 1 | 0 | 30,000 |
| | 0 | 1 | 1 | 80,000 |
| 2 | 1 | 0 | 0 | 20,000 |
| | 1 | 0 | 1 | 70,000 |
| 3 | 1 | 1 | 0 | 50,000 |
| | 1 | 1 | 1 | 100,000 |

TABLE 5.6. Cash Flow Profiles for Four Mutually Exclusive
Investment Alternatives

| End of year, $t$ | Net Cash Flows for Alternatives | | | |
|---|---|---|---|---|
| | $A_{0t}$ | $A_{1t}$ | $A_{2t}$ | $A_{3t}$ |
| 0 | 0 | −$50,000 | −$20,000 | −$50,000 |
| 1 | 0 | − 5,000 | − 4,000 | 0 |
| 2 | 0 | 10,000 | 2,000 | 8,000 |
| 3 | 0 | 25,000 | 8,000 | 16,000 |
| 4 | 0 | 40,000 | 14,000 | 24,000 |
| 5 | 0 | 10,000 | 25,000 | 45,000 |

cash flows associated with it. In principle, one should forecast the cash flows that will result if the present method is continued and compare the cash flows with those associated with other alternatives.

In most, if not all, economic evaluations, it is not necessary to develop a detailed forecast of all items of cost, revenue, and investment associated with an alternative. Costs and revenues that will be the same regardless of the alternative selected can be omitted. If a cost reduction alternative will not affect sales revenues, no forecast of such revenues need be developed. Attention is focused on the items of cost and revenue that will be affected by the alternative selected.

Since economic analyses are to be performed to judge the merits of investment alternatives for the future, the data required differ from those normally provided by the accounting system. As indicated in Chapter Two, the fundamental and historical purpose of accounting is to maintain a consistent *historical* record of the financial results of the operations of the organization. Accounting figures are based on definitions derived consistent with this objective. Accounting methods are not designed to determine the economic worth of *future* alternative courses of action.

Depending on the organization, specially designed forms are often provided for aiding the analyst in developing cash flow profiles and conducting the analysis. Sample forms used by one major industrial organization are provided in Figures 5.2 to 5.4. This particular company compares investment alternatives using the rate of return method; they refer to it as the discounted cash flow rate of return method. The next section of this text deals with comparison of alternatives.

## 5.5 COMPARING THE INVESTMENT ALTERNATIVES

We are now ready to begin comparing mutually exclusive investment alternatives on the basis of economic considerations. This chapter has dealt with structuring mutually exclusive alternatives, establishing a planning horizon, and defining cash flow profiles. The time value of money and one or more methods of measuring the worth of investments, both developed in Chapter Four, may now be applied to select the preferred alternative. Recall that the measures of investment worth include (1) present worth, (2) annual worth, (3) future worth, (4) internal rate of return, (5) external rate of return, (6) savings/investment ratio, (7) payback period, and (8) capitalized worth. The first six of these measures are stressed in the remainder of this section. The payback period and capitalized worth measures are a bit different and will be treated individually toward the end of this section.

Up to this point, there has been little mention of income taxes, depreciation, investment credits, and so forth. Certainly, these affect the cash flow patterns of our alternatives. In order to simplify the comparison of investment alternatives, however, we will assume that either the cash flows are after-tax cash flows or a before-tax study is desired. In Chapter Six we address the subject of income taxes and their effect on the preference among investment alternatives.

There are two basic concepts used in comparing mutually exclusive alternatives when any of the six consistent measures of investment worth are employed. One involves use of the *total cash flow approach* in which the *total* cash flows associated with each alternative are considered individually. The other involves use of the *incremental cash flow approach* in which alternatives are compared pairwise, and only the incremental (or differences in) cash flows are considered. Both methods are correct and will yield consistent decisions *if performed correctly*. Note that the alternative to do nothing should be included when it is feasible.

Example 5.6 ————————————————————————

To illustrate the use of the various methods of comparing investment alternatives, consider the four mutually exclusive alternatives having the cash flow profiles given in Table 5.7 for a planning horizon of 5 years. Note that to do

TABLE 5.7. Cash Flow Profiles for Four Mutually Exclusive Investment Alternatives

| | Net Cash Flows for Alternatives | | | |
|---|---|---|---|---|
| End of Year, $t$ | $A_{0t}$ | $A_{1t}$ | $A_{2t}$ | $A_{3t}$ |
| 0 | $0 | $ 0 | -$50,000 | -$75,000 |
| 1 | 0 | 4,500 | 20,000 | 20,000 |
| 2 | 0 | 4,500 | 20,000 | 25,000 |
| 3 | 0 | 4,500 | 20,000 | 30,000 |
| 4 | 0 | 4,500 | 20,000 | 35,000 |
| 5 | 0 | 4,500 | 20,000 | 40,000 |

NET CASH FLOW SCHEDULE

ALTERNATE  **A** OR **B**
CIRCLE ONE

TITLE (PROJECT) _____
TITLE (THIS ALTERNATE) _____
PROJECT NO. _____          DATE _____

| LINE | EXPENSES * | 0 | 1 | 2 | 3 | PERIOD 4 | 5 | 6 | 7 | 8 | TOTAL |
|---|---|---|---|---|---|---|---|---|---|---|---|
| 1 | NET BOOK VALUE (BEGINNING OF YEAR) | ✕ | | | | | | | | | |
| 2 | DEPRECIATION (____ YEARS LIFE) | | | | | | | | | | |
| 3 | RENTAL / LEASE COST | | | | | | | | | | |
| 4 | IN-PLANT LABOR WITH FRINGE BENEFITS @ ____ % | | | | | | | | | | |
| 5 | PURCHASED LABOR | | | | | | | | | | |
| 6 | REWORK / SCRAP | | | | | | | | | | |
| 7 | TOOLING / ANCILLARY EQUIPMENT | | | | | | | | | | |
| 8 | MAINTENANCE & REPAIR | | | | | | | | | | |
| 9 | POWER & FUEL / UTILITIES EXPENSE | | | | | | | | | | |
| 10 | MATERIALS & SUPPLIES | | | | | | | | | | |
| 11 | PRODUCT SHIPPING COSTS | | | | | | | | | | |
| 12 | PERSONAL PROPERTY TAX | | | | | | | | | | |
| 13 | REAL ESTATE TAX | | | | | | | | | | |
| 14 | IMPLEMENTING COST | | | | | | | | | | |
| 15 | | | | | | | | | | | |
| 16 | | | | | | | | | | | |
| 17 | TOTAL CASH EXPENSES (LINES 2 THRU 16) | | | | | | | | | | |
| 18 | AFTER TAX ANNUAL COST ** (50% x LINE 17) | | | | | | | | | | |
| 19 | INVESTMENT/EXISTING ASSET VALUE | | | | | | | | | | |
| 20 | INVESTMENT TAX CREDIT/REBATE | | | | | | | | | | |
| 21 | SALVAGE | | | | | | | | | | |
| 22 | | | | | | | | | | | |
| 23 | NET CASH FLOW (ALGEBRAIC SUM OF LINES 18 THRU 22 AND ADD BACK DEPRECIATION) | | | | | | | | | | |

* DO NOT SHOW COSTS WHICH ARE IDENTICAL IN BOTH ALTERNATIVES.
** SHOW ALGEBRAIC SIGNS ON LINES 18 THRU 23.

FIGURE 5.2.

DISCOUNTED CASH FLOW SUMMARY

| PERIOD | "A" NCF | "B" NCF | "A"-"B" △ NCF 0% INTEREST | CUMULATIVE CASH FLOW BACK AMOUNT | % | 15% INTEREST FACTOR | PW | 25% INTEREST FACTOR | PW | 40% INTEREST FACTOR | PW | 60% INTEREST FACTOR | PW |
|---|---|---|---|---|---|---|---|---|---|---|---|---|---|
| −1 | | | | | | 1.150 | | 1.250 | | 1.400 | | 1.600 | |
| 0 | | | | | | 1.000 | | 1.000 | | 1.300 | | 1.000 | |
| TOTAL "X" (DISBURSEMENTS) | | | | | | | | | | | | | |
| 1 | | | | | | .870 | | .800 | | .714 | | .625 | |
| 2 | | | | | | .756 | | .640 | | .510 | | .391 | |
| 3 | | | | | | .658 | | .512 | | .364 | | .244 | |
| 4 | | | | | | .572 | | .409 | | .260 | | .153 | |
| 5 | | | | | | .497 | | .328 | | .186 | | .095 | |
| 6 | | | | | | .432 | | .262 | | .133 | | .060 | |
| 7 | | | | | | .376 | | .210 | | .095 | | .037 | |
| 8 | | | | | | .327 | | .168 | | .068 | | .023 | |
| 9 | | | | | | .284 | | .134 | | .048 | | .015 | |
| 10 | | | | | | .247 | | .107 | | .035 | | .009 | |
| 11 | | | | | | .215 | | .086 | | .025 | | .006 | |
| 12 | | | | | | .187 | | .069 | | .018 | | .004 | |

TOTAL "Y" (RECEIPTS)

RATIO "X" / "Y"

PROPOSED INVESTMENT _____

DCF / ROR _____ %

PAYOUT @ 0% _____ YRS.

DIVISION _____

PROJECT _____

PROJECT NO. _____ J.O. NO. _____

PREPARED BY _____ DATE _____

APPROVED BY _____ DATE _____

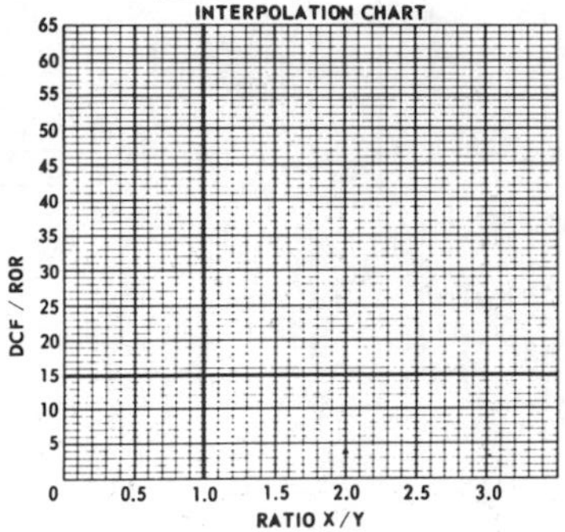

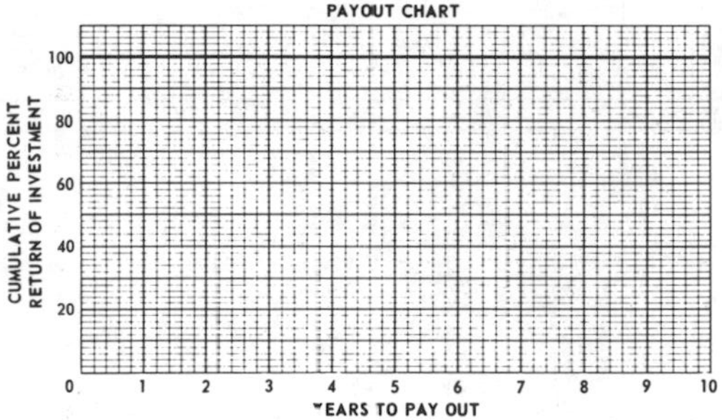

INTERPOLATION CHART

DCF / ROR
65 60 55 50 45 40 35 30 25 20 15 10 5 0
RATIO X/Y
0 0.5 1.0 1.5 2.0 2.5 3.0

PAYOUT CHART

CUMULATIVE PERCENT RETURN OF INVESTMENT
0 20 40 60 80 100
YEARS TO PAY OUT
1 2 3 4 5 6 7 8 9 10

FIGURE 5.3.

## DCF / ROR DATA SHEET

PROJECT TITLE _____ PROJECT NO. _____ DATE _____

PROJECT LIFE IS _____ YEARS, DETERMINED BY: _____

_____

_____

OTHER ALTERNATIVES CONSIDERED AND REASONS FOR REJECTION: _____

_____

_____

_____

| EXPLANATIONS AND CALCULATIONS | LINE REFERENCE | ALTERNATE A  TITLE _____  METHOD DESCRIPTION _____ | LINE REFERENCE | ALTERNATE B  TITLE _____  METHOD DESCRIPTION _____ |
|---|---|---|---|---|
| DEPRECIATION  DESCRIBE EXISTING / PROPOSED EQUIPMENT AND EXPLAIN DEPRECIATION | 2 | | 2 | |
| EXPENSES (LINES 3–16)  NOTE ZERO YEAR EXPENSES SUCH AS IMPLEMENTATION AND TRAINING; MAINTENANCE/MAJOR OVERHAUL; OR ANY UNUSUAL EXPENSES.  NOTE IF ONLY DELTA COSTS OF ONE ALTERNATIVE ARE BEING SHOWN. | | | | |
| INVESTMENT  SHOW DERIVATION OF EXISTING ASSET VALUE AND PROPOSED CAPITAL INVESTMENT. IDENTIFY ANY GOVT. EQUIP. STATE IF OLD EQUIP TO BE KEPT. | 19 | | 19 | |
| INVESTMENT TAX CREDIT OR REBATE  SHOW PERCENT USED. | 20 | | 20 | |
| SALVAGE  CALCULATE RESIDUAL AFTER TAX VALUE OF ASSETS | 21 | | 21 | |
| OTHER CONSIDERATIONS  INCLUDES INTANGIBLES, RISK AND LIKELIHOOD, OTHER ANALYSES AND RECOMMENDATION. | | | | |

FIGURE 5.4.

184 • COMPARISON OF ALTERNATIVES

nothing is presented here as an explicit alternative. A minimum attractive rate of return of 15% will be used in subsequent analyses.

## 5.5.1 Total Cash Flow Approach

With the *total cash flow approach,* we calculate for each alternative the measure of investment worth resulting from investment of the *entire budget* available.

Example 5.7 ──────────────────────────────

Consider Alternative 2 in Example 5.6, Table 5.7. Note that an initial investment of $50,000 is required; however, we evidently have a budget of (at least) $75,000 because that is the amount required by our most costly alternative. Let us calculate the present worth and internal rate of return measures of investment worth for Alternative 2.

In order to calculate either the *PW* or *IRR* measures resulting from investment of the *entire budget,* we must establish the cash flows during each year. The cash flows directly due to Alternative 2 are known to be −$50,000 in year 0 and $20,000/year in years 1 to 5. The additional $25,000 will be invested in year 0 at *MARR* = 15% and result in equivalent cash flows of $25,000(A|P 15,5) = $25,000(.2983) = $7457.50/year in years 1 to 5. Table 5.8 summarizes the total cash flow if Alternative 2 is selected.

*Present Worth Measure.* The present worth of the *total cash flow associated with Alternative 2* is given by:

$$PW_2(15\%) = -\$75,000 + \$27,457.50(P|A\ 15,5)$$
$$= -\$75,000 + \$27,457.50(3.3522)$$
$$= \$17,043.03$$

As a matter of interest, we can calculate the present worth measure for *just those cash flows directly due to Alternative 2.*

TABLE 5.8. Calculation of Total Cash Flows Associated with Alternative 2

| End of Year, t | Cash Flow Directly Due to Alternative 2, $A_{2t}$ | Equivalent Cash Flow from Investment at the MARR = 15% | Total Cash Flow Associated with Alternative 2 |
|---|---|---|---|
| 0 | −$50,000 | −$25,000.00 | −$75,000.00 |
| 1 | 20,000 | 7,457.50 | 27,457.50 |
| 2 | 20,000 | 7,457.50 | 27,457.50 |
| 3 | 20,000 | 7,457.50 | 27,457.50 |
| 4 | 20,000 | 7,457.50 | 27,457.50 |
| 5 | 20,000 | 7,457.50 | 27,457.50 |

$$PW_2(15\%) = -\$50,000 + \$20,000(P|A \ 15,5)$$
$$= -\$50,000 + \$20,000(3.3522)$$
$$= \$17,044.00$$

Aside from negligible round-off error, these present worths are the same! This should have been expected since the present worth (and annual worth and future worth) of an investment at the MARR is zero.

*Internal Rate of Return Measure.* The internal rate of return measure on the *total cash flow associated with Alternative 2* can be found by solving for $i$ as follows:

$$FW_2(i) = 0 = -\$75,000(F|P \ i,5) + \$27,457.50(F|A \ i,5)$$

$@i = 20\%$    $0 \neq -\$75,000(2.4883) + \$27,457.50(7.4416) = \$17,705.23$

$@i = 25\%$    $0 \neq -\$75,000(3.0518) + \$27,457.50(8.2070) = -\$3541.30$

Interpolating, $i = 24.17\%$. Now, if we were to calculate the internal rate of return measure for *just those cash flows directly due to Alternative 2*, the *IRR* would be 28.73%. Note that these are different! This is because only $50,000 is invested at 28.73%, while the remaining $25,000 is invested at *MARR* = 15%. The total $75,000 is, therefore, invested at a weighted average of 24.17%. It is the latter *IRR* = 24.17% that should be used for comparing alternatives when the total cash flow approach is used.

---

In general, *when the total cash flow approach is used and the measure of investment worth is present worth, annual worth, or future worth, we can simply calculate the measure based upon cash flows directly due to each alternative.* There is no need to consider the remaining portion of the budget since it is assumed to be invested at MARR and will add zero to the *PW*, *AW*, or *FW* measures. *When the measure of investment worth is internal rate of return, external rate of return, or savings/investment ratio, the total cash flows must be considered.* This includes the cash flows directly due to the alternative, plus the equivalent cash flows from investment of the remaining budget at the MARR. Figure 5.5 presents the total cash flow approach as described above. The *total cash flow approach* is commonly used when the measure of investment worth is *PW*, *AW*, or *FW*; with the *IRR*, *ERR*, or *SIR* measures, it is more common to use the *incremental cash flow approach*.

### 5.5.2 Incremental Cash Flow Approach

With the *incremental cash flow approach*, we calculate the measure of investment worth for the *incremental* or *difference* cash flows between pairs of investment alternatives. The set of mutually exclusive alternatives is usually ordered from lowest to highest initial investment. Then, the first alternative is compared against the "do nothing" alternative and one is found to be more desirable, while the other is eliminated from further consideration. The more

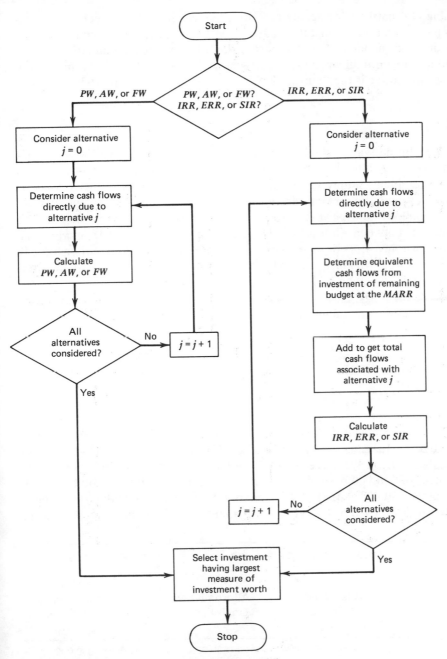

FIGURE 5.5. Flow chart of total cash flow approach.

desirable alternative is designated as the "current best." The second alternative is then compared against the "current best" from the first comparison. The better one is designated "current best"; the other one is eliminated. This pattern continues until all successive pairwise comparisons have been made; the winning alternative is the "current best" as a result of the last pairwise comparison.

## Example 5.8

Consider Alternatives 1 and 2 in Example 5.6, Table 5.7. We will assume that Alternative 1 has already been compared against Alternative 0 and has been found preferable to simply investing at the *MARR* of 15%. In fact, it is obvious that Alternative 1 should be preferred since it requires no initial investment but returns $4500/year during years 1 to 5. Now, let us perform an incremental cash flow analysis between Alternatives 1 and 2 using the present worth and internal rate of return measures of investment worth. Before calculating either the *PW* or *IRR* measures, we must establish the incremental cash flows between Alternatives 1 and 2 as given in Table 5.9.

*Present Worth Measure.* The present worth of the incremental cash flows between Alternatives 1 and 2 is given by

$$PW_{2-1}(15\%) = -\$50,000 + \$15,500(P|A\ 15,5)$$
$$= -\$50,000 + \$15,500(3.3522)$$
$$= \$1959.10$$

Since this present worth of incremental cash flows exceeds zero, we designate Alternative 2 as the "current best." Table 5.9 can help us understand why. The cash flows for Alternative 2 $(A_{2t})$ can be decomposed into the cash flows for Alternative 1 $(A_{1t})$ plus the incremental cash flows $(A_{2t} - A_{1t})$. That is, $A_{2t} = A_{1t} + (A_{2t} - A_{1t})$. We have a clear choice between using the incremental investment of $50,000 to employ Alternative 2, or selecting Alternative 1 and investing the $50,000 at *MARR*. Presumably, we already know that the Alternative 1 cash flows $(A_{1t})$ are a better investment than doing nothing. Therefore,

TABLE 5.9. Calculation of Incremental Cash Flows Between Alternatives 1 and 2

| End of Year, $t$ | Alternative 1 Cash Flow, $A_{1t}$ | Alternative 2 Cash FLow, $A_{2t}$ | Incremental Cash Flows Between Alternatives 1 and 2, $A_{2t} - A_{1t}$ |
|---|---|---|---|
| 0 | $ 0 | −$50,000 | −$50,000 |
| 1 | 4,500 | 20,000 | 15,500 |
| 2 | 4,500 | 20,000 | 15,500 |
| 3 | 4,500 | 20,000 | 15,500 |
| 4 | 4,500 | 20,000 | 15,500 |
| 5 | 4,500 | 20,000 | 15,500 |

we need only concern ourselves with deciding whether the incremental investment $(A_{2t} - A_{1t})$ is better than investing at the $MARR$. Since the $PW$ measure is positive, we will select Alternative 2 and eliminate Alternative 1 from further consideration.

*Internal Rate of Return Measure.* The internal rate of return measure on the incremental cash flows between Alternatives 1 and 2 can be found by solving for $i$ as follows:

$$FW_{2-1}(i) = 0 = -\$50,000(F|P\ i,5) + \$15,500(F|A\ i,5)$$

$$@i = 15\% \quad 0 \neq -\$50,000(2.0114) + \$15,500(6.7424) = \$3937.20$$

$$@i = 20\% \quad 0 \neq -\$50,000(2.4883) + \$15,500(7.4416) = -\$9070.20$$

Interpolating, $i = 16.51\%$ which is better than the $MARR$ of 15%. Therefore, again, Alternative 2 is the "current best" alternative. We would normally now compare Alternative 3 against Alternative 2.

---

In general, when the incremental cash flow approach is used, any of the present worth, annual worth, future worth, internal rate of return, external rate of return, or savings/investment ratio measures can be used following the same basic procedure. The incremental cash flow approach is presented in Figure 5.6.

## 5.5.3 Present Worth Method

The *present worth method* is probably the most common and trouble-free approach for analyzing a set of mutually exclusive alternatives.

### Example 5.9 _____

Using the four cash flow profiles detailed in Example 5.6, let us select the preferred alternative using both the total and incremental cash flow approaches under the present worth method. The $MARR$ is 15% and Table 5.10 repeats the cash flow profiles for convenience.

*Total Cash Flow Approach*

For Alternative 0,

$$PW_0(15\%) = \$0$$

For Alternative 1,

$$PW_1(15\%) = \$4500(P|A\ 15,5)$$
$$= \$4500(3.3522)$$
$$= \$15,084.90$$

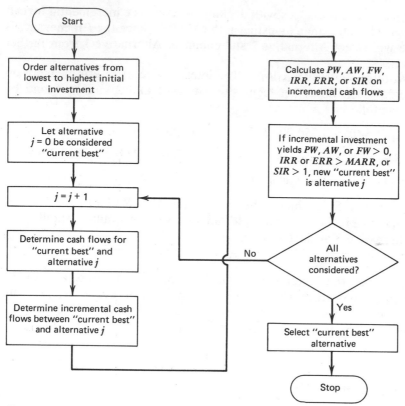

FIGURE 5.6. Flowchart of incremental cash flow approach.

For Alternative 2,

$$PW_2(15\%) = -\$50,000 + \$20,000(P|A\ 15,5)$$
$$= -\$50,000 + \$20,000(3.3522)$$
$$= \$17,044$$

TABLE 5.10. Cash Flow Profiles for Four Mutually
Exclusive Investment Alternatives

| End of Year, $t$ | $A_{0t}$ | $A_{1t}$ | $A_{2t}$ | $A_{3t}$ |
|---|---|---|---|---|
| | | Net Cash Flows for Alternatives | | |
| 0 | $0 | $ 0 | -$50,000 | -$75,000 |
| 1 | 0 | 4,500 | 20,000 | 20,000 |
| 2 | 0 | 4,500 | 20,000 | 25,000 |
| 3 | 0 | 4,500 | 20,000 | 30,000 |
| 4 | 0 | 4,500 | 20,000 | 35,000 |
| 5 | 0 | 4,500 | 20,000 | 40,000 |

For Alternative 3,

$$PW_3(15\%) = -\$75,000 + \$20,000(P|A\ 15,5) + \$5000(P|G\ 15,5)$$
$$= -\$75,000 + \$20,000(3.3522) + \$5000(5.7751)$$
$$= \$20,919.50$$

We prefer Alternative 3 since it provides the highest $PW$ when evaluated at the MARR.

*Incremental Cash Flow Approach*

Between Alternatives 1 and 0,

$$PW_{1-0}(15\%) = \$4500(P|A\ 15,5)$$
$$= \$4500(3.3522)$$
$$= \$15,084.90 \quad \text{(Prefer 1 over 0)}$$

Between Alternatives 2 and 1,

$$PW_{2-1}(15\%) = -\$50,000 + \$15,500(P|A\ 15,5)$$
$$= -\$50,000 + \$15,500(3.3522)$$
$$= \$1959.10 \quad \text{(Prefer 2 over 1)}$$

Between Alternatives 3 and 2,

$$PW_{3-2}(15\%) = -\$25,000 + \$5000(P|G\ 15,5)$$
$$= -\$25,000 + \$5000(5.7751)$$
$$= \$3875.50 \quad \text{(Prefer 3 over 2)}$$

We prefer Alternative 3 since it is the winner in pairwise comparisons using the present worth of incremental cash flows between alternatives.

## 5.5.4 Annual Worth Method

The *annual worth method* is also straightforward and useful for those analysts and decision makers who prefer to think in terms of annual worths or dollars per year.

### Example 5.10

Let us again select the preferred alternative from those presented in Example 5.9, Table 5.10. We will use the annual worth method and both the total and incremental cash flow approaches.

*Total Cash Flow Approach*

For Alternative 0,

$$AW_0(15\%) = \$0/\text{year}$$

For Alternative 2,

$$AW_1(15\%) = \$4500/year$$

For Alternative 2,

$$AW_2(15\%) = -\$50,000(A|P\ 15,5) + \$20,000$$
$$= -\$50,000(.2983) + \$20,000$$
$$= \$5085/year$$

For Alternative 3,

$$AW_3(15\%) = -\$75,000(A|P\ 15,5) + \$20,000 + \$5000(A|G\ 15,5)$$
$$= -\$75,000(.2983) + \$20,000 + \$5000(1.7228)$$
$$= \$6241.50/year$$

Alternative 3 is preferred since it provides the highest $AW$ when evaluated at the *MARR*.

*Incremental Cash Flow Approach*

Between Alternatives 1 and 0,

$$AW_{1-0}(15\%) = \$4500/year \qquad \text{(Prefer 1 over 0)}$$

Between Alternatives 2 and 1,

$$AW_{2-1}(15\%) = -\$50,000(A|P\ 15,5) + \$15,500$$
$$= -\$50,000(.2983) + \$15,500$$
$$= \$585/year \qquad \text{(Prefer 2 over 1)}$$

Between Alternatives 3 and 2,

$$AW_{3-2}(15\%) = -\$25,000(A|P\ 15,5) + \$5000(A|G\ 15,5)$$
$$= -\$25,000(.2983) + \$5000(1.7228)$$
$$= \$1156.50/year \qquad \text{(Prefer 3 over 2)}$$

Alternative 3 is preferred since it is the winner in pairwise comparisons using the annual worth of incremental cash flows between alternatives.

---

## 5.5.5 Future Worth Method

The *future worth method* is easily used, but not often selected for economic analyses of alternatives.

## Example 5.11

Using the future worth method and the total and incremental cash flow approaches, let us select the preferred alternative from those given in Example 5.9, Table 5.10.

*Total Cash Flow Approach*

For Alternative 0,

$$FW_0(15\%) = \$0$$

For Alternative 1,

$$FW_1(15\%) = \$4500(F|A\ 15,5)$$
$$= \$4500(6.7424)$$
$$= \$30,340.80$$

For Alternative 2,

$$FW_2(15\%) = -\$50,000(F|P\ 15,5) + \$20,000(F|A\ 15,5)$$
$$= -\$50,000(2.0114) + \$20,000(6.7424)$$
$$= \$34,278.00$$

For Alternative 3,

$$FW_3(15\%) = -\$75,000(F|P\ 15,5) + \$20,000(F|A\ 15,5)$$
$$+ \$5000(P|G\ 15,5)(F|P\ 15,5) = -\$75,000(2.0114)$$
$$+ \$20,000(6.7424) + \$5000(5.7751)(2.0114) = \$42,073.18$$

We prefer Alternative 3 on the basis that it has the highest $FW$ measure.

*Incremental Cash Flow Approach*

Between Alternatives 1 and 0,

$$FW_{1-0}(15\%) = \$4500(F|A\ 15,5)$$
$$= \$4500(6.7424)$$
$$= \$30,340.80 \qquad \text{(Prefer 1 over 0)}$$

Between Alternatives 2 and 1,

$$FW_{2-1}(15\%) = -\$50,000(F|P\ 15,5) + \$15,500(F|A\ 15,5)$$
$$= -\$50,000(2.0114) + \$15,500(6.7424)$$
$$= \$3937.20 \qquad \text{(Prefer 2 over 1)}$$

Between Alternatives 3 and 2,

$$FW_{3-2}(15\%) = -\$25,000(F|P\ 15,5) + \$5000(P|G\ 15,5)(F|P\ 15,5)$$
$$= -\$25,000(2.0114) + \$5000(5.7751)(2.0114)$$
$$= \$7795.18 \qquad \text{(Prefer 3 over 2)}$$

Again, we prefer Alternative 3 since it wins the series of pairwise comparisons using the future worth of incremental cash flows between alternatives.

## 5.5.6 Internal Rate of Return Method

The *internal rate of return method* is one of the most popular, yet misused, methods of comparing alternatives.

Example 5.12_____

We will compare the alternatives of Example 5.9, Table 5.10 using the internal rate of return method and both total and incremental cash flows. Table 5.10 gives us the cash flows directly due to the various alternatives. In order to use the total cash flow approach, we now need the *total* cash flows associated with each alternative. This is given in Table 5.11.

TABLE 5.11. Calculation of Total Cash Flows Associated with Each Alternative

| Alternative $j$ | End of Year, $t$ | Cash Flows Directly Due to Alternative $j$, $A_{jt}$ | Equivalent Cash Flows from Investment at the MARR = 15% | Total Cash Flows Associated with Alternative $j$ |
|---|---|---|---|---|
| 0 | 0 | $ 0 | −$75,000.00 | −$75,000.00 |
|   | 1 | 0 | 22,372.50 | 22,372.50 |
|   | 2 | 0 | 22,372.50 | 22,372.50 |
|   | 3 | 0 | 22,372.50 | 22,372.50 |
|   | 4 | 0 | 22,372.50 | 22,372.50 |
|   | 5 | 0 | 22,372.50 | 22,372.50 |
| 1 | 0 | $  0 | −$75,000.00 | −75,000.00 |
|   | 1 | 4,500 | 22,372.50 | 26,872.50 |
|   | 2 | 4,500 | 22,372.50 | 26,872.50 |
|   | 3 | 4,500 | 22,372.50 | 26,872.50 |
|   | 4 | 4,500 | 22,372.50 | 26,872.50 |
|   | 5 | 4,500 | 22,372.50 | 26,872.50 |
| 2 | 0 | −$50,000 | −$25,000.00 | −$75,000.00 |
|   | 1 | 20,000 | 7,457.50 | 27,457.50 |
|   | 2 | 20,000 | 7,457.50 | 27,457.50 |
|   | 3 | 20,000 | 7,457.50 | 27,457.50 |
|   | 4 | 20,000 | 7,457.50 | 27,457.50 |
|   | 5 | 20,000 | 7,457.50 | 27,457.50 |
| 3 | 0 | −$75,000 | $ 0 | −$75,000.00 |
|   | 1 | 20,000 | 0 | 20,000.00 |
|   | 2 | 25,000 | 0 | 25,000.00 |
|   | 3 | 30,000 | 0 | 30,000.00 |
|   | 4 | 35,000 | 0 | 35,000.00 |
|   | 5 | 40,000 | 0 | 40,000.00 |

*Total Cash Flow Approach*

For Alternative 0,

$$FW_0(i) = \$0 = -\$75,000(F|P\ i,5) + \$22,372.50(F|A\ i,5)$$

$$@i = 15\% \quad \$0 = -\$75,000(2.0114) + \$22,372.50(6.7424) = -\$10.66$$

Therefore, $i_0 = 15\%$.

The relatively minor discrepancy between $0 and −$10.66 is due to round-off error in the interest tables. Recall that Alternative 0 earns the *MARR* = 15%; therefore, the IRR could have simply been stated without the above calculation.

For Alternative 1,

$$FW_1(i) = \$0 = -\$75,000(F|P\ i,5) + \$26,872.50(F|A\ i,5)$$

$$@i = 20\% \quad \$0 \neq -\$75,000(2.4883) + \$26,872.50(7.4416) = \$13,351.90$$

$$@i = 25\% \quad \$0 \neq -\$75,000(3.0518) + \$26,872.50(8.2070) = -\$8342.39$$

Interpolating, $i_1 = 23.08\%$.

For Alternative 2,

$$FW_2(i) = \$0 = -\$75,000(F|P\ i,5) + \$27,457.50(F|A\ i,5)$$

$$@i = 20\% \quad \$0 \neq -\$75,000(2.4883) + \$27,457.50(7.4416) = \$17,705.23$$

$$@i = 25\% \quad \$0 \neq -\$75,000(3.0518) + \$27,457.50(8.2070) = -\$3541.30$$

Interpolating, $i_2 = 24.17\%$.

For Alternative 3,

$$FW_3(i) = \$0 = -\$75,000(F|P\ i,5) + \$20,000(F|A\ i,5)$$
$$+ \$5000(P|G\ i,5)(F|P\ i,5)$$

$$@i = 20\% \quad \$0 \neq -\$75,000(2.4883) + \$20,000(7.4416)$$
$$+ \$5000(4.9061)(2.4883)$$
$$= \$23,248.74$$

$$@i = 25\% \quad \$0 \neq -\$75,000(3.0518) + \$20,000(8.2070)$$
$$+ \$5000(4.235)(3.0518)$$
$$= -\$603.79$$

Interpolating, $i_3 = 24.87\%$.

We prefer Alternative 3 on the basis that it yields the highest *IRR* on the *total cash flow* associated with each alternative.

*Incremental Cash Flow Approach*

Between Alternatives 1 and 0, there is no incremental investment, but there are incremental returns of $4500 in each of years 1 to 5. The incremental IRR is therefore $i_{1-0} = \infty\%$ (Prefer 1 over 0).

Between Alternatives 2 and 1,

$$FW_{2-1}(i) = \$0 = -\$50,000(F|P\ i,5) + \$15,500(F|A\ i,5)$$

$@i = 15\%$  $\$0 \neq -\$50,000(2.0114) + \$15,500(6.7424) = \$3937.20$

$@i = 20\%$  $\$0 \neq -\$50,000(2.4883) + \$15,500(7.4416) = -\$9070.20$

Interpolating, $i_{2-1} = 16.51\%$ (Prefer 2 over 1).

Between Alternatives 3 and 2,

$$FW_{3-2}(i) = \$0 = -\$25,000(F|P\ i,5) + \$5000(P|G\ i,5)(F|P\ i,5)$$

$@i = 15\%$  $\$0 \neq -\$25,000(2.0114) + \$5000(5.7751)(2.0114) = \$7795.18$

$@i = 20\%$  $\$0 \neq -\$25,000(2.4883) + \$5000(4.9061)(2.4883) = -\$1168.26$

Interpolating, $i_{3-2} = 19.35\%$ (Prefer 3 over 2).

We again prefer Alternative 3 since it prevailed in the pairwise comparison of alternatives in terms of *IRR* on their incremental cash flows. Note that Alternative 3 did not have the highest incremental *IRR*! That is not important; what *is* important is that in the final pairwise comparison its rate exceeded the *MARR* of 15%.

---

As mentioned earlier, decision makers often use the *IRR* method incorrectly. They try to use a total cash flow approach but use only the cash flows directly due to each alternative, with no consideration of the other component of investment at the *MARR*. If we were to do this, Alternatives 1, 2, and 3 would have *IRR* measures of $\infty\%$, 28.73%, and 24.90%, respectively. Note that Alternatives 1, 2, and 3 would then be *incorrectly* ranked in reverse order of desirability!

The incremental cash flow approach is recommended when using the *IRR* method. It is not only correct, but much easier to use than the total cash flow approach.

### 5.5.7 External Rate of Return Method

The *external rate of return method* is rarely used, but it has advantages as described in Chapter Four over the *IRR* method.

Example 5.13 ——————————————————————

Let us now compare the alternatives of Example 5.9, Table 5.10 using the external rate of return method and both the total and incremental cash flows approaches. The positive net cash flows will be assumed to be reinvested at the $MARR = 15\%$ and Equation (4.15) will be employed.

## Total Cash Flow Approach

Like with the IRR method, we must use *total cash flows* associated with each alternative as developed in Table 5.11.

For Alternative 0,

$$\$22,372.50(F|A \ 15,5) = \$75,000(1 + i')^5$$
$$\$22,372.50(6.7424) = \$75,000(1 + i')^5$$

Solving, $i_0' = 15\%$.

For Alternative 1,

$$\$26,872.50(F|A \ 15,5) = \$75,000(1 + i')^5$$
$$\$26,872.50(6.7424) = \$75,000(1 + i')^5$$

Solving, $i_1' = 19.29\%$.

For Alternative 2,

$$\$27,457.50(F|A \ 15,5) = \$75,000(1 + i')^5$$
$$\$27,457.50(6.7424) = \$75,000(1 + i')^5$$

Solving, $i_2' = 19.81\%$.

For Alternative 3,

$$\$20,000(F|A \ 15,5) + \$5000(P|G \ 15,5)(F|P \ 15,5) = \$75,000(1 + i')^5$$
$$\$20,000(6.7424) + \$5000(5.7751)(2.0114) = \$75,000(1 + i')^5$$

Solving, $i_3' = 20.80\%$.

We prefer Alternative 3 since it provides the largest ERR on the total cash flows.

## Incremental Cash Flow Approach

Between Alternatives 1 and 0, there is no incremental investment, but there are incremental returns of $4500 in each of years 1 to 5. The incremental *ERR* is therefore $i_{1-0}' = \infty\%$ (Prefer 1 over 0).

Between Alternatives 2 and 1,

$$\$15,500(F|A \ 15,5) = \$50,000(1 + i')^5$$
$$\$15,500(6.7424) = \$50,000(1 + i')^5$$

Solving, $i_{2-1}' = 15.89\%$ (Prefer 2 over 1).

Between Alternatives 3 and 2,

$$\$5000(P|G\ 15,5)(F|P\ 15,5) = \$25,000(1 + i')^5$$

$$\$5000(5.7751)(2.0114) = \$25,000(1 + i')^5$$

Solving, $i' = 18.36\%$ (Prefer 3 over 2).

Again, Alternative 3 is preferred on the basis of its winning the series of pairwise comparisons using the *ERR* method and the incremental cash flow approach. Note that Alternative 3 did not have the highest ERR on its incremental investment; its *ERR* was, however, higher than the *MARR* in the final pairwise comparison. The incremental cash flow approach is recommended when using the *ERR*, since it is easily applied.

## 5.5.8 Savings/Investment Ratio Method

The *savings/investment ratio method* is most often used in public sector projects; it is also known as the benefit/cost ratio.

## Example 5.14_____

We will now analyze the four alternatives of Example 5.9, Table 5.10 using the *SIR* formulation given in Equation (4.16). Again, the *MARR* = 15%.

*Total Cash Flow Approach*

Like with the *IRR* and *ERR* methods, we must use *total cash flows* associated with each alternative as developed in Table 5.11.

For Alternative 0,

$$SIR_0(15\%) = \frac{\$22,372.50(P|A\ 15,5)}{\$75,000}$$

$$= \frac{\$22,372.50(3.3522)}{\$75,000}$$

$$= 1.00$$

For Alternative 1,

$$SIR_1(15\%) = \frac{\$26,872.50(P|A\ 15,5)}{\$75,000}$$

$$= \frac{\$26.872.50(3.3522)}{\$75,000}$$

$$= 1.20$$

For Alternative 2,

$$SIR_2(15\%) = \frac{\$27,457.50(P|A\ 15,5)}{\$75,000}$$

$$= \frac{\$27,457.50(3.3522)}{\$75,000}$$

$$= 1.23$$

For Alternative 3,

$$SIR_3(15\%) = \frac{\$20,000(P|A\ 15,5) + \$5000(P|G\ 15,5)}{\$75,000}$$

$$= \frac{\$20,000(3.3522) + \$5000(5.7751)}{\$75,000}$$

$$= 1.28$$

Alternative 3 is preferred since it yields the highest *SIR* value based upon *total cash flows* associated with each alternative.

### Incremental Cash Flow Approach

Between Alternatives 1 and 0, there is no incremental investment, but there are incremental returns of \$4500 in each of years 1 to 5. The incremental *SIR* is therefore $SIR_{1-0}(15\%) = \infty$ (Prefer 1 to 0).
Between Alternatives 2 and 1,

$$SIR_{2-1}(15\%) = \frac{\$15,500(P|A\ 15,5)}{\$50,000}$$

$$= \frac{\$15,500(3.3522)}{\$50,000}$$

$$= 1.04 \qquad \text{(Prefer 2 to 1)}$$

Between Alternatives 3 and 2,

$$SIR_{3-2}(15\%) = \frac{\$5000(P|G\ 15,5)}{\$25,000}$$

$$= \frac{\$5000(5.7751)}{\$25,000}$$

$$= 1.16 \qquad \text{(Prefer 3 to 2)}$$

Alternative 3 is preferred since it won the final pairwise comparison by having an *SIR* greater than 1.00 on the incremental investment. The incremental cash flow approach is preferred when using the *SIR* method due to its ease of use.

### 5.5.9 Payback Period Method

The *payback period method* is not consistent with the previous six methods presented, and as such may yield entirely different selections or rankings of alternatives. Neither the total nor the incremental cash flow approach as presented is applicable. The *PBP* method is recommended for use only as an auxilliary or secondary criterion in alternative selection.

**Example 5.15**

Let us now use the payback period method to analyze the alternatives presented in Example 5.9, Table 5.10.

For Alternative 0, it is not meaningful to talk about payback. Since Alternative 1 requires no investment, its payback is instantaneous.

For Alternative 2.

$$\sum_{i=1}^{2} R_{2t} = \$20,000 + \$20,000 = \$40,000 < C_{20} = \$50,000$$

$$\sum_{i=1}^{3} R_{2t} = \$20,000 + \$20,000 + \$20,000 = \$60,000 > C_{20} = \$50,000$$

The payback period is 2½ (or 3) years.

For Alternative 3,

$$\sum_{i=1}^{2} R_{3t} = \$20,000 + \$25,000 + \$30,000 = \$75,000 = C_{30}$$

The payback period is exactly 3 years.

Note that the ranking based upon PBP for our Alternatives is 1, 2, 3 which is exactly in reverse order of preference dictated by the six consistent methods.

---

### 5.5.10 Capitalized Worth Method

The *capitalized worth method* is applicable only if there is reason to believe that a series of cash flows will repeat indefinitely into the future. As such, it is not a method comparable to those used previously in this section. Neither the total nor the incremental cash flow approach as presented is applicable.

**Example 5.16**

Let us assume that the cash flows as presented in Example 5.9, Table 5.10 actually do repeat indefinitely into the future. We will continue using the *MARR* of 15%. Also, we will make use of the annual worths calculated in Example 5.10.

For Alternative 0,

$$CW_0(15\%) = \$0$$

For Alternative 1,

$$CW_1(15\%) = \frac{AW_1(15\%)}{.15}$$

$$= \frac{\$4500}{.15}$$

$$= \$30,000$$

For Alternative 2,

$$CW_2(15\%) = \frac{AW_2(15\%)}{.15}$$

$$= \frac{\$5085}{.15}$$

$$= \$33,900$$

For Alternative 3,

$$CW_3(15\%) = \frac{AW_3(15\%)}{.15}$$

$$= \frac{\$6241.50}{.15}$$

$$= \$41,610$$

We prefer Alternative 3 on the basis that it has the highest capitalized worth. Had we been calculating the capitalized cost where costs are positive, we would have preferred the smallest value.

---

## 5.6 PERFORMING SUPPLEMENTARY ANALYSES

The fifth step in performing an economic evaluation of investment alternatives is the performance of subsequent analyses. A sensitivity analysis consists of an exploration of the behavior of the measure of investment worth to changes in the values of the parameters for an investment alternative. The parameters subject to change might include the planning horizon, the discount rate, and any or all of the cash flows. This step recognizes that the process of providing estimates of cash flows and decisions concerning the planning horizon and discount rate is not a precise, errorless process. An examination of breakeven, sensitivity, and risk analyses as they relate to economic analyses is reserved for Chapter Eight. For now, we assume that the values assigned to the parameters are neither inaccurate nor subject to change.

## 5.7 SELECTING THE PREFERRED ALTERNATIVE

The final step in performing a comparison of investment alternatives is the selection of the preferred alternative. Our discussion in this chapter has concentrated solely on the economic factor; we have been concerned with determining the most economical alternative. The final decision may be based on a host of criteria instead of on the single criterion of economics. In Chapter Nine we examine the decision process in the face of multiple objectives.

The selection process is complicated not only by the presence of multiple objectives, but also by the risks and uncertainties associated with the future. The analysis of Chapter Eight is extended in Chapter Nine to include decision making in the face of risk or uncertainty.

As pointed out in Chapter One, the selection or rejection of the recommended solution is heavily dependent on the sales ability of the individual presenting the recommendation to management. Since the corporate decision makers are normally presented with many more investment alternatives than can be funded, it is important to communicate effectively in order to compete favorably for the company's limited capital. Toward this end, Klausner [13] provides four specific suggestions for the engineer or systems analyst.

1. He should recognize that the decision-makers' perspective is broad and develop his proposal accordingly. The project's capital requirements should be related to previous and estimated future capital requirements on similar investments. The project's Discounted Cash Flow return or Net Present Value should be compared to other projects with which the decision-makers are familiar. The proposal should be shown to fit in with long-range corporate plans and support short-range objectives. Comparison should be made with similar investments by competition and competitive advantage (if any) shown. Ancillary marketing, public relations and/or political benefits that the company will derive from the investment should be pointed out. The effect that the investment will have on other functional activities of the company should be noted and overall benefits stressed. In other words, the investment proposal should be related to the well-being of the total enterprise.

2. He should recognize that this investment proposal is only one of several that the decision-makers are reviewing and that not all proposals will be accepted. The engineer should know how the decision-makers classify investments and what competition there is for available capital resources. he should know the relative strengths and weaknesses of his investment proposal vis-a-vis competing investment proposals and deal with each in the presentation. If the proposed investment is relatively risk-free, that point should be made strongly to possibly offset less desirable aspects of the proposal (e.g., low return on investment).

3. He should know the decision-makers and tailor the investment proposal accordingly. The engineer should, for example, know and use the measure of merit

they prefer, and support it with any other measures of merit that may be necessary or helpful. For example, if the decision-makers favor Discounted Cash Flow return on investment and the proposal has one that barely meets minimum return standards but does have a large Net Present Value, the engineer must make sure that the potential contribution to profit is clearly pointed out. If it is known that the decision-makers still calculate the payout period of each investment proposal, the investment's payout period should be stated—either to support the proposal or to point out its irrelevance in hopes of minimizing its impact on the ultimate investment decision. The engineer should always use technical and economical terms that the decision-makers will understand and relate to.

The engineer should become familiar with the values that the decision-makers attach to different aspects of investment proposals—particularly economic uncertainty. Low risk should be emphasized in proposals to decision-makers who tend to be risk-avoiders; potential economic gain should be highlighted for the risk-taking decision-maker. This, of course, does not suggest that the engineer should be less than honest in his proposal, but only makes the point that the communication of the investment proposal should be developed with the decision-maker in mind and emphasis varied accordingly.

4. He should not oversell the technical engineering aspects of the investment proposal. It should be remembered that decision-makers are primarily interested in the economic aspects of the proposal. The engineer must resist the temptation to overstate the complicated and sophisticated technology that might underlie a proposal. The less decision-makers understand about the technical/engineering aspects of a proposal, the more uneasy they become. This apprehension becomes part of the uncertainty which the decision-makers subjectively assign to the proposal and the result could be its rejection in favor of a proposal with which they are more familiar, or at least, more comfortable. The engineer should keep in mind the background and interests of the decision-makers and use technical and commercial terms with which they are familiar.

# 5.8 ANALYZING ALTERNATIVES WITH NO POSITIVE CASH FLOWS

The previous analysis of an investment decision involved four mutually exclusive alternatives, including the "do nothing" alternative, which included positive valued cash flows. However, there exist situations in which no positive valued cash flows are present.

Example 5.17_____

Consider a case in which two new cost reduction alternatives have been proposed. The present method, which we refer to as the do nothing alternative, is also a feasible alternative. Thus, three mutually exclusive alternatives are considered. Cash flow profiles for the alternatives are given in Table 5.12.

Alternative 0 is the "do nothing" alternative in which the present expenditure of $12,000/year is continued. Alternative 1 involves an initial investment of

TABLE 5.12. Cash Flow Profiles for
Three Mutually Exclusive Investment
Alternatives with No Positive Cash Flows

| $t$ | $A_{0t}$ | $A_{1t}$ | $A_{2t}$ |
|---|---|---|---|
| 0 | 0 | $-\$10,000$ | $-\$15,000$ |
| 1 | $-\$12,000$ | $-\ 9,000$ | $-\ 9,000$ |
| 2 | $-\ 12,000$ | $-\ 9,000$ | $-\ 8,000$ |
| 3 | $-\ 12,000$ | $-\ 9,000$ | $-\ 7,000$ |
| 4 | $-\ 12,000$ | $-\ 9,000$ | $-\ 6,000$ |
| 5 | $-\ 12,000$ | $-\ 9,000$ | $-\ 5,000$ |

$10,000 in order to reduce the annual expenditures by $3000 over the 5-year period. Alternative 2 requires an initial investment of $15,000 in order to obtain decreasing annual expenditures. The minimum attractive rate of return is specified to be 10%.

The *PW, AW,* or *FW* methods using the total cash flow approach, or any measure of investment worth using the incremental cash flow approach, may be applied to this problem. Let us analyze these alternatives in three different ways.

*Annual Worth Method—Total Cash Flow Approach*

$$AW_0(10\%) = -\$12,000/\text{year}$$

$$AW_1(10\%) = -\$10,000(A|P\ 10,5) - \$9000$$
$$= -\$10,000(.2638) - \$9000$$
$$= -\$11,638/\text{year}$$

$$AW_2(10\%) = -\$15,000(A|P\ 10,5) - \$9000 + \$1000(A|G\ 10,5)$$
$$= -\$15,000(.2638) - \$9000 + \$1000(1.8101)$$
$$= -\$11,146.90/\text{year}$$

Alternative 2 is preferred since it has the highest annual worth (or lowest annual cost).

*Annual Worth Method—Incremental Cash Flow Approach*

$$AW_{1-0}(10\%) = -\$10,000(A|P\ 10,5) + \$3000$$
$$= -\$10,000(.2638) + \$3000$$
$$= \$362/\text{year} \qquad (\text{Prefer 1 over 0})$$

$$AW_{2-1}(10\%) = -\$5000(A|P\ 10,5) + \$1000(A|G\ 10,5)$$
$$= \$5000(.2638) + \$1000(1.8101)$$
$$= \$491.10/\text{year} \qquad (\text{Prefer 2 over 1})$$

Again, Alternative 2 is preferred.

*Internal Rate of Return Method–Incremental Cash Flow Approach*

$$FW_{1-0}(i) = \$0 = -\$10,000(F|P\ i,5) + \$3000(F|A\ i,5)$$

$$@i = 15\% \quad \$0 \neq -\$10,000(2.0114) + \$3000(6.7424) = \$113.20$$

$$@i = 20\% \quad \$0 \neq -\$10,000(2.4883) + \$3000(7.4416) = -\$2558.20$$

Interpolating, $i_{1-0} = 15.21\%$ (Prefer 1 over 0).

$$FW_{2-1}(i) = \$0 = -\$5000(F|P\ i,5) + \$1000(P|G\ i,5)(F|P\ i,5)$$

$$@i = 15\% \quad \$0 \neq -\$5000(2.0114) + \$1000(5.7751)(2.0114) = \$1559.04$$

$$@i = 20\% \quad \$0 \neq -\$5000(2.4883) + \$1000(4.9061)(2.4883) = -\$233.65$$

Interpolating, $i_{2-1} = 19.35\%$ (Prefer 2 over 1).

Again, Alternative 2 is preferred.

The *IRR* method deserves additional explanation. Since the sum of the cash flows (ignoring the time value of money) is negative for each alternative, positive valued rates of return among the three alternatives do not exist. However, even though Alternative 1 does not have a rate of return by itself, in comparison with Alternative 0, it is clear that the investment of the $10,000 yields annual savings of $3000 in annual expenditures. Thus, on an incremental basis, the $10,000 incremental investment produces positive valued cash flows of $3000/year for each of the 5 years.

The return on the incremental investment, using the internal rate-of-return method, is found to be approximately 15.21%. Since we would do better to invest the $10,000 in Alternative 1 and earn 15.21% than to choose Alternative 0 and only earn the minimum attractive rate of return on the remaining money available for investment, Alternative 1 is preferred to Alternative 0.

Of course, we have $15,000 to invest (otherwise Alternative 2 would be unfeasible). Thus, at this point we are willing to invest $10,000 in Alternative 1 and earn 15.21% and invest the remaining $5000 in some other opportunity and earn the minimum attractive rate of return of 10%. The question now considered is, "Should we use the additional $5000 and pool it with the $10,000 to invest in Alternative 2?" The rate of return on the incremental investment required to obtain Alternative 2 is calculated to be approximately 19.35%.

At this point, the question is "Should we invest $10,000 in Alternative 1 and invest $5000 elsewhere at the *MARR*, or should we invest $15,000 in Alternative 2?" Investing in Alternative 2 yields a return of 15.21% on the $10,000 increment and 19.35% on the $5000 increment. Consequently, Alternative 2 would be preferred to the investment of $10,000 in Alternative 1 and investing the remaining $5000 to earn only 10%.

## 5.9 MORE ON DEALING WITH UNEQUAL LIVES

Recall that in Section 5.3 we recommended that a planning horizon be specified, cash flows over the planning horizon be given explicitly, and the evaluation be performed over the common planning horizon time frame. While this method is becoming better known and accepted, many still compare alternatives having unequal lives on the basis of the individual life cycles. This approach either implicitly assumes a least common multiple of lives approach, or it implicitly establishes a salvage value which may or may not be reasonable. Enough background material has now been presented to permit us to clarify these points and demonstrate their drawbacks.

## Example 5.18

Consider two mutually exclusive alternatives having cash flow profiles as depicted in Figure 5.7. Using a minimum attractive rate of return of 15%, the following annual worths are obtained:

$$AW_1(15\%) = -\$5000(A|P\ 15,5) - \$2000 + \$1000(A|F\ 15,5)$$
$$= -\$5000(.2983) - \$2000 + \$1000(.1483)$$
$$= -\$3343.20$$

$$AW_2(15\%) = -\$7000(A|P\ 15,6) - \$1500 + \$1500(A|F\ 15,6)$$
$$= -\$7000(.2642) - \$1500 + \$1500(.1142)$$
$$= -\$3178.10$$

Implicit in the comparison of these alternatives using annual worths based on individual life cycles is the assumption that a 30-year planning horizon is being used. As shown in Table 5.13, if the present worths are computed for a 30-year period, the following values will be obtained:

$$PW_1(15\%) = -\$21,952.10$$
$$PW_2(15\%) = -\$20,868.30$$

Converting the present worths to annual worths yields

$$AW_1(15\%) = -\$21,952.10(A|P\ 15,30)$$
$$= -\$21,952.10(.1523)$$
$$= -\$3343.30$$

$$AW_2(15\%) = -\$20,868.30(A|P\ 15,30)$$
$$= -\$20,868.30(.1523)$$
$$= -\$3178.24$$

which are the values obtained (except for round-off error) using individual life cycles.

An alternative assumption that could be made is that a 5-year planning horizon is being used and the salvage value for Alternative 2 is such that an annual worth of $-\$3178.10$ will still be obtained. Hence, $S_5$, the salvage value at the

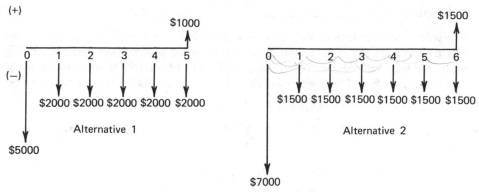

FIGURE 5.7. Cash flow diagrams for two mutually exclusive alternatives.

end of the fifth year, for Alternative 2 must be

$$AW_2(15\%) = -\$7000(A|P\ 15,5) - \$1500 + S_5(A|F\ 15,5)$$
$$-\$3178.10 = -\$7000(.2983) - \$1500 + S_5(.1483)$$
$$S_5 = \$2764.67$$

This salvage value may or may not have any relationship to reality.

TABLE 5.13. Cash Flows for Two Mutually
Exclusive Alternatives, with a 30-year
Planning Horizon

| EOY | CF(1) | CF(2) |
|---|---|---|
| 0 | −$ 5,000 | −$ 7,000 |
| 1–4 | − 2,000 | − 1,500 |
| 5 | − 6,000 | − 1,500 |
| 6 | − 2,000 | − 7,000 |
| 7–9 | − 2,000 | − 1,500 |
| 10 | − 6,000 | − 1,500 |
| 11 | − 2,000 | − 1,500 |
| 12 | − 2,000 | − 7,000 |
| 13–14 | − 2,000 | − 1,500 |
| 15 | − 6,000 | − 1,500 |
| 16–17 | − 2,000 | − 1,500 |
| 18 | − 2,000 | − 7,000 |
| 19 | − 2,000 | − 1,500 |
| 20 | − 6,000 | − 1,500 |
| 21–23 | − 2,000 | − 1,500 |
| 24 | − 2,000 | − 7,000 |
| 25 | − 6,000 | − 1,500 |
| 26–29 | − 2,000 | − 1,500 |
| 30 | − 1,000 | 0 |
| Present worth | −$21,952.10 | −$20,868.30 |

## Example 5.19

Consider the alternatives depicted in Figure 5.8. They will be used in a more glaring example to illustrate the shortcoming of an assumption that the salvage value for unused portions of an asset's life will be such that the annual worth will be unchanged. The annual worths for individual life cycles are found to be:

$$AW_1(15\%) = -\$5000(A|P\ 15,5) - \$3000 + \$1000(A|F\ 15,5)$$
$$= -\$5000(.2983) - \$3000 + \$1000(.1483)$$
$$= -\$4343.20$$

$$AW_2(15\%) = -\$6000(A|P\ 15,6) - \$1000 - \$1000(A|G\ 15,6)$$
$$+ \$1000(A|F\ 15,6)$$
$$= -\$6000(.2642) - \$1000 - \$1000(2.0972) + \$1000(.1142)$$
$$= -\$4568.20$$

If a 5-year planning horizon is used, to obtain an annual worth of $-\$4568.20$ for Alternative 2 requires that the salvage value at the end of the fifth year, $F_5$, be such that

$$-\$4568.20 = -\$6000(A|P\ 15,5) - \$1000 - \$1000(A|G\ 15,5) + S_5(A|F\ 15,5)$$

$$-\$4568.20 = -\$6000(.2983) - \$1000 - \$1000(1.7228) + S_5(.1483)$$

$$S_5 = \$374.92$$

Thus, instead of having a $1000 salvage value at the end of the sixth year, a salvage value of $-\$374.92$ must exist at the end of the fifth year to yield the annual worth obtained by treating individual life cycles.

---

We do not recommend that alternatives having unequal lives be blindly compared on the basis of the annual worths for individual life cycles. Such an

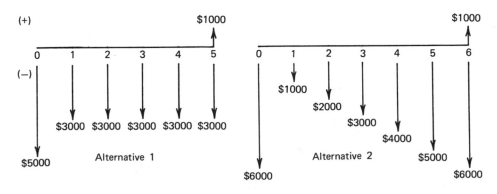

*MARR* = 15%

FIGURE 5.8. Cash flow diagrams for two mutually exclusive alternatives.

approach assumes *implicitly* that either a least common multiple of lives planning horizon is appropriate or the salvage values for unused portions of an asset's life are such that the annual worth is unchanged. We prefer to make *explicit* all assumptions concerning the planning horizon and the salvage values.

## 5.10 REPLACEMENT ANALYSIS

*Replacement analysis* is one of the most important and most common types of alternative comparisons encountered in practice. In a replacement analysis, one of the feasible alternatives involves maintaining the status quo; the remaining alternatives provide various replacement options that are available. In some organizations, replacement analyses are performed routinely in an effort to ensure that the best equipment and facilities are in use, compared to their possible successors.

The reasons for considering replacement are numerous. First, the current asset (defender) may have a number of deficiencies including high set-up cost, excessive maintenance, declining production efficiency, heavy energy consumption, and physical impairment. For example, when you are confronted with a car that is expensive to operate and maintain, or one soon to need a major overhaul, you begin to consider replacing the car.

Second, potential replacement assets (challengers) may take advantage of new technology and be easily set up, maintained at low cost, high in output, energy efficient, and possessing increased capabilities, perhaps at a vastly reduced cost. For example, some new generation computer-controlled manufacturing equipment has rendered many old machines economically obsolete. Also, we can relate to the phenomenal accomplishments with which calculators and personal computers have resulted in increased capabilities, vastly lower prices, and economic obsolescence for equipment only a few years old.

Finally, the environment affects replacement decisions. Consumer demand preferences change, and present equipment may be unable to adapt to the new designs dictated by consumers. Demand levels may also cause equipment capacity to be relatively high or low, resulting in inefficiency or inability to perform. Rental firms specializing in equipment, automobiles, furniture, and so forth, may affect ownership decisions by offering lease options. Also, outside contractors may be able to perform some tasks cheaper and better than in-house facilities. All of the above considerations and more may affect the decision to keep or replace an asset.

Replacement analyses are basically just another type of alternative comparison. That is, they follow the same systematic six-step approach outlined and developed in this chapter. All of the six consistent measures of investment worth are also applicable. Unfortunately, replacement decisions are often confounded by our nearness to the existing asset. That is, we may have an "emotional" attachment to present equipment, particularly if we are the ones who recommended its installation previously. This attachment often results in attempts to recover *sunk costs,* identified in Chapter Two as past costs that are unrecoverable. Sunk costs should not be included in a replacement analysis (or any other alternative evaluation); to include sunk costs penalizes or burdens the potential replacement asset and unfairly favors the current asset.

Two approaches are commonly used in replacement analyses. One is the *cash flow approach* in which actual cash flows associated with keeping, purchasing, or leasing an asset are used directly. The other is the *outsider viewpoint approach* in which the cash flow profiles faced by an objective outsider are used. Which approach is used in a replacement analysis is strictly a matter of preference. Both are mathematically equivalent and yield consistent decisions.

## 5.10.1 Cash Flow Approach

The *cash flow approach* might properly be called the *insider viewpoint approach*. Often, there is no additional capital cost if the present asset is kept. If a replacement is purchased, there is often a trade-in allowance given for the present asset. In order to develop the cash flow profiles, the decision maker should ask "How much money will be spent and received if I adopt this alternative?" As noted above, past costs should be viewed from the proper perspective; unrecoverable past costs are *sunk costs* and are not to be included in economy studies that deal with the future, except as those sunk costs may affect income taxes if a present asset is disposed.

The planning horizon to be used is again at the discretion of the decision maker. As in any alternative evaluation, the current asset and its proposed replacements must be evaluated over a common planning horizon, with cash flow profiles extending throughout but not beyond that horizon for each alternative. Since the remaining life of the present asset is usually shorter than that of a new asset, the *shortest life among alternatives* is often chosen.

Example 5.20 _____

To illustrate a replacement analysis, consider a situation involving a chemical plant that owns a filter press that was purchased 3 years ago for $30,000. Actual operating and maintenance (O & M) expenses (excluding labor) for the press have been $4000, $5000, and $6000 each of the past 3 years, as depicted in Table 5.14. It is anticipated that the filter press can be used for 5 more years and salvaged for $2000 at that time. The undepreciated worth of the press as indicated on our accounting books is $12,600; however, technological developments of the past 3 years have resulted in excellent new competitive filter presses. As a result, the current market value for the used filter press is only $9000. If the old filter press is retained, annual operating and maintenance costs are anticipated to be as shown in Table 5.14.

TABLE 5.14. Data for a Replacement Alternative

| | Alternative 1 | | Alternative 2 | | |
|---|---|---|---|---|---|
| End of Year, $t$ | Operating and Maintenance Costs | End of Year, $t$ | Operating and Maintenance Costs | Salvage Value, $S_t$ |
| -3 | — | 0 | — | $36,000 |
| -2 | − $4,000 | 1 | — | 30,000 |
| -1 | − 5,000 | 2 | −$1,000 | 24,600 |
| 0 | − 6,000 | 3 | − 2,000 | 19,800 |
| 1 | − 7,000 | 4 | − 3,000 | 15,600 |
| 2 | − 8,000 | 5 | − 4,000 | 12,000 |
| 3 | − 9,000 | 6 | − 5,000 | 9,000 |
| 4 | − 10,000 | 7 | − 6,000 | 6,600 |
| 5 | − 11,000 | 8 | − 7,000 | 4,800 |
| | | 9 | − 8,000 | 3,600 |
| | | 10 | − 9,000 | 3,000 |

A new filter press is available and can be purchased for $36,000. It has an anticipated life of 10 years and is expected to have annual operating and maintenance costs as also given in Table 5.14. Based upon historical data concerning salvage values of filter presses, estimated salvage values ($S_t$) for the new press are also given in the table. Alternative 1 is defined to be "keep the old press," while Alternative 2 is defined to be "replace the old press with the new press."

Since the "old" press can only be used for 5 more years, a planning horizon of 5 years is specified; the salvage value for the "new" press is estimated to be $12,000 in 5 years. Cash flows for each alternative are given in Table 5.15. Note that the $9000 market value of the existing press is applied as a positive cash flow for the challenger since the old unit will be sold if the new unit is purchased.

TABLE 5.15. Cash Flows for a Replacement Alternative—
Cash Flow Approach

| End of Year, $t$ | Alternative 1 Net Cash Flows, $A_{1t}$ | Alternative 2 Net Cash Flows, $A_{2t}$ | Difference in Cash Flows, $A_{2t} - A_{1t}$ |
|---|---|---|---|
| 0 | $0 | −$36,000 + 9,000 | −$27,000 |
| 1 | −$7,000 | 0 | 7,000 |
| 2 | − 8,000 | −1,000 | 7,000 |
| 3 | − 9,000 | −2,000 | 7,000 |
| 4 | −10,000 | −3,000 | 7,000 |
| 5 | −11,000 + 2,000 | −4,000 + 12,000 | 17,000 |

Computing the annual worths for each alternative using a minimum attractive rate of return of 15% yields the following results:

$$AW_1(15\%) = -\$7000 - \$1000(A|G\ 15,5) + \$2000(A|F\ 15,5)$$
$$= -\$7000 - \$1000(1.7228) + \$2000(.1483)$$
$$= -\$8426.20/\text{year}$$

$$AW_2(15\%) = -\$27,000(A|P\ 15,5) - \$1000(A|G\ 15,5) + \$12,000(A|F\ 15,5)$$
$$= -\$27,000(.2983) - \$1000(1.7228) + \$12,000(.1483)$$
$$= -\$7997.30/\text{year}$$

On the basis of the annual worth difference of $428.90/year, it is recommended that the new filter press be purchased and the old press be sold.

---

In listing the cash flows for each alternative in Table 5.15, we ignored the operating and maintenance costs that will occur after the fifth year if the new filter press is purchased. The argument against including these operating and maintenance costs is based on the specification of a planning horizon of 5 years. The 5-year planning horizon was based on the maximum useful life for the old filter press. If the old press is retained, then it will have to be replaced in 5 years (if not before, because of the possible development of attractive replacement alternatives in the future). Consequently, in 5 years we might have available an alternative that will yield even greater operating and maintenance savings than the filter press currently being considered. Once the planning horizon has been established, it is not fair to include any cash flows that might occur later for one alternative without including similar estimates for the other alternative(s).

The above interpretation defines the planning horizon as a window through which only the cash flows that occur during the planning horizon can be seen. This window should *include a terminal (salvage) value for any alternative having a life longer than the planning horizon,* even though the alternative might not be physically replaced at that time. The end of the planning horizon defines a point in time at which another replacement study is planned. At that time, the future savings and costs will be compared against other available replacement candidates.

It should also be noted that neither the $30,000 first cost nor the $12,600 book value of the existing press appears as a cash flow. The $30,000 is a past cost which is now irrelevant (except as it affects taxes as seen in Chapter Six), and the $12,600 was never a cash flow. The difference between the value on our accounting books ($12,600) and the market value ($9000) is $3600. This $3600 is a sunk cost that analysts are often tempted to add to the first cost of the challenger in an attempt to recover it. To do so is absolutely incorrect, and biases the decision maker against the proposed replacement item. For example, if we burdened Alternative 2 by adding a sunk cost of $3600 to its first cost, its annual worth would be decreased by $-\$3600(A|P\ 15,5) = -\$3600(.2983) =$ $-\$1073.88/\text{year}$ to a total of $-\$7997.30/\text{year} - \$1073.88/\text{year} = -\$9071.18/$

year. On this basis, we would mistakenly select Alternative 1 and continue to operate an economically inferior filter press.

## Example 5.21

If, in Example 5.20, a 10-year planning horizon is desired, we recommend that consideration be given to the replacement of the "old" filter press in 5 years. Based on a forecast of the growth of filter press technology, suppose we anticipate that at the end of 5 years a filter press will be available at a cost of $31,000. Net operating and maintenance costs, and a terminal salvage value are anticipated to be as depicted in Table 5.16.

Calculating the annual worth of these alternatives results in:

$$
\begin{aligned}
AW_1(15\%) &= [-\$7,000(P|A\ 15,5) - \$1000(P|G\ 15,5) - \$29,000(P|F\ 15,5) \\
&\quad -\$1000(P|G\ 15,5)(P|F\ 15,5) + \$15,000(P|G\ 15,10)](A|P\ 15,10) \\
&= (-\$7000(3.3522) - \$1000(5.7751) - \$29,000(.4972) \\
&\quad -\$1000(5.7751)(.4972) + \$15,000(.2472)](.1993) \\
&= -\$8534.56/\text{year}
\end{aligned}
$$

$$
\begin{aligned}
AW_2(15\%) &= -\$27,000(A|P\ 15,10) - \$1000(A|G\ 15,10) + \$3000(A|F\ 15,10) \\
&= -\$27,000(.1993) - \$1000(3.3832) + \$3000(.0493) \\
&= -\$8616.40/\text{year}
\end{aligned}
$$

These computations establish that the difference in annual worths for the two alternatives is $81.84/year, with Alternative 1 being more economic. Thus, on the basis of technological forecasts of filter press alternatives in 5 years, it would appear advantageous to postpone the replacement. However, the degree of uncertainty in our forecast of the cash flows for a projected future replace-

TABLE 5.16. Cash Flows for a Replacement Alternative Using a 10-Year Planning Horizon-Cash Flow Approach

| End of Year, t | Alternative 1 Net Cash Flows, $A_{1t}$ | Alternative 2 Net Cash Flows, $A_{2t}$ |
|---|---|---|
| 0 | $ 0 | -$36,000 + $9,000 |
| 1 | -$ 7,000 | 0 |
| 2 | - 8,000 | -1,000 |
| 3 | - 9,000 | -2,000 |
| 4 | -10,000 | -3,000 |
| 5 | -11,000 + 2,000 - 31,000 | -4,000 |
| 6 | 0 | -5,000 |
| 7 | - 1,000 | -6,000 |
| 8 | - 2,000 | -7,000 |
| 9 | - 3,000 | -8,000 |
| 10 | -4,000 + 15,000 | -9,000 + 3,000 |

TABLE 5.17. Cash Flows for Several Replacement Alternatives—
Cash Flow Approach

| End of Year, $t$ | Alternative 1 Net Cash Flows, $A_{1t}$ | Alternative 2 Net Cash Flows, $A_{2t}$ | Alternative 3 Net Cash Flows, $A_{3t}$ | Alternative 4 Net Cash Flows, $A_{4t}$ |
|---|---|---|---|---|
| 0 | 0 | −$36,000 + $10,000 | −$40,000 + 12,000 | −$7,500 + 9,000 |
| 1 | −$ 7,000 | 0 | −  500 | −7,500 |
| 2 | −  8,000 | −1,000 | −1,000 | − 7,500 −  800 |
| 3 | −  9,000 | −2,000 | −1,500 | − 7,500 − 1,600 |
| 4 | − 10,000 | −3,000 | −2,000 | − 7,500 − 2,400 |
| 5 | −11,000 + 2,000 | −4,000 + 12,000 | −  2,500 + 13,000 | −3,200 |

*(handwritten annotations above Alternative 3 and Alternative 4 columns: "9000 + 1000" and "9000 + 3000")*

ment candidate would cause us to question the merits of postponing the replacement because of a difference of $81.84/year. These kinds of considerations are explored more fully in Chapters Eight and Nine.

Often, there is a discrepancy between trade-in allowances if several potential replacement assets are being considered. Further, these trade-in values have little relationship to the true market value that could be realized if the existing asset were sold separately. In this case, when using the cash flow approach, the appropriate question to ask is still, "How much money will be spent and received if I adopt this alternative?"

Example 5.22 _____

In Example 5.20, Alternative 1 was to keep our existing filter press having a market value of $9000. Alternative 2 included selling the current press and buying a new one for $36,000. Suppose, however, that the dealer has offered to allow a $10,000 trade-in value for the old press. Also two additional alternatives have been identified. Alternative 3 is a new asset costing $40,000, having a salvage value after 5 years of $13,000, and having annual operating and maintenance costs as given in Table 5.17. A trade-in value of $12,000 is offered for the current press in Alternative 3. Alternative 4 is to lease a press for $7500/year payable at the beginning of each year during the 5-year horizon. If the lease is taken, the existing press will be sold on the open market. The complete cash flow profile for each alternative is given in Table 5.17.

Calculating the annual worths of these alternatives at a *MARR* of 15% results in

$$AW_1(15\%) = -\$7000 - \$1000(A|G\ 15,5) + \$2000(A|F\ 15,5)$$
$$= -\$7000 - \$1000(1.7228) + \$2000(.1483)$$
$$= -\$8426.20/\text{year}$$

$$AW_2(15\%) = -\$26{,}000(A|P\ 15{,}5) - \$1000(A|G\ 15{,}5) + \$12{,}000(A|F\ 15{,}5)$$
$$= -\$26{,}000(.2983) - \$1000(1.7228) + \$12{,}000(.1483)$$
$$= -\$7699.00/\text{year}$$

$$AW_3(15\%) = -\$28{,}000(A|P\ 15{,}5) - \$500 - \$500(A|G\ 15{,}5)$$
$$+ \$13{,}000(A|F\ 15{,}5) = -\$28{,}000(.2983) - \$500 - \$500(1.7228)$$
$$+ \$13{,}000(.1483) = -\$7785.90/\text{year}$$

$$AW_4(15\%) = -\$7500(F|P\ 15{,}1) + \$9000(A|P\ 15{,}5) - \$800(A|G\ 15{,}5)$$
$$= -\$7500(1.15) + \$9000(.2983) - \$800(1.7228)$$
$$= -\$7318.54/\text{year}$$

Based upon these calculations, the lease alternative appears to be economically most favorable.

---

### 5.10.2 Outsider Viewpoint Approach

The *outsider viewpoint approach* is preferred by many because it forces the decision maker to view both the existing asset and its challengers from an objective point of view as would an "outsider." That is, the outsider is assumed to have no existing asset. The outsider is then free to choose either a used asset (the defender) available for the price of its market value, or any of the potential replacement assets (the challengers).

In essence, the outsider viewpoint approach considers the salvage value of the existing asset to be its investment cost if it is retained in service. Such an approach is consistent with the opportunity cost concept described in Chapter Two. Since the retention of the defender is equivalent to a decision to forego the receipt of its salvage value, then an opportunity cost is assigned to the defender.

Example 5.23 ───────────────────────────────

Let us again consider Example 5.20, this time using the outsider viewpoint. Recall that the present filter press has a market value of $9000, a life of 5 years, and a salvage value of $2000 at that time. The challenger has a cost of $36,000, a life of 10 years, and estimated salvage values at any point in time as given in Table 5.14. A 5-year planning horizon is to be used and the *MARR* is 15%. Cash flows from the outsider's point of view are as given in Table 5.18.

We can see by comparing Tables 5.15 and 5.18 that the differences in cash flows between alternatives are the same; therefore, the cash flow approach and the outsider viewpoint approach are equivalent. As further verification, we can calculate the annual worth of each alternative.

TABLE 5.18. Cash Flows for a Replacement Alternative—Outsider's Viewpoint Approach

| End of Year, $t$ | Alternative 1 Net Cash Flows, $A_{1t}$ | Alternative 2 Net Cash Flows, $A_{2t}$ | Difference in Cash Flows, $A_{2t} - A_{1t}$ |
|---|---|---|---|
| 0 | −$ 9,000 | −$36,000 | −$27,000 |
| 1 | − 7,000 | 0 | 7,000 |
| 2 | − 8,000 | −1,000 | 7,000 |
| 3 | − 9,000 | −2,000 | 7,000 |
| 4 | − 10,000 | −3,000 | 7,000 |
| 5 | −11,000 + 2,000 | −4,000 + 12,000 | 17,000 |

$$AW_1(15\%) = -\$9000(A|P\ 15,5) - \$7000 - \$1000(A|G\ 15,5) + \$2000(A|F\ 15,5)$$
$$= -\$9000(.2983) - \$7000 - \$1000(1.7228) + \$2000(.1483)$$
$$= -\$11,110.90/\text{year}$$

$$AW_2(15\%) = -\$36,000(A|P\ 15,5) - \$1000(A|G\ 15,5) + \$12,000(A|F\ 15,5)$$
$$= -\$36,000(.2983) - \$1000(1.7228) + \$12,000(.1483)$$
$$= -\$10.682.00/\text{year}$$

Note that the difference in annual worths is $428.90/year in favor of the new filter press. This is the identical conclusion reached in Example 5.20.

In using the outsider viewpoint approach, we must be careful to use "rational" first costs for each alternative. Example 5.22 showed that the trade-in allowances may be different between alternatives, and that there may be little correspondence between the trade-in value and an asset's true market value. Usually, a high trade-in value indicates the seller is using an inflated selling price. In reality, this inflated selling price would most likely be decreased if a straight purchase with no trade-in were made. Therefore, for analysis purposes, when using the outsider's viewpoint approach, *we will decrease the inflated selling price of the asset by the difference between the trade-in value and a lower market value*. Presumably, no adjustment is necessary if the trade-in allowance is less than the market value since we would simply choose to dispose of the defender separately at the market value price.

Example 5.24_____

Now, let us consider Example 5.22 from the outsider viewpoint. Recall that Alternatives 2 and 3 cost $36,000 and $40,000, respectively. Also, recall that the market value for the current asset is $9000 and the trade-in allowances under Alternatives 2 and 3 are $10,000 and $12,000, respectively. Since these trade-in allowances are inflated by $1000 and $3000, respectively, the corre-

TABLE 5.19. Cash Flows for Several Replacement Alternatives—Outsider's Viewpoint Approach

| End of Year, $t$ | Alternative 1 Net Flows, Cash $A_{1t}$ | Alternative 2 Net Flows, Cash $A_{2t}$ | Alternative 3 Net Flows, Cash $A_{3t}$ | Alternative 4 Net Flows, Cash $A_{4t}$ |
|---|---|---|---|---|
| 0 | −$ 9,000 | −$35,000 | −$37,000 | −$7,500 |
| 1 | −$ 7,000 | 0 | − 500 | − 7,500 |
| 2 | − 8,000 | − 1,000 | − 1,000 | −7,500 − 800 |
| 3 | − 9,000 | − 2,000 | − 1,500 | −7,500 − 1,600 |
| 4 | − 10,000 | − 3,000 | − 2,000 | −7,500 − 2,400 |
| 5 | −11,000 + 2,000 | −4,000 + 12,000 | −2,500 + 13,000 | − 3,200 |

sponding costs of Alternatives 2 and 3 will be reduced to $35,000 and $37,000. A planning horizon of 5 years and a *MARR* of 15% are still in effect. Table 5.19 presents the cash flow profiles from the outsider's point of view.

Calculating the annual worths of these alternatives results in

$$AW_1(15\%) = -\$9000(A|P\ 15,5) - \$7000 - \$1000(A|G\ 15,5) + \$2000(A|F\ 15,5)$$
$$= -\$9000(.2983) - \$7000 - \$1000(1.7228) + \$2000(.1483)$$
$$= -\$11,110.90/\text{year}$$

$$AW_2(15\%) = -\$35,000(A|P\ 15,5) - \$1000(A|G\ 15,5) + \$12,000(A|F\ 15,5)$$
$$= -\$35,000(.2983) - \$1000(1.7228) + \$12,000(.1483)$$
$$= -\$10,383.70/\text{year}$$

$$AW_3(15\%) = -\$37,000(A|P\ 15,5) - \$500 - \$500(A|G\ 15,5) + \$13,000(A|F\ 15,5)$$
$$= -\$37,000(.2983) - \$500 - \$500(1.7228) + \$13,000(.1483)$$
$$= -\$10,470.60/\text{year}$$

$$AW_4(15\%) = -\$7500(F|P\ 15,1) - \$800(A|G\ 15,5)$$
$$= -\$7500(1.15) - \$800(1.7228)$$
$$= -\$10,003.24/\text{year}$$

Again, the lease alternative is preferred based upon annual worth calculations. Note that the above annual worth differences between alternatives using the outsider viewpoint approach are *exactly* the same as the annual worth differences between alternatives in Example 5.22 in which the cash flow approach was used. This further substantiates the equivalence of the two approaches.

## 5.10.3 Optimum Replacement Interval

Although logical arguments can be given for treating a replacement decision as just another economic investment alternative, many firms fail to subject their existing equipment to careful scrutiny on a periodic basis to determine if replacement is required. Despite the fact that replacement studies can yield significant reductions in costs, it still remains that many firms postpone replacing assets beyond the "optimum" time for replacement, perhaps because a decision to replace an asset involves a change, and resistance to change is inherent in most individuals. For example, an engineer who 2 years ago successfully argued that compressor Z should be replaced by compressor Y may now find that compressor X is more economical than compressor Y. If X is now championed it may be viewed by management as an admission that the wrong compressor was selected as a replacement for Z.

Some reasons for delaying the replacement of assets beyond the economic replacement time are:

1. The firm is making a profit with its present equipment.
2. The present equipment is operational and is producing an acceptable quality product.
3. There is risk or uncertainty associated with predicting the expenses of a new machine, whereas one is relatively certain about the expenses of the current machine.
4. A decision to replace equipment is a stronger commitment, for a period of time into the future, than keeping the existing equipment.
5. Management tends to be conservative in decisions regarding the replacement of costly equipment.
6. There may be a limitation on funds available for purchasing new equipment, but no limitation on funds for maintaining existing equipment.
7. There may be considerable uncertainty concerning the future demands for the services of the equipment in question.
8. Sunk costs psychologically affect decisions to replace equipment.
9. An anticipation that technological improvements in the future might render obsolete equipment available currently; a wait-and-see attitude prevails.
10. Reluctance to be a pioneer in adopting new technology; instead of replacing now, wait for the competition to act.

As a unit of equipment ages, operating and maintenance costs increase. At the same time, the capital recovery cost decreases with prolonged use of the equipment. The combination of decreasing capital recovery costs and increasing annual operating and maintenance cost results in the equivalent uniform annual cost taking on a form similar to that depicted in Figure 5.9.

By forecasting the operating and maintenance costs for each year of service, as well as the anticipated salvage values for various replacement ages, one can determine the "optimum replacement interval" for equipment.

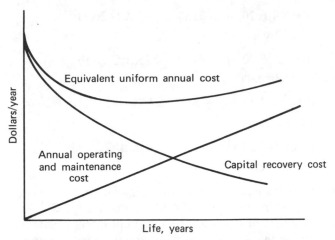

FIGURE 5.9. Portrayal of components of equivalent annual costs.

## Example 5.25

Suppose a small compressor can be purchased for $1000; the salvage value for the compressor is assumed to be negligible, regardless of the replacement interval. Annual operating and maintenance costs are expected to increase by $75/year, with the first year's cost anticipated to be $150. Using a minimum attractive rate of return of 20%, the following equivalent uniform annual costs are obtained:

$$EUAC(n = 1) = \$1000(A|P\ 20,1) + \$150 + \$75(A|G\ 20,1)$$
$$= \$1350.00/\text{year}$$

$$EUAC(n = 2) = \$1000(A|P\ 20,2) + \$150 + \$75(A|G\ 20,2)$$
$$= \$838.59/\text{year}$$

$$EUAC(n = 3) = \$1000(A|P\ 20,3) + \$150 + \$75(A|G\ 20,3)$$
$$= \$690.63/\text{year}$$

$$EUAC(n = 4) = \$1000(A|P\ 20,4) + \$150 + \$75(A|G\ 20,4)$$
$$= \$631.87/\text{year}$$

$$EUAC(n = 5) = \$1000(A|P\ 20,5) + \$150 + \$75(A|G\ 20,5)$$
$$= \$607.44/\text{year}$$

$$EUAC(n = 6) = \$1000(A|P\ 20,6) + \$150 + \$75(A|G\ 20,6)$$
$$= \$599.11/\text{year}$$

$$EUAC(n = 7) = \$1000(A|P\ 20,7) + \$150 + \$75(A|G\ 20,7)$$
$$= \$599.17/\text{year}$$

$$EUAC(n = 8) = \$1000(A|P\ 20,8) + \$150 + \$75(A|G\ 20,8)$$
$$= \$603.77/\text{year}$$

$$EUAC(n = 9) = \$1000(A|P\ 20,9) + \$150 + \$75(A|G\ 20,9)$$
$$= \$610.83/\text{year}$$

$$EUAC(n = 10) = \$1000(A|P\ 20,10) + \$150 + \$75(A|G\ 20,10)$$
$$= \$619.04/\text{year}$$

$$EUAC(n = 11) = \$1000(A|P\ 20,11) + \$150 + \$75(A|G\ 20,11)$$
$$= \$627.80/\text{year}$$

As $n$ increases beyond 6 years, the equivalent uniform annual cost increases. Hence, for this example, a replacement interval of 6 years is indicated.

Certain assumptions are inherent in optimum replacement interval calculations. First, we implicitly assume that the planning horizon is an integer multiple of the replacement interval selected. (Recall our discussion of the use of the annual worth based on one life cycle and a planning horizon of the least common multiple of lives.) Second, we have assumed that each time the compressor is replaced it will be replaced with a compressor having an *identical cash flow profile*. If neither assumption is valid, then the above approach is not valid.

---

### Example 5.26

Suppose in the previous case an 11-year planning horizon is appropriate. For simplicity, we will continue to assume that replacements will have identical cash flow profiles. If the original compressor is kept for 11 years, the present worth equivalent will be

$$PW(n = 11) = -\$627.80(P|A\ 20,11)$$
$$= -\$2716.55$$

If the compressor is to be replaced at some intermediate point during the planning horizon, say after $k$ years, and the replacement is to be kept until the end of the planning horizon, the following present worth calculations result for $k = 6$ or 7:

$$PW(k = 6) = -\$599.11(P|A\ 20,6) - \$607.44(P|A\ 20,5)(P|F\ 20,6)$$
$$= -\$2600.72$$

$$PW(k = 7) = -\$599.17(P|A\ 20,7) - \$631.87(P|A\ 20,4)(P|F\ 20,7)$$
$$= -\$2616.30$$

Thus, the compressor should be replaced after 6 years and the replacement should be kept until the end of the planning horizon. Why is it unnecessary to consider the remaining values of $k$ (i.e., $k = 5, 4, 3, 2,$ and 1)? In order to answer this, consider the case of $k = 5$ and compare the cash flow profiles over

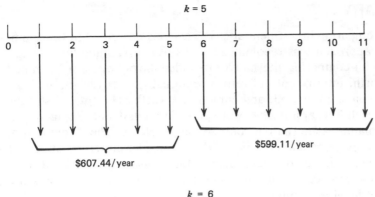

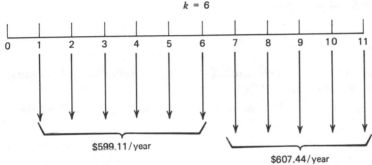

FIGURE 5.10. Comparison of replacement strategies involving replacement at the end of year 5 versus the end of year 6.

the 11-year planning horizon using $k = 5$ and $k = 6$. As shown in Figure 5.10, a consideration of the time value of money eliminates the possibility of $k = 5$ having a lower present worth than $k = 6$. Similar comparisons can be used to eliminate the cases of $k = 1, 2, 3,$ and 4. Likewise, $k = 8, 9,$ and 10 may be eliminated since costs have already begun to increase from $k = 6$ to $k = 7$.

For this particular example failure to replace the equipment at precisely the optimum time will not result in a significantly lower annual *worth* or present *worth*. This is often the case with many replacement situations; the resulting measure of effectiveness is relatively insensitive to deviations from the optimum strategy. Hence, the firm might establish an operating policy of reviewing the *actual* operating and maintenance costs for the equipment at the anticipated "optimum" time of replacement. Then, perform a replacement study at that time, using the six-step procedure for comparing investment alternatives.

---

The subject of replacement analysis is popular among engineering economists. In fact, sufficient literature is available on the subject to devote an entire book to it. Such books do exist, e.g. [31]. For further treatment of replacement analysis, you may wish to consult the *Engineering Economist* publication, Canada [5] Morris [20], and Terborgh [31], among others.

## 5.11 SUMMARY

The process of comparing investment alternatives was described in this chapter. A cash flow approach was recommended, as was a planning horizon approach. Methods for comparing alternatives included the present worth, annual worth, future worth, internal rate of return, external rate of return, savings/investment ratio, payback period, and capitalized worth. The total cash flow and incremental cash flow approaches were considered in detail. Replacement problems were treated as alternative comparison problems. The discussion of the comparison of investment alternatives did not consider either the effects of income taxes, the public sector requirements for measuring benefits in economic units, or the effects of risk and uncertainty; such considerations are reserved for the following chapters.

## BIBLIOGRAPHY

1. American Telephone and Telegraph Company, *Engineering Economy,* Third Edition, McGraw-Hill, 1977.
2. Baldwin, R. H., "How to Assess Investment Proposals," *Harvard Business Review, 37* (3), 1959, pp. 98–99.
3. Bernhard, R. H., "Discount Methods for Expenditure Evaluation—A Clarification of Their Assumptions," *The Journal of Industrial Engineering, 13* (1), 1962, pp. 19–27.
4. Bussey, L. E., *The Economic Analysis of Industrial Projects,* Prentice-Hall, 1978.
5. Canada, J. R., *Intermediate Economic Analysis for Management and Engineers,* Prentice-Hall, 1971.
6. DeGarmo, Paul E., Canada, John R., and Sullivan, William G., *Engineering Economy,* Sixth Edition, Macmillan, 1979.
7. *Economic Analysis Handbook,* Publication P-442, Department of the Navy, Naval Facilities Engineering Command, Washington, D.C., 1971.
8. Estes, C. B. and Jalali-Yazdi, A., "Now Replace Equipment on Actual Cost Basis," *Industrial Engineering, 10* (11), November 1978, pp. 34–37.
9. Fleischer, G. A., "Two Major Issues Associated with the Rate of Return Method for Capital Allocation: The 'Ranking Error' and 'Preliminary Selection,'" *The Journal of Industrial Engineering, 17* (4), 1966, pp. 202–208.
10. Grant, E. L., Ireson, W. G., and Leavenworth, R. S., *Principles of Engineering Economy,* Seventh Edition, Wiley, 1982.
11. Heebink, D. V., "Rate of Return, Reinvestment and the Evaluation of Capital Expenditures," *The Journal of Industrial Engineering, 13* (1), 1962, pp. 48–49.
12. Hirshleifer, J., "On the Theory of Optimal Investment Decision," *Journal of Political Economy, 66* (5), 1958, pp. 329–352.

13. Klausner, R. F., "Communicating Investment Proposals to Corporate Decision-Makers," *The Engineering Economist, 17* (1), 1971, pp. 45–55.

14. Kulonda, D. J., "Replacement Analysis with Unequal Lives—The Study Period Method," *The Engineering Economist, 23* (3), 1978, pp. 171–180.

15. Leautaud, J. L. L., "On the Fundamentals of Economic Evaluation," *The Engineering Economist, 19* (2), 1974, pp. 105–126.

16. Lutz, R. P., "How to Justify the Purchase of Material Handling Equipment to Management," *Proceedings of the 1976 MHI Material Handling Seminar,* The Material Handling Institute, Pittsburgh, Pa., 1976.

17. Mao, J. C. T., "The Internal Rate of Return as a Ranking Criterion," *The Engineering Economist, 11* (4), 1966, pp. 1–13.

18. Mao, J. C. T., "An Analysis of Criteria for Investment and Financing Decisions Under Certainty: A Comment," *Management Science, 13* (3), 1966, pp. 289–291.

19. Mao, J. C. T., *Quantitative Analysis of Financial Decisions,* Macmillan, 1969.

20. Morris, W. T., *Engineering Economic Analysis,* Reston Publishing Company, 1976.

21. Newnan, D. G., *Engineering Economic Analysis,* Engineering Press, 1976.

22. Oakford, R. V., *Capital Budgeting: A Quantitative Evaluation of Investment,* Ronald Press, 1970.

23. Renshaw, E., "A Note on the Arithmetic of Capital Budgeting Decisions," *The Journal of Business, 30* (3), 1957, pp. 193–204.

24. Riggs, J. L., *Engineering Economics,* Second Edition, McGraw-Hill, 1982.

25. Shore, B., "Replacement Decisions Under Capital Budgeting Constraints," *The Engineering Economist, 20* (4), 1975, pp. 243–256.

26. Solomon, E., "The Arithmetic of Capital Budgeting Decisions," *The Journal of Business, 29* (2), 1956, pp. 124–129.

27. Stevens, G. T., Jr., *Economic and Financial Analysis of Capital Investments,* Wiley, 1979.

28. Szonyi, A. J., Fenton, R. G., White, J. A., Agee, M. H., and Case, K. E., *Principles of Engineering Economic Analysis,* Canadian Edition, Wiley Canada Limited, 1982.

29. Tarquin, A. J., and Blank, L. T., *Engineering Economy,* McGraw-Hill, Second Edition, 1983.

30. Teichroew, D., Robicheck, A. A., and Montalbano, M., "An Analysis of Criteria for Investment and Financing Decisions Under Certainty," *Management Science, 12* (3), 1965, pp. 151–179.

31. Terborgh, G., *Business Investment Management,* Machinery and Allied Products Institute, 1958.

**32.** Thuesen, H. G., Fabrycky, W. J., and Thuesen, G. J., *Engineering Economy,* Fifth Edition, Prentice-Hall, 1977.

## PROBLEMS

**1.** Marlow Electronics has identified four investment proposals, A, B, C, and D. Proposals A and B are mutually exclusive and Proposals A and D are mutually exclusive. Proposal C is contingent on either Proposal A or B. To "do nothing" is not a viable alternative. Clearly specify all possible alternatives that should be considered. (5.2)

**2.** The ETG Corporation Capital Budget Committee has identified three major proposals, A, B, and C, to enhance customer service. Proposals A and C are mutually exclusive; Proposal C is contingent on Proposal B. The initial investments required for Proposals A, B, and C are $70,000, $80,000, and $35,000, respectively. A total of $120,000 is available for investment. Clearly specify all feasible alternatives to be considered. (5.2)

**3.** Greasy Petroleum Company has available four proposals, A, B, C, and D. Proposal B is contingent on the acceptance of either Proposal A or Proposal D. Also, Proposal D is contingent on Proposal A, while Proposal A is contingent on either Proposal B or Proposal C. The initial investments for these proposals are $420,000, $60,000, $45,000, and $300,000, respectively. If Greasy has a budget limitation of $600,000, clearly identify all feasible alternatives to be considered. (5.2)

**4.** Consider the net cash flows (NCF) and salvage values (SV) for each of Alternatives 1, 2, and 3 having lives of 3, 8, and 4 years, respectively.

| EOY | Alternative 1 | | Alternative 2 | | Alternative 3 | |
|---|---|---|---|---|---|---|
| | $NCF_1$ | $SV_1$ | $NCF_2$ | $SV_2$ | $NCF_3$ | $SV_3$ |
| 0 | −$60,000 | $60,000 | −$100,000 | $100,000 | −$140,000 | $140,000 |
| 1 | 35,000 | 30,000 | 30,000 | 60,000 | 65,000 | 80,000 |
| 2 | 35,000 | 10,000 | 30,000 | 40,000 | 65,000 | 40,000 |
| 3 | 35,000 | 0 | 30,000 | 25,000 | 65,000 | 20,000 |
| 4 | | | 30,000 | 15,000 | 65,000 | 10,000 |
| 5 | | | 30,000 | 10,000 | | |
| 6 | | | 30,000 | 5,000 | | |
| 7 | | | 30,000 | 5,000 | | |
| 8 | | | 30,000 | 5,000 | | |

Assume each alternative can be renewed indefinitely with the same NCF and SV profiles.

**(a)** If a least common multiple of lives approach is to be used, specify the planning horizon and the complete set of cash flows for each alternative. (5.3)

**(b)** Repeat part (a) using the shortest life among alternatives. (5.3)

**(c)** Repeat part (a) using the longest life among alternatives. (5.3)

**(d)** Repeat part (a) using a planning horizon of 2 years. (5.3)

**(e)** Repeat part (a) using a planning horizon of 6 years. (5.3)

**(f)** Repeat part (a) using a (tentative) planning horizon of 30 years. (5.3)

**5.** Alternatives 1, 2, and 3 have lives of 3, 4, and 6 years, respectively. Their net cash flow (NCF) and salvage value (SV) profiles are as follows:

| EOY | Alternative 1 | | Alternative 2 | | Alternative 3 | |
|-----|---------------|---------------|---------------|---------------|---------------|---------------|
| | $NCF_1$ | $SV_1$ | $NCF_2$ | $SV_2$ | $NCF_3$ | $SV_3$ |
| 0 | -$20,000 | — | -$40,000 | $40,000 | -$70,000 | $70,000 |
| 1 | 8,000 | — | 20,000 | 30,000 | 30,000 | 50,000 |
| 2 | 8,000 | — | 20,000 | 20,000 | 30,000 | 30,000 |
| 3 | 28,000 | — | 20,000 | 10,000 | 30,000 | 20,000 |
| 4 | | | 20,000 | 0 | 30,000 | 10,000 |
| 5 | | | | | 30,000 | 5,000 |
| 6 | | | | | 30,000 | 2,000 |

The NCF profile for Alternative 1 is due to a $20,000/year lease payable at the beginning of each year, plus an end-of-year net revenue of $28,000. This lease arrangement may be renewed indefinitely; however, premature cancellation of the lease costs $10,000. All other cash flows and salvage values are expected to repeat indefinitely as shown.

**(a)** If a least common multiple of lives approach is to be used, specify the planning horizon and the complete set of cash flows for each alternative. (5.3)

**(b)** Repeat part (a) using the shortest life among alternatives. (5.3)

**(c)** Repeat part (a) using the longest life among alternatives. (5.3)

**(d)** Repeat part (a) using a planning horizon of 2 years. (5.3)

**(e)** Repeat part (a) using a planning horizon of 5 years. (5.3)

**(f)** Repeat part (a) using a planning horizon of 9 years. (5.3)

**(g)** Repeat part (a) using a (tentative) planning horizon of 18 years. (5.3)

**6.** Two alternatives have the following net cash flow (NCF) and salvage value (SV) profiles *for the first cycle of each.*

| EOY | Alternative 1 | | Alternative 2 | |
|-----|---------------|---------------|---------------|---------------|
| | $NCF_1$ | $SV_1$ | $NCF_2$ | $SV_2$ |
| 0 | -$50,000 | $50,000 | -$80,000 | $80,000 |
| 1 | 25,000 | 25,000 | 35,000 | 50,000 |
| 2 | 30,000 | 10,000 | 45,000 | 20,000 |
| 3 | 35,000 | 0 | 50,000 | 10,000 |
| 4 | | | 55,000 | 0 |
| 5 | | | 60,000 | 0 |

Alternatives 1 and 2 can both repeat; however, the cash flow profiles and salvage values will change. Renewal of Alternative 1 will cost 40% more for the initial investment, increasing to $70,000. Alternative 1 salvage values will decrease in the same proportion to the initial investment as shown above (e.g., $70,000, $35,000,

$14,000, $0). Further renewals of Alternative 1 will cause initial investments and salvage values to increase 40% over the previous cycle's values, and salvage values will again decrease in the same proportions as illustrated in the table. Annual net revenues are projected to continue increasing at $5000/year (e.g., $40,000/year, $45,000/year, etc.) indefinitely. Renewal of Alternative 2 will cost 60% more for the initial investment with each renewal, and salvage values will decrease in proportion to those shown. Also, annual net revenues will continue increasing at $5000/year.

(a) Specify the complete set of cash flows for each alternative if a planning horizon of 2 years is to be used. (5.3)

(b) Repeat part (a) using a planning horizon of 3 years. (5.3)

(c) Repeat part (a) using a planning horizon of 4 years. (5.3)

(d) Repeat part (a) using a planning horizon of 5 years. (5.3)

(e) Repeat part (a) using a planning horizon of 10 years. (5.3)

7. Three investment proposals have been selected for further evaluation. Proposals A and C are mutually exclusive; Proposal A is contingent on Proposal B; and Proposal B is contingent on Proposal C. A budget limitation of $200,000 exists. The cash flow profiles for the three investment proposals are given below for the 4-year planning horizon.

| EOY | NCF(A) | NCF(B) | NCF(C) |
|-----|--------|--------|--------|
| 0 | −$60,000 | −$40,000 | −$80,000 |
| 1–4 | 20,000 | 10,000 | 15,000 |
| 4 | — | 15,000 | 20,000 |

(a) Specify the mutually exclusive alternatives to be considered. (5.2)

(b) Clearly show the net cash flow profile associated with each alternative. (5.4)

8. Four investment proposals, A, B, C, and D, are available. Proposals A and B are mutually exclusive and Proposals A and C are mutually exclusive. Proposal D is contingent on either Proposal A or B. Funds available for investment are limited to $200,000.

| EOY | NCF(A) | NCF(B) | NCF(C) | NCF(D) |
|-----|--------|--------|--------|--------|
| 0 | −$80,000 | −$125,000 | −$100,000 | −$80,000 |
| 1–5 | 30,000 | 40,000 | 35,000 | 32,000 |

(a) Specify the mutually exclusive alternatives to be considered. (5.2)

(b) Clearly show the net cash flow profile associated with each alternative. (5.4)

9. A firm has available four proposals, A, B, C, and D. Proposal A is contingent on acceptance of either Proposal C or Proposal D. In addition, Proposal C is contingent on Proposal D, while Proposal D is contingent on either Proposal A or Proposal B. The firm has a budget limitation of $500,000. The cash flows are as follows:

| EOY | NCF(A) | NCF(B) | NCF(C) | NCF(D) |
|---|---|---|---|---|
| 0 | −$250,000 | −$350,000 | −$50,000 | −$38,000 |
| 1 | 85,000 | 119,000 | 10,000 | 3,000 |
| 2 | 85,000 | 122,000 | 10,000 | 8,000 |
| 3 | 85,000 | 125,000 | 10,000 | 13,000 |
| 4 | 85,000 | 128,000 | 10,000 | 18,000 |
| 5 | 150,000 | 170,000 | 10,000 | 23,000 |

(a) Specify the mutually exclusive alternatives to be considered. (5.2)

(b) Clearly show the net cash flow profile associated with each alternative. (5.4)

10. Metal Salvage, Inc. has available three investment proposals A, B, and C, having the cash flow profiles shown below. Proposals B and C are mutually exclusive and Proposal C is contingent on Proposal A being chosen.

|  | NCF(A) | NCF(B) | NCF(C) |
|---|---|---|---|
| Initial investment | $400,000 | $600,000 | $300,000 |
| Life | 8 years | 12 years | 6 years |
| Annual receipts | $320,000 | $380,000 | $400,000 |
| Annual disbursements | $230,000 | $240,000 | $300,000 |
| Salvage value | $100,000 | $200,000 | $100,000 |

The firm is willing to use a planning horizon of 24 years and accepts the assumption of identical cash flow profiles for successive life cycles of a proposal.

(a) Determine the set of feasible alternatives. (5.2)

(b) Determine the net cash flow profile for each alternative. (5.4)

11. Four investment proposals, W, X, Y, and Z, are being considered by the Ajax Corporation. Proposals X and Z are mutually exclusive. Proposal Y is contingent on either X or Z. Proposals W and Y are mutually exclusive. A budget limitation of $400,000 exists. Either Proposal X or Proposal Z must be included in the alternative selected.

| EOY | NCF(W) | NCF(X) | NCF(Y) | NCF(Z) |
|---|---|---|---|---|
| 0 | −$200,000 | −$250,000 | −$180,000 | −$200,000 |
| 1–8 | 60,000 | 100,000 | 50,000 | 40,000 |
| 8 | 80,000 | 20,000 | 240,000 | 200,000 |

(a) Determine the set of feasible mutually exclusive alternatives. (5.2)

(b) Determine the net cash flow for each alternative. (5.4)

12. Charles Machine Works is faced with three investment proposals, A, B, and C, having the cash flow profiles shown below over the planning horizon of 5 years.

Proposals B and C are mutually exclusive, and Proposal C is contingent on Proposal A being selected. A budgetary limitation of $125,000 exists on the amount that can be invested initially.

| EOY | NCF(A) | NCF(B) | NCF(C) |
|---|---|---|---|
| 0 | −$50,000 | −$25,000 | −$10,000 |
| 1 | 10,000 | 10,000 | 3,000 |
| 2 | 15,000 | 10,000 | 3,000 |
| 3 | 20,000 | 10,000 | 3,000 |
| 4 | 25,000 | 10,000 | 3,000 |
| 5 | 30,000 | 10,000 | 7,500 |

Specify the cash flow profiles for all mutually exclusive, feasible investment alternatives. (5.4)

13. Two mutually exclusive alternatives are to be evaluated and the economically better alternative specified. The cash flow profiles for the alternatives are shown below.

| EOY | NCF(1) | NCF(2) |
|---|---|---|
| 0 | −$10,000 | −$15,000 |
| 1 | 1,000 | 1,500 |
| 2 | 1,500 | 2,000 |
| 3 | — | 500 |
| 4 | 2,500 | 3,000 |
| 5 | 3,500 | 4,000 |
| 6 | — | 500 |
| 7 | 4,500 | 5,000 |
| 8 | 6,000 | 11,500 |

(a) Assuming $15,000 is available to invest, determine the total cash flow associated with each alternative over each year of the planning horizon. Any excess capital is invested at $MARR = 15\%$, earning equal annual returns over the planning horizon of 8 years. (5.5.1)

(b) Determine the incremental cash flow profile between Alternatives 2 and 1 for each year of the planning horizon. (5.5.2)

14. Consider the following investment decision.

| | Machine 1 | Machine 2 |
|---|---|---|
| First cost | $15,000 | $20,000 |
| Estimated life | 5 years | 10 years |
| Estimated annual revenues | $12,000 | $14,000 |
| Estimated annual operating costs | $ 6,000 | $ 8,000 |
| Estimated salvage value at end of 5 years | $ 1,500 | $ 5,000 |

(a) Assuming $20,000 is available to invest, determine the total cash flow associated with each machine over each year of the planning horizon. Any excess capital is invested at *MARR* = 20% earning equal annual returns over the planning horizon of 5 years. (5.5.1)

(b) Determine the incremental cash flow profile between Machine 2 and Machine 1 for each year of the planning horizon of 5 years. (5.5.2)

15. Two mutually exclusive proposals, each with a life of 5 years, are under consideration. *MARR* is 15%. Each proposal has the following cash flow profile:

| EOY | NCF(A) | NCF(B) |
|-----|--------|--------|
| 0 | -$30,000 | -$42,000 |
| 1 | 9,300 | 12,625 |
| 2 | 9,300 | 12,625 |
| 3 | 9,300 | 12,625 |
| 4 | 9,300 | 12,625 |
| 5 | 9,300 | 12,625 |

(a) Clearly specify the mutually exclusive alternatives available and their net cash flow profiles. (5.4)

(b) Determine which alternative the decision maker should select. Use a *total cash flow approach* and the present worth method. (5.5.3)

(c) Determine which alternative the decision maker should select. Use an *incremental cash flow approach* and the present worth method. (5.5.3)

(d) Repeat part (b) using the annual worth method. (5.5.4)

(e) Repeat part (c) using the annual worth method. (5.5.4)

(f) Repeat part (b) using the future worth method. (5.5.5)

(g) Repeat part (c) using the future worth method. (5.5.5)

(h) Repeat part (b) using the internal rate of return method. (5.5.6)

(i) Repeat part (c) using the internal rate of return method. (5.5.6)

(j) Repeat part (b) using the external rate of return method. (5.5.7)

(k) Repeat part (c) using the external rate of return method. (5.5.7)

(l) Repeat part (b) using the savings/investment ratio method. (5.5.8)

(m) Repeat part (c) using the savings/ investment ratio method. (5.5.8)

16. The Belt Company is considering three investment proposals, A, B, and C. Proposals A and B are mutually exclusive, and Proposal C is contingent on Proposal B. The cash flow data for the investments over a 10-year planning horizon are given below. The Belt Company has a budget limit of $1,000,000 for investments of the type being considered currently. *MARR* = 25%.

| | NCF(A) | NCF(B) | NCF(C) |
|---|--------|--------|--------|
| Initial investment | $600,000 | $800,000 | $470,000 |
| Life | 10 years | 10 years | 10 years |
| Salvage values | $ 70,000 | $130,000 | $ 65,000 |
| Annual receipts | $400,000 | $600,000 | $260,000 |
| Annual disbursements | $130,000 | $270,000 | $ 70,000 |

**(a)** Clearly specify the mutually exclusive alternatives available and their net cash flow profiles. (5.4)
**(b)** Determine which alternative Belt's decision maker should select. Use a *total cash flow approach* and the present worth method. (5.5.3)
**(c)** Determine which alternative Belt's decision maker should select. Use an *incremental cash flow approach* and the present worth method. (5.5.3)
**(d)** Repeat part (b) using the annual worth method. (5.5.4)
**(e)** Repeat part (c) using the annual worth method. (5.5.4)
**(f)** Repeat part (b) using the future worth method. (5.5.5)
**(g)** Repeat part (c) using the future worth method. (5.5.5)
**(h)** Repeat part (b) using the internal rate of return method. (5.5.6)
**(i)** Repeat part (c) using the internal rate of return method. (5.5.6)
**(j)** Repeat part (b) using the external rate of return method. (5.5.7)
**(k)** Repeat part (c) using the external rate of return method. (5.5.7)
**(l)** Repeat part (b) using the savings/investment ratio method. (5.5.8)
**(m)** Repeat part (c) using the savings/investment ratio method. (5.5.8)

**17.** A firm has available two mutually exclusive investment proposals, A and B. Their net cash flows are as shown in the table over a 10-year planning horizon. *MARR* is 20%.

| EOY | NCF(A) | NCF(B) |
|-----|--------|--------|
| 0 | -$40,000 | -$30,000 |
| 1 | 8,000 | 9,000 |
| 2 | 8,000 | 8,500 |
| 3 | 8,000 | 8,000 |
| 4 | 8,000 | 7,500 |
| 5 | 8,000 | 7,000 |
| 6 | 8,000 | 6,500 |
| 7 | 8,000 | 6,000 |
| 8 | 8,000 | 5,500 |
| 9 | 8,000 | 5,000 |
| 10 | 8,000 | 4,500 |

**(a)** Clearly specify the mutually exclusive alternatives available and their net cash flow profiles. (5.4)
**(b)** Determine which alternative the decision maker should select. Use a *total cash flow approach* and the present worth method. (5.5.3)
**(c)** Determine which alternative the decision maker should select. Use an *incremental cash flow approach* and the present worth method. (5.5.3)
**(d)** Repeat part (b) using the annual worth method. (5.5.4)
**(e)** Repeat part (c) using the annual worth method. (5.5.4)
**(f)** Repeat part (b) using the future worth method. (5.5.5)
**(g)** Repeat part (c) using the future worth method. (5.5.5)
**(h)** Repeat part (b) using the internal rate of return method. (5.5.6)
**(i)** Repeat part (c) using the internal rate of return method. (5.5.6)
**(j)** Repeat part (b) using the external rate of return method. (5.5.7)

(k) Repeat part (c) using the external rate of return method. (5.5.7)
(l) Repeat part (b) using the savings/investment ratio method. (5.5.8)
(m) Repeat part (c) using the savings/investment ratio method. (5.5.8)

18. A firm is faced with four investment proposals, A, B, C, and D, having the cash flow profiles shown below. Proposals A and C are mutually exclusive, and Proposal D is contingent on Proposal B being chosen. Currently, $750,000 is available for investment, and the firm has stipulated a *MARR* of 20%.

|  | NCF(A) | NCF(B) | NCF(C) | NCF(D) |
|---|---|---|---|---|
| Initial investment | $400,000 | $400,000 | $600,000 | $300,000 |
| Planning horizon | 10 years | 10 years | 10 years | 10 years |
| Annual receipts | $205,000 | $215,000 | $260,000 | $230,000 |
| Annual disbursements | $110,000 | $125,000 | $120,000 | $150,000 |
| Salvage value | $ 50,000 | $ 50,000 | $100,000 | $ 50,000 |

(a) Clearly specify the mutually exclusive alternatives available and their net cash flow profiles. (5.4)
(b) Determine which alternative the decision maker should select. Use a *total cash flow approach* and the present worth method. (5.5.3)
(c) Determine which alternative the decision maker should select. Use an *incremental cash flow approach* and the present worth method. (5.5.3)
(d) Repeat part (b) using the annual worth method. (5.5.4)
(e) Repeat part (c) using the annual worth method. (5.5.4)
(f) Repeat part (b) using the future worth method. (5.5.5)
(g) Repeat part (c) using the future worth method. (5.5.5)
(h) Repeat part (b) using the internal rate of return method. (5.5.6)
(i) Repeat part (c) using the internal rate of return method. (5.5.6)
(j) Repeat part (b) using the external rate of return method. (5.5.7)
(k) Repeat part (c) using the external rate of return method. (5.5.7)
(l) Repeat part (b) using the savings/investment ratio method. (5.5.8)
(m) Repeat part (c) using the savings/investment ratio method. (5.5.8)

19. Rich N. Smug is considering two mutually exclusive investment alternatives, each requiring an investment of $5000. Alternative 1 returns $1000 after 1 year and $6000 after 2 years. This alternative is not renewable. Alternative 2 returns $1000/year for the first 3 years and $5000 after 4 years. Mr. Smug has established a *MARR* of 10%. He is committed to make one of the two investments.
(a) Clearly specify the mutually exclusive alternatives available and their net cash flow profiles. (5.4)
(b) Determine which alternative Mr. Smug should select. Use a *total cash flow approach* and the present worth method. (5.5.3)
(c) Determine which alternative Mr. Smug should select. Use an *incremental cash flow approach* and the present worth method. (5.5.3)
(d) Repeat part (b) using the annual worth method. (5.5.4)
(e) Repeat part (c) using the annual worth method. (5.5.4)
(f) Repeat part (b) using the future worth method. (5.5.5)

**(g)** Repeat part (c) using the future worth method. (5.5.5)
**(h)** Repeat part (b) using the internal rate of return method. (5.5.6)
**(i)** Repeat part (c) using the internal rate of return method. (5.5.6)
**(j)** Repeat part (b) using the external rate of return method. (5.5.7)
**(k)** Repeat part (c) using the external rate of return method. (5.5.7)
**(l)** Repeat part (b) using the savings/investment ratio method. (5.5.8)
**(m)** Repeat part (c) using the savings/investment ratio method. (5.5.8)

**20.** Insulation is being considered for installation on a steam pipe. Either 1-inch or 2-inch insulation is being figured. The annual heat loss from the pipe at present is estimated to be $2.50 per foot of pipe. The 1-inch insulation costs $1/foot and will reduce the heat loss by 65%; the 2-inch insulation costs $1.90/foot and will reduce the heat loss by 79%. The insulation will last 5 years with no salvage value. *MARR* is 25%.

**(a)** Clearly specify the mutually exclusive alternatives available and their net cash flow profiles. (5.4)
**(b)** Determine which kind of insulation the decision maker should select. Use a *total cash flow approach* and the present worth method. (5.5.3)
**(c)** Determine which kind of insulation the decision maker should select. Use an *incremental cash flow approach* and the present worth method. (5.5.3)
**(d)** Repeat part (b) using the annual worth method. (5.5.4)
**(e)** Repeat part (c) using the annual worth method. (5.5.4)
**(f)** Repeat part (b) using the future worth method. (5.5.5)
**(g)** Repeat part (c) using the future worth method. (5.5.5)
**(h)** Repeat part (b) using the internal rate of return method. (5.5.6)
**(i)** Repeat part (c) using the internal rate of return method. (5.5.6)
**(j)** Repeat part (b) using the external rate of return method. (5.5.7)
**(k)** Repeat part (c) using the external rate of return method. (5.5.7)
**(l)** Repeat part (b) using the savings/investment ratio method. (5.5.8)
**(m)** Repeat part (c) using the savings/investment ratio method. (5.5.8)

**21.** Three mutually exclusive investment alternatives have the cash flow profiles shown below. A *MARR* of 30% is used in the selection of a preferred alternative.

| EOY | NCF(1) | NCF(2) | NCF(3) |
|-----|--------|--------|--------|
| 0 | −$40,000 | −$25,000 | $0 |
| 1 | 16,500 | 5,000 | 0 |
| 2 | 16,500 | 10,000 | 0 |
| 3 | 16,500 | 15,000 | 0 |
| 4 | 16,500 | 20,000 | 0 |
| 5 | 16,500 | 25,000 | 0 |

**(a)** Clearly specify the mutually exclusive alternatives available and their net cash flow profiles. (5.4)
**(b)** Determine which alternative the decision maker should select. Use a *total cash flow approach* and the present worth method. (5.5.3)

(c) Determine which alternative the decision maker should select. Use an *incremental cash flow approach* and the present worth (5.5.3)

(d) Repeat part (b) using the annual worth method. (5.5.4)

(e) Repeat part (c) using the annual worth method. (5.5.4)

(f) Repeat part (b) using the future worth method. (5.5.5)

(g) Repeat part (c) using the future worth method. (5.5.5)

(h) Repeat part (b) using the internal rate of return method. (5.5.6)

(i) Repeat part (c) using the internal rate of return method. (5.5.6)

(j) Repeat part (b) using the external rate of return method. (5.5.7)

(k) Repeat part (c) using the external rate of return method. (5.5.7)

(l) Repeat part (b) using the savings/investment ratio method. (5.5.8)

(m) Repeat part (c) using the savings/investment ratio method. (5.5.8)

22. In Problem 19, Rich N. Smug has two alternatives to consider. What are the payback periods for each alternative? On the basis of payback period, which alternative does Mr. Smug prefer? (5.5.9)

23. Using the payback period, evaluate Alternatives 1 and 2 in Problem 21. (5.5.9)

24. Consider the net cash flows of the following three alternatives. None is renewable. *MARR* is 20%.

| EOY | NCF(1) | NCF(2) | NCF(3) |
|-----|--------|--------|--------|
| 0 | −$3,000 | −$4,000 | −$6,000 |
| 1 | 1,000 | 4,000 | 3,000 |
| 2 | 1,500 | 2,000 | 3,000 |
| 3 | 0 | 0 | 3,000 |
| 4 | 4,000 | 0 | 0 |
| 5 | 6,000 | 0 | 0 |

(a) Compare these alternatives on the basis of the present worth method. (5.5.3)

(b) Compare these alternatives on the basis of the payback period method. (5.5.9)

25. Two alternatives for a recreational facility are being considered. Their cash flow profiles are as follows for lives of 5 and 3 years, respectively.

| EOY | NCF(1) | NCF(2) |
|-----|--------|--------|
| 0 | −$11,000 | −$5,000 |
| 1 | 5,000 | 2,000 |
| 2 | 4,000 | 3,000 |
| 3 | 3,000 | 4,000 |
| 4 | 2,000 | |
| 5 | 1,000 | |

Each alternative is expected to repeat forever with roughly the same net cash flow profile. Using the capitalized worth method and a *MARR* of 10%, select the preferred alternative. (5.5.10)

26. Three bridge designs are under consideration for a Payne County road. Wood, steel, and concrete underlie the three designs. The wood design costs $25,000 installed, with $2000/year maintenance and $15,000 rework at the end of every 10 years. Steel costs $35,000 installed, with $1500/year maintenance. Concrete costs $40,000 installed, but costs only $1000/year in maintenance. If *MARR* is 8% which alternative should be selected on the basis of capitalized cost, assuming cost profiles remain constant indefinitely? (5.5.10)

27. A consulting engineer is designing an irrigation system and is considering two pumps to meet a pumping demand of 15,000 gallons/minute at 15 feet total dynamic head. The specific gravity of the fertilizer mixture being pumped is 1.25. Pump A operates at 78% efficiency and costs $11,500; Pump B operates at 86% efficiency and costs $16,250. Energy costs $0.060/kilowatt-hour. Continuous pumping for 200 days/year is required (i.e., 24 hours/day). *MARR* is 15%, the planning horizon is 5 years, and equal salvage values are estimated for the two pumps. (*Note:* Dynamic head × gallons/minute × specific gravity × 0.746 ÷ 3960 = kilowatts.) (5.8)
    (a) Clearly specify the net cash flow profiles for each mutually exclusive alternative. (5.8)
    (b) Determine which pump should be selected. Use a *total cash flow approach* and the present worth method. (5.8)
    (c) Repeat part (b) using the annual worth method. (5.8)
    (d) Repeat part (b) using the future worth method. (5.8)
    (e) Determine which pump should be selected. Use an *incremental cash flow approach* and the internal rate of return method. (5.8)
    (f) Repeat part (e) using the external rate of return method. (5.8)
    (g) Repeat part (e) using the savings/investment ratio method. (5.8)

28. The motor on a numerically controlled machine tool must be replaced. Two different 20-horsepower (output) electric motors are being considered. Motor U sells for $1360 and has an efficiency rating of 90%; Motor V sells for $300 and has a rating of 84%. The cost of energy is $0.055/kilowatt-hour. An 8-year planning horizon is used, and zero salvage values are assumed for both motors. Annual usage of the motor averages approximately 4800 hours. A *MARR* of 20% is to be used. Assume the motor selected will be loaded to capacity. (*Note:* 0.746 kilowatt = 1 horsepower.)
    (a) Clearly specify the net cash flow profiles for each mutually exclusive alternative. (5.8)
    (b) Determine which motor should be selected. Use a *total cash flow approach* and the present worth method. (5.8)
    (c) Repeat part (b) using the annual worth method. (5.8)
    (d) Repeat part (b) using the future worth method. (5.8)
    (e) Determine which motor should be selected. Use an *incremental cash flow approach* and the internal rate of return method. (5.8)
    (f) Repeat part (e) using the external rate of return method. (5.8)
    (g) Repeat part (e) using the savings/investment ratio method. (5.8)

29. A chemical plant is considering the installation of a storage tank for water. The tank is estimated to have an initial cost of $213,000, annual costs for maintenance are estimated to be $3200/year. As an alternative, a holding pond can be provided some distance away at an initial cost of $90,000 for the pond, plus $45,000 for pumps and

piping; annual operating and maintenance costs for the pumps and holding pond are estimated to be $8500. The planning horizon is 20 years with neither alternative having any salvage value. *MARR* is 15%.

(a) Clearly specify the net cash flow profiles for each mutually exclusive alternative. (5.8)

(b) Determine which alternative should be selected. Use a *total cash flow approach* and the present worth method. (5.8)

(c) Repeat part (b) using the annual worth method. (5.8)

(d) Repeat part (b) using the future worth method. (5.8)

(e) Determine which alternative should be selected. Use an *incremental cash flow approach* and the internal rate of return method. (5.8)

(f) Repeat part (e) using the external rate of return method. (5.8)

(g) Repeat part (e) using the savings/investment ratio method. (5.8)

30. The state civil defense organization wishes to establish a communications network to cover the state. It is desired to maintain a specified minimum signal strength at all points in the state. Two alternatives have been selected for detailed consideration. Design I involves the installation of five transmitting stations of low power. The investment at each installation is estimated to be $40,000 in structure and $25,000 in equipment. Design II involves the installation of two transmitting stations of much higher power. Investment in structure will be approximately $130,000/installation; equipment at each installation will cost $180,000. Annual operating and maintenance costs per installation are anticipated to be $25,000 for Design I and $55,000 for Design II. Structures are anticipated to last for 20 years; equipment life is estimated to be 10 years, with replacements assumed to be identical in costs and performance. The planning horizon is 20 years, salvage values are nil at $t = 20$, and *MARR* is 12%.

(a) Clearly specify the net cash flow profiles for each mutually exclusive alternative. (5.8)

(b) Determine which design should be selected. Use a *total cash flow approach* and the present worth method. (5.8)

(c) Repeat part (b) using the annual worth method. (5.8)

(d) Repeat part (b) using the future worth method. (5.8)

(e) Determine which design should be selected. Use an *incremental cash flow approach* and the internal rate of return method. (5.8)

(f) Repeat part (e) using the external rate of return method. (5.8)

(g) Repeat part (e) using the savings/investment ratio method. (5.8)

31. Two compressors are being considered by the Ajax Company. Compressor A can be purchased for $9700; annual operating and maintenance costs are estimated to be $3500. Alternatively, Compressor B can be purchased for $7500; annual operating and maintenance costs are estimated to be $4200. An 8-year planning horizon is to be used; salvage values are estimated to be 15% of the original purchase price; a *MARR* of 25% is to be used.

(a) Clearly specify the net cash flow profiles for each mutually exclusive alternative. (5.8)

(b) Determine which compressor should be selected. Use a *total cash flow approach* and the present worth method. (5.8)

(c) Repeat part (b) using the annual worth method. (5.8)

(d) Repeat part (b) using the future worth method. (5.8)

**(e)** Determine which compressor should be selected. Use an *incremental cash flow approach* and the internal rate of return method. (5.8)

**(f)** Repeat part (e) using the external rate of return method. (5.8)

**(g)** Repeat part (e) using the savings/investment ratio method. (5.8)

**32.** The Ajax Manufacturing Company wishes to choose one of the following machines:

|  | Machine 1 | Machine 2 |
|---|---|---|
| First cost | $10,000 | $12,000 |
| Planning horizon | 6 years | 6 years |
| Salvage value | $ 1,000 | $ 1,000 |
| Operating and maintenance cost for year $k$ | $800 + 80k$ | $200 + 100k$ |

*MARR* is 12%.

**(a)** Clearly specify the net cash flow profiles for each mutually exclusive alternative. (5.8)

**(b)** Determine which machine should be selected. Use a *total cash flow approach* and the present worth method. (5.8)

**(c)** Repeat part (b) using the annual worth method. (5.8)

**(d)** Repeat part (b) using the future worth method. (5.8)

**(e)** Determine which machine should be selected. Use an *incremental cash flow approach* and the internal rate of return method. (5.8)

**(f)** Repeat part (e) using the external rate of return method. (5.8)

**(g)** Repeat part (e) using the savings/investment ratio method. (5.8)

**33.** The manager of the distribution center for the southeastern territory of a major pharmaceutical manufacturer is contemplating installing an improved material handling system linking receiving and storage, as well as storage and shipping. Two designs are being considered. The first consists of a conveyor system that is tied into an automated storage/retrieval system. Such a system is estimated to cost $1,125,000 initially, have annual operating and maintenance costs of $72,000, and a salvage value of $75,000 at the end of the 10-year planning horizon.

The second design consists of manually operated narrow aisle, high stacking lift trucks. To provide service comparable to that provided by the alternative design, an initial investment of $490,000 is required. Annual operating and maintenance costs of $160,000 are anticipated. An estimated salvage value of $30,000 is expected at the end of the planning horizon. *MARR* is 20%.

**(a)** Clearly specify the net cash flow profiles for each mutually exclusive alternative. (5.8)

**(b)** Determine which alternative should be selected. Use a *total cash flow approach* and the present worth method. (5.8)

**(c)** Repeat part (b) using the annual worth method. (5.8)

**(d)** Repeat part (b) using the future worth method. (5.8)

**(e)** Determine which alternative should be selected. Use an *incremental cash flow approach* and the internal rate of return method. (5.8)

**(f)** Repeat part (e) using the external rate of return method. (5.8)

**(g)** Repeat part (e) using the savings/investment ratio method. (5.8)

**34.** A manufacturing plant in Minnesota has been contracting snow removal at a cost of $400/day. The past 4 years have yielded unusually heavy snowfalls, resulting in the cost of snow removal being of concern to the plant manager. The plant engineer has found that a snow-removal machine can be purchased for $53,000; it is estimated to have a useful life of 10 years, and a zero salvage value at that time. Annual costs for operating and maintaining the equipment are estimated to be $15,000. *MARR* is 20% and the estimated demand for the equipment is equal to 80 days/year.

(a) Clearly specify the net cash flow profiles for each mutually exclusive alternative. (5.8)

(b) Determine which alternative should be selected. Use a *total cash flow approach* and the present worth method. (5.8)

(c) Repeat part (b) using the annual worth method. (5.8)

(d) Repeat part (b) using the future worth method. (5.8)

(e) Determine which alternative should be selected. Use an *incremental cash flow approach* and the internal rate of return method. (5.8)

(f) Repeat part (e) using the external rate of return method. (5.8)

(g) Repeat part (e) using the savings/investment ratio method. (5.8)

**35.** Two submergible pumps are under consideration. Their cost data follow:

|  | Pump 1 | Pump 2 |
|---|---|---|
| First cost | $9,000 | $14,000 |
| Salvage value | 0 | $1,000 |
| Annual operating cost | $1,600 | $1,100 |
| Life | 5 years | 10 years |

Either pump is needed over a 10-year planning horizon. The company uses a *MARR* of 15%.

(a) Clearly specify the net cash flow profiles for each mutually exclusive alternative. (5.8)

(b) Determine which pump should be selected. Use a *total cash flow approach* and the present worth method. (5.8)

(c) Repeat part (b) using the annual worth method. (5.8)

(d) Repeat part (b) using the future worth method. (5.8)

(e) Determine which pump should be selected. Use an *incremental cash flow approach* and the internal rate of return method. (5.8)

(f) Repeat part (e) using the external rate of return method. (5.8)

(g) Repeat part (e) using the savings/investment ratio method. (5.8)

**36.** A firm is considering two compressors. One must be chosen; the relevant data follow:

|  | Compressor 1 | Compressor 2 |
|---|---|---|
| Initial investment | $15,000 | $30,000 |
| Life | 6 years | 7 years |
| Salvage value | $ 1,000 | $ 7,500 |
| Annual disbursements | $ 8,000 | $ 4,500 |

*MARR* is 20% and the cash flow profiles are believed to be repeatable over a planning horizon of 42 years.

(a) Clearly specify the net cash flow profiles for each mutually exclusive alternative. (5.8)

(b) Determine which compressor should be selected. Use a *total cash flow approach* and the present worth method. (5.8)

(c) Repeat part (b) using the annual worth method. (5.8)

(d) Repeat part (b) using the future worth method. (5.8)

(e) Determine which compressor should be selected. Use an *incremental cash flow approach* and the internal rate of return method. (5.8)

(f) Repeat part (e) using the external rate of return method. (5.8)

(g) Repeat part (e) using the savings/investment ratio method. (5.8)

37. Two numerically controlled drill presses are being considered by the production department of a major corporation; one must be selected. Both machines meet the quality and safety standards of the firm. Comparative data are as follows. *MARR* is 15%.

|  | Drill Press X | Drill Press Y |
|---|---|---|
| Initial investment | $26,000 | $39,000 |
| Estimated life | 6 years | 6 years |
| Salvage value | $ 4,000 | $ 7,000 |
| Annual operating cost | $12,000 | $ 7,500 |
| Annual maintenance cost | $ 3,000 | $ 5,000 |

(a) Clearly specify the net cash flow profiles for each mutually exclusive alternative. (5.8)

(b) Determine which alternative should be selected. Use a *total cash flow approach* and the present worth method. (5.8)

(c) Repeat part (b) using the annual worth method. (5.8)

(d) Repeat part (b) using the future worth method. (5.8)

(e) Determine which alternative should be selected. Use an *incremental cash flow approach* and the internal rate of return method. (5.8)

(f) Repeat part (e) using the external rate of return method. (5.8)

(g) Repeat part (f) using the savings/investment ratio method. (5.8)

38. Machine 1 initially costs $160,000. Its resale value at the end of the $k$th year of service equals $160,000 - 12,000k$ for $k = 1, 2, \ldots, 10$. Operating and maintenance expenses equal $35,000/year. Machine 2 initially costs $76,000. Its resale value at the end of the $k$th year, $S_k$, equals $76,000(0.80)^k$ for $k = 1, 2, \ldots, 5$. Operating and maintenance expenses equal $40,000/year. Using the annual worth method and a *MARR* of 15%, determine which machine is preferred assuming machine cash flows are repeatable.

(a) Use a planning horizon of 5 years. (5.8)

(b) Use a planning horizon of 20 years. (5.8)

**39.** The Ajax Manufacturing Company wishes to choose one of the following machines.

|  | Machine 1 | Machine 2 | Machine 3 |
|---|---|---|---|
| First cost | $10,000 | $15,000 | $18,000 |
| Planning horizon | 5 years | 5 years | 5 years |
| Salvage value | — | $ 2,000 | $ 3,000 |
| Operating and maintenance cost for year $k$, $k = 1, \ldots, n$ | $600(1.10)^{k-1}$ | $400(1.08)^{k-1}$ | $200 + 100k$ |

*MARR* is 15% and the planning horizon is 5 years. Based upon the annual worth method, determine the preferred machine. (5.8)

**40.** A firm will either lease an office copier at an end-of-year cost of $10,000 for a 6-year period, or they will purchase the copier at an initial cost of $66,117. If purchased, the copier will have a zero salvage value at the end of its 6-year life. No other costs are to be considered. *MARR* is 12%. Should the firm lease or buy? Justify your answer on the basis of the following.
(a) The present worth method. (5.8)
(b) The internal rate of return method. (5.8)

**41.** A company can construct a new warehouse for $1,000,000, or it can lease an equivalent building for $100,000/year for 25 years with the option of purchasing the building for $1,000,000 at the end of the 25-year period. Lease payments are due at the *beginning* of each year. The company can earn 20%/year before taxes on its invested capital. Indicate the preferred alternative based upon the following.
(a) The annual worth method. (5.8)
(b) The savings/investment ratio method. (5.8)

**42.** In Problem 41, for what beginning-of-year lease payment will the company be indifferent between leasing and buying? (5.8)

**43.** E-M Systems Company must decide if they should purchase a small computer or lease the computer. The computer costs $48,000 initially and will last 5 years, having a $5000 salvage value at that time. If the computer is purchased, all maintenance costs must be paid by E-M. Maintenance costs are $2,500/year over the life of the equipment. The E-M Company uses an interest rate of 15% in evaluating investment alternatives. For what beginning-of-year annual leasing charge is the firm indifferent between purchasing and leasing over the 5-year period?
(a) Use the annual worth method. (5.8)
(b) Use the external rate of return method. (5.8)

**44.** A firm is considering either leasing or buying a minicomputer system. If purchased, the initial cost will be $250,000; annual operating and maintenance costs will be $80,000/year. Based on a 6-year planning horizon, it is anticipated the computer will have a salvage value of $30,000 at that time. If the computer is leased, annual operating and maintenance costs in excess of the annual lease payment will be

$60,000/year. Based on an interest rate of 20%, what annual beginning-of-year lease payment will make the firm be indifferent between leasing and buying?

**(a)** Use the present worth method. (5.8)

**(b)** Use the internal rate of return method. (5.8)

**45.** Thermo-D, Inc. is considering investing $14,000 in a heat exchanger. The heat exchanger will last 6 years, at which time it will be sold for $2000. Maintenance costs for the exchanger are estimated to increase by $300/year over its life. The maintenance cost for the first year is estimated to be $2000. As an alternative, the company may lease the equipment for $X/year, including maintenance. For what value of $X$ should the company lease the heat exchanger? The company expects to earn 20% on its investments. Assume beginning-of-year lease payments.

**(a)** Use the future worth method. (5.8)

**(b)** Use the external rate of return method. (5.8)

**46.** Cowboy Creek in Stillwater has a history of flooding during heavy rainfalls. Two improvement alternatives are under consideration by the city engineer.

1. Leave the existing 36-inch corrugated steel culvert in place and install another of the same size alongside at an installed cost of $7000.

2. Remove the culvert currently in place and replace it with a single 50-inch culvert at an installed cost of $11,000.

    If Alternative 1 is adopted, the existing 36-inch culvert will need to be replaced in 10 years. It is assumed that a replacement at that time will cost $12,000. A newly installed culvert will last for 20 years; replacement prior to the end of 20 years will result in a zero salvage value. Use a planning horizon of 20 years and a 10% *MARR* to determine the preferred alternative. (5.10)

**47.** Pear Computers is considering replacing its material handling system and either purchasing or leasing a new system. The old system has an annual operating and maintenance cost of $32,000, a remaining life of 8 years, and an estimated salvage value of $5000 at that time.

A new system can be purchased for $280,000; it will be worth $60,000 in 8 years; and it will have annual operating and maintenance costs of $18,000/year. If the new system is purchased, the old system can be traded in for $20,000.

Leasing a new system will cost $26,000/year, payable at the beginning of the year, plus operating costs of $9000/year, payable at the end of the year. If the new system is leased, the old system will be sold for $10,000.

*MARR* is 15%. Compare the annual worths of keeping the old system, buying a new system, and leasing a new system based upon a planning horizon of 8 years.

**(a)** Use the cash flow approach. (5.10.1)

**(b)** Use the outsider viewpoint approach. (5.10.2)

**48.** Fluid Dynamics Company owns a pump that it is contemplating replacing. The old pump has annual operating and maintenance costs of $8000/year, it can be kept for 4 years more and will have a zero salvage value at that time.

The old pump can be traded in on a new pump. The trade-in value is $4000, with the purchase price of the new pump being $18,000. The new pump will have a value of $9000 in 4 years and will have annual operating and maintenance costs of $4500/year.

Using a *MARR* of 20%, evaluate the investment alternative based upon the present worth method and a planning horizon of 4 years.
(a) Use the cash flow approach. (5.10.1)
(b) Use the outsider viewpoint approach. (5.10.2)

49. A company owns a 5-year old turret lathe that has a book value of $20,000. The present market value for the lathe is $16,000. The expected decline in market value is $2000/year to a minimum market value of $4000. Maintenance plus operating costs for the lathe equal $4200/year.
A new turret lathe can be purchased for $45,000 and will have an expected life of 8 years. The market value for the turret lathe is expected to equal $45,000(0.70)$^k$ at the end of year $k$. Annual maintenance and operating cost is expected to equal $1600. Based on a 15% before-tax *MARR*, should the old lathe be replaced now? Use an equivalent uniform annual cost comparison, a planning horizon of 7 years, and the outsider viewpoint approach. (5.10.2)

50. Esteez Construction Company has an overhead crane that has an estimated remaining life of 7 years. The crane can be sold for $14,000. If the crane is kept in service it must be overhauled immediately at a cost of $6000. Operating and maintenance costs will be $5000/year after the crane is overhauled. After overhauling it, the crane will have a zero salvage value at the end of the 7-year period. A new crane will cost $36,000, will last for 7 years, and will have a $8000 salvage value at that time. Operating and maintenance costs are $2500 for the new crane. Esteez uses an interest rate of 15% in evaluating investment alternatives. Should the company buy the new crane based upon an annual cost analysis?
(a) Use the cash flow approach. (5.10.1)
(b) Use the outsider viewpoint approach. (5.10.2)

51. A firm is considering replacing a compressor that was purchased 4 years ago for $70,000. Currently, the compressor has a book value of $40,000, based on straight-line depreciation. If the compressor is retained for 4 years more, it is estimated to have a salvage value of $10,000. If the old compressor is retained for 8 years more, it is estimated that its salvage value will be negligible at that time. Operating and maintenance costs for the compressor have been increasing at a rate of $750/year, with the cost during the past year being $11,000.
A new compressor can be purchased for $60,000. It is estimated to have uniform annual operating and maintenance costs of $6000/year. The salvage value for the compressor is estimated to be $40,000 after 4 years and $15,000 after 8 years. If a new compressor is purchased, the old compressor will be traded in for $25,000.
Using a before-tax analysis, a *MARR* of 20%, and an annual worth comparison, determine the preferred alternative using a planning horizon of (a) 4 years, and (b) 8 years. Use the cash flow approach. (5.10.1)

52. A small foundry is considering the replacement of a No. 1 Whiting cupola furnace that is capable of melting gray iron only with a reverberatory-type furnace that has gray iron and nonferrous metals melting capability. Both furnaces have approximately the same melting rates for gray iron in pounds per hour. The foundry company plans to use the reverberatory furnace, if purchased, primarily for melting gray iron, and the total quantity melted is estimated to be about the same with either

furnace. Annual raw material costs would therefore be about the same for each furnace. Available information and cost estimates for each furnace is given below. *Cupola furnace*. Purchased used and installed 8 years ago for a cost of $20,000. The present market value is determined to be $8000. Estimated remaining life is somewhat uncertain but, with repairs, the furnace should remain functional for 7 years more. If kept 7 years more, the salvage value is estimated as $2000 and average annual expenses expected are:

| | |
|---|---|
| Fuel | $35,000 |
| Labor (including maintenance) | $40,000 |
| Payroll taxes | 10% of direct labor costs |
| Taxes and insurance on furnace | 1% of purchase price |
| Other | $16,000 |

*Reverberatory furnace*. This furnace costs $32,000. Expenses to remove the cupola and install the reverberatory furnace are about $2400. The new furnace has an estimated salvage value of $3000 after 7 years of use and annual expenses are estimated as:

| | |
|---|---|
| Fuel | $29,000 |
| Labor (operating) | $30,000 |
| Payroll taxes | 10% of direct labor costs |
| Taxes and insurance on furnace | 1% of purchase price |
| Other | $16,000 |

In addition, this furnance must be relined every 2 years at a cost of $4000/occurrence. If the foundry presently earns an average of 20% on invested capital before income taxes, should the cupola furnace be replaced by the reverberatory furnace?
(a) Use the cash flow approach. (5.10.1)
(b) Use the outsider viewpoint approach. (5.10.2)

53. Metallic Peripherals, Inc. has received a production contract for a new product. The contract lasts for 5 years. To do the necessary machining operations, the firm can use one of its own lathes, which was purchased 3 years ago at a cost of $16,000. Today the lathe can be sold for $8000. In 5 years the lathe will have a zero salvage value. Annual operating and maintenance costs for the lathe are $4000/year. If the firm uses its own lathe it must also purchase an additional lathe at a cost of $12,000; its value in 5 years will be $3000. The new lathe will have annual operating and maintenance costs of $3500/year.
As an alternative, the presently owned lathe can be traded in for $10,000 and a new lathe of larger capacity purchased for a cost of $24,000; its value in 5 years is estimated to be $8000, and its annual operating and maintenance costs will be $6000/year.
An additional alternative is to sell the presently owned lathe and subcontract the work to another firm. Company X has agreed to do the work for the 5-year period at an annual cost of $12,000/end-of-year.
Using a 15% interest rate, determine the least cost alternative for performing the required production operations.
(a) Use the cash flow approach. (5.10.1)
(b) Use the outsider viewpoint approach. (5.10.2)

54. A machine was purchased 5 years ago for $12,000. At that time, its estimated life was 10 years with an estimated end-of-life salvage value of $1200. The average annual operating and maintenance costs have been $14,000 and are expected to continue at this rate for the next 5 years. However, average annual revenues have been and are expected to be $20,000. Now, the firm can trade in the old machine for a new machine for $5000. The new machine has a list price of $15,000, an estimated life of 10 years, annual operating plus maintenance costs of $7500, annual revenues of $13,500, and salvage values at the end of the $j$th year according to

$$S_j = \$15,000 - \$1500j, \quad \text{for } j = 0, 1, 2, 3, 4, 5, 6, 7, 8, 9, 10$$

Determine whether to replace or not by the annual worth method using a *MARR* equal to 15% compounded annually. Use a 5-year planning horizon and the outsider viewpoint approach. (5.10.2)

55. Kwik-Kleen Car Wash has been experiencing difficulties in keeping its equipment operational. The owner is faced with the alternative of overhauling the present equipment or replacing it with new equipment. The cost of overhauling the present equipment is $8500. The present equipment has annual operating and maintenance costs of $7500. If it is overhauled, the present equipment will last for 5 years more and be scrapped at zero value. If it is not overhauled, it has a trade-in value of $3200 toward the new equipment.
New equipment can be purchased for $28,000. At the end of 5 years the new equipment will have a resale value of $12,000. Annual operating and maintenance costs for the new equipment will be $3000.
Using a *MARR* of 12%, what is your recommendation to the owner of the car wash? Base your recommendation on a present worth comparison and the outsider viewpoint approach. (5.10.2)

56. A building supplies distributor purchased a gasoline-powered forklift truck 4 years ago for $8000. At that time, the estimated useful life was 8 years with a salvage value of $800 at the end of this time. The truck can now be sold for $2500. For this truck, average annual operating expenses for year $j$ have been

$$C_j = \$2000 + \$400(j - 1)$$

Now the distributor is considering the purchase of a smaller battery-powered truck for $6500. The estimated life is 10 years, with the salvage value decreasing by $600 each year. Average annual operating expenses are expected to be $1200. If a *MARR* of 20% is assumed and a 4-year planning horizon is adopted, should the replacement be made now? (5.10.2)

57. National Chemicals has an automatic chemical mixer that it has been using for the past 4 years. The mixer originally cost $18,000. Today the mixer can be sold for $10,000. The mixer can be used for 10 years more and will have a $2500 salvage value at that time. The annual operating and maintenance costs for the mixer equal $6000/year.
Because of an increase in business, a new mixer must be purchased. If the old mixer is retained, a new mixer will be purchased at a cost of $25,000 and have a $4000 salvage value in 10 years. This new mixer will have annual operating and maintenance costs equal to $5000/year.

The old mixer can be sold and a new mixer of larger capacity purchased for $32,000. This mixer will have a $6000 salvage value in 10 years and will have annual operating and maintenance costs equal to $8000/year.

Based on a *MARR* of 15%, what do you recommend? (5.10.2)

58. A firm is contemplating replacing a computer they purchased 3 years ago for $400,000. It will have a salvage value of $20,000 in 4 years. Operating and maintenance costs have been $75,000/year. Currently the computer has a trade-in value of $100,000 toward a new computer that costs $300,000 and has a life of 4 years, with a salvage value of $50,000 at that time. The new computer will have annual operating and maintenance costs of $80,000.

If the current computer is retained, another small computer will have to be purchased in order to provide the required computing capacity. The smaller computer will cost $150,000, has a salvage value of $20,000 in 4 years, and has annual operating and maintenance costs of $30,000.

Using an annual worth comparison before taxes, with a *MARR* of 25%, determine the preferred course of action. (5.10.2)

59. The Ajax Specialty Items Corporation has received a 5-year contract to produce a new product. To do the necessary machining operations, the company is considering two alternatives.

Alternative A involves continued use of the currently owned lathe. The lathe was purchased five years ago for $20,000. Today the lathe is worth $8000 on the used machinery market. If this lathe is to be used, special attachments must be purchased at a cost of $3500. At the end of the 5-year contract, the lathe (with attachments) can be sold for $2000. Operating and maintenance costs will be $7000/year if the old lathe is used.

Alternative B is to sell the currently owned lathe and buy a new lathe at a cost of $25,000. At the end of the 5-year contract, the new lathe will have a salvage value of $13,000. Operating and maintenance costs will be $4000/year for the new lathe.

Using an annual worth analysis, should the firm use the currently owned lathe or buy a new lathe? Base your analysis on a minimum attractive rate of return of 20%. (5.10.2)

60. The Telephone Company of America purchased a numerically controlled production machine 5 years ago for $300,000. The machine currently has a trade-in value of $70,000. If the machine is continued in use, another machine, X, must be purchased to supplement the old machine. Machine X costs $200,000, has annual operating and maintenance costs of $40,000, and will have a salvage value of $30,000 in 10 years. If the old machine is retained, it will have annual operating and maintenance costs of $55,000 and will have a salvage value of $15,000 in 10 years.

As an alternative to retaining the old machine, it can be replaced with Machine Y. Machine Y costs $400,000, has anticipated annual operating and maintenance costs of $70,000, and has a salvage value of $140,000 in 10 years.

Using a *MARR* of 20% and a present worth comparison, determine the preferred economic alternative. (5.10.2)

61. A highway construction firm purchased an item of earth-moving equipment 3 years ago for $125,000. The salvage value at the end of 8 years was estimated to be 35% of first cost. The firm earns an average annual gross revenue of $105,000 with the equipment and the average annual operating costs have been and are expected to be $65,000.

The firm now has the opportunity to sell the equipment for $70,000 and subcontract the work normally done by the equipment over the next 5 years. If the subcontracting is done, the average annual gross revenue will remain $105,000 but the subcontractor charges $85,000/end-of-year for these services.

If a 25% rate of return before taxes is desired, determine by the annual worth method whether or not the firm should subcontract. (5.10.2)

62. Bumps Unlimited, a highway contractor, must decide whether to overhaul a tractor and scraper or replace it. The old equipment was purchased 5 years ago for $130,000; it had a projected life of 12 years, with a $15,000 salvage value at that time. If traded in on a new tractor and scraper, it can be sold for $60,000. Overhauling the old equipment will cost $20,000. If overhauled, operating and maintenance costs will be $18,000/year, and the overhauled equipment will have a newly projected salvage value of $0 in 7 years.

A new tractor and scraper can be purchased for $150,000; it will have annual operating and maintenance costs of $10,000, and will have a salvage value of $40,000 in 7 years.

Using an annual worth comparison with a MARR of 15%, should the equipment be overhauled or replaced? (5.10.2)

63. A firm is presently using a machine that has a market value of $11,000 to do a specialized production job. The requirement for this operation is expected to last only 6 years more, after which it will no longer be done. The predicted costs and salvage values for the present machine are:

| Year | 1 | 2 | 3 | 4 | 5 | 6 |
|---|---|---|---|---|---|---|
| Operating cost | $1,500 | $1,800 | $2,100 | $2,400 | $2,700 | $3,000 |
| Salvage value | 8,000 | 6,000 | 5,000 | 4,000 | 3,000 | 2,000 |

A new machine has been developed that can be purchased for $17,000 and has the following predicted cost performance.

| Year | 1 | 2 | 3 | 4 | 5 | 6 |
|---|---|---|---|---|---|---|
| Operating cost | $ 1,000 | $ 1,100 | $ 1,200 | $1,300 | $1,400 | $1,500 |
| Salvage value | 13,000 | 11,000 | 10,000 | 9,000 | 8,000 | 7,000 |

If interest is at 0%, when should the new machine be purchased? (5.10.3)

64. A particular unit of production equipment has been used by a firm for a period of time sufficient to establish very accurate estimates of its operating and maintenance costs. Replacements can be expected to have identical cash flow profiles in successive life cycles if constant worth dollar estimates are used. The appropriate discount rate is 30%. Operating and maintenance costs for a unit of equipment in its $t$th year

of service, denoted by $C_t$, are as follow.

| $t$ | $C_t$ | $t$ | $C_t$ |
|---|---|---|---|
| 1 | $ 6,000 | 6 | $15,000 |
| 2 | 7,500 | 7 | 17,250 |
| 3 | 9,150 | 8 | 19,650 |
| 4 | 10,950 | 9 | 22,200 |
| 5 | 12,900 | 10 | 24,900 |

Each unit of equipment costs $45,000 initially. Because of its special design, the unit of equipment cannot be disposed of at a positive salvage value following its purchase; hence, a zero salvage value exists, regardless of the replacement interval used.

(a) Determine the optimum replacement interval assuming an infinite planning horizon. (Maximum feasible interval = 10 years.)

(b) Determine the optimum replacement interval assuming a finite planning horizon of 15 years, with $C_{t+1} = C_t + \$1500 + \$150(t - 1)$ for $t = 10, 11, \ldots$

(c) Solve parts (a) and (b) using a discount rate of 0%.

(d) Based on the results obtained, what can you conclude concerning the effect the discount rate has on the optimum replacement interval? (5.10.3)

65. Given an infinite planning horizon, identical cash flow profiles for successive life cycles, and the following functional relationships for $C_t$, the operating and maintenance cost for the $t$th year of service for the unit of equipment in current use, and $F_n$, the salvage value at the end of $n$ years of service:

$$C_t = \$4000(1.10)^t \qquad t = 1, 2, \ldots, 12$$

$$F_n = \$44,000(0.50)^n \qquad n = 0, 1, 2, \ldots, 12$$

Determine the optimum replacement interval assuming a *MARR* of (a) 0%, (b) 15%. (Maximum life = 12 years.) (5.10.3)

66. Solve Problem 65 given the following functional relationships. (5.10.3)

$$C_t = \$4000(1.40)^t \qquad t = 1, 2, \ldots, 12$$

$$F_n = \$44,000(0.90)^n \qquad n = 0, 1, 2, \ldots, 12$$

# CHAPTER SIX
# DEPRECIATION AND INCOME TAX CONSIDERATIONS

## 6.1 INTRODUCTION

Depreciation and taxes are particularly important in engineering economic analyses. Although depreciation allowances are not actually cash flows, their magnitudes and timing do affect taxes. Tax dollars are cash flows and therefore it makes sense to include them in an economic analysis, just as wages, equipment, materials, and energy are included. Proper knowledge and application of tax laws can make the economic difference between accepting and rejecting a project, as well as between profit and loss on the corporate bottom line.

The Economic Recovery Tax Act of 1981 and the Tax Equity and Fiscal Responsibility Act of 1982 virtually rewrote the book on depreciation and income tax treatment. This chapter presents the Accelerated Cost Recovery System (ACRS), several depreciation methods replaced by the ACRS, and some other still-allowable depreciation strategies. It also illustrates corporate income tax law, including the effects of depreciation, borrowed money, capital gains and losses, investment credits, and expense deductions on after-tax cash flow profiles. No attempt is made to discuss the minute details of tax law. Experts devote entire careers trying to keep abreast of the latest laws and judgments. Instead, the objective here is to communicate the major aspects of corporate tax treatment such that you can perform economic analyses on an after-tax basis and recognize when to seek corporate legal assistance.

## 6.2 THE MEANING OF DEPRECIATION

Most property decreases in value with use and time. That is, it *depreciates*. The law permits a deduction of a reasonable allowance for the exhaustion and wear and tear of property used in a trade or business, or of property held for the production of income. The amount of the allowance depends upon several factors, including date first placed in service, type of property, and cost basis or investment. Under the ACRS, sums are set aside or recovered each year such that the total of these sums throughout the asset's recovery period will equal the original cost.

The federal income tax is based on net income that results by deducting certain items such as expenses from gross income. For tax purposes, investment in a depreciable asset is treated as a prepaid expense, and the depreciation deduction allocates that expense over time. It is important to note that cost recovery or depreciation is not an actual cash flow, but is merely treatable as an expense for income tax purposes. A larger deduction in a year decreases net taxable income and, hence, income taxes, making more money available for reinvestment by the firm. Because of the time value of money, it is generally desirable to take larger allowable deductions in the early years and lesser allowances in the later years of an asset's life. Often, but not always, the ACRS is more favorable to the corporate taxpayer than even the best of the depreciation strategies that preceded it. That is, it permits more cost recovery, more quickly, such that the present worth of after-tax cash flows is higher.

## 6.3 SOME DEFINITIONS USED IN DEPRECIATION ACCOUNTING

*Depreciable property* meets these requirements:

1. It must be used in business or held for production of income.
2. It must have a useful life that can be determined, and its useful life must be longer than 1 year.
3. It must be something that wears out, decays, gets used up, becomes obsolete, or loses value from natural causes.

In general, if property does not meet all three of these conditions, it is not depreciable [1].

Depreciable property may be tangible or intangible. *Tangible property* can be seen or touched. *Intangible property,* such as a copyright or franchise, is not tangible. Depreciable property may also be real or personal. *Personal property* includes machinery or equipment that is not real estate. *Real property* is land and generally anything that is erected on, growing on, or attached to land. Land itself, however, is never depreciable. Most engineering economic analyses involve tangible personal property [1].

Depreciation allowance is determined under the Accelerated Cost Recovery System (ACRS) unless the property does not qualify as recovery property.

*Recovery property* is tangible depreciable property that is placed in service after 1980. It does not include property depreciated on a basis other than years, such as the *units of production* method, nor does it include some public utility property. The cost basis of the property is recovered over the recovery period of the asset. The *cost basis* is essentially the taxpayer's investment. In most cases, this is the cost of the property plus the cost of additions to that property, including installation cost. It is also known as the *unadjusted basis*. The cost basis less capital recovered such as depreciation allowances is known as the *unrecovered investment, adjusted basis,* or, more often, *book value.* The *recovery period* is simply the time over which the cost basis can be recovered, and is 3, 5, 10, or 15 years depending upon the type of property [2]. It is usually shorter than the inherent physical life of the asset.

Depreciation on assets placed in service before 1981 and assets that are not recovery property is not determined using the ACRS. Rather, the cost basis, less salvage value, is recovered during the useful life of the property, usually in accordance with straight-line, sum of the years' digits, or declining balance depreciation methods. *Salvage value* is an estimate of the market value at the end of the asset's useful life. *Useful life* refers to an estimate of time during which the asset will actually be used by the taxpayer. It is often shorter than the inherent physical life of the asset. Neither the salvage value nor the useful life is used explicitly in the ACRS.

Most current engineering economic analyses involving depreciable property make use of the ACRS. Regular straight-line depreciation still applies in the case of intangible property; ACRS straight-line deductions may be used in lieu of the regular ACRS allowances. Other depreciation methods may be needed to determine cash flows on assets first placed in service prior to 1981, or when depreciation is based upon something other than years. As such, the ACRS and ACRS straight-line depreciation will be treated in considerable detail; other methods will be treated more briefly. Anyone desiring a more thorough treatment of depreciation strategy for assets placed in service prior to 1981 is referred to the first edition of this text.

## 6.4 ACCELERATED COST RECOVERY SYSTEM (ACRS)

Capital costs for most depreciable tangible property placed in service after 1980 must be recovered using the ACRS. Under the ACRS, cost recovery methods and periods are the same whether the property is new or used. Salvage value is disregarded, and need not be estimated. The ACRS is actually simpler, more uniform, and often more economically beneficial to the taxpayer than its predecessors.

### 6.4.1 Recovery Property Classes

An explanation of *recovery property* was given earlier (see Sec. 6.3). Recovery property can be further defined as either *Section 1245 class property* or *Section*

*1250 class property*. Definitions of these properties are presented later in this chapter (see Sec. 6.15). For now, however, think of Section 1245 class property as depreciable tangible property such as equipment, machinery, tooling, office furniture, or vehicles used in trade or business. Also, many research facilities are Section 1245 class property. Section 1250 class property may be thought of as depreciable tangible real property such as buildings, fixtures, fences, and parking lots.

Each item of recovery property may be assigned to one of the following classes:

*3-Year Property*—Included is Section 1245 class property having a present class life[1] of 4 years or less. Also included is Section 1245 class property used in connection with research and experimentation. Examples of items having present class lives of 4 years or less are automobiles, taxis, light general-purpose trucks, over-the-road tractor units, and certain special tools and devices for manufacturing.

*5-Year Property*—Included is Section 1245 class property that is not 3- or 10-year property, or 15-year public utility property. This recovery property class contains the bulk of equipment, machinery, tooling, and office furniture used in engineering economic analyses. Also included are single-purpose agricultural structures, storage facilities used for petroleum distribution, and public utility property having a present class life of 5 to 18 years, inclusive.

*10-Year Property*—Included is public utility property, other than Section 1250 class property or 3-year property, having a present class life of more than 18 but not more than 25 years. Also included is Section 1250 class property having a present class life of 12.5 years or less (theme and amusement parks). Residential manufactured homes, including mobile homes, belong to this class.

*15-Year Public Utility Property*—Included is public utility property, other than Section 1250 class property or 3-year property, having a present class life of more than 25 years.

*15-Year Real Property*—Included is Section 1250 class property with a present class life of more than 12.5 years. This covers all real property, such as buildings, fixtures, fences, and parking lots not specifically designated in an earlier class. This real property class is also frequently used in engineering economic analyses.

These five recovery property classes dictate the percentage of the cost basis to be recovered each year under the ACRS.

---

[1] "Present class life" is currently interpreted as class life guideline under the old Class Life Asset Depreciation Range (CLADR) system.

## 6.4.2 Calculating ACRS Deductions

The ACRS deduction is calculated by applying the appropriate percentage ($d_t \times 100\%$), for a specified recovery property class and year, to the cost basis ($P$). These percentages are found in Tables 6.1 and 6.2. The allowable deduction in year $t$, ($D_t$), is given by

$$D_t = d_t P \tag{6.1}$$

Note that the percentages in Tables 6.1 and 6.2 add up to 100% for each recovery property class. This means the entire cost basis of a property can be recovered over its recovery property class life, even if its useful life is much longer. The unrecovered investment ($B_t$) at the end of year $t$ is given by

$$B_t = P - \sum_{j=1}^{t} D_j = P\left(1 - \sum_{j=1}^{t} d_j\right) \tag{6.2}$$

This represents the amount of the cost basis yet to be recovered.

### Example 6.1

A company is purchasing a stand-alone computer with colorgraphics display, printer, plotter, and interfacing equipment for real-time monitoring and control of manufacturing equipment. First cost including installation will be $82,000 with an estimated salvage value of $5000 at the end of a projected useful life of 7 years. The allowable ACRS deduction and unrecovered investment for each year are given in Table 6.3. Note that this is 5-year recovery property.

---

TABLE 6.1. ACRS Percentages ($d_t \times 100\%$) for 3-, 5-, and 10-Year Property, and 15-Year Public Utility Property

| Year, t | 3-Year Property | 5-Year Property | 10-Year Property | 15-Year Public Utility Property |
|---|---|---|---|---|
| 1 | 25 | 15 | 8 | 5 |
| 2 | 38 | 22 | 14 | 10 |
| 3 | 37 | 21 | 12 | 9 |
| 4 | | 21 | 10 | 8 |
| 5 | | 21 | 10 | 7 |
| 6 | | | 10 | 7 |
| 7 | | | 9 | 6 |
| 8 | | | 9 | 6 |
| 9 | | | 9 | 6 |
| 10 | | | 9 | 6 |
| 11 | | | | 6 |
| 12 | | | | 6 |
| 13 | | | | 6 |
| 14 | | | | 6 |
| 15 | | | | 6 |

TABLE 6.2. ACRS Percentages ($d_t \times 100\%$) for 15-Year Real Property Except Low-Income Housing

| Year, | Month in Taxable Year Property Placed in Service | | | | | | | | | | | |
|---|---|---|---|---|---|---|---|---|---|---|---|---|
| t | 1 | 2 | 3 | 4 | 5 | 6 | 7 | 8 | 9 | 10 | 11 | 12 |
| 1 | 12 | 11 | 10 | 9 | 8 | 7 | 6 | 5 | 4 | 3 | 2 | 1 |
| 2 | 10 | 10 | 11 | 11 | 11 | 11 | 11 | 11 | 11 | 11 | 11 | 12 |
| 3 | 9 | 9 | 9 | 9 | 10 | 10 | 10 | 10 | 10 | 10 | 10 | 10 |
| 4 | 8 | 8 | 8 | 8 | 8 | 8 | 9 | 9 | 9 | 9 | 9 | 9 |
| 5 | 7 | 7 | 7 | 7 | 7 | 7 | 8 | 8 | 8 | 8 | 8 | 8 |
| 6 | 6 | 6 | 6 | 6 | 7 | 7 | 7 | 7 | 7 | 7 | 7 | 7 |
| 7 | 6 | 6 | 6 | 6 | 6 | 6 | 6 | 6 | 6 | 6 | 6 | 6 |
| 8 | 6 | 6 | 6 | 6 | 6 | 6 | 5 | 6 | 6 | 6 | 6 | 6 |
| 9 | 6 | 6 | 6 | 6 | 5 | 6 | 5 | 5 | 5 | 6 | 6 | 6 |
| 10 | 5 | 6 | 5 | 6 | 5 | 5 | 5 | 5 | 5 | 5 | 6 | 5 |
| 11 | 5 | 5 | 5 | 5 | 5 | 5 | 5 | 5 | 5 | 5 | 5 | 5 |
| 12 | 5 | 5 | 5 | 5 | 5 | 5 | 5 | 5 | 5 | 5 | 5 | 5 |
| 13 | 5 | 5 | 5 | 5 | 5 | 5 | 5 | 5 | 5 | 5 | 5 | 5 |
| 14 | 5 | 5 | 5 | 5 | 5 | 5 | 5 | 5 | 5 | 5 | 5 | 5 |
| 15 | 5 | 5 | 5 | 5 | 5 | 5 | 5 | 5 | 5 | 5 | 5 | 5 |
| 16 | — | — | 1 | 1 | 2 | 2 | 3 | 3 | 4 | 4 | 4 | 5 |

Example 6.1 shows that the estimated salvage value does not enter into the ACRS calculations in any way. We will have to account for the actual salvage value received, but this will be illustrated later. Further, the entire cost basis was recovered during the class life of 5 years, rather than over the 7-year useful life. Now, let us consider a slightly more complicated example which makes use of Table 6.2.

Example 6.2 _____

A firm having a taxable year of January 1 to December 31 will place a business real estate property in service on May 1. A total of $220,000 will be paid for the

TABLE 6.3. ACRS Deduction and Unrecovered Investment on Computer

| End of Year, t | ACRS Deduction, $D_t$ | Unrecovered Investment, $B_t$ |
|---|---|---|
| 0 | — | $82,000.00 |
| 1 | $12,300.00 | 69,700.00 |
| 2 | 18,040.00 | 51,660.00 |
| 3 | 17,220.00 | 34,440.00 |
| 4 | 17,220.00 | 17,220.00 |
| 5 | 17,220.00 | 0.00 |
| 6 | 0.00 | 0.00 |
| 7 | 0.00 | 0.00 |

TABLE 6.4. ACRS Deduction and Unrecovered Investment
on Real Estate

| End of year, $t$ | ACRS Deduction, $D_t$ | Unrecovered Investment, $B_t$ |
|---|---|---|
| 0 | — | $220,000.00 |
| 1 | $17,600.00 | 202,400.00 |
| 2 | 24,200.00 | 178,200.00 |
| 3 | 22,000.00 | 156,200.00 |
| 4 | 17,600.00 | 138,600.00 |
| 5 | 15,400.00 | 123,200.00 |
| 6 | 15,400.00 | 107,800.00 |
| 7 | 13,200.00 | 94,600.00 |
| 8 | 13,200.00 | 81,400.00 |
| 9 | 11,000.00 | 70,400.00 |
| 10 | 11,000.00 | 59,400.00 |
| 11 | 11,000.00 | 48,400.00 |
| 12 | 11,000.00 | 37,400.00 |
| 13 | 11,000.00 | 26,400.00 |
| 14 | 11,000.00 | 15,400.00 |
| 15 | 11,000.00 | 4,400.00 |
| 16 | 4,400.00 | 0.00 |

property. The allowable ACRS deductions and the unrecovered investment for each year are given in Table 6.4.

In Example 6.2, recovery takes place over a period of 16 taxable years. This is the case whenever real estate in this class is placed in service after the second month of a taxable year. Also, note that the entire cost basis is recovered. This would have been true even if an estimated salvage value had been declared.

In Examples 6.1 and 6.2, it is implicitly assumed that the assets are kept throughout the recovery period. If they are sold or otherwise disposed of, the recovery allowance in the year of disposition may be affected. For personal property in the 3-, 5-, and 10-year property, and 15-year public utility property classes covered by Table 6.1, no recovery deduction is allowed in the year of disposition. An ACRS deduction is permitted for real property covered in Table 6.2, reflecting the number of months the property was in service in the year of disposition.

Example 6.3

In Example 6.1, the ACRS deduction in year 4 is $17,220.00. What would be the year 4 deduction if our computer were sold sometime late in year 4? Immediately after the end of year 4?
   Answers: $0.00 and $17,220.00, respectively.

## Example 6.4

In Example 6.2, the ACRS deduction in year 8 is $13,200.00. What would be the year 8 deduction if the real estate were sold on October 1?

Answer: $9/12 \times \$13,200.00 = \$9,900.00$.

### 6.4.3 Straight-Line Election

A taxpayer may elect to claim ACRS straight-line deductions instead of the regular allowances presented in the previous section. The optional straight-line recovery periods are given in Table 6.5. In general, the applicable deduction rate in each year $t$ for a recovery period of $n$ years is given by

$$d_t = 1/n \tag{6.3}$$

Note that the rate is the same for each year of an $n$-year recovery period. This does not imply, however, that the allowable deduction will be the same each year.

During the first year that 3-, 5-, and 10-year property, and 15-year public utility property is placed in service, half the amount of recovery that would be allowed for a full year may be deducted, regardless of when during the year the property was placed in service. If the property is held for at least the entire recovery period, another half-year's deduction may be allowed following the end of the recovery period. That is,

$$D_t = .5d_t P \qquad t = 1, n + 1 \tag{6.4}$$

and

$$D_t = d_t P \qquad t = 2, 3, \ldots, n \tag{6.5}$$

if the property is kept throughout the recovery period. If the property is sold during the recovery period, standard ACRS rules apply in that no deduction is allowable in the year of sale.

TABLE 6.5. Optional ACRS Straight-Line Recovery Periods

| Recovery Property Class | Optional Recovery Periods in Years, $n$ |
|---|---|
| 3-Year | 3, 5, or 12 |
| 5-Year | 5, 12, or 25 |
| 10-Year | 10, 25, or 35 |
| 15-Year public utility | 15, 35, or 45 |
| 15-Year real | 15, 35, or 45 |

**TABLE 6.6. ACRS Straight-Line Recovery on Stamping Machine ($n = 12$)**

| End of Year, $t$ | Straight-Line ACRS Deduction, $D_t$ |
| --- | --- |
| 0 | — |
| 1 | $25,000/12 \times 1/2 = \$1,041.67$ |
| 2–12 | $25,000/12 \quad = \$2,083.33$ |
| 13 | $25,000/12 \times 1/2 = \$1,041.67$ |

## Example 6.5

A sheet metal stamping machine is purchased for $25,000 on December 1. It is classified as 5-year recovery property. The firm elects to use the ACRS straight-line option with a recovery period of 12 years, in order to defer depreciation deductions. The firm's allowable recovery during each year is given in Table 6.6. Note that the entire $25,000 is recovered.

For 15-year real property, the first year's deduction and the year of disposition deduction must be prorated for the number of months in use. The half-year convention does not apply to 15-year real property.

## Example 6.6

Reconsider the real estate purchase for $220,000 described in Example 6.2. Assume the purchase is made on May 1 and we elect to use the ACRS straight-line option over a 35-year recovery period. If the property is eventually sold on October 1 of year 8, the schedule of allowable deductions is as presented in Table 6.7.

For the 3-, 5-, 10-year property, and 15-year public utility property classes, the same method and recovery period must be used for all property in the same class that is placed in service in the same tax year. For example, if four items of 5-year recovery property are placed in service in the same tax year and ACRS straight-line recovery with $n = 12$ is elected for one item, the same method must be used for all four items. Different methods and periods may be used for

**TABLE 6.7. ACRS Straight-Line Recovery on Real Estate ($n = 35$)**

| End of year, $t$ | ACRS Straight-Line Deduction, $D_t$ |
| --- | --- |
| 0 | — |
| 1 | $220,000/35 \times 8/12 = \$4,190.48$ |
| 2–7 | $220,000/35 \quad = \$6,285.71$ |
| 8 | $220,000/35 \times 9/12 = \$4,714.29$ |

property in different classes, or for property in the same class if placed in service in different tax years. For 15-year real property, a different ACRS straight-line recovery period can be chosen for each item. Unlike the other classes of property, the choice is made on a property-by-property basis [1].

## 6.5 OTHER DEPRECIATION METHODS BASED ON YEARS

Use of the ACRS is mandatory for most tangible depreciable property placed in service after 1980. However, post-1980 depreciation on assets first placed in service before 1981 is to be computed according to the method selected at that time. Prior to 1981, the three most popular depreciation strategies included the (1) straight-line, (2) sum of the years' digits, and (3) declining balance methods. All three methods utilize a useful life ($n$) instead of a recovery period. They also require an estimate of the salvage value ($F$) at the end of the useful life. Under each of these methods, total depreciation taken over the useful life of the property may not exceed the cost basis less the salvage value ($P - F$). Knowledge of these methods may be of use in estimating after-tax cash flows of existing assets being considered for replacement. Also, a knowledge of these methods will permit comparison to the ACRS in terms of desirability for the taxpayer and investor.

### 6.5.1 Straight-Line Depreciation

The straight-line method provides for the uniform write-off of an asset. The depreciation allowed at the end of each year $t(D_t)$ is equal throughout the property's useful life and is given by

$$D_t = \frac{P - F}{n} \tag{6.6}$$

The unrecovered investment, often referred to as the book value, at the end of year $t(B_t)$ is given by

$$B_t = P - \left(\frac{P - F}{n}\right)t \tag{6.7}$$

The straight-line method was very popular for property placed in service prior to 1981, and those assets will continue to be depreciated as originally set up. It is still to be used for depreciating intangible property. Also, when elected for use under the ACRS, depreciation is computed as shown above, except that a recovery period is used instead of a useful life, salvage value is not taken into account ($F = 0$), and the half-year convention applies.

### Example 6.7

Reconsider the stand-alone computer of Example 6.1 in which the cost basis is $82,000 with an estimated salvage value of $5000 and a useful life of 7 years.

TABLE 6.8. Straight-Line Depreciation and Book Value

| End of Year, $t$ | Depreciation, $D_t$ | Book Value, $B_t$ |
|---|---|---|
| 0 | — | $82,000.00 |
| 1 | $11,000.00 | 71,000.00 |
| 2 | 11,000.00 | 60,000.00 |
| 3 | 11,000.00 | 49,000.00 |
| 4 | 11,000.00 | 38,000.00 |
| 5 | 11,000.00 | 27,000.00 |
| 6 | 11,000.00 | 16,000.00 |
| 7 | 11,000.00 | 5,000.00 |

Had it been placed in service prior to 1981, the straight-line depreciation and book value for each year would be as given in Table 6.8.

## 6.5.2 Sum of the Years' Digits Depreciation

The *sum of the years' digits method* is known for its accelerated write-off of assets. That is, it provides relatively high depreciation allowances in the early years and lower allowances throughout the rest of the property's useful life. The name "sum of the years' digits" comes from the fact that the sum

$$1 + 2 + \cdots + (n - 1) + n = \frac{n(n + 1)}{2}$$

is used directly in the calculation of depreciation. The depreciation allowance during any year $t$ is expressed as

$$D_t = \frac{n - (t - 1)}{n(n + 1)/2} (P - F) \tag{6.8}$$

The book value at the end of each year $t$ is given by

$$B_t = P - \sum_{j=1}^{t} \frac{n - (j - 1)}{n(n + 1)/2} (P - F)$$

which reduces to

$$B_t = (P - F) \frac{(n - t)(n - t + 1)}{n(n + 1)} + F \tag{6.9}$$

Sum of the years' digits depreciation is allowed if it was elected for pre-1981 tangible property having a useful life of 3 years or more and that was new in use (special rules apply to depreciable realty).

## Example 6.8

Again let us consider the computer of Example 6.1 having $P = \$82,000$, $F = \$5000$, and $n = 7$. Had it been placed in service prior to 1981, the sum of the

TABLE 6.9. Sum of the Years' Digits Depreciation and Book Value

| End of Year, $t$ | Value of $\dfrac{n - (t - 1)}{n(n + 1)/2}$ | Depreciation, $D_t$ | Book Value, $B_t$ |
|---|---|---|---|
| 0 | — | — | $82,000.00 |
| 1 | 7/28 | $19,250.00 | 62,750.00 |
| 2 | 6/28 | 16,500.00 | 46,250.00 |
| 3 | 5/28 | 13,750.00 | 32,500.00 |
| 4 | 4/28 | 11,000.00 | 21,500.00 |
| 5 | 3/28 | 8,250.00 | 13,250.00 |
| 6 | 2/28 | 5,500.00 | 7,750.00 |
| 7 | 1/28 | 2,750.00 | 5,000.00 |

years' digits depreciation and book value for each year would be as given in Table 6.9.

### 6.5.3 Declining Balance Depreciation

The *declining balance method,* like sum of the years' digits depreciation, is known for its accelerated write-off of assets. In this method, the depreciation allowed at the end of each year $t$ is a constant fraction ($p$) of the book value at the end of the previous year. That is,

$$D_t = pB_{t-1} \tag{6.10}$$

The book value at the end of each year $t$ is given by

$$B_t = P(1 - p)^t \tag{6.11}$$

Substituting Equation 6.11 into Equation 6.10 allows us to calculate the year $t$ depreciation directly as

$$D_t = pP(1 - p)^{t-1} \tag{6.12}$$

Note that in the declining balance method of depreciation the estimated salvage value need not come into play in figuring the deduction; however, the book value must never fall below the salvage value.

Twice the straight-line rate, or $2/n$, is the maximum constant fraction permissible under law. When $p = 2/n$, this method is known as *double declining balance* or *200% declining balance* depreciation. It can be used only if it was elected for pre-1981 tangible property having a useful life of 3 years or more and that was new in use (special rules apply to depreciable realty). For used tangible property the maximum permissible rate is $p = 1.5/n$, known as *150% declining balance* depreciation.

TABLE 6.10. Double Declining Balance Depreciation and Book Value

| End of Year, $t$ | Depreciation, $D_t$ | Book Value, $B_t$ |
|---|---|---|
| 0 | — | $82,000.00 |
| 1 | $23,428.57 | 58,571.43 |
| 2 | 16,734.69 | 41,836.74 |
| 3 | 11,953.35 | 29,883.39 |
| 4 | 8,538.11 | 21,345.28 |
| 5 | 6,098.65 | 15,246.63 |
| 6 | 4,356.18 | 10,890.45 |
| 7 | 3,111.56 | 7,778.89 |

Example 6.9_____

Let us again work with the computer of Example 6.1 in which $P = \$82,000$, $F = \$5000$, and $n = 7$. Had it been placed in service prior to 1981, the double declining balance depreciation and book value for each year would be as given in Table 6.10. The value of $p$ is 2/7.

_____

In Example 6.9, the book value at the end of year 7 is $7778.89. Since the declining balance rate does not directly use the estimate of a salvage value, the book value in the last year need not be the same as for the other methods. If the book value falls below the salvage value, depreciation should be truncated such that book value just equals salvage value.

The Internal Revenue Service (IRS) also allows switching from either the 200% or 150% declining balance method to straight-line depreciation.[2] This may be desirable in order to present a greater depreciation charge, resulting in lower taxes in the current year, deferring taxes until later years, and thus providing a present worth tax advantage. In this case, the switch should take place whenever straight-line depreciation on the undepreciated portion of the asset exceeds the declining balance allowance. That is, we should switch to straight line at the first year for which

$$\frac{B_{t-1} - F}{n - (t - 1)} > pB_{t-1} \tag{6.13}$$

The estimated salvage value is used in determining the straight-line depreciation component, even though it is neglected in the double declining balance method. Switching to straight-line depreciation will never be desirable if the estimated salvage value $F$ exceeds the declining balance book value for the last year $B_n$, causing depreciation to be truncated.

_____

[2] Switching from sum of the years' digits to straight-line depreciation is also allowed, but normally cannot be justified economically.

TABLE 6.11. Double Declining Balance Switching to Straight-Line Depreciation, and Book Value

| End of Year, t | Double Declining Balance Depreciation, $D_t$ | Straight-Line Depreciation on Remaining Life, $D_t$ | Book Value, $B_t$ |
|---|---|---|---|
| 0 | — | — | $82,000.00 |
| 1 | $23,428.57* | $11,000.00 | 58,571.43 |
| 2 | 16,734.69* | 8,928.57 | 41,836.73 |
| 3 | 11,953.35* | 7,367.35 | 29,883.38 |
| 4 | 8,538.11* | 6,220.85 | 21,345.27 |
| 5 | 6,098.65* | 5,448.42 | 15,246.62 |
| 6 | 4,356.18 | 5,123.31* | 10,123.31 |
| 7 | 3,111.56 | 5,123.31* | 5,000.00 |

* Indicates depreciation allowance actually used. Switching to straight line occurs in year 6.

## Example 6.10

In Example 6.9, we see that the last year's book value of $B_7 = \$7778.89$ exceeds the estimated salvage value of $5000. We therefore have reason to believe that switching to straight-line depreciation is desirable. First let us consider year 5. Straight-line depreciation for the last 3 years would be

$$\frac{21,345.27 - 5000.00}{3} = \$5448.42$$

which does not exceed the double declining balance depreciation of $6098.65. Trying year 6, the straight-line depreciation for the last two years would be

$$\frac{15,246.62 - 5000.00}{2} = \$5123.31$$

This does exceed the $4356.18 allowance under double declining balance depreciation and, consequently, in year 6 we will switch to straight line. The results are illustrated in Table 6.11.

## 6.6 COMPARISON OF DEPRECIATION METHODS

ACRS, straight-line (SL), sum of the years' digits (SOYD), double declining balance (DDB), and double declining balance switching to straight-line (DDB/SL) depreciation methods can be compared in several meaningful ways. The comparisons give some interesting results as well as insights into the importance of depreciation strategy.

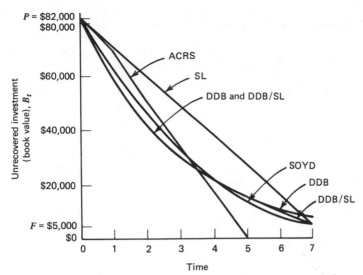

FIGURE 6.1. Value-time curves of five different depreciation methods.

## 6.6.1 Value-Time Curves

Value-time curves are plotted in Figure 6.1 using the stand-alone computer data from Examples 6.1, 6.7, 6.8, 6.9, and 6.10. Note that the method of depreciation determines the path of unrecovered investments (or book values), which can vary widely, particularly in the intermediate years.

## 6.6.2 Depreciation and the EUAC

Calculation of the equivalent uniform annual cost ($EUAC$) is presented in previous chapters. The $EUAC$ represents the equivalent uniform annual cost of capital recovered plus a return on the unrecovered investment. Using the stand-alone computer having $P = \$82,000$, $F = \$5000$, $n = 7$, and interest rate $i = 15\%$, the annual cost is[3]

$$EUAC = P(A|P\ 15,7) - F(A|F\ 15,7)$$
$$= \$82,000\ (.2404) - \$5000\ (.0904)$$
$$= \$19,257.75/year$$

Note that $EUAC$ is calculated here without respect to any depreciation method.

Table 6.12 presents the straight-line capital recovered plus return on the unrecovered investment from Example 6.7, covering the 7-year useful life of the computer from Example 6.7.

[3] In Chapter Six, some calculations are performed using interest factors determined and retained within a calculator for reasons of precision. Use of the interest tables in the Appendix will provide answers which differ insignificantly.

TABLE 6.12. Straight-Line Capital Recovery Plus Return

| End of Year, $t$ | SL Depreciation (Capital Recovered), $D_t$ | Book Value (Unrecovered Investment), $B_t$ | Return on Unrecovered Investment, $iB_{t-1}$ | Capital Recovered Plus Return, $D_t + iB_{t-1}$ |
|---|---|---|---|---|
| 0 | — | $82,000.00 | — | — |
| 1 | $11,000.00 | 71,000.00 | $12,300.00 | $23,300.00 |
| 2 | 11,000.00 | 60,000.00 | 10,650.00 | 21,650.00 |
| 3 | 11,000.00 | 49,000.00 | 9,000.00 | 20,000.00 |
| 4 | 11,000.00 | 38,000.00 | 7,350.00 | 18,350.00 |
| 5 | 11,000.00 | 27,000.00 | 5,700.00 | 16,700.00 |
| 6 | 11,000.00 | 16,000.00 | 4,050.00 | 15,050.00 |
| 7 | 11,000.00 | 5,000.00 | 2,400.00 | 13,400.00 |

Calculating the uniform annual cost of capital recovered plus return at $i = 15\%$ yields

$$A = [23,300(P|F\ 15,1) + 21,650(P|F\ 15,2) + \cdots + 13,400(P|F\ 15,7)]$$
$$[A|P\ 15,7]$$
$$= [23,300(.8696) + 21,650(.7561) + \cdots + 13,400(.3759)][.2404]$$
$$= \$19,257.75/year$$

This is the same as the *EUAC*! Had the *SOYD* or *DDB/SL* methods of Examples 6.8 and 6.10 been used, the result would have been the same, $A = \$19,257.75/year$.

Even using the ACRS method of Example 6.1 will yield the same result if we are careful. Remember, the ACRS allows us to recover the entire cost basis $P = \$82,000$, rather than just $P - F = \$77,000$. This $5000 over-recovery will eventually be lost in the form of "depreciation recapture" (discussed in Sec. 6.15) at the time of sale. If this $5000 is subtracted from ACRS Deduction ($D_t$) and capital recovered plus return ($D_t + iB_{t-1}$) in year 7, the uniform annual cost of capital recovered plus return will be $A = \$19,257.75/year$. Identical results can be obtained with the *DDB* method from Example 6.9 if a similar adjustment is made for the $7778.89 - \$5000 = \$2778.89$ under-recovery in year 7.

In summary, the uniform annual cost of an asset before taxes remains the same, regardless of the depreciation method used. It is most easily calculated using the formula for *EUAC*.

## 6.7 OTHER DEPRECIATION METHODS NOT BASED ON YEARS

Most depreciation methods for both pre-1981 and post-1980 are tied to the passage of time. Occasionally, however, recovery of cost over an ACRS recovery period, for example, would have little relation to an asset's use or its production of income. That is why the IRS allows taxpayers to utilize other consistent methods of depreciation. Some recognized methods are briefly presented below. Assets that we elect to depreciate in any of these ways are specifically excluded from the definition of recovery property to which the ACRS applies.

### 6.7.1 Units of Production Method

This procedure allows equal depreciation per each unit of output, regardless of the lapse of time involved. The allowance for year $t$ is equal to the total depreciable amount $(P - F)$ times the ratio of units produced during the year $(U_t)$ to the total units that are expected to be produced during the useful life of the asset $(U)$. That is,

$$D_t = (P - F) \frac{U_t}{U} \tag{6.14}$$

This method has proven suitable for depreciating equipment used in mining, oil and gas production, and the forest products industry.

### 6.7.2 Operating Day (Hour) Method

This is similar to the previous method in that year $t$ depreciation is based on the ratio of days (hours) used during the year $(Q_t)$ to total days (hours) expected in a useful life $(Q)$. Depreciation is expressed as

$$D_t = (P - F) \frac{Q_t}{Q} \tag{6.15}$$

### 6.7.3 Income Forecast Method

This method is applicable to depreciate the cost of rented property such as motion picture films. The ratio of year $t$ rental income $(R_t)$ to the total useful life income $(R)$ is multiplied by the total lifetime depreciation, or

$$D_t = (P - F) \frac{R_t}{R} \tag{6.16}$$

## 6.8 TAX CONCEPTS

The taxes paid by a corporation represent a real cost of doing business and, consequently, affect the cash flow profile. For this reason, it is wise to perform economic analyses on an *after-tax* basis. After-tax analysis procedures are identical to the before-tax evaluation procedures studied already; however, the cash flows are adjusted for taxes paid or saved.

There are numerous kinds of taxes including *ad valorem* (property), *sales, excise* (a tax or duty on the manufacture, sale, or consumption of various commodities), and *income taxes*. Income taxes are usually the only significant taxes to be considered in an economic analysis. Income taxes are assessed on gross income less certain allowable deductions, incurred both in the normal course of business as well as on gains resulting from the disposal of property.

Federal and state income tax regulations are not only detailed and intricate, but they are subject to change over time. During periods of recession and inflation, there is a tendency for the tax laws to be changed in order to improve the state of the economy. For this reason, only the general concepts and procedures for calculating after-tax cash flow profiles and performing after-tax analyses are emphasized here. Furthermore, only federal income tax will be considered because of the diversity of state laws. Practitioners involved in an actual analysis should seek the assistance of corporate legal counsel regarding any uncertainties about tax laws in effect at that time.

## 6.9 CORPORATE INCOME TAX—ORDINARY INCOME

Corporate income tax is not limited to business organizations that have actually incorporated. Engineers, lawyers, doctors, and other professionals may be treated as corporations if they have formally organized as professional corporations or associations. Associations, business trusts, joint stock companies, insurance firms, and certain limited partnerships are also taxed as corporations.

Ordinary federal income tax imposed on these corporations is presented in Table 6.13.

The tax rates shown are those current as of 1983 for tax years beginning in 1983 and beyond.[4] Companies such as many small businesses with taxable incomes less than $100,000 are subject to significantly reduced tax rates. Note, however, that every dollar of taxable income over $100,000 is taxed at the full rate of 46%. For illustration and consistency throughout the text and problems

TABLE 6.13. Corporate Income Tax
Rates for Tax Years Beginning in
1983 and Beyond

| Taxable Income, in Dollars | Tax Rate, Percent |
|---|---|
| $0 < TI \le 25,000$ | 15 |
| $25,000 < TI \le 50,000$ | 18 |
| $50,000 < TI \le 75,000$ | 30 |
| $75,000 < TI \le 100,000$ | 40 |
| $100,000 < TI$ | 46 |

[4] If tax rates for years prior to 1983 are required, references such as [1] or [2] for earlier years are excellent.

of this chapter, an applicable tax rate of 46% will be used unless otherwise specified. This implicitly assumes we are dealing with a company already having over $100,000 in taxable income.

## Example 6.11 _____

Our small company is currently forecasting a taxable income of $40,000 for this year. We are considering another investment that will increase taxable income by $30,000. If we take on the project, what will be our increased income tax liability? What would it be if we were currently forecasting a taxable income of $140,000 for this year?

Answer 1: Without the project the tax is .15(25,000) + .18(15,000) = $6450. With the project the tax is .15(25,000) + .18(25,000) + .30(20,000) = $14,250. The tax increase is $14,250 − $6450 = $7800, or 26% of the $30,000.

Answer 2: Without the project the tax is .15(25,000) + .18(25,000) + .30(25,000) + .40(25,000) + .46(40,000) = $44,150. With the project the tax is .15(25,000) + .18(25,000) + .30(25,000) + .40(25,000) + .46(70,000) = $57,950. The tax increase is $57,950 − $44,150 = $13,800, or 46% of the $30,000.

---

Taxable income must first be determined before any tax rate can be applied. Basically, *taxable income is gross income less allowable deductions. Gross income* is income in a general sense less any monies specifically exempt from tax liability. Corporate deductions are subtracted from gross income and commonly include items such as salaries, wages, repairs, rent, bad debts, taxes (other than income), charitable contributions, casualty losses, interest, and depreciation (including cost recovery). Interest and depreciation are of particular interest, since we can control them to some extent through financing arrangements and accounting procedures.

*Taxable income* is represented pictorially in Figure 6.2, which shows that taxable income for any year is what is left after deductions, including interest on borrowed money and depreciation allowance, are subtracted from gross income. These components are not all cash flows, since the depreciation allowance is simply treated as an expense in determining taxable income.

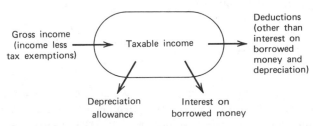

FIGURE 6.2. Pictorial representation of taxable income.

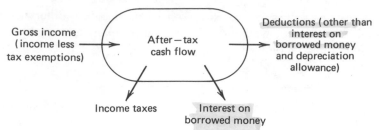

FIGURE 6.3. Pictorial representation of after-tax cash flow.

## 6.10 AFTER-TAX CASH FLOW

We have now looked at the basic elements needed to calculate after-tax cash flows. These elements are summarized in Figure 6.3. This shows that the after-tax cash flow is the amount remaining after income taxes and deductions, including interest but excluding depreciation allowance, are subtracted from gross income.

In many of the tables to follow, we simplify our terminology by speaking of before-tax cash flows. The term *before-tax cash flow* is used when no borrowed money is involved, and it equals gross income less deductions, not including depreciation. *Before-tax and loan cash flow* is used when borrowed money is involved, and it equals gross income less deductions, not including either depreciation or principal or interest on the loan.

Example 6.12_____

Let us again consider the stand-alone computer of Example 6.1 for which $P = \$82,000$, $F = \$5000$, and $n = 7$. It will be monitoring and controlling manufacturing equipment, and we expect it to reduce operating expenses by \$26,000/year. Therefore, before-tax cash flow will increase by \$26,000 during each year of the computer's useful life. We plan to depreciate it using the ACRS, and desire to know the after-tax cash flows. We will then apply the present worth method as an effectiveness measure to see whether the alternative to purchase the computer yields at least a 15% after-tax rate of return. The cash flow calculations are given in Table 6.14.

Calculating the present worth of after-tax cash flows, we have

$$PW(15) = -\$82,000 + \$19,698(P|F\ 15,1) +$$
$$\$22,338.40(P|F\ 15,2)$$
$$+ \cdots + \$14,040(P|F\ 15,7) + \$2700(P|F\ 15,7)$$
$$= \$2297.63$$

Year 7 is listed twice, once to record the normal cash flows and once to show the salvage value transaction. Note that the \$5000 salvage value is considered a "depreciation recapture" and taxed as ordinary income. This is because the

TABLE 6.14. After-Tax Cash Flow Profile Using the ACRS

| End of Year, A | Before-Tax Cash Flow, B | ACRS Deduction, C | Taxable Income B − C, D | Tax D × .46, E | After-Tax Cash Flow B − E, F |
|---|---|---|---|---|---|
| 0 | −$82,000.00 | — | — | — | −$82,000.00 |
| 1 | 26,000.00 | $12,300.00 | $13,700.00 | $ 6,302.00 | 19,698.00 |
| 2 | 26,000.00 | 18,040.00 | 7,960.00 | 3,661.60 | 22,338.40 |
| 3 | 26,000.00 | 17,220.00 | 8,780.00 | 4,038.80 | 21,961.20 |
| 4 | 26,000.00 | 17,220.00 | 8,780.00 | 4,038.80 | 21,961.20 |
| 5 | 26,000.00 | 17,220.00 | 8,780.00 | 4,038.80 | 21,961.20 |
| 6 | 26,000.00 | 0.00 | 26,000.00 | 11,960.00 | 14,040.00 |
| 7 | 26,000.00 | 0.00 | 26,000.00 | 11,960.00 | 14,040.00 |
| 7 | 5,000.00 (salvage) | — | 5,000.00 (depreciation recapture) | 2,300.00 | 2,700.00 |

entire cost basis ($P$ = $82,000) had already been recovered through ACRS deductions. Depreciation recapture is covered in Section 6.15 later in this chapter.

Example 6.12 shows that the after-tax cash flow profile is nothing more than lumps of money flowing in or out at different points in time over a planning horizon. Although the present worth effectiveness measure was used in the example, a rate of return, future worth, or other equivalent analysis could have been applied. Since the present worth of the discounted cash flows in our example was positive ($2297.63), we are able to conclude that the alternative as described indeed yields more than a 15% after-tax rate of return.

When we look at the taxable income and after-tax cash flow representations in Figures 6.2 and 6.3, it is natural to wonder about the tax effects of different depreciation methods, recovery periods, and interest on borrowed money. We will show that these factors have a substantial effect on taxes and thus on cash flow profiles.

## 6.11 EFFECT OF DEPRECIATION METHOD AND RECOVERY PERIOD

We have seen how some depreciation methods provide for a higher depreciation allowance during the early years of an asset's life and a correspondingly lower allowance in later years. This places a lower tax burden on the asset during the early years followed by a higher yearly burden. *In most cases the total undiscounted ordinary income tax dollars paid will be the same*, regardless of the depreciation method used. This is always true when the tax rate remains constant and the total depreciation allowance for each method is the same or an adjustment for under- or over-recovery is made at ordinary income tax rates as was done in Example 6.12.

The ACRS does not give much flexibility for selecting depreciation strategies on post-1980 property. In order to illustrate the effects of depreciation method and recovery period, two ACRS straight-line elections, plus two pre-1981 depreciation strategies will be used.

Example 6.13_____

Let us evaluate the stand-alone computer using ACRS straight-line depreciation and an $n = 5$ recovery period. Recall that the computer has $P = \$82,000$, $F = \$5,000$, and a useful life of 7 years. Also, it will reduce operating expenses (and therefore increase before-tax cash flow) by \$26,000/year. The cash flow calculations are given in Table 6.15.

Calculating the present worth of the after-tax cash flows at $i = 15\%$, we have

$$PW(15) = -\$82,000 + \$17,812(P|F\ 15,1) + \$21,584(P|F\ 15,2)$$
$$+ \cdots + \$14,040(P|F\ 15,7) + \$2700(P|F\ 15,7)$$
$$= \$1066.72$$

Note that the half-year convention on ACRS straight-line depreciation is applied in years 1 and 6. Also, the entire cost basis $P = \$82,000$ is recovered, necessitating taxation of the \$5000 salvage value in year 7.

---

TABLE 6.15. After-Tax Cash Flow Profile Using ACRS Straight-Line Depreciation

| End of Year, A | Before-Tax Cash Flow, B | ACRS SL Deduction, C | Taxable Income B − C, D | Tax D × .46, E | After-Tax Cash Flow B − E, F |
|---|---|---|---|---|---|
| 0 | −$82,000.00 | — | — | — | −$82,000.00 |
| 1 | 26,000.00 | $ 8,200.00 | $17,800.00 | $ 8,188.00 | 17,812.00 |
| 2 | 26,000.00 | 16,400.00 | 9,600.00 | 4,416.00 | 21,584.00 |
| 3 | 26,000.00 | 16,400.00 | 9,600.00 | 4,416.00 | 21,584.00 |
| 4 | 26,000.00 | 16,400.00 | 9,600.00 | 4,416.00 | 21,584.00 |
| 5 | 26,000.00 | 16,400.00 | 9,600.00 | 4,416.00 | 21,584.00 |
| 6 | 26,000.00 | 8,200.00 | 17,800.00 | 8,188.00 | 17,812.00 |
| 7 | 26,000.00 | 0.00 | 26,000.00 | 11,960.00 | 14,040.00 |
| 7 | 5,000.00 (salvage) | — | 5,000.00 (depreciation recapture) | 2,300.00 | 2,700.00 |

Comparing Tables 6.14 and 6.15, we can see that the standard ACRS deductions are higher in the early years (1 to 5), with correspondingly lower taxes and higher after-tax cash flows. The time value of money effect is seen in the present worth of after-tax cash flows. Merely electing one depreciation method over another can make a difference of $2297.63 − $1066.72 = $1230.91 in the present worth of after-tax cash flows at the *MARR* of 15%.

## Example 6.14

Now, let us consider another legal option, ACRS straight-line depreciation having a recovery period of $n = 12$ years. Recall that the computer has $P = \$82,000$, $F = \$5000$, a useful life of 7 years, and before-tax cash flows of $26,000/year. The cash flow calculations are given in Table 6.16.
  Calculating the present worth of the after-tax cash flows at $i = 15\%$, we have

$$PW(15) = -\$82,000 + \$15,611.67(P|F\ 15,1) + \$17,183.33(P|A\ 15,5)(P|F\ 15,1)$$
$$+ \$14,040(P|F\ 15,7) + \$23,131.67(P|F\ 15,7)$$
$$= -\$4362.44$$

Table 6.16 is different from any previous table in several respects. The computer is assumed to be sold toward the end of year 7, before the recovery period is completed. According to the ACRS rules, no deduction is permitted in the year of sale. Also, only $3416.67 + $6833.33(5) = $37,583.32 is recovered through ACRS deductions, leaving $44,416.68 unrecovered. Another $5000 is

TABLE 6.16. After-Tax Cash Flow Profile Using ACRS Straight-Line Depreciation Over 12-Year Recovery Period

| End of Year, A | Before-Tax Cash Flow, B | ACRS SL Deduction, C | Taxable Income B − C, D | Tax D × .46, E | After-Tax Cash Flow B − E, F |
|---|---|---|---|---|---|
| 0 | −$82,000.00 | — | — | — | −$82,000.00 |
| 1 | 26,000.00 | $3,416.67 | $22,583.33 | $10,388.33 | 15,611.67 |
| 2 | 26,000.00 | 6,833.33 | 19,166.67 | 8,816.67 | 17,183.33 |
| 3 | 26,000.00 | 6,833.33 | 19,166.67 | 8,816.67 | 17,183.33 |
| 4 | 26,000.00 | 6,833.33 | 19,166.67 | 8,816.67 | 17,183.33 |
| 5 | 26,000.00 | 6,833.33 | 19,166.67 | 8,816.67 | 17,183.33 |
| 6 | 26,000.00 | 6,833.33 | 19,166.67 | 8,816.67 | 17,183.33 |
| 7 | 26,000.00 | 0.00 | 26,000.00 | 11,960.00 | 14,040.00 |
| 7 | 5,000.00 (salvage) | 44,416.68 (unrecovered through ACRS deductions) | − 39,416.68 (Section 1231 loss) | − 18,131.67 | 23,131.67 |

recovered through salvage. The remaining $82,000 − $37,583.32 − $5000 = $39,416.68 is considered a Section 1231 loss and is here assumed to be applied against Section 1231 gains elsewhere in the company, resulting in a $39,416.68(.46) = $18,131.67 tax savings or negative tax. Section 1231 gains and losses are covered later in this chapter (see Sec. 6.15).

---

A quick look at the −$4362.44 present worth of after-tax cash flows indicates that the purchase of a stand-alone computer should be rejected. Yet, this is the same computer found to be desirable in Examples 6.12 and 6.13; only the depreciation method and/or recovery period have changed! Extending the recovery period, and thus deferring capital recovery, increases taxes and reduces after-tax cash flows in the early years, reducing the economic desirability of the investment.

In Example 6.14, we saw that an alternative's taxable income and tax can be negative for a particular year. It was assumed that those values were used to offset positive tax liabilities from other corporate activities. When there are insufficient positive taxable incomes to be offset, the taxable income losses can be carried back 3 years or forward 15 years in order to reduce positive taxable income. That is, if an operating loss occurs in 1987, we can reopen books beginning in 1984, then 1985 and 1986. If the loss is not fully absorbed against positive taxable incomes, the remainder may be used in 1988, 1989, and so on for as many as 15 years in the future. We will continue to assume that negative taxable incomes are used to offset positive values for the year in question.

## Example 6.15

Let us compare our post-1980 depreciation methods of Examples 6.12, 6.13, and 6.14 with the most widely used pre-1981 method, straight-line depreciation. Many pre-1981 assets will continue throughout the 1980s using straight-line depreciation. Again, $P = \$82,000$, $F = \$5000$, $n = 7$, and before-tax cash flow is $26,000/year. The cash flow calculations are given in Table 6.17.

Calculating the present worth of the after-tax cash flows at i = 15%, we have

$$PW(15) = -\$82,000 + \$19,100(P|A\ 15,7) + \$5000(P|F\ 15,7)$$
$$= -\$656.30$$

In this example, the $5000 salvage value is not taxed. The total straight-line depreciation is only $P − F = \$77,000$ and the last $5000 is recovered through the salvage value. Since there is no "over-" or "under-recovery," no extra tax treatment is necessary.

---

Had we been considering the computer for installation prior to 1981, an after-tax analysis using straight-line depreciation would have resulted in a decision to not invest due to the negative present worth of after-tax cash flows. Again, nothing about the computer is different; only the depreciation strategy has

TABLE 6.17. After-Tax Cash Flow Profile Using Pre-1981
Straight-Line Depreciation

| End of Year, A | Before-Tax Cash Flow, B | SL Depreciation, C | Taxable Income B − C, D | Tax D × .46, E | After-Tax Cash Flow B − E, F |
|---|---|---|---|---|---|
| 0 | −$82,000.00 | — | — | — | −$82,000.00 |
| 1 | 26,000.00 | $11,000.00 | $15,000.00 | $6,900.00 | 19,100,00 |
| 2 | 26,000.00 | 11,000.00 | 15,000.00 | 6,900.00 | 19,100.00 |
| 3 | 26,000.00 | 11,000.00 | 15,000.00 | 6,900.00 | 19,100.00 |
| 4 | 26,000.00 | 11,000.00 | 15,000.00 | 6,900.00 | 19,100.00 |
| 5 | 26,000.00 | 11,000.00 | 15,000.00 | 6,900.00 | 19,100.00 |
| 6 | 26,000.00 | 11,000.00 | 15,000.00 | 6,900.00 | 19,100.00 |
| 7 | 26,000.00 | 11,000.00 | 15,000.00 | 6,900.00 | 19,100.00 |
| 7 | 5,000.00 (salvage) | 5,000.00 (recovered through salvage) | — | — | 5,000.00 |

changed. By now, it should be coming clear why depreciation methods are of such concern to corporate taxpayers.

Example 6.16

Let us now evaluate the computer using another pre-1981 depreciation method, double declining balance switching to straight line. This method was often elected by taxpayers prior to 1981 due to its accelerated depreciation, resulting in higher present worths of after-tax cash flows. Many pre-1981 assets will continue in use throughout the 1980s under this depreciation method. Again, $P = \$82,000$, $F = \$5000$, $n = 7$, and before-tax cash flow is $26,000/year. The cash flow calculations are given in Table 6.18.

Calculating the present worth of the after-tax cash flows at $i = 15\%$, we have

$$PW(15) = -\$82,000 + \$24,817.14(P|F\ 15,1) + \$21,737.96(P|F\ 15,2)$$
$$+ \cdots + \$16,396.72(P|F\ 15,7) + \$5000(P|F\ 15,7)$$
$$= \$2644.75$$

Had the computer been considered for installation prior to 1981, this after-tax analysis would have resulted in a decision to invest, due to the positive present worth of after-tax cash flows. In fact, $2644.75 is the highest present worth we have seen for any depreciation method thus far. It is $2644.75 − $2297.63 = $347.12 higher than the ACRS method yields in Example 6.12.

Examples 6.12 to 6.16 illustrate the important points that the method of depreciation and the recovery period affect cash flows and, therefore, the

TABLE 6.18. After-Tax Cash Flow Profile Using Double Declining Balance Switching to Straight-Line Depreciation

| End of Year, A | Before-Tax Cash Flow, B | DDB/SL Depreciation, C | Taxable Income B − C, D | Tax D × .46, E | After-Tax Cash Flow B − E, F |
|---|---|---|---|---|---|
| 0 | −$82,000.00 | — | — | — | −$82,000.00 |
| 1 | 26,000.00 | $23,428.57 | $ 2,571.43 | $1,182.86 | 24,817.14 |
| 2 | 26,000.00 | 16,734.69 | 9,265.31 | 4,262.04 | 21,737.96 |
| 3 | 26,000.00 | 11,953.35 | 14,046.65 | 6,461.46 | 19,538.54 |
| 4 | 26,000.00 | 8,538.11 | 17,461.89 | 8,032.47 | 17,967.53 |
| 5 | 26,000.00 | 6,098.65 | 19,901.35 | 9,154.62 | 16,845.38 |
| 6 | 26,000.00 | 5,123.31 | 20,876.69 | 9,603.28 | 16,396.72 |
| 7 | 26,000.00 | 5,123.31 | 20,876.69 | 9,603.28 | 16,396.72 |
| 7 | 5,000.00 (salvage) | 5,000.00 (recovered through salvage) | — | — | 5,000.00 |

economic desirability of a project. It is interesting to note that the total capital recovery ($82,000), total taxes ($48,300), and total undiscounted after-tax cash flows ($56,700) are the same in each of Tables 6.14 to 6.18. Can we make any generalized statements *from these examples* about the preferability of depreciation methods and recovery periods with respect to effects on taxes and after-tax cash flows? No, because these examples are worked only for a particular set of conditions. It can, however, be shown mathematically that certain depreciation strategies are superior to others in that they provide a higher present worth of tax savings, assuming that the effective tax rate remains the same from year to year. For example, for post-1980 property, the regular ACRS percentage as given in Tables 6.1 and 6.2 are always preferable to the ACRS straight-line alternative. Further, if the ACRS straight-line method is used, the shortest recovery period is always superior to the longer alternatives. For pre-1981 property, both double declining balance switching to straight-line and sum of the years' digits are preferable to straight-line depreciation; however, double declining balance switching to straight line is not always better than the sum of the years' method. Although not illustrated, the shortest useful life legally permitted is superior to selection of a longer useful life.

Quite often the ACRS is better than any of the pre-1981 depreciation methods, especially when the recovery period is significantly shorter than the useful life. In the computer example, double declining balance switching to straight line slightly outperformed the ACRS, mainly due to the closeness between recovery period (5 years) and useful life (7 years). In general, however, the ACRS provides corporate taxpayers with reasonably attractive depreciation allowances that should be useful in stimulating investment. More on comparisons of post-1980 and pre-1981 depreciation methods can be found in [3] and [4].

## 6.12 EFFECT OF INTEREST ON BORROWED MONEY

Investment alternatives may be financed using equity (owner's) funds or debt (borrowed) funds. Until now we have implicitly assumed all financing to be through equity, although many companies use a mix of debt and equity for financing plant and equipment. Borrowed funds, including both principal and interest, must be repaid. The interest repaid each year affects taxable income and, consequently, taxes. Both the principal and interest payments affect after-tax cash flows.

There are four common (and many less common) ways in which money can be repaid. First is the periodic payment of interest over the stipulated repayment period with the entire principal being repaid at the end of that time. The second requires a periodic payment that uniformly repays the principal and also covers the periodic interest. In this method, the payments decrease as the interest on the unrepaid principal decreases. Third is the method requiring a uniform periodic payment for the sum of principal plus interest. In each payment the proportion of principal gradually increases as the proportion of interest decreases. The fourth method repays nothing, neither interest nor principal, until the end of a specified period.

Loan interest rates have varied widely in recent years. For example, during 1 calendar year, the prime interest rate charged by banks to their best corporate customers varied in the range from about 11% to 22%. Since neither the banks nor the customers want to get "stuck" in an unfavorable long-term agreement, loans with adjustable interest rates have become commonplace. For example, a typical small business loan might span 1 or 2 years, with interest adjusted to the current market rate and paid monthly, and with the entire principal due at the end of the loan period.

Example 6.17_____

Let us illustrate the four basic plans for repaying principal and interest on borrowed money. Assume that a business borrows \$40,000 to be used in financing an alternative, and the interest rate on this loan is 18% compounded annually. The stipulated repayment period is 5 years.

A summary of all relevant components of our example is presented in Table 6.19. In method 1, the interest equals \$40,000(.18) = \$7200/year. Only the interest is paid, and only the principal of \$40,000 is owed after each year's payment. Method 2 repays the principal in equal amounts of \$40,000/5 = \$8000 as well as the interest for year $t$, which is given by [\$40,000 − \$8000($t$ − 1)]0.18. Clearly, the interest payment, total payment, and total money owed after yearly payments are decreasing gradient series. Method 3 requires equal annual total payments. This annual payment is equal to \$40,000($A|P$ 18,5) =

TABLE 6.19. Illustration of Four Common Methods of Principal and Interest Repayment

| End of Year | Interest Accrued During Year | Total Money Owed Before Yearly Payment | Interest Payment | Principal Payment | Total Payment | Total Money Owed After Yearly Payment |
|---|---|---|---|---|---|---|
| **Method 1** | | | | | | |
| 0 | | | | | | $40,000.00 |
| 1 | $ 7,200.00 | $47,200.00 | $ 7,200.00 | $      0 | $ 7,200.00 | 40,000.00 |
| 2 | 7,200.00 | 47,200.00 | 7,200.00 | 0 | 7,200.00 | 40,000.00 |
| 3 | 7,200.00 | 47,200.00 | 7,200.00 | 0 | 7,200.00 | 40,000.00 |
| 4 | 7,200.00 | 47,200.00 | 7,200.00 | 0 | 7,200.00 | 40,000.00 |
| 5 | 7,200.00 | 47,200.00 | 7,200.00 | 40,000.00 | 47,200.00 | 0 |
| **Method 2** | | | | | | |
| 0 | | | | | | $40,000.00 |
| 1 | $ 7,200.00 | $47,200.00 | $ 7,200.00 | $ 8,000.00 | $15,200.00 | 32,000.00 |
| 2 | 5,760.00 | 37,760.00 | 5,760.00 | 8,000.00 | 13,760.00 | 24,000.00 |
| 3 | 4,320.00 | 28,320.00 | 4,320.00 | 8,000.00 | 12,320.00 | 16,000.00 |
| 4 | 2,880.00 | 18,880.00 | 2,880.00 | 8,000.00 | 10,880.00 | 8,000.00 |
| 5 | 1,440.00 | 9,440.00 | 1,440.00 | 8,000.00 | 9,440.00 | 0 |
| **Method 3** | | | | | | |
| 0 | | | | | | $40,000.00 |
| 1 | $ 7,200.00 | $47,200.00 | $7,200.00 | $ 5,591.11 | $12,791.11 | 34,408.89 |
| 2 | 6,193.60 | 40,602.49 | 6,193.60 | 6,597.51 | 12,791.11 | 27,811.37 |
| 3 | 5,006.05 | 32,817.42 | 5,006.05 | 7,785.07 | 12,791.11 | 20,026.31 |
| 4 | 3,604.73 | 23,631.04 | 3,604.73 | 9,186.38 | 12,791.11 | 10,839.93 |
| 5 | 1,951.19 | 12,791.11 | 1,951.19 | 10,839.93 | 12,791.11 | 0 |
| **Method 4** | | | | | | |
| 0 | | | | | | $40,000.00 |
| 1 | $ 7,200.00 | $47,200.00 | $      0 | $      0 | $      0 | 47,200.00 |
| 2 | 8,496.00 | 55,696.00 | 0 | 0 | 0 | 55,696.00 |
| 3 | 10,025.28 | 65,721.28 | 0 | 0 | 0 | 65,721.28 |
| 4 | 11,829.83 | 77,551.11 | 0 | 0 | 0 | 77,551.11 |
| 5 | 13,959.20 | 91,510.31 | 51,510.31 | 40,000.00 | 91,510.31 | 0 |

$12,791.11. The principal component of this annual payment for year $t$ can be found quickly as $12,791.11(P|F\ 18,5 - t + 1)$, using the method given in Chapter Three. It is interesting to note that the principal payment is an 18% increasing compound series. Also, the interest payment and the total money owed after yearly payment decrease each year, and the amount by which they decrease is an 18% increasing compound series. In method 4, the interest accrued during each year is added to the principal such that the total amount owed after $t$ years is $40,000(1.18)^t$. When payment is made at the end of year 5, everything over $40,000 is considered interest.

---

We have seen that the interest on borrowed money is deductible for tax purposes, whereas the principal repayment does not enter into taxable income. In addition, both the interest and principal portions of a payment are real and must be taken into account when calculating cash flows.

### Example 6.18

Let us illustrate the effect of borrowed money by recalling the stand-alone computer having $P = \$82,000$, $F = \$5000$, a useful life of 7 years, and before-tax and loan cash flows of $26,000. Assume that $40,000 of the $82,000 paid for the computer is through debt funding. The loan is to be repaid in equal annual installments (method 3) at 18% over 5 years. The remaining $42,000 will be equity money. Applicable depreciation and loan activity (method 3) have previously been calculated in Examples 6.1 and 6.17, respectively. The resulting after-tax cash flow profile is detailed in Table 6.20.

Calculating the present worth of the after-tax cash flows at $i = 15\%$, we have

$$
\begin{aligned}
PW(15) = & -\$42,000 + \$10,218.89(P|F\ 15,1) + \$12,396.34(P|F\ 15,2) \\
& + \$11,472.87(P|F\ 15,3) + \$10,828.26(P|F\ 15,4) \\
& + \$10,067.67(P|F\ 15,5) + \$14,040(P|F\ 15,6) \\
& + \$14,040(P|F\ 15,7) + \$2700(P|F\ 15,7) \\
= & -\$42,000 + \$10,218.89(.8696) + \$12,396.34(.7561) \\
& + \$11,472.87(.6575) + \$10,828.26(.5718) \\
& + \$10,067.67(.4972) + \$14,040(.4323) \\
& + \$14,040(.3759) + \$2700(.3759) \\
= & \$7362.55
\end{aligned}
$$

The value of our effectiveness measure jumped to $7362.55 for this particular example. Note that the present worth calculation on cash flows was made using a discount rate equal to our 15% *MARR*. The 18% loan rate was used only in sizing the loan repayments. If we actually need to borrow the $40,000 to implement the project, our estimates indicate that a handsome monetary return will be received on the $42,000 equity investment. If, on the other hand, we really

TABLE 6.20. After-Tax Cash Flow Profile Using ACRS Depreciation Deductions and $40,000 Borrowed Money at 18%

| End of Year, A | Before-Tax and Loan Cash Flow, B | Loan Principal Payment, C | Loan Interest Payment, D | ACRS Deduction, E | Taxable Income B − D − E, F | Taxes F × .46, G | After-Tax Cash Flow B − C − D − G, H |
|---|---|---|---|---|---|---|---|
| 0 | −$82,000.00 | −$40,000.00 | | | | | −$42,000.00 |
| 1 | 26,000.00 | 5,591.11 | $7,200.00 | $12,300.00 | $ 6,500.00 | $ 2,990.00 | 10,218.89 |
| 2 | 26,000.00 | 6,597.51 | 6,193.60 | 18,040.00 | 1,766.40 | 812.54 | 12,396.35 |
| 3 | 26,000.00 | 7,785.07 | 5,006.05 | 17,220.00 | 3,773.95 | 1,736.02 | 11,472.86 |
| 4 | 26,000.00 | 9,186.38 | 3,604.73 | 17,220.00 | 5,175.27 | 2,380.62 | 10,828.27 |
| 5 | 26,000.00 | 10,839.93 | 1,951.19 | 17,220.00 | 6,828.81 | 3,141.25 | 10,067.63 |
| 6 | 26,000.00 | 0.00 | 0.00 | 0.00 | 26,000.00 | 11,960.00 | 14,040.00 |
| 7 | 26,000.00 | 0.00 | 0.00 | 0.00 | 26,000.00 | 11,960.00 | 14,040.00 |
| 7 | 5,000.00 (salvage) | | | | 5,000.00 (Section 1231 Gain) | 2,300.00 | 2,700.00 |

have at least the other $40,000 available, borrowing allows us to invest that money as equity capital in another alternative that will earn a return at least equal to the *MARR*. This, of course, depends on the availability of investments that will yield a rate at least equal to the *MARR*.

---

A relatively high rate of interest, 18%, was used for the loan in Example 6.18 for two reasons. First, it appears illogical for company management to borrow at 18% when they require only a 15% *MARR*. Remember though, the 15% *MARR* is an *after-tax* rate of return. Since loan interest is a deductible item, it is being paid with *before-tax* dollars. The after-tax interest rate or "cost of capital" for the loan is 18% × (1 − tax rate) = 18(1 − .46) = 9.72%. The second reason is to illustrate that borrowed money, even at relatively high rates, can often "leverage" equity capital such that a nice return is earned as shown in Example 6.18.

We cannot conclude from the example that borrowing money is always favorable. The desirability of borrowed funds depends on the terms of the loan, including method of repayment, interest, and repayment period. Furthermore, collateral, which may be lost (after legal action) if the principal and interest cannot be paid on schedule, is frequently required. In summary, each alternative investment and financing strategy should be compared on its own merits.

We have used the same basic example to illustrate the implications of depreciation and financing strategies thus far throughout the chapter. The present worth of after-tax cash flows was seen to vary from −$4362.44 using equity financing under the ACRS straight-line option over 12 years, to $7362.55 using a more sophisticated mix of debt and equity financing and the ACRS with a 5-year recovery period. Nothing about the basic incomes or costs relating to the asset were changed. By now the reader should have a good idea how to calculate and assess after-tax cash flows under differing depreciation schedules and financing arrangements, and should be motivated to apply these techniques in the analysis of alternative investments.

## 6.13 CAPITAL GAINS AND LOSSES

Many tax experts agree that there are no provisions in the internal revenue laws more difficult to understand and apply than those pertaining to capital gains and losses. Assets receiving capital treatment are important for tax purposes because they receive preferential tax rates, being less severely taxed than ordinary income. Congress originally desired to lessen the blow of ordinary income tax on gains caused by appreciation of items held in a risk situation over a substantial period of time. Over the years, however, taxpayers have maneuvered to assure that certain items are classified as eligible for capital gains treatment. As a result, unforeseen loopholes have been closed by the IRS, increasing the complexity of the tax provisions. Many of the common capital gain and loss situations faced in economic analyses will be presented in this chapter; however, any uncertainty should be satisfied by legal counsel.

A capital asset is any property held by a taxpayer *except:*

1. Stock in trade or other property properly included in inventory.
2. Property held *primarily* for sale to customers in the ordinary course of the taxpayer's trade or business.
3. Depreciable property used in a trade or business.
4. Real property used in the taxpayer's trade or business.
5. A copyright, a literary, musical, or artistic composition, a letter or memorandum, or similar property (but not a patent) held by a taxpayer whose efforts created the property; or by a taxpayer for whom a letter, memorandum, or similar property was prepared or produced; or by a taxpayer acquiring (e.g., by gift) the basis of a person who created the property or for whom it was prepared or produced.
6. Accounts or notes receivable acquired in the ordinary course of trade or business for services rendered, or from the sale of stock in trade, inventory or property held for sale to customers in the ordinary course of trade or business.
7. Certain short-term government obligations issued on a discount basis.
8. U.S. Government publications acquired free or for less than the public sales price [1].

Shares of stock, stock rights, bonds, notes, debentures, and similar securities are considered capital assets unless they fall under one of the exceptions listed. Real property and property held for the production of income, not used in a trade or business, are capital assets. Patents held for investment, inventions in the hands of the inventor, life estates, inherited jewelry, bank accounts, and cotton acreage allotments, for example, are also considered capital assets. Gold, silver, stamps, coins, gems, and so on, expand the list.

Eligibility is the key difficulty in complying with IRS capital gains codes. For example, the courts have used purpose of acquisition and disposal, frequency and extent of sales, nature of the taxpayer's business, the taxpayer's sales efforts, and so forth, to determine whether or not capital treatment is allowable. For example, any items sold to customers in the ordinary course of trade or business may not be considered capital assets by the seller, and they are therefore subject to ordinary gains or losses.

Before applying capital treatment, the extent of the capital gain or loss must be determined. This requires a balancing of long- and short-term gains and losses. A capital gain or loss may occur when there is a sale or exchange of a capital asset. The excess of selling price over cost basis is a *capital gain.* If the selling price is less than the cost basis, then a *capital loss* has occurred. If the period during which the taxpayer holds the asset is less than or equal to 1 year, the gain or loss is referred to as a *short-term gain or loss*. If the asset is held longer than 1 year, a *long-term gain or loss* has occurred. A net short-term gain (loss) is the sum of short-term gains minus the sum of short-term losses. Similarly, a net long-term gain (loss) is the sum of long-term gains minus long-term losses.

Example 6.19

A corporation has the following capital gains and losses during a tax year. The net short- and long-term gains and losses are summarized as follows:

| | |
|---|---|
| Short-term capital gains | $19,300 |
| Short-term capital losses | ($27,600) |
| Net short-term capital loss | ($ 8,300) |
| | |
| Long-term capital gains | $77,500 |
| Long-term capital losses | ($18,750) |
| Net long-term capital gain | $58,750 |

## 6.14 TAX TREATMENT OF CAPITAL ASSETS

Short- and long-term capital gains may be consolidated for tax purposes. The result may be a net capital gain, or a net capital loss. The consolidated result dictates whether the tax treatment will be capital gains, ordinary income, or a loss to be carried back or over. The consolidation rules and tax treatments are summarized in Table 6.21.

Example 6.20

In Example 6.19 we determined that a corporation has a net short-term capital loss of $8300 and a net long-term capital gain of $58,750. Consolidating these in accordance with Table 6.21, we have a net capital gain of $58,750 − $8300 = $50,450, which may be taxed as a capital gain.

There are two basic ways of computing taxes when capital gains and losses are involved. First is the method whereby all taxable income including capital gains are treated as ordinary income. An alternative method excludes the excess of net long-term capital gains over net short-term capital losses. The ordinary income is then taxed as usual and the excluded net capital gain is taxed at the rate of 28%. The taxpayer may select either of these two methods.

TABLE 6.21. Results and Corporate Tax Treatment of Consolidated Long- and Short-Term Capital Gains and Losses*

| Net Short-Term Capital Gains and Losses | Net Long-Term Capital Gains and Losses | | |
|---|---|---|---|
| | None | Long-Term Gain | Long-Term Loss |
| None | No capital treatment | Net capital gain taxed as capital gain | Net capital loss may be carried over |
| Short-term gain | Net capital gain taxed as ordinary income | Net capital gain. Long-term gain taxed as capital gain *and* short-term gain taxed as ordinary income | If a net capital loss, may be carried back or over. If a net capital gain, taxed as ordinary income |
| Short-term loss | Net capital loss may be carried back or over | If a net capital gain, taxed as capital gain. If a net capital loss, may be carried back or over | Net capital loss may be carried back or over |

* Capital losses carried back and over are treated as short-term capital losses.

Example 6.21 _____

Now, suppose the corporation has $170,450 taxable income of which we found $50,450 to be a net capital gain taxable as a capital gain. Therefore $120,000 would be ordinary income. Computing the federal taxes by both the regular and alternative methods we have:

*Regular Method*

$$\text{Ordinary income tax} = \$25,000(.15) + \$25,000(.18) + \$25,000(.30)$$
$$+ \$25,000(.40) + \$70,450(.46) = \$58,157$$

*Alternate Method*

*Note:* Separate $50,450 net capital gain income from $120,000 ordinary income.

$$\text{Ordinary income tax} = \$25,000(.15) + \$25,000(.18) + \$25,000(.30)$$
$$+ \$25,000(.40) + \$20,000(.46) = \$34,950$$

$$\text{Capital gains tax} = \$50,450(.28) = \$14,126$$

$$\text{Total tax} = \$34,950 + \$14,126 = \$49,076$$

The alternative method for capital gains would save over $9000 in taxes for the company.

## Example 6.22

A small business has taxable income of $22,000, including $5000 of net long-term capital gains *and* $3000 of net short-term capital gains. Table 6.21 shows that there is a net capital gain, and the long-term gains may be taxed as capital gains and the short-term gains as ordinary income. Computing the taxes using the regular method, we have:

*Regular Method*

$$\text{Ordinary income tax} = \$22,000(.15) = \$3300$$

*Alternate Method*

*Note:* Separate $5000 long-term capital gain from other $17,000 taxable as ordinary income.

$$\text{Ordinary income tax} = \$17,000(.15) = \$2550$$

$$\text{Capital gains tax} = \$5000(.28) = \$1400$$

$$\text{Total tax} = \$2550 + \$1400 = \$3950$$

We can see that a company would not elect to use the alternate procedure here, since only a 15% tax rate is applied to its taxable income.

When capital losses are substracted from capital gains and the result is negative, Table 6.21 shows that the net loss may be carried back or over. The carry-back period is 3 years and the carry-forward period is 5 years, during which time the capital loss is treated as a short-term capital loss and can be used to offset capital gains. Capital losses cannot be used to offset ordinary income.

## 6.15 DISPOSITION OF DEPRECIABLE PROPERTY

The definition of capital assets in the previous section specifically excluded depreciable business property and business real estate. Yet, tax law contains a special benefit provision that, under certain conditions, allows assets classified as "Section 1231" properties to receive favored capital asset treatment upon disposal. Also, whether or not the property qualifies for capital asset treatment, there may be a recapture of excess depreciation that is taxable as ordinary income. In general, we need to know how to handle income from the sale of an asset. First, a few definitions are needed.

### 6.15.1 Section 1245, 1250, and 1231 Properties

*Recovery property* was previously described as tangible depreciable property placed in service after 1980. It does not include property depreciated on a basis other than years, such as the units of production method. Nor does it include some public utility property. Basically, tangible (things you can touch) property depreciated under the ACRS is recovery property.

*Section 1245 property* is depreciable property that is either:

1. Personal property (tangible or intangible).
2. Other tangible property (not including a building or its structural components) that
   a. is used in connection with manufacturing, production, extraction, or furnishing transportation, communications, or utilities,
   b. constitutes a research facility connected with anything in (a), or
   c. includes certain other specific facilities.
3. Livestock.
4. Single purpose agricultural or horticultural structures.
5. Storage facilities used in connection with the distribution of petroleum.
6. Elevators and escalators.
7. Certain specific portions of some real property.

*Section 1245 class property* includes the tangible portion of Section 1245 property as described above, less (1) elevators and escalators, and (2) certain specific portions of some real property. *For most engineering economic analyses, Section 1245 class property involves depreciable tangible property such as equipment, machinery, tooling, office furniture, or vehicles used in trade or business.*

*Section 1250 property* is depreciable property that includes all intangible real property (such as leases of land), buildings and their structural components, and all tangible real property except Section 1245 property. *Section 1250 class property* includes the tangible portion of 1250 property as described above, plus elevators and escalators. *For most engineering economic analyses, Section 1250 class property involves tangible real property such as buildings, fixtures, fences, and parking lots.*

*Section 1231 property* includes all Section 1245 and Section 1250 properties described above, plus some other assets. Examples of these assets include land used in trade or business, mining property, timber, and royalties for timber, coal, and iron ore. In general, property must be held longer than 1 year to qualify under Section 1231 for the favorable capital asset tax treatment. Figure 6.4 illustrates the relationship between Section 1245, 1250, and 1231 properties.

## 6.15.2 Tax Treatment of Depreciable Properties

A gain or loss is usually realized when a depreciable asset is sold or in some way disposed. The nature of this gain or loss determines the tax treatment which is very favorable for Section 1231 property. First, all Section 1231 gains and losses are grouped. Then, if the gains exceed the losses, an overall gain is realized and *each individual* gain and loss is treated as if it were from the sale of a long-term capital asset. If gains do not exceed losses, *each individual* gain and loss is treated as if it were ordinary income. Thus, long-term capital gain

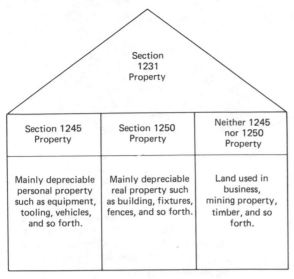

FIGURE 6.4. Relationship between Section 1245, 1250, and 1231 Properties.

benefits are extended where there is a net gain, and ordinary loss benefits are available where there is a net loss.

Since it is impossible to know the performance of all Section 1231 assets in advance, a reasonable rule is necessary in order to treat anticipated disposals of property in engineering economic analyses. Therefore, unless better information is available, Section 1231 gains are treated as capital gains and Section 1231 losses are treated as ordinary losses for tax purposes.

To help understand the tax treatment of Section 1231 property, three figures are presented in which the following notation is used:

$P$ = cost basis or original investment.

$B_t$ = unrecovered investment at time $t$; this is the cost basis less allowable ACRS deductions.

$S_t$ = unrecovered investment at time $t$ using ACRS straight line depreciation over a 15-year recovery period.

$F_t$ = salvage value actually received at time $t$.

$t$ = any time, during or beyond the recovery period.

In conjunction with Figures 6.5, 6.6, and 6.7, Table 6.22 summarizes the rules for tax treatment of post-1980 depreciable property.

Rules for the tax treatment upon disposal of pre-1981 property are very similar. Pre-1981 Section 1245 and 1250 properties follow the rules of Parts A and B of Table 6.22, respectively. Even the notation used in Figures 6.5 and 6.6 remains the same with the exception of $B_t$ and $S_t$. Notation $B_t$ is the unrecovered investment using one of the pre-1981 depreciation methods such as described in Section 6.5; notation $S_t$ relates to pre-1981 straight-line depreciation.

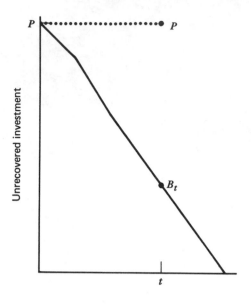

FIGURE 6.5. Unrecovered investment versus time for Section1245 recovery property and Section 1250 nonresidential recovery property except when ACRS straight-line depreciation is used.

## Example 6.23

In Example 6.12, the stand-alone computer's cost basis $P$ = \$82,000 was fully recovered under the ACRS by the end of year 5. How should the \$5000 salvage value received at year 7 be treated for tax purposes?

The computer is Section 1245 recovery property. Referring to Tables 6.14 and 6.22 and Figure 6.5, $B_7$ = \$0 and $F_7$ = \$5000. Therefore, $F_7 - B_7$ = \$5000 is

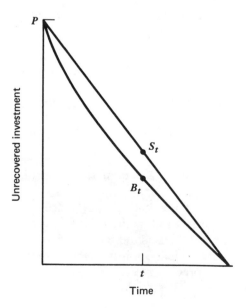

FIGURE 6.6. Unrecovered investment versus time for Section 1250 residential (rental) recovery property except when ACRS straight-line depreciation is used.

TABLE 6.22. Rules for Tax Treatment of Depreciable Properties

A. Property Types

Section 1245 recovery property
Section 1250 nonresidential recovery property except when ACRS straight-line depreciation is used.

Rules (see Figure 6.5)

| If | Then |
|---|---|
| $F_t > P$ | $F_t - P$ goes into Section 1231 group as a *gain* |
| | and |
| | $P - B_t$ depreciation recapture treated as ordinary income |
| $B_t \leq F_t \leq P$ | $F_t - B_t$ depreciation recapture treated as ordinary income |
| $F_t < B_t$ | $B_t - F_t$ goes into Section 1231 group as a *loss* |

B. Property Type

Section 1250 residential (rental) recovery property except when ACRS straight-line depreciation is used

Rules (see Figure 6.6)

| If | Then |
|---|---|
| $F_t > S_t$ | $F_t - S_t$ goes into Section 1231 groups as a *gain* |
| | and |
| | $S_t - B_t$ depreciation recapture treated as ordinary income |
| $B_t \leq F_t \leq S_t$ | $F_t - B_t$ depreciation recapture treated as ordinary income |
| $F_t < B_t$ | $B_t - F_t$ goes into Section 1231 group as a *loss* |

C. Property Type

Section 1250 nonresidential or residential (rental) recovery property on which ACRS straight-line depreciation is used

Rules (see Figure 6.7)

| If | Then |
|---|---|
| $F_t > S_t$ | $F_t - S_t$ goes into Section 1231 group as a *gain* |
| $F_t \leq S_t$ | $S_t - F_t$ goes into Section 1231 group as a *loss* |

depreciation recapture and should be taxed as ordinary income in the amount of $5000(.46) = \$2300$. This agrees with the treatment in Example 6.12.

Example 6.24

Let us now reconsider Example 6.14 in which we used ACRS straight-line depreciation over a 12-year recovery period. Since the useful life of the computer was 7 years, only $37,583.32 of the cost basis $P = \$82,000$ had been recovered through depreciation. How should disposal of the computer in year 7 have been handled?

The computer is Section 1245 recovery property. Referring to Tables 6.16 and 6.22 and Figure 6.5, $B_7 = \$82,000 - \$37,583.32 = \$44,416.68$, and $F_7 = \$5000$. Therefore, $B_7 - F_7 = \$44,416.68 - \$5000 = \$39,416.68$ goes into the

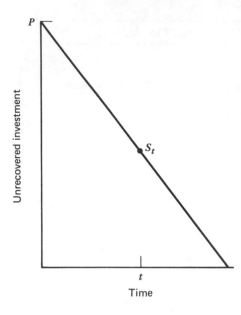

FIGURE 6.7. Unrecovered investment versus time for Section 1250 nonresidential or residential (rental) recovery property on which ACRS straight-line depreciation is used.

Section 1231 group as a loss. Since we have no information regarding other Section 1231 gains and losses, this loss will be assumed to apply against ordinary income. As such, there should be a tax of $-\$39,416.68(.46) = -\$18,131.67$ or a negative tax to apply against other positive taxes in year seven. This agrees with the treatment in Example 6.14.

---

## Example 6.25

Although unrealistic, suppose the computer having $P = \$82,000$, a useful life of 7 years, and ACRS recovery over 5 years is sold at $t = 7$ for $\$100,000$. How should this salvage value be treated for tax purposes?

The computer is Section 1245 recovery property. Referring to Table 6.22 and Figure 6.5, $B_7 = \$0$ and $F_7 = \$100,000$. Therefore, $F_7 - P = \$100,000 - \$82,000 = \$18,000$ goes into the Section 1231 group as a gain. Since we have no information on other Section 1231 gains and losses, this gain will be assumed to be taxed as a capital gain. The amount $P - B_7 = \$82,000$ represents depreciation recapture and is taxed as ordinary income. Thus, the total tax should be $\$18,000(.28) + \$82,000(.46) = \$42,760$.

---

## Example 6.26

Consider the business real estate property of Example 6.2 in which $P = \$220,000$. After 18 years, the property is sold for $\$170,000$. Assume the property is nonresidential. How should this sale be taxed?

This is Section 1250 nonresidential recovery property. Referring to Tables 6.4 and 6.22 and Figure 6.5, $B_{18} = \$0$ and $F_{18} = \$170,000$. Therefore, $F_{18} - B_{18} = \$170,000$ is depreciation recapture taxed as ordinary income in the amount of $\$170,000(.46) = \$78,200$.

---

Example 6.27 _____

Suppose the scenario of Example 6.26 remains the same, but the property is residential. How should the sale be taxed?

This is Section 1250 residential recovery property. Referring to Tables 6.4 and 6.22 and Figure 6.6, $B_{18} = \$0$ and $F_{18} = \$170,000$. Also, $S_{18} = 0$ since ACRS straight-line depreciation would have fully recovered the $P = \$220,000$. Therefore the $F_{18} - S_{18} = \$170,000$ is treated as a Section 1231 gain. Lacking information about other Section 1231 property, we will tax the $\$170,000$ as a capital gain, resulting in $\$170,000(.28) = \$47,600$ tax.

---

Example 6.28 _____

Suppose the residential rental property $P = \$220,000$ is placed in service on May 1 and sold exactly 8 years later (this will be during the ninth tax year) for $\$170,000$. How should the sale be taxed?

Total depreciation under both the ACRS and ACRS straight line are presented in Table 6.23.

Referring to Tables 6.22 and 6.23 and Figure 6.6, $P = \$220,000$, $B_9 = \$220,000 - \$142,266.67 = \$77,733.33$, $S_9 = \$220,000 - \$117,333.33 = \$102,666.67$, and $F_9 = \$170,000$. Therefore, $F_9 - S_9 = \$170,000 - \$102,666.67 = \$67,333.33$ goes into the Section 1231 group as a gain, and $S_9 - B_9 = \$102,666.67 - \$77,733.33 = \$24,933.34$ is depreciation recapture taxable as

---

TABLE 6.23. ACRS and ACRS Straight-Line Deduction on Real Estate

| End of Year, $t$ | ACRS Deduction, $D_t$ | ACRS SL Deduction, $D_t$ |
|---|---|---|
| 0 | — | — |
| 1 | $17,600.00 | 8/12 × $14,666.67 = $ 9,777.78 |
| 2 | 24,200.00 | 14,666.67 |
| 3 | 22,000.00 | 14,666.67 |
| 4 | $17,600.00 | 14,666.67 |
| 5 | 15,400.00 | 14,666.67 |
| 6 | 15,400.00 | 14,666.67 |
| 7 | 13,200.00 | 14,666.67 |
| 8 | 13,200.00 | 14,666.67 |
| 9 | 4/12 × 11,000 = 3,666.67 | 4/12 × 14,666.67 = 4,888.89 |
| Total Deductions | $142,266.67 | $117,333.33 |

ordinary income. In the absence of further information about Section 1231 gains and losses, total tax will be $67,333.33(.28) + $24,933.34(.46) = $30,322.67.

---

Example 6.29

Consider the following results in a given taxable year (all items held longer than 1 year):

1. Stock—gain of $10,000
2. Bonds—loss of $3000
3. Factory building—gain above depreciation recapture $20,000
4. Machinery—loss of $5000

What is the total tax liability of these assets?

Stocks and bonds are capital assets with a net capital gain of $10,000 − $3000 = $7000. The building and machinery are Section 1231 property having a net group gain of $20,000 − $5000 = $15,000. Since a group gain was realized, the building and machinery are *each* treated as capital assets for a total tax liability of ($10,000 − $3000 + $20,000 − $5000)(.28) = $6160.

---

Example 6.30

Suppose the scenario in Example 6.29 were switched such that the building lost $20,000 but the machinery had a gain of $5000. What would be the new tax liability?

The Section 1231 group would have a net loss of $15,000 and *each* asset would be treated as ordinary income. Total tax liability would be ($10,000 − $3000)(.28) + (−$20,000 + $5000)(.46) = −$4940.

---

## 6.16 INVESTMENT CREDIT AND SECTION 179 EXPENSE DEDUCTION

The investment credit (*IC*) and the Section 179 expense deduction are designed to stimulate investment by providing reduced taxation in the first year of an asset's life. The IC has had a turbulent history dating back to 1962. It has been on-again, off-again during that time, with varying rules, percentages, and so forth. The Section 179 expense deduction was introduced in 1981 to begin with tax year 1982. It replaces the pre-1981 additional first-year depreciation that amounted to 20% of up to $10,000 of qualifying investment. Due to the way in which lawmakers continually modify these investment stimuli, the *IC* and the Section 179 expense deduction bear checking with tax experts prior to incorporating them into an engineering economic analysis.

## 6.16.1 Investment Credit

The investment credit (*IC*) permits taxpayers to claim a credit, not just a deduction, against income taxes for qualified investment in certain new or used depreciable property. The regular credit allowable is 10% of the eligible investment. The *IC* is limited to $25,000 plus 85% of the tax liability over $25,000 for tax years beginning in 1983 and beyond. Any unused credit may be carried back 3 years or carried forward 15 years. If an asset is disposed of prior to completion of its recovery period, or its useful life if nonrecovery property, the *IC* may need to be refigured. In this case, we should consult the tax law.

The *IC* is available only for qualifying property, known as Section 38 property. Qualifying property must:

1. Be depreciable as recovery property under the ACRS, or as other depreciable property having a useful life of at least 3 years.
2. Be tangible personal property (other than an air-conditioning or space-heating unit), or other tangible property (except buildings and their structural components), used as an integral part of manufacturing, production, extraction, and so forth.
3. Be placed in service during the year in a trade or business or for the production of income, and be used predominantly in the United States [1].

The amount of qualified investment in Section 38 property is summarized in Table 6.24.

The amount of qualifying new property is not limited. Used property, however, is limited to a total investment of $125,000 in 1983 and earlier, and $150,000 in 1984 and beyond.

If the full allowable *IC* is taken, law requires that half the amount of the *IC* be used to reduce the cost basis of the asset, thereby lessening the allowable cost recovery. For example, if the 10% *IC* is claimed, the cost basis of the asset must be reduced immediately by 5%. For 3-year recovery property, the *IC* is

TABLE 6.24. Amount of Investment Eligible for *IC*

| ACRS Recovery Property Class | Proportion Eligible |
| --- | --- |
| 3-year | 60% |
| 5-year | 100% |
| 10-year | 100% |
| 15-year public utility | 100% |
| 15-year real | None |

| Non-ACRS Recovery Property Life | Proportion Eligible |
| --- | --- |
| 0, 1, 2 | None |
| 3, 4 | 1/3 |
| 5, 6 | 2/3 |
| 7 or more | 100% |

only 6%, and the basis need only be reduced by 3%. As an alternative, if the taxpayer is willing to reduce the *IC* from 10 to 8% or from 6 to 4%, the cost basis need not be reduced at all.

## Example 6.31

A firm invests $1,500,000 in elaborate robotics equipment, $740,000 in an additional building and associated real property, and $40,000 in a fleet of automobiles. Tax liability for the year is $120,000. Applying the investment credit, what is the reduction in tax liability?

The three properties are ACRS 5-year, 15-year real, and 3-year recovery property. The 15-year real property does not qualify for the *IC*.

$$\text{Investment credit} = \$1,500,000(.10) + \$40,000(.06) = \$152,400$$

$$\text{Investment credit used} = \$25,000 + (\$120,000 - \$25,000).85$$
$$= \$105,750.$$

Note that the excess tax credit of $152,400 - $105,750 = $46,650 can be carried backward or forward to apply against taxes in other years.

## 6.16.2 Election to Expense Certain Depreciable Assets

Taxpayers may elect to treat the cost of certain property as an expense rather than as a capital expenditure. This property, called Section 179 property, must be recovery property acquired by purchase for use in a trade or business and that also qualifies for the investment credit. The election to expense must take place in the year the property is placed in service. The maximum deduction limits are $5000 in 1982 and 1983, $7500 in 1984 and 1985, and $10,000 in 1986 and beyond. These limits apply to each taxpayer, not to each asset. The property's cost basis, $P$, must be reduced by the amount expensed. Also, no investment credit is allowable for any portion of the cost that is expensed under Section 179. We can see that the maximum expense deduction for a company in any year is a few thousand dollars, and this will have its greatest effect in assisting small businesses.

## Example 6.32

Let us now apply both the Section 179 expense deduction and the investment credit to the stand-alone computer of Example 6.12. Recall that $P = \$82,000$, $F = \$5000$, useful life is 7 years, before-tax cash flow is $26,000/year, and $i = 15\%$. Assume that this is the company's only asset purchased and placed in service in 1984. What will be the after-tax cash flow profits?

We must begin with the Section 179 treatment. The maximum expense deduction in year 1 is $7500. The cost basis is therefore reduced to $82,000 − $7500 = $74,500. In addition, the *IC* will be $74,500(.10) = $7450. The cost

TABLE 6.25. After-Tax Cash Flows Using the Investment Credit and the Section 179 Expense Deduction

| End of Year A | Before-Tax Cash Flow B | Section 179 Expense Deduction C | ACRS Deduction D | Taxable Income B − C − D E | Tax E × .46 F | Investment Credit G | After-Tax Cash Flow B − F + G H |
|---|---|---|---|---|---|---|---|
| 0 | −$82,000.00 | — | — | — | — | — | −$82,000.00 |
| 1 | 26,000.00 | $7,500.00 | $10,616.25 | $ 7,883.75 | $ 3,626.53 | $7,450.00 | 29,823.47 |
| 2 | 26,000.00 | 0.00 | 15,570.50 | 10,429.50 | 4,797.57 | 0.00 | 21,202.43 |
| 3 | 26,000.00 | 0.00 | 14,862.75 | 11,137.25 | 5,123.14 | 0.00 | 20,876.86 |
| 4 | 26,000.00 | 0.00 | 14,862.75 | 11,137.25 | 5,123.14 | 0.00 | 20,876.86 |
| 5 | 26,000.00 | 0.00 | 14,862.75 | 11,137.25 | 5,123.14 | 0.00 | 20,876.86 |
| 6 | 26,000.00 | 0.00 | 0.00 | 26,000.00 | 11,960.00 | 0.00 | 14,040.00 |
| 7 | 26,000.00 | 0.00 | 0.00 | 26,000.00 | 11,960.00 | 0.00 | 14,040.00 |
| 7 | 5,000.00 (salvage) | — | — | 5,000.00 (depreciation recapture) | 2,300.00 | — | 2,700.00 |

basis must be further reduced to $74,500(.95) = $70,775$. Table 6.25 summarizes the cash flow calculations.

Calculating the present worth of after-tax cash flows, we have

$$PW(15) = -\$82,000 + \$29,823.47(P|F\ 15,1) + \$21,202.43(P|F\ 15,2)$$
$$+ \cdots + \$14,040(P|F\ 15,7) + \$2700(P|F\ 15,7)$$
$$= \$8371.39$$

The investment credit and the Section 179 expense deduction increased the present worth of after-tax cash flows from $2297.63 to $8371.39. The investment credit will usually be a major economic analysis factor for companies of any size. In the problems at the end of the chapter, it should be applied at the end of year 1. The Section 179 expense deduction will only be a major factor for small businesses. Do not apply the deduction to the problems at the end of the chapter, unless specifically told otherwise, and then apply it at the end of year 1.

## 6.17 CARRY-BACK AND CARRY-FORWARD RULES

In Examples 6.14 and 6.24 we saw that an alternative's taxable income and tax can be negative for a particular year. It was implicitly assumed that those values were used to offset positive tax liabilities from other corporate activities. When there are insufficient incomes or taxes to be offset, we can carry back or carry forward ordinary income losses, capital losses, and the investment credit to past or future years' returns.

Ordinary income losses can be carried back 3 years, beginning with the earliest. If not entirely used to offset income in that year, it is carried to the second year back and so on. After applying losses to all 3 back years, remaining losses can be carried forward to as many as 15 years following the loss year. Capital losses may be carried back 3 years and forward 5 years. They are treated as short-term capital losses and used to offset capital gains. They cannot be used to offset ordinary income. The investment credit can be carried back 3 years and forward 15 years to be applied directly against taxes on those returns.

## 6.18 LEASE–BUY CONSIDERATIONS

It is often attractive to lease property as opposed to owning it. When an asset is leased, a schedule of payments over time is agreed upon. Lease charges, being expenses, apply directly against taxable income during the year in which they occur. Thus, each dollar spent on a leased item has the effect of costing only $0.54 (assuming a .46 tax rate). This is not true for dollars spent on a purchased item, because this cost must be written off over the recovery period of the asset. Thus, the effective cost per dollar spent on a purchased item is more than $.54.

## Example 6.33

The stand-alone computer has an $82,000 first cost, salvage value of $5000, life of 7 years, and cost-savings of $26,000/year. It may be purchased, with cost recovery under the ACRS with a recovery period of 5 years. This case was illustrated in Example 6.12, Table 6.14, and had a $2,297.63 present worth of after-tax cash flows at a *MARR* of 15%.

A leasor offers to lease the computer for $26,784 − $1500($t$ − 1) during the $t$th year. This decreasing gradient is equivalent to 120% of the pretax equivalent uniform annual cost of the computer. The after-tax cash flows for this lease alternative are presented in Table 6.26.

Calculating the present worth of after-tax cash flows, we have

$$PW(15) = [-\$423.36 + \$810.00(A|G\ 15,7)](P|A\ 15,7)$$
$$= [-\$423.36 + \$810.00(2.4498)](4.1604)$$
$$= \$6494.32$$

Thus, leasing appears to be more favorable than purchasing. However, if we are eligible for the investment tax credit and Section 179 expense deduction, we have already seen in Example 6.32, Table 6.25 that the present worth of after-tax cash flows will be $8371.39.

## 6.19 DEPLETION

Depletion is a gradual reduction of minerals, gas and oil, timber, and natural deposits. In a sense, depletion is closely akin to depreciation. The difference is that while a depleting asset is losing value by actually being removed and sold,

TABLE 6.26. After-Tax Cash Flow for Lease Alternative

| End of Year A | Before-Tax Cash Flow B | Cost to Lease C | Taxable Income, B − C D | Tax D × .46 E | After-Tax Cash Flow B − C − E F |
|---|---|---|---|---|---|
| 0 | — | | | | |
| 1 | $26,000.00 | $26,784.00 | −$ 784.00 | −$ 360.64 | −$ 423.36 |
| 2 | 26,000.00 | 25,284.00 | 716.00 | 329.36 | 386.64 |
| 3 | 26,000.00 | 23,784.00 | 2,216.00 | 1,019.36 | 1,196.64 |
| 4 | 26,000.00 | 22,284.00 | 3,716.00 | 1,709.36 | 2,006.64 |
| 5 | 26,000.00 | 20,784.00 | 5,216.00 | 2,399.36 | 2,816.64 |
| 6 | 26,000.00 | 19,284.00 | 6,716.00 | 3,089.36 | 3,626.64 |
| 7 | 26,000.00 | 17,784.00 | 8,216.00 | 3,779.36 | 4,436.64 |

TABLE 6.27. Depletion Percentage for Some Minerals and Similar Resources*

| Type | Depletion Percentage |
|---|---|
| Oil and gas (percentage depletion is not allowed, except for certain gas and oil production) | |
| a. Oil and gas—independent producers and royalty owners, with some limitations | 15 (1984 and beyond) |
| b. Natural gas—regulated natural gas and gas sold under a fixed contract, with some limitations | 22 |
| Sulphur and uranium; and, if from U.S. deposits, asbestos, bauxite, chromite, graphite, mica, quartz crystals, cadmium, cobalt, lead, mercury, nickel, tin, zinc, and certain other ores and minerals | 22 |
| Gold, silver, oil shale, copper, and iron ore if from U.S. deposits | 15 |
| Nonmetallic minerals such as diatomaceous earth, granite, limestone, marble, and so forth | 14 |
| Coal, lignite, perlite, and sodium chloride | 10 |
| Clay and shale used for sewer pipe and brick | $7\frac{1}{2}$ |
| Clay (for roofing tile, flower pots, etc.), gravel, peat, pumice, sand, and stone | 5 |

* As of 1983.

a depreciable asset is losing value through wear, tear, and obsolescence in the manufacture of goods to be sold. Money recovered through the depletion allowance is likely to be used in the exploration and development of depletable assets, just as depreciation reserves are reinvested for new equipment. As in most tax-related matters, laws relating to lessors, lessees, royalties, and sales

TABLE 6.28. Depletion Allowance, Tax, and

| End of Year A | Barrels Sold B | Gross Income C | Operating Cost D | Net Income Before Depletion C − D E | 50% of Net Income E × .5 F |
|---|---|---|---|---|---|
| 0 | | −240,000 | | | |
| 1 | 15,000 | 510,000 | $105,000 | $405,000 | $202,500 |
| 2 | 13,000 | 442,000 | 139,750 | 302,250 | 151,125 |
| 3 | 10,000 | 340,000 | 145,000 | 195,000 | 97,500 |
| 4 | 6,000 | 204,000 | 109,500 | 94,500 | 47,250 |
| 5 | 3,000 | 102,000 | 66,000 | 36,000 | 18,000 |
| 6 | 1,000 | 34,000 | 25,750 | 8,250 | 4,125 |

* Indicates depletion allowance used.

are complex and will probably require expert assistance. However, the importance of depletion and the alternative methods that can be used for figuring the allowance will be illustrated here.

*Cost depletion* is the basic method of computing the depletion deduction. First, an estimate of the number of units (tons, barrels, board-feet, etc.) that make up the deposit is necessary. Then, the cost basis is divided by the number of units. This quotient, multiplied by the number of units sold during the year, determines the cost depletion allowance. It is expressed as

$$\text{Cost depletion deduction} = \frac{\text{Cost basis} \times \text{Number of units sold during year}}{\text{Number of units in property}}$$

The calculated deduction may or may not be used, depending on whether or not the alternative depletion allowance method provides a larger deduction.

*Percentage depletion* provides an allowance equal to a percentage of the gross income from the property. The applicable percentage depends on the type of property depleted. Table 6.27 gives some recent depletion percentages. The percentage depletion deduction may not exceed 50% of the taxable income before allowance for depletion. Also, if cost depletion results in a greater depletion, cost depletion must be used.

Example 6.34_____

We own a small independent oil company and have purchased the rights, drilled, and developed a 48,000-barrel oil well on NW/4 35/19N/1W, Payne County Oklahoma, for $240,000. Operating expenses, based on past experience, are equal to $7 + $3.75(t - 1)/barrel where t is the year in which the oil is removed and sold. This increase in cost per barrel over time indicates the increased difficulty of recovering the oil as the field is depleted. Our geologist

| After-Tax Cash Flow for Oil Well Example | | | | |
|---|---|---|---|---|
| Cost Depletion at $5.00/barrel $B \times \$5.00$ $G$ | Percentage Depletion at 15% of Gross Income $C \times .15$ $H$ | Taxable Income $E\text{-max}\{G,[\min(F,H)]\}$ $I$ | Income Tax $I \times .46$ $J$ | After-Tax Cash Flow $E - J$ $K$ |
| | | | | −$240,000 |
| $75,000 | $76,500* | $328,500 | $151,110 | 253,890 |
| 65,000 | 66,300* | 235,950 | 108,537 | 193,713 |
| 50,000 | 51,000* | 144,000 | 66,240 | 128,760 |
| 30,000 | 30,600* | 63,900 | 29,394 | 65,106 |
| 15,000 | 15,300* | 20,700 | 9,522 | 26,478 |
| 5,000* | 5,100 | 3,250 | 1,495 | 6,755 |

expects the field to last 6 years, yielding 15,000, 13,000, 10,000, 6000, 3000, and 1000 barrels of oil over that time. Each barrel will be worth $34 to our company. Using the cost method, $240,000/48,000 barrels or $5.00/barrel will be our depletion allowance. From Table 6.27, our depletion percentage is seen to be 15%, assuming we meet all the qualifications that allow a small producer to claim a percentage depletion. In Table 6.28 we calculate each year's estimated depletion, tax, and after-tax cash flow.

Note that in year 6 we chose the cost method because the percentage depletion was too high, being over 50% of net income. When working with depletion problems, it is always advisable to calculate the allowance both ways in order to assure the greatest tax advantage.

## 6.20 SUMMARY

This chapter has presented the most important elements of depreciation and income tax law as they pertain to economic analyses. It is clear that depreciation (or depletion) method, recovery period, financing, and tax credits can have significant effects on the desirability of making an investment. Many of these factors are law-related, and changes are being made daily. Therefore, in cases of uncertainty regarding depreciation and tax treatment, it is wise to seek competent legal advice.

Capital gains and losses are often not important in engineering economic analyses. However, it is wise to know when a capital gain or loss has occurred and when to apply capital gains treatment versus ordinary income treatment.

Finally, unless there are overriding considerations, lease versus buy alternatives should be analyzed closely. Quite often, a lease alternative will compare with surprising favor against an equivalent purchase alternative.

## BIBLIOGRAPHY

1. Department of the Treasury, Internal Revenue Service, *Tax Guide for Small Business,* Publication 334, Rev. Nov. 82.
2. Commerce Clearing House, Inc., *1983 U.S. Master Tax Guide,* November 1982.
3. Blank, Leland T., and Donald R. Smith, "A Comparative Analysis of the Accelerated Cost Recovery System as Enacted by the 1981 Economic Recovery Tax Act," *The Engineering Economist, 28* (1), 1983, pp. 1–30.
4. Canada, John R., "How the New Tax Laws Increase Attractiveness of Investment Projects," *Industrial Engineering, 14* (5) May 1982, pp. 76–81.

# PROBLEMS

1. For each of the following assets, state whether the asset is tangible/intangible property, personal/real property, and depreciable/nondepreciable property. (6.3)
   (a) A patent.          (f) Land.
   (b) Office furniture.   (g) A bond.
   (c) A taxi.            (h) A mobile home.
   (d) A lathe.           (i) A building.
   (e) A computer.        (j) A fence.

2. For each of the assets named in Problem 1, state whether it is Section 1245 class property, Section 1250 class property, or neither, and give the recovery property class if applicable. (6.4.1)

3. A unit of machinery (5-year recovery property) is purchased at the beginning of the fiscal year for $15,000. The estimated salvage value is $1000 after 8 years. Calculate the ACRS deduction and the unrecovered investment during each year of the asset's life.
   (a) Use the regular ACRS allowances. (6.4.2)
   (b) Use the ACRS straight-line election and a recovery period of 5 years. (6.4.3)

4. A small truck (3-year recovery property) is purchased for $11,000. The truck is expected to be of use to the company for 4 years, after which it will be sold for $2500. Calculate the ACRS deduction and the unrecovered investment during each year of the asset's life.
   (a) Use the regular ACRS allowances. (6.4.2)
   (b) Use the ACRS straight-line election and a recovery period of 3 years. (6.4.3)

5. An electrostatic precipitator is purchased on April 1 by a calendar-year taxpayer for $160,000. It is expected to last 10 years and have a salvage value of $15,000. Calculate the ACRS recovery (deduction) during years 1, 3, and 6.
   (a) If the regular ACRS allowances are used. (6.4.2)
   (b) If the ACRS straight-line election and the minimum permissible number of years in the recovery period are used. (6.4.3)

6. A smart terminal and peripherals are purchased in December by a calendar-year taxpayer for $40,000. The terminal will be used on a space-tracking project for 7 years and be worth $5000 at that time. Calculate the ACRS recovery (deduction) during years 1, 4, and 7.
   (a) If regular ACRS allowances are used. (6.4.2)
   (b) If the ACRS straight-line election and the minimum permissible number of years in the recovery period are used. (6.4.3)

7. Material handling equipment is purchased and installed for $220,000. It is placed in service in the middle of the tax year. If it is removed just *before* the end of the tax year approximately $4\frac{1}{2}$ years from the date placed in service, determine the ACRS deduction during each of the tax years involved.
   (a) Use the regular ACRS allowances. (6.4.2)
   (b) Use the ACRS straight-line election and the minimum permissible number of years in the recovery period. (6.4.3)

**8.** Repeat Problem 7 if the material handling equipment is removed just *after* the end of the tax year.
   **(a)** Use the regular ACRS allowances. (6.4.2)
   **(b)** Use the ACRS straight-line election and the minimum permissible number of years in the recovery period. (6.4.3)

**9.** A business building is purchased by a calendar-year taxpayer on August 1 for $150,000. Calculate the ACRS deduction and the unrecovered investment assuming the building is kept beyond the recovery period.
   **(a)** Use the regular ACRS allowances. (6.4.2)
   **(b)** Use the ACRS straight-line election and the minimum permissible number of years in the recovery period. (6.4.3)

**10.** A rental apartment complex is purchased on May 1 for $180,000 by a calendar-year taxpayer. If the apartments are kept for 5 years and 2 months (sold on July 1), determine the ACRS deduction during each of the six tax years involved.
   **(a)** Use the regular ACRS allowances. (6.4.2)
   **(b)** Use the ACRS straight-line election and a recovery period of 35 years. (6.4.3)

**11.** A water treatment system is needed in order to supply water to a series of baths for a bonding operation. It costs $250,000, has a useful life of 12 years, and a salvage value predicted to be $20,000 at that time. Determine the allowable deduction and the unrecovered investment for years 1, 5, 6, and 9.
   **(a)** Use the regular ACRS allowances. (6.4.2)
   **(b)** Use the ACRS straight-line election and the minimum permissible number of years in the recovery period. (6.4.3)
   **(c)** Use straight-line depreciation. (6.5.1)
   **(d)** Use sum of the years' digits depreciation. (6.5.2)
   **(e)** Use double declining balance switching to straight-line depreciation. (6.5.3)

**12.** A firm purchases a computer for $140,000. It has a useful life of 7 years and a terminal salvage value of $15,000 at that time. Determine the allowable deduction for year 4 and the unrecovered investment at the *beginning* of year 4. Assume post-1980 placement in service.
   **(a)** Use the regular ACRS allowances. (6.4.2)
   **(b)** Use ACRS straight-line allowances (minimum permissible recovery period). (6.4.3)
   Assume pre-1981 placement in service.
   **(c)** Use straight-line depreciation. (6.5.1)
   **(d)** Use sum of the years' digits depreciation. (6.5.2)
   **(e)** Use double declining balance switching to straight-line depreciation. (6.5.3)

**13.** A furnace has a first cost of $50,000 and a salvage value after 5 years of $0. Calculate the allowable deduction each year. Assume post-1980 placement in service.
   **(a)** Use regular ACRS allowances. (6.4.2)
   **(b)** Use ACRS straight-line allowances (minimum permissible recovery period). (6.4.3)
   Assume pre-1981 placement in service.
   **(c)** Use straight-line depreciation. (6.5.1)
   **(d)** Use sum of the years' digits depreciation. (6.5.2)
   **(e)** Use double declining balance switching to straight-line depreciation. (6.5.3)

14. Determine the recovery and unrecovered investment for each year of a public utility property having a present class life of 15 years. (Be careful on your selection of the recovery period!) The asset has a cost of $100,000, a salvage value of $10,000, and a useful life of 20 years. It is placed in service after 1980.
   (a) Use regular ACRS allowances. (6.4.1)
   (b) Use ACRS straight-line allowances and the minimum permissible recovery period. (6.4.2)

15. Repeat Problem 14, but assume the property is sold just before the end of the fifth year.
   (a) Use regular ACRS allowances. (6.4.1)
   (b) Use ACRS straight-line allowances and the minimum permissible recovery period. (6.4.2)

16. A compressor is purchased for $6000 and has an estimated salvage value of $500 after a useful life of 6 years. Interest is 15%. Determine for each year the capital recovered, the return on capital unrecovered, and the capital recovered plus return. Also find the annual equivalent capital recovered plus return.
   (a) Use regular ACRS allowances. (6.4.1)
   (b) Use straight-line depreciation. (6.5.1)
   (c) Use sum of the years' digits depreciation. (6.5.2)
   (d) Then compare the annual equivalent capital recovered plus return values above to the capital recovery cost given by $(P - F)(A|Pi,n) + Fi$.

17. A firm purchases a distillation column for $300,000 which they anticipate will be used for 8 years and have a salvage value at that time of $60,000. Determine for each year the capital recovered, the return on capital unrecovered, and the capital recovery plus return. Also, find the annual equivalent capital recovered plus return. Use a MARR of 15%
   (a) Use regular ACRS allowances. (6.4.1)
   (b) Use ACRS straight-line allowances (minimum permissible recovery period). (6.4.2)
   (c) Use straight-line depreciation. (6.5.1)
   (d) Double declining balance switching to straight-line depreciation. (6.5.3)

18. An automatic control mechanism is estimated to provide 3500 hours of service during its life. The mechanism costs $7600 and has a salvage value of $300 after 3500 hours of use. Its use is projected over a 4-year period as follows:

| Year | Hours of Use |
| --- | --- |
| 1 | 1,000 |
| 2 | 1,500 |
| 3 | 700 |
| 4 | 300 |

Calculate the depreciation charge for each year using a method similar to the "operating day" method, but based on operating hours. (6.7.1)

19. A utility trailer costs $2500 and is rented out by the hour, day, or week. It is expected to depreciate to zero by the time it has been rented out for a total of $12,500 in gross income. If its forecasted annual revenues are $4000, $4500, $2500, and $1500, what will be the appropriate annual depreciation charges? (6.7.1)

20. What is the federal income tax for each of the following corporate taxable incomes?
  (a) $12,500.    (d) $89,000.
  (b) $36,000.    (e) $150,000.
  (c) $64,000.    (f) $1,000,000.
  Neatly plot a graph of federal income tax versus taxable income. Be sure to label key values on both scales. (6.9)

21. A consulting engineer is seeking a project which will increase taxable income significantly. Determine the tax increase and the percentage of new taxable income which goes to pay the tax increase. Assume the present and additional new taxable incomes are, respectively: (a) $40,000 and $36,000 and (b) $110,000 and $36,000. (6.9)

22. A dust collector with a first cost of $50,000 is purchased. The collector will be used for 8 years, will yield an annual gross income less operating expenses of $15,000, and will have no salvage value. The effective tax rate is 46%. Determine the after-tax cash flow for each year.
  (a) Use regular ACRS allowances. (6.10)
  (b) Use ACRS straight-line allowances (minimum permissible recovery period). (6.10)
  (c) Use straight-line depreciation. (6.10)
  (d) Use sum of the years' digits depreciation. (6.10)
  (e) Use double declining balance switching to straight-line depreciation. (6.10)

23. An automatic welding robot costs $100,000 and has a salvage value of $12,000 after a useful life of 8 years. The robot generates a net savings of $33,000/year. Find the after-tax cash flow for each year.
  (a) Use regular ACRS allowances. (6.10)
  (b) Use ACRS straight-line allowances (minimum permissible recovery period). (6.10)
  (c) Use straight-line depreciation. (6.10)
  (d) Use sum of the years' digits depreciation. (6.10)
  (e) Use double declining balance switching to straight-line depreciation. (6.10)

24. A firm purchases a centrifugal separator for $18,000; it is estimated to have a life of 10 years. Operating and maintenance costs increase by $1000/year, with the cost for the first year being $1000. A $1000 salvage value is anticipated. Assuming a 46% income tax rate, and a 15% after-tax *MARR*, determine the annual equivalent cost of the separator.
  (a) Use regular ACRS allowances. (6.10)
  (b) Use ACRS straight-line allowances (minimum permissible recovery period). (6.10)
  (c) Use straight-line depreciation. (6.10)
  (d) Use sum of the years' digits depreciation. (6.10)
  (e) Use double declining balance switching to straight-line depreciation. (6.10)

25. An investment has the following net cash flows before taxes.

| EOY | 0 | 1 | 2 | 3 | 4 |
|---|---|---|---|---|---|
| | −$70,000 | 42,000 | 44,000 | 28,000 | 23,000 + 20,000 (salvage value) |

The effective tax rate is 46% and the after-tax *MARR* is 20%. Assume the company is profitable in its other activities. Find the after-tax cash flows and the present worth of those cash flows.

(a) Use regular ACRS allowances. (6.10)
(b) Use ACRS straight-line allowances (minimum permissible recovery period). (6.10)
(c) Use straight-line depreciation. (6.10)
(d) Use sum of the years' digits depreciation. (6.10)
(e) Use double declining balance switching to straight-line depreciation. (6.10)

26. An earth-moving company invests $35,000 in equipment that will have a useful life of 6 years with a $5000 salvage value at that time. Increases in the firm's income from cut and fill contracts due to the investment total $15,000/year over 6 years. The company pays taxes at a rate of 46%. With an after-tax *MARR* of 15%, should the firm have undertaken the investment?

(a) Use regular ACRS allowances. (6.10)
(b) Use ACRS straight-line allowances (minimum permissible recovery period). (6.10)
(c) Use straight-line depreciation. (6.10)
(d) Use sum of the years' digits depreciation. (6.10)
(e) Use double declining balance switching to straight-line depreciation. (6.10)

27. An investment proposal is described by the following tabulation:

|  | Year | | |
|---|---|---|---|
|  | 1 | 2 | 3 |
| Gross income during year | $ 8,000 | $16,000 | $20,000 |
| Investment at *beginning* of year | 24,000 | 0 | 0 |
| Operating cost | 2,000 | 3,000 | 4,000 |
| Depreciation charge (ACRS) | 6,000 | 9,120 | 8,880 |

The company considering this proposal is profitable in its other activities. Determine the taxable income, tax, and after-tax cash flow for years 0, 1, 2, and 3. Assume an effective tax rate of 46%. (6.10)

28. An investment opportunity has the following financial outlook over a 5-year period.

| EOY | Before-Tax Cash Flow | Depreciation Charges | Taxes |
|---|---|---|---|
| 0 | −$30,000 |  |  |
| 1 | 10,000 | $4,500 | $2,530 |
| 2 | 10,000 | 6,600 | 1,564 |
| 3 | 10,000 | 6,300 | 1,702 |
| 4 | 10,000 | 6,300 | 1,702 |
| 5. | 10,000 | 6,300 | 1,702 |

If our firm requires a 15% rate of return after taxes, should we invest? (6.10)

29. A firm may either invest $20,000 in a numerically controlled lathe that will last for 6 years and have a zero salvage value at that time, or invest $X in a methods improvement study (this is not depreciable). Both investment alternatives yield an increase in income of $10,000/year for 6 years. With regular ACRS allowances for the investment, and with a 46% tax rate, for what value of $X$ will the firm be indifferent between the two investment alternatives? Assume an after-tax $MARR$ of 15%. (6.10)

30. An automatic copier is being considered that will cost $24,000, have a life of 5 years, and no prospective salvage value at the end of that time. It is estimated that because of this venture, there will be a yearly gross income of $22,000. The operating cost during the first year will be $6000, increasing by $500 each year thereafter. The company contemplating the purchase of this asset has an effective tax rate of 46%. Determine the after-tax cash flow for each year and the annual worth of this investment if $MARR$ = 15%.
   (a) Use regular ACRS allowances. (6.10)
   (b) Use ACRS straight-line allowances (minimum permissible recovery period). (6.10)

31. Two mutually exclusive alternatives, A and B, are available. Alternative A requires an original investment of $80,000, has a useful life of 6 years, zero annual operating costs, and a salvage value in year $k$ given by $80,000 (0.80)^k$. Alternative B requires an original investment of $120,000, has a life of 8 years, annual operating costs of $2000, and a salvage value in year $k$ given by $120,000 (0.85)^k$. The after-tax $MARR$ is 15%; a 46% tax rate is applicable. Regular ACRS allowances are used. Perform an annual worth comparison and recommend the least cost alternative.
   (a) Use a planning horizon of 6 years.
   (b) Use a planning horizon of 8 years.
   (c) Use a planning horizon of 24 years.
   If replacements are required, assume they have identical cash flow profiles. (6.10)

32. An automatic soldering machine is purchased for $54,000. Installation cost is $9000 because of the extreme care and provisions needed. It is estimated that the asset can be sold for $7500 after a useful life of 10,000 hours of operation. However, before selling, it must be removed at a cost of $1500. Inspection and maintenance costs $18/operating hour. It requires one-half hour to solder one unit on the machine, and a total of 10,000 units are soldered per year. The after-tax $MARR$ is 15%, and the effective tax rate is 46%. Depreciation is based on operating hours.
   (a) What is the equivalent annual after-tax cost of the soldering machine? (6.10)
   (b) What is the cost of soldering per unit? (6.10)

33. Suppose a back-hoe is purchased for $36,000 with an estimated salvage value of $6000 just before the end of the eighth year. Annual revenues and annual operating costs, excluding depreciation, are $38,000 and $23,000, respectively. Assume that $i$ = 20% and the firm's tax rate is 46%.
   (a) Determine the present worth of taxes paid for the 8-year period (1) if regular ACRS allowances are used, and (2) if ACRS straight-line allowances are used (12-year recovery period).

**(b)** Determine the present worth of the recovery allowances (deductions) for the 8-year period (1) if regular ACRS allowances are used and (2) if ACRS straight-line allowances are used (12-year recovery period).

Do your answers to parts (a) and (b) suggest that the present worth of taxes is minimized if the present worth of recovery allowances is maximized? (6.11)

34. The ABC Company is considering the purchase of a computer-controlled printing press that will cost $40,000, have a life of 12 years, and an estimated salvage value of zero. The press will develop a gross income of $46,000/year. The annual cost connected with the press (exclusive of depreciation) will amount to $32,000/year. The company will use ACRS straight-line recovery and the effective tax rate is 46%. If the *MARR* is 15% after taxes, determine the present worth of the after-tax cash flows.

**(a)** Use a recovery period of 5 years.

**(b)** Use a recovery period of 12 years.

Do your answers to parts (a) and (b) suggest that the present worth of after-tax cash flows is maximized with a shorter recovery period? (6.11)

35. A company has the opportunity to invest in a water purification system requiring $250,000 capital. The firm has only $150,000 available and must borrow the additional $100,000 at an interest rate of 15%/year. The system has a life of 6 years and an estimated salvage value at the end of its life of $0. The before-tax and loan cash flow for each of years 1 to 6 is $75,000. Only the loan interest is paid each year and the entire principal is repaid at the end of year 3. Regular ACRS recovery is used; the applicable tax rate is 46%; and the firm is actually able to sell the system for $25,000 at the end of year 6. Construct a table showing each of the following for each of the 6 years: (6.12)

**(a)** Before-tax and loan cash flow.

**(b)** Loan principal payment.

**(c)** Loan interest payment.

**(d)** ACRS deduction.

**(e)** Taxable income.

**(f)** Taxes.

**(g)** After-tax cash flow.

36. Rework Problem 35 assuming that the loan is paid back as follows.

**(a)** In equal annual amounts.

**(b)** The principal in equal annual amounts plus yearly interest. (6.12)

37. Precision Components, Inc. purchases of new piece of computer-controlled, highly versatile equipment for machining blocks of solid metal. The machine costs $500,000, and has an expected life of 15 years. It will replace several old manually operated machines for a cost savings of $150,000/year. It is financed at 16% for $210,000; the remaining $290,000 is equity money. Repayment occurs over 3 years with three equal principal amounts of $70,000, plus interest on the unrecovered balance. Just before the end of the fifth year, the machine is sold for $200,000 to purchase a new and better one. Clearly and completely set up a table and determine the after-tax cash flow for each appropriate year. Be sure to include everything that makes up or helps determine the ATCF. The tax rate is 46% and regular ACRS allowances are used. (6.12)

**38.** Complete the partial table below.

| EOY | Before-Tax and Loan Cash Flow | Loan Principal Payment | Loan Interest Payment | ACRS Deduction | Taxable Income | Taxes (Rate = 0.46) | After-Tax Cash Flow |
|---|---|---|---|---|---|---|---|
| 0 | −$100,000 | −$40,000 | | | | | −$60,000 |
| 1 | 20,000 | | $4,000 | $15,000 | ? | ? | ? |
| 2 | 20,000 | | 4,000 | 22,000 | ? | ? | ? |
| 3 | ? | | ? | 21,000 | ? | $ 0 | ? |
| 4 | ? | | ? | ? | ? | 2,300 | ? |
| 5 | 40,000 | | 4,000 | ? | ? | 6,900 | ? |
| 6 | 20,000 | | 4,000 | ? | ? | ? | 8,640 |
| 7 | ? | | 4,000 | ? | $19,000 | ? | ? |
| 8 | ? | | 4,000 | ? | 31,000 | ? | ? |
| 9 | 45,000 | | 4,000 | ? | ? | ? | ? |
| 10 | 30,000 | | 4,000 | ? | ? | ? | ? |
| 10 | | 40,000 | | | | | − 40,000 |

*Note 1.* There is no uniformity to the before-tax and loan cash flows.
*Note 2.* The effective income tax rate is 46%.

**39.** A firm purchases a heat exchanger by borrowing the $10,000 purchase price. The loan is to be repaid with three equal annual payments at an annual compound rate of 12%. It is anticipated that the exchanger will be used for 6 years and then be sold for $1000. Annual operating and maintenance expenses are estimated to be $5000/year. Assume ACRS depreciation, a 46% tax rate, and an after-tax *MARR* of 15%. Compute the after-tax, equivalent uniform annual cost for the heat exchanger. (6.12)

**40.** A firm borrows $20,000, paying back $5000/year on the principal plus 15% on the unpaid balance at the end of each year. The $20,000 is used to purchase a digital frequency meter that lasts 5 years and has a $2000 salvage value at that time. The firm has obtained IRS approval to use ACRS recovery under a 3-year recovery period. It requires a 15% after-tax *MARR* in such analyses and has a 46% tax rate on taxable income. Determine the after-tax equivalent uniform annual cost for the meter. (6.12)

**41.** A company has the following capital gains and losses and ordinary taxable incomes:

| | |
|---|---|
| Short-term capital gains | $ 25,000 |
| Short-term capital losses | 4,000 |
| Long-term capital gains | 103,000 |
| Long-term capital losses | 8,000 |
| Ordinary taxable income | 80,000 |

Determine the federal taxes to be paid by the company.
**(a)** Use the regular method. (6.14)
**(b)** Use the alternate method. (6.14)

**42.** Our company has short term capital gains of $14,000, short-term capital losses of $10,000, long-term capital gains of $40,000, and long-term capital losses of $8000. Ordinary taxable income is $90,000. How much tax do we have to pay? (6.14)

**43.** Suppose our company in Problem 42 had short-term capital losses of $25,000 (the other information stays the same). How much tax do we have to pay now? (6.14)

**44.** Determine the amount of tax (or other appropriate handling) for each of the following situations (assume a tax rate of 46% on ordinary income): (6.14)

| | |
|---|---|
| **(a)** Net short-term capital gains | $30,000 |
| Net long-term capital gains | $40,000 |
| **(b)** Net short-term capital gains | $20,000 |
| Net long-term capital losses | $12,000 |
| **(c)** Net short-term capital losses | $26,000 |
| Net long-term capital losses | $14,000 |
| **(d)** Net short-term capital losses | $15,000 |
| Net long-term capital gains | $26,000 |
| **(e)** Net short-term capital gains | $18,000 |
| Net long-term capital losses | $27,000 |
| **(f)** Net short-term capital losses | $21,000 |
| Net long-term capital gains | $10,000 |

**45.** Consider an asset that is Section 1245 5-year recovery property. The purchase price is $150,000, and the asset is sold just before the end of the ninth year for $20,000. How should the salvage value received be treated for tax purposes if the following method are used?
**(a)** Regular ACRS allowances. (6.15.2)
**(b)** ACRS straight-line allowances over a 12-year recovery period. (6.15.2)

**46.** Reconsider the asset in Problem 45. Assume the asset is sold for $20,000 just before the end of the third year.
**(a)** Use regular ACRS allowances. (6.15.2)
**(b)** Use ACRS straight-line allowances over a 12-year recovery period. (6.15.2)

**47.** A Section 1245 5-year recovery property is purchased for $100,000 and recovered using regular ACRS allowances. Assuming that we do not know the status of other Section 1231 assets of the company, determine the tax treatment of the asset's salvage value in the following cases. (6.15.2)
**(a)** It is sold during year 3 for $108,000.
**(b)** It is sold during year 3 for $80,000.
**(c)** It is sold during year 3 for $10,000.

**48.** Rework Problem 47 assuming ACRS straight-line allowances were used to recover the asset. (6.15.2)

**49.** A Section 1250 residential 15-year recovery property is purchased for $250,000 and recovered using regular ACRS allowances. It is purchased in the first month of the tax year. Assuming that we do not know the status of other Section 1231 assets of the company, determine the tax treatment of the asset's salvage value in the following cases. (6.15.2)
**(a)** It is sold just before the end of year 5 for $300,000.
**(b)** It is sold just before the end of year 5 for $225,000.
**(c)** It is sold just before the end of year 5 for $175,000.
**(d)** It is sold just before the end of year 5 for $125,000.

**50.** A Section 1250 15-year recovery property is purchased for $700,000 and recovered using ACRS straight-line allowances. It is purchased at the beginning of the fourth month of the tax year. Assuming that we do not know the status of other Section

1231 assets of the company, determine the tax treatment of the asset's salvage value in the following cases. (6.15.2)

(a) It is sold just before the end of year 5 for $1,000,000.
(b) It is sold just before the end of year 5 for $600,000.
(c) It is sold just before the end of year 5 for $300,000.

51. Consider an asset that is Section 1250 residential recovery property. The purchase price was $240,000, and the asset was sold after 24 years for $150,000. How should this sale be treated for tax purposes if:

(a) We have used regular ACRS allowances? (6.15.2)
(b) We have used ACRS straight-line allowances (minimum permissible number of years)? (6.15.2)

52. An electronic multiaxis wheel balancer is purchased for $10,000. Its estimated life is 8 years, with an estimated salvage value of $1000. The balancer wears badly, and will be sold toward the end of year 3 for $1000. The effective tax rate is 0.46, the capital gains tax rate is 0.28, and regular ACRS allowances are being used. Operating expenses are $4000/year and gross income is $8000/year. Determine the aftertax cash flow for each year of the balancer's life. (6.15.2)

53. A firm purchases a crane for $100,000. Originally, it was estimated that the crane would be retained for 9 years and sold for $10,000. However, the crane is used and sold just before the end of the sixth year for $2000. Income less operating and maintenance costs for the crane have been $30,000/year. Using a 46% income tax rate and regular ACRS allowances, compute the after-tax cash flows for the 6-year period. Incorporate the Section 179 expense deduction and the investment credit in year 1. The company annually has a large amount of long-term capital gains. (6.16.2)

54. Rework Problem 53 assuming the crane is sold just before the end of the fifth year for $2000. (6.16.2)

55. A $39,000 investment in an automatic control panel is proposed. It is anticipated that this investment will cause a reduction in net annual operating disbursements of $12,000/year for 8 years. The investment will be depreciated for income tax purposes by the ACRS straight line method. The effective tax rate is 46%. Include the Section 179 expense deduction. An investment credit applies at the end of year 1. With a required after-tax return of 12%, what is the equivalent present worth cost for the investment? (6.16.2)

56. A sum of $10,000 is borrowed at 15% compounded annually and paid back with equal annual payments over a 2-year period. The $10,000 is combined with $15,000 of equity funds to purchase a transformer that has a service life of 8 years and a terminal salvage value of $3400. Annual operating and maintenance costs are anticipated to be $5000/year. Section 179 and the investment credit applies in year 1, along with an income tax rate of 46%. Regular ACRS allowances are used. The after-tax MARR is 20%. Determine the present worth of costs for this machine. (6.16.2)

57. A $50,000 heat recovery incinerator is expected to cause a reduction in net out-of-pocket costs of $20,000/year for 6 years. The incinerator will be depreciated for income tax purposes using regular ACRS allowances and no salvage value is ex-

pected. The investment credit is allowed in year 1. If the tax rate is 46% and the after-tax *MARR* is 20%, determine whether or not the incinerator should be purchased. (6.16.2)

58. A firm is considering investing $765,000 in a material handling system that will reduce annual operating costs by $150,000/year over a 15-year planning horizon. Perform a before-tax analysis using a *MARR* of 20%. Is the investment justified? Suppose further analysis indicates $640,000 of the investment is for equipment; the remaining $125,000 is for expense items. A 46% tax rate, the investment credit in year 1, regular ACRS allowances (5-year recovery period), and 15% after-tax *MARR* are to be used. Is the investment justified on an after-tax basis? (6.16.2)

59. A tractor originally costs $20,000. Operating costs for year $j$ are equal to $4000 + $500($j - 1$). A 46% tax rate, investment credit in year 1, regular ACRS allowances, zero salvage value at all times, 15% minimum attractive rate of return after taxes, and indefinite planning horizon are to be assumed. Determine the optimum replacement interval. (6.15.2)

60. A firm is contemplating either purchasing or renting a computer. If purchased, the computer will cost $400,000 and have annual operating and maintenance costs of $75,000. Because of the obsolescence rate on computing equipment, a 5-year study period is chosen. Just before the end of 5 years, it is anticipated the computer will have a value of $50,000. The same computer can be leased, with *beginning of year* lease charges of $170,000, which includes maintenance. If leased, the firm will have end-of-year operating costs of $50,000. Using regular ACRS allowances, a 46% tax rate, and an after-tax *MARR* of 20%, perform an annual worth comparison to determine if the computer should be leased. (6.18)

61. A firm is considering purchasing an analog computer for $200,000. The computer is estimated to have a life of 6 years and a salvage value of $20,000. Operating and maintenance costs are estimated to be $45,000/year. Regular ACRS allowances are to be used and a 10% investment credit applies in year 1. Alternatively, the computer can be leased for $X at the *beginning* of each year for a 6-year period. If leased, the annual operating and maintenance cost to be paid by the firm reduces to $35,000. Using a 46% tax rate, an after-tax *MARR* of 15%, and an annual worth comparison, determine the value of X that will yield equivalent annual costs between buying and leasing. (6.18)

62. A firm is considering purchasing a digital computer for $500,000. The computer is estimated to have a life of 5 years and a salvage value of $50,000. End-of-year operating and maintenance costs are estimated to be $60,000/year. ACRS straight-line allowances are used. The investment credit applies in year 1. Alternatively, the computer can be leased for $X/year for the 5-year period with the lease payment due at the *beginning* of the year. If leased, the annual end-of-year operating and maintenance costs are estimated to be $40,000. Using a 46% tax rate, an after-tax *MARR* of 15%, and an annual worth comparison, determine the value of X that will yield equivalent annual costs between buying and leasing. (6.18)

63. The Smart Construction Company has been paying a subcontractor $30,000/year at the end of each year to perform a particular job. Annual revenues received by Smart associated with this particular job have averaged $50,000. The agreement with the

subcontractor has now expired but could be renewed for another 5 years. The same revenues and subcontract charges are expected to continue.

The Smart Company is considering the purchase of equipment to handle this same job and terminate the subcontracting. The initial cost of the equipment plus installation expenses is $50,000. The economic life of the equipment is established as 10 years with a salvage value of $5000 at that time. However, the company is only interested in a planning horizon of 5 years at which time a salvage value of $20,000 is anticipated. With the equipment, annual revenues are expected to remain $50,000 but average annual operating expenses expected are $20,000.

The company uses regular ACRS allowances. If the company's annual tax rate is 46%, a 15% after-tax rate of return is desired, and a 5-year planning horizon is assumed, should the equipment be purchased or the subcontract agreement be renewed? (6.18)

**64.** A gold mine that is expected to produce 30,000 ounces of gold is purchased for $2,000,000. The gold can be sold for $450/ounce; however, it costs $275/ounce for mining and processing costs. If 3500 ounces are produced this year, what will be the depletion allowance for (a) unit depletion and (b) percentage depletion? (6.19)

**65.** A West Virginia coal mine having 6 million tons of coal has a first cost of $25 million. The gross income for this coal is $45/ton. Operating costs are $30/ton. If, during the first 2 years of operation, the mine yields 150,000 tons and 200,000 tons, respectively, determine the after-tax cash flow for each year, assuming that this mine is the owner's only venture. Use the better method of depletion. (6.19)

## CHAPTER SEVEN

# ECONOMIC ANALYSIS OF PROJECTS IN THE PUBLIC SECTOR

## 7.1 INTRODUCTION

Knowing how to evaluate and select projects to be approved, paid for, and operated by the government is at least as important to today's engineer as a similar knowledge relating to the private sector. Fortunately, analysis methods for public and private projects are very similar, even though there are some basic, significant differences between the two. The methods considered in this chapter are those most frequently used in evaluating government (national, state, or local) projects—*benefit-cost* and *cost-effectiveness* analysis. More emphasis will be spent on benefit-cost methods, which require that benefits and costs be evaluated on a monetary basis. Cost effectiveness requires a numerical measure of effectiveness; however, that measure need not be in terms of money.

## 7.2 THE NATURE OF PUBLIC PROJECTS

There are many types of government projects and many agencies involved. Four classes reasonably cover the spectrum of projects entered into by government. They include cultural development, protection, economic services, and natural resources. *Cultural development* is enhanced through education, recreation, historic, and similar institutions or preservations. *Protection* is achieved through military services, police and fire protection, and the judicial system. *Economic services* include transportation, power generation, and housing loan programs. *Natural resource* projects might entail wildland management, pollution control, and flood control. Although these are obviously incomplete project lists in each class, it is not so obvious that some projects belong in more than one area. For example, flood control is certainly a form of protection for some, as well as being related to natural resources.

Government projects have a number of interesting characteristics that set them apart from projects in the private sector. Many government projects are huge, having first costs of tens of millions of dollars. They tend to have extremely long lives, such as 50 years for a bridge or a dam. The multiple-use concept is common, as in wildland management projects, where economic (timber), wildlife preservation (deer, squirrel), and recreation projects (camping, hiking) are each considered uses of import for the land. The benefits or enjoyment of government projects are often completely out of proportion to the financial support of individuals or groups. Also, there are almost always multiple government agencies that have an interest in a project. Also, public sector projects are not easily evaluated, since it may be many years before their benefits are realized. Finally, there is usually not a clear-cut measure of success or failure in the public sector such as the rate of return or net present worth criteria.

## 7.3 OBJECTIVES IN PROJECT EVALUATION

If large, complex, lengthy, multiple-use projects of interest to several groups are to be evaluated for their desirability, the criteria for the evaluation must first be agreed on. The setting for modern evaluation of government projects dates back to the River and Harbor Act of 1902, which "required a board of engineers to report on the desirability of Army Corps of Engineers' river and harbor projects, taking into account the amount of commerce *benefited* and the *cost*." Even more applicable to today is the criterion specified in the Flood Control Act of June 22, 1936, which stated ". . . that the Federal Government should improve or participate . . . if the *benefits to whomsoever they may accrue* are in excess of the *estimated costs* . . ." [13]. Obviously, the idea of benefit-cost analysis has been around for a long time, even though only in recent years has there been much written about it.

Prest and Turvey [10] give a short, reasonable definition of benefit-cost analysis: It is "a practical way of assessing the desirability of projects where it is important to take a long view (in the sense of looking at repercussions in the further, as well as the nearer, future) and a wide view (in the sense of allowing for side-effects of many kinds on many persons, industries, regions, etc.) that is, it implies the enumeration and evaluation of all the relevant costs and benefits." The "long view" is nothing more than considering the entire planning horizon which, granted, is generally far longer for projects in the public sector than for those in the private sector. The "wide view" and the notion of evaluating all "relevant costs and benefits" probably spell out the greatest single difference between government and private economic evaluation. That is, government projects often affect many individuals, groups, and things, either directly or indirectly, for better or for worse. In evaluating these projects, the analyst tries to capture these effects on the public, quantify them and, where possible, make the measures in monetary terms. Positive effects are referred to as *benefits,* while negative effects are *disbenefits.* In contrast, the private sector "effects" of primary importance are those that relate to income being returned to the organization. *Costs* of construction, financing, operation, and maintenance are estimated in much the same way in both the public and private sectors.

## 7.4 BENEFIT-COST ANALYSIS

The notion of benefit-cost analysis is simple in principle. It follows the same systematic approach used in selecting between economic investment alternatives, including the following six steps adopted from Chapter Five.

1. Define the set of feasible, mutually exclusive, public sector alternatives to be compared.
2. Define the planning horizon to be used in the benefit-cost study.
3. Develop the cost-savings and benefit-disbenefit profiles in monetary terms for each alternative.
4. Compare the alternatives using a specified time value of money and measure(s) of merit.
5. Perform supplementary analyses.
6. Select the preferred alternative.

These will be discussed from the viewpoint of a benefit-cost analysis.

The heart of many engineering problems is in identifying all of the *alternatives* available to achieve a particular goal. Some alternatives can be excluded from further consideration immediately. Each remaining alternative should be described thoroughly. This involves specifying all of the good and bad effects that a project will have on the public. This includes effects directly on people, land values, and the environment. In addition, all aspects of project development, operation, maintenance, and eventual salvage must be stated.

Defining the *planning horizon* is essential in order to define the period over which the best project(s) is to be selected. If the planning horizon is longer than the life of a nonrenewable public project, there may be several years during which no benefits or costs are considered. If the planning horizon is shorter than the life of a project, residual values (same concept as salvage values) may be estimated and used.

Quantifying all *costs and savings* refers to governmental expenditures and incomes received relating to a project over the planning horizon. Disbursements include all first and continuing costs of a project, while income may result from tolls, fees, or other charges to the user public. Residual or salvage values may exist if the project is believed to still be in operation at the end of the planning horizon, and these are treated as savings or negative costs at that time. Costs, unlike benefits, are quantified for public projects in a way similar to private projects.

Each *benefit or disbenefit* during the planning horizon must also be quantified monetarily. Benefits, or the positive effects of a government investment, refer to desirable consequences on the public instead of on any governmental body. Disbenefits are the negative effects on the public instead of on any government organization. Unfortunately, placing dollar values on benefits received by a diverse public is often not an easy task. Neither is deciding on the interest rate to be used.

A base measure such as annual equivalent benefits and costs or present worths of benefits and costs is then established, and the *measure of merit* is chosen. Benefit-cost analyses frequently use the benefit-cost ratio ($B/C$) or, to a lesser extent, a measure of benefits less costs ($B - C$). If

$B_{jt}$ = public benefits associated with project $j$ during year $t$,
  $t = 1, 2, \ldots, n$
$C_{jt}$ = governmental costs associated with project $j$
  during year $t$, $t = 0, 1, 2, \ldots, n$

and

  $i$ = appropriate interest rate

then the $B/C$ criterion may be expressed mathematically, using a present worth base measure, as

$$B/C_j(i) = \frac{\sum_{t=1}^{n} B_{jt}(1 + i)^{-t}}{\sum_{t=0}^{n} C_{jt}(1 + i)^{-t}} \tag{7.1}$$

Note that the $B/C$ ratio is an alternative name for the savings/investment ratio treated in Chapter Four. The $B - C$ criterion is expressed as

$$(B - C)_j(i) = \sum_{t=0}^{n} (B_{jt} - C_{jt})(1 + i)^{-t} \tag{7.2}$$

which is similar to the present worth method described in Chapter Four.

When two or more project *alternatives are being compared* using a $B/C$ ratio, the analysis should be done on an *incremental basis*. That is, first the alternatives should be ordered from lowest to highest cost (present worth, annual equivalent, etc.). Then, the incremental benefits of the second alternative over the first, $\Delta B_{2-1}(i)$, are divided by the incremental costs of the second over the first, $\Delta C_{2-1}(i)$. That is,

$$\Delta B/C_{2-1}(i) = \frac{\Delta B_{2-1}(i)}{\Delta C_{2-1}(i)} = \frac{\sum_{t=1}^{n} (B_{2t} - B_{1t})(1 + i)^{-t}}{\sum_{t=0}^{n} (C_{2t} - C_{1t})(1 + i)^{-t}} \tag{7.3}$$

Note that if the first alternative is "do nothing," the incremental $B/C$ ratio is also the straight $B/C$ ratio for the second alternative. As long as $\Delta B/C_{2-1}(i)$ exceeds 1.0, Alternative 2 is preferable to Alternative 1. Otherwise, Alternative 1 is preferred to Alternative 2. The winner of these is then compared on an incremental basis with the next most costly alternative. These pairwise comparisons continue until all alternatives have been exhausted, and only one "best" project remains. The procedure used is very similar to that specified for the rate-of-return method in Chapter Four.

With the $B - C$ criterion, an incremental basis may be used following the some rules as for the $B/C$ ratio, but preferring Alternative 2 to Alternative 1 as long as the following condition holds:

$$\Delta(B - C)_{2-1}(i) = \Delta B_{2-1}(i) - \Delta C_{2-1}(i)$$

$$= \sum_{t=1}^{n} [B_{2t}(i) - B_{1t}(i)](1 + i)^{-t}$$

$$- \sum_{t=0}^{n} [C_{2t}(i) - C_{1t}(i)](1 + i)^{-t} \geq 0 \tag{7.4}$$

Where benefits and costs are known directly, the value of $(B - C)_j$ for each Alternative $j$ may be calculated and the maximum value selected.

Apparently straightforward, there are a number of potential pitfalls in the evaluation of government projects. That is, economic analyses of public projects can be easily biased unknowingly by a project evaluator. The more significant of these pitfalls are discussed in subsequent sections.

Next, *supplementary analyses* in the form of risk analysis, sensitivity analysis, and break-even analysis may be performed. These are often useful in determining how critical various inputs can be and how close various alternatives are economically, and they otherwise quantify much other information that would likely remain intangible.

The final step is *selecting the preferred alternative*. Not only is it to be selected, but all quantitative and qualitative supporting considerations should be recorded in detail. This is particularly true in the case of public sector projects.

These seven major steps in conducting a benefit-cost analysis, as well as the $B/C$ and $B - C$ relationship to the present worth, annual equivalent, and rate of return criteria, are illustrated in the remainder of this section.

## Example 7.1

We are given the task of deciding between three highway alternatives to replace a winding, old, dangerous road. As the crow drives, the length of the current route is 26 miles. Alternative A is to overhaul and resurface the old road at a cost of $3,000,000. Resurfacing will then be required at a cost of $2.5 million at the end of each 10-year period. Annual maintenance for Alternative A will cost $10,000/mile. Alternative B is to cut a new road following the terrain; it will be only 22 miles long. Its first cost will be $10,000,000, and surface renovation will be required every 10 years at a total cost of $2,250,000. Annual maintenance will be $10,000/mile. Alternative C also involves a new highway which, for practical considerations, will be built along a 20.5-mile straight line. Its first cost, however, will be $18,000,000, because of the extensive additional excavating necessary along this route. It, too, will require resurfacing every 10 years at a cost of $2,250,000. Annual maintenance will be $18,000/mile. This increase over route B is due to the additional roadside bank retention efforts that will be required.

Our task is to select one of these alternatives, considering a planning horizon of 30 years with negligible residual value for each of the highways at that time. One of these alternatives is required, since the old road has deteriorated below acceptable standards. We can calculate the annual equivalent first cost and maintenance cost of each alternative using an interest rate of 8%.

Construction and resurfacing cost:

Route A:  [$3,000,000 + $2,500,000($P|F$ 8,10)
    + $2,500,000($P|F$ 8,20)]($A|P$ 8,30)
    = [$3,000,000 + $2,500,000(0.4632) + $2,500,000(0.2145)](0.0888)
    = $416,849/year

Route B:  [$10,000,000 + $2,250,000($P|F$ 8,10)
    + $2,250,000($P|F$ 8,20)]($A|P$ 8,30)
    = [$10,000,000 + $2,250,000(0.4632) + $2,250,000(0.2145)](0.0888)
    = $1,023,404/year

Route C:  [$18,000,000 + $2,250,000($P|F$ 8,10)
    + $2,250,000($P|F$ 8,20)]($A|P$ 8,30)
    = [$18,000,000 + $2,250,000(0.4632) + $2,250,000(0.2145)](0.0888)
    = $1,733,804/year

Maintenance cost:

Route A: $\left(\$10,000 \dfrac{\$}{\text{mile} - \text{year}}\right)$ (26 miles) = $260,000/year

Route B: ($10,000)(22) = $220,000/year
Route C: ($18,000)(20.5) = $369,000/year

Clearly, route A costs less than route B, which itself costs less than route C. Can we now conclude that route A should be selected based on cost to the government? Absolutely not, according to the criterion that seeks to maximize benefits to the public as a whole, minus costs. In fact, we have analyzed only one side of the problem. Now, we must attempt to quantify the benefits along each of the routes.

Traffic density along each of the three routes will fluctuate widely from day to day, but will average 4000 vehicles/day throughout the year. This volume is composed of 350 light commercial trucks, 250 heavy trucks, 80 motorcycles, and the remainder are automobiles. The average cost per mile of operation for these vehicles is $0.35, $0.50, $0.10, and $0.20, respectively.

There will be a time savings because of the different distances along each of the routes, as well as the different speeds which each of the routes will sustain. Route A will allow heavy trucks to average 35 miles/hour, while other traffic can maintain an average speed of 45 miles/hour. Routes B and C will allow heavy trucks to average 40 miles/hour, and the rest of the vehicles can average 50 miles/hour. The cost of time for all commercial traffic is valued at $15/vehicle/hour, and for noncommercial traffic, $5/vehicle/hour. Twenty-five percent of the automobiles and all of the trucks are considered commercial.

Finally, there is a significant safety factor that should be included. Along the old winding road, there has been an excessive number of accidents per year. Route A will reduce the number of vehicles involved in accidents to 105, and routes B and C are expected to involve only 75 and 70 vehicles in accidents, respectively, per year. The average cost per vehicle in an accident is estimated to be $7500, considering actual physical property damages, lost wages because of injury, medical expenses, and other relevant costs.

We now set about to analyze the various benefits in monetary terms. We have considered savings in vehicle operation, time, and accident prevention. The costs incurred by the public for these items are calculated in the following steps.

Operational costs

Route A: $\left(350 \dfrac{\text{light trucks}}{\text{day}}\right) \left(26 \dfrac{\text{miles}}{\text{light truck}}\right) \left(0.35 \dfrac{\$}{\text{mile}}\right) \left(365 \dfrac{\text{days}}{\text{year}}\right)$

$+ \left(250 \dfrac{\text{heavy trucks}}{\text{day}}\right) \left(26 \dfrac{\text{miles}}{\text{heavy truck}}\right) \left(0.50 \dfrac{\$}{\text{mile}}\right) \left(365 \dfrac{\text{days}}{\text{year}}\right)$

$+ \left(80 \dfrac{\text{motorcycles}}{\text{day}}\right) \left(26 \dfrac{\text{miles}}{\text{motorcycle}}\right) \left(0.10 \dfrac{\$}{\text{mile}}\right) \left(365 \dfrac{\text{days}}{\text{year}}\right)$

$+ \left(3320 \dfrac{\text{automobiles}}{\text{day}}\right) \left(26 \dfrac{\text{miles}}{\text{automobile}}\right) \left(0.20 \dfrac{\$}{\text{mile}}\right) \left(365 \dfrac{\text{days}}{\text{year}}\right)$

$= \$8,726,055/\text{year}$

Route B:   $[350(\$0.35) + 250(\$0.50) + 80(\$0.10) + 3320(\$0.20)](22)(365)$
$= \$7,383,585/\text{year}$

Route C:   $[350(\$0.35) + 250(\$0.50) + 80(\$0.10) + 3320(\$0.20)](20.5)(365)$
$= \$6,880,159/\text{year}$

Time costs

Route A:   $\left(350\ \dfrac{\text{light trucks}}{\text{day}}\right)\left(26\ \dfrac{\text{miles}}{\text{light trucks}}\right)\left(\dfrac{1\ \text{hour}}{45\ \text{miles}}\right)\left(365\ \dfrac{\text{days}}{\text{year}}\right)$

$\times \left(15\ \dfrac{\$}{\text{hour}}\right) + \left(250\ \dfrac{\text{heavy trucks}}{\text{day}}\right)\left(26\ \dfrac{\text{miles}}{\text{heavy truck}}\right)\left(\dfrac{1\ \text{hour}}{35\ \text{miles}}\right)$

$\times \left(365\ \dfrac{\text{days}}{\text{year}}\right)\left(15\ \dfrac{\$}{\text{hour}}\right) + \left(80\ \dfrac{\text{motorcycles}}{\text{day}}\right)\left(26\ \dfrac{\text{miles}}{\text{motorcycle}}\right)$

$\times \left(\dfrac{1\ \text{hour}}{45\ \text{miles}}\right)\left(365\ \dfrac{\text{days}}{\text{year}}\right)\left(5\ \dfrac{\$}{\text{hour}}\right) + \left(3320\ \dfrac{\text{automobiles}}{\text{day}}\right)$

$\times \left(26\ \dfrac{\text{miles}}{\text{automobile}}\right)\left(\dfrac{1\ \text{hour}}{45\ \text{miles}}\right)\left(365\ \dfrac{\text{days}}{\text{year}}\right)$

$\times \left(0.25 \times 15\ \dfrac{\$}{\text{hour}} + 0.75 \times 5\ \dfrac{\$}{\text{hour}}\right) = \$7,459,441/\text{year}$

Route B:   $\left[\dfrac{350}{50}\ (\$15) + \dfrac{250}{40}\ (\$15) + \dfrac{80}{50}\ (\$5) + \dfrac{3320}{50}\ (0.25 \times \$15 + 0.75 \times \$5)\right]$

$\times (22)(365) = \$6,688,078/\text{year}$

Route C:   $\left[\dfrac{350}{50}\ (\$15) + \dfrac{250}{40}\ (\$15) + \dfrac{80}{50}\ (\$5)\right.$

$\left. + \dfrac{3320}{50}\ (0.25 \times \$15 + 0.75 \times \$5)\right](20.5)(365)$

$= \$5,273,292/\text{year}$

Safety costs

Route A: $\left(105\ \dfrac{\text{vehicles}}{\text{year}}\right)\left(7500\ \dfrac{\$}{\text{vehicle}}\right) = \$787,500/\text{year}$

Route B: $(75)(\$7500) = \$562,500/\text{year}$
Route C: $(70)(\$7500) = \$525,000/\text{year}$

All relevant government and public costs are summarized in Table 7.1.

We can desire to compare these three alternative routes using benefit-cost criteria. Let our first criterion be the popular benefit-cost ratio. Since one of these alternatives must be selected, we will assume the lowest government cost alternative, route A, will be selected unless the extra expenditures for routes B or C prove more worthy. Since we have not defined "benefits" per se, user benefits will be taken as the incremental reduction in user costs from the less expensive to the more expensive alternatives. Since we are looking at incre-

TABLE 7.1. Summary of Annual Equivalent Government and Public Costs

| | Route A | Route B | Route C |
|---|---|---|---|
| Government first cost of highway | $ 416,849/year | $ 1,023,404/year | $ 1,733,804/year |
| Government cost of highway maintenance | 260,000/year | 220,000/year | 369,000/year |
| Public operational costs | 8,726,055/year | 7,383,585/year | 6,880,159/year |
| Public time costs | 7,459,444/year | 6,688,078/year | 5,273,292/year |
| Public safety costs | 787,500/year | 562,500/year | 525,000/year |
| Total government costs | 676,849/year | 1,243,404/year | 2,102,804/year |
| Total public costs | 16,972,996/year | 14,634,163/year | 12,678,451/year |

mental benefits, it makes sense to compare these against the respective incremental costs needed to achieve these additional benefits.

The incremental benefits and costs for route B as compared to route A for $i = 8\%$ are given as follows:

$$\Delta B_{B-A}(8) = \text{public costs}_A(8) - \text{public costs}_B(8)$$
$$\$16,972,996 - \$14,634,163 = \$2,338,833/\text{year}$$

$$\Delta C_{B-A}(8) = \text{government costs}_B(8) - \text{government costs}_A(8)$$
$$\$1,243,404 - \$676,849 = \$566,555/\text{year}$$

That is, for an incremental expenditure of $566,555/year, the government can provide added benefits of $2,338,833/year for public. The appropriate benefit cost ratio is then

$$\Delta B | C_{B-A}(8) = \frac{\Delta B_{B-A}(8)}{\Delta C_{B-A}(8)} = \frac{\$2,338,833}{\$566,555} = 4.13$$

This clearly indicates that the additional funds for route B are worthwhile, and we desire route B over route A.

Using a similar analysis, we now calculate the benefits, costs, and $\Delta B | C$ ratio to determine whether or not route C is preferable to route B.

$$\Delta B_{C-B}(8) = \$14,634,163 - \$12,678,451 = \$1,955,712/\text{year}$$

$$\Delta C_{C-B}(8) = \$2,102,804 - \$1,243,404 = \$859,400/\text{year}$$

$$\Delta B / C_{C-B}(8) = \frac{\$1,955,712}{\$859,400} = 2.28$$

This benefit-cost ratio, being greater than 1.00, indicates that the additional expenditure of $859,400/year to build and maintain route $C$ would provide commensurate benefits to the public. In fact, the user savings would be $1,955,712/year. Of the three alternative routes, route C is preferred.

The next benefit-cost criterion takes advantage of the fact that if $\Delta B/C > 1$, then $\Delta(B - C) > 0$. That is, the difference in incremental benefits and costs may be used in place of the incremental benefit-cost ratio.

## Example 7.2

Applying this measure to the routes in the previous example results in the following calculations for comparing routes A and B. We know that

$$\Delta B_{B-A}(8) = \$2,338,833/\text{year}$$

$$\Delta C_{B-A}(8) = \$\ \ 566,555/\text{year}$$

This results in

$$\Delta(B - C)_{B-A}(8) = \Delta B_{B-A}(8) - \Delta C_{B-A}(8) = \$2,338,833 - \$566,555$$
$$= \$1,772,278/\text{year}$$

Thus, we again conclude that route B is preferred to route A. Similarly,

$$\Delta(B - C)_{C-B}(8) = \Delta B_{C-B}(8) - \Delta C_{C-B}(8)$$
$$= \$1,955,712 - \$859,400 = \$1,096,312/\text{year}$$

which indicates that route C is worthy of the additional expenditure required; hence route C should be constructed.

---

The benefit-cost criteria are consistent with the methods of alternative evaluation presented in Chapter Five. For example, suppose it is desired to base a decision on the present worth criterion, minimizing the sum of government construction and maintenance costs as well as public user costs.

## Example 7.3

The present worth of all costs for each highway alternative is given by the following formula:

$$PW(\text{total}) = PW_{\text{total government costs}}(i) + PW_{\text{total public costs}}(i)$$

Then,

Route A: $PW_A(8) = \$676,849(P|A\ 8,30) + \$16,972,996(P|A\ 8,30)$
$\qquad\qquad = \$676,849(11.2578) + \$16,972,996(11.2578)$
$\qquad\qquad = \$198,698,425$

Route B: $PW_B(8) = \$1,243,404(P|A\ 8,30) + \$14,634,163(P|A\ 8,30)$
$\qquad\qquad = \$1,243,404(11.2578) + \$14,634,163(11.2578)$
$\qquad\qquad = \$178,746,474$

Route C: $PW_C(8) = \$2,102,804(P|A\ 8,30) + \$12,678,451(P|A\ 8,30)$

$\qquad\qquad = \$2,102,804(11.2578) + \$12,678,451(11.2578)$

$\qquad\qquad = \$166,404,413$

Since the present worth of all costs for route C is smallest, we again see that this alternative is preferred.

---

The annual equivalent cost is calculated by simply multiplying a constant times the present worth. Thus, the benefit-cost criteria are also comparable to the annual worth measure of merit.

Example 7.4_____

Calculating the equivalent uniform annual cost using

$$AW_{total}(i) = AW_{total\ government\ costs}(i) + AW_{total\ public\ costs}(i)$$

we have

Route A: $AW_A(8) = \$676,849 + \$16,972,996 = \$17,649,845/\text{year}$

Route B: $AW_B(8) = \$1,243,404 + \$14,634,163 = \$15,877,567/\text{year}$

Route C: $AW_C(8) = \$2,102,804 + \$12,678,451 = \$14,781,255/\text{year}$

Again route C is preferred.

It is interesting and logical to note that the differences in equivalent uniform annual costs for routes A,B and B,C are the same as the differences in incremental benefits minus incremental costs for routes A,B and B,C. That is,

$$AW_A(8) - AW_B(8) = \$17,649,845 - \$15,877,567 = \$1,772,278/\text{year}$$

$$\Delta B_{B-A}(8) - \Delta C_{B-A}(8) = \$2,338,833 - \$566,555 = \$1,772,278/\text{year}$$

and

$$AW_B(8) - AW_C(8) = \$15,877,567 - \$14,781,255 = \$1,096,312/\text{year}$$

$$\Delta B_{C-B}(8) - \Delta C_{C-B}(8) = \$1,955,712 - \$859,400 = \$1,096,312/\text{year}$$

---

Finally, the benefit-cost criteria can even be related to the internal rate-of-return approach discussed in Chapter Four. That is, the interest rate is found that causes the annual equivalent incremental benefits less incremental costs to equal zero.

Example 7.5_____

The internal rate of return approach may be used by finding the value of the interest rate $i$ that equates the incremental benefits of Alternative $k$ over Alternative $j$ to their incremental costs. That is,

$$AW_{k-j}(i) = 0 = \Delta B_{k-j}(i) - \Delta C_{k-j}(i)$$

Using this expression to consider route B versus route A, we have

$$\Delta B_{B-A}(i) = \$16,972,996 - \$14,634,163 = \$2,338,833/\text{year}$$

$$
\begin{aligned}
\Delta C_{B-A}(i) = &[\$10,000,00 + \$2,250,000(P|F\ i,10) \\
&+ \$2,250,000(P|F\ i,20)](A|P\ i,30) + \$220,000 \\
&- \{[\$3,000,000 + \$2,500,000(P|F\ i,10) + \$2,500,000(P|F\ i,20)] \\
&[A|P\ i,30] + \$260,000\}
\end{aligned}
$$

Searching for the value of $i$ yielding $AW_{B-A}(i) = \Delta B_{B-A}(i) - \Delta C_{B-A}(i) = 0$, we find

$$AW_{B-A}(30) = \$283,970/\text{year}$$

$$AW_{B-A}(35) = -\$66,903/\text{year}$$

The interest rate we are seeking is between 30 and 35%, or approximately 34%. In any event, $i$ easily exceeds 8% and we prefer route B to route A.

Using the rate-of-return approach, we can also compare routes B and C.

$$\Delta B_{C-B}(i) = \$14,634,163 - \$12,678,451 = \$1,955,712/\text{year}$$

$$
\begin{aligned}
\Delta C_{C-B}(i) = &[\$18,000,000 + \$2,250,000(P|F\ i,10) + \$2,250,000(P|F\ i,20)] \\
&(A|P\ i,30) + \$369,000 \\
&- \{[\$10,000,000 + \$2,250,000(P|F\ i,10) + \$2,250,000(P|F\ i,20)] \\
&(A|P\ i,30) + \$220,000\}
\end{aligned}
$$

The rate of interest causing $AW_{C-B}(i) = \Delta B_{C-B}(i) - \Delta C_{C-B}(i) = 0$ is between 20 and 25%, since

$$AW_{C-B}(20) = \$200,312/\text{year}$$

$$AW_{C-B}(25) = -\$195,688/\text{year}$$

The solving value of $i$ is approximately 22.5%; hence, we prefer route C to route B.

---

Several different approaches to the same problem have been illustrated and shown to be consistent project evaluators. Typically, where government projects are involved, one of the benefit-cost criteria ($B/C$ or $B - C$) are used. More often than not, the criterion is the benefit-cost ratio. This is unfortunate because, just as in rate of return analyses in the private sector, the benefit-cost ratio is easy to misuse and misinterpret, and it is very sensitive to the classification of problem elements as "benefits" or "costs." These problems will be discussed subsequently.

The present worth, annual worth, and rate-of-return methods are seldom used in government analyses. Even though the mechanics underlying these approaches are perfectly suitable, their underlying philosophy is more attuned to return on investment, discounted net profits, and minimum attractive rate of return, implying measures of return to *investors* based on capital investment *by the same investors*. Government investments, however, do not necessarily result in *any* monetary incomes to the government and are seldom evaluated solely on the basis of monetary return. Also, government employees and politicians prefer benefit-cost analyses to enable them to speak intelligently of the "benefits" derived by the public as a result of government's expenditures of the public's tax, assessment, and other money.

## 7.5 IMPORTANT CONSIDERATIONS IN EVALUATING PUBLIC PROJECTS

In the section on benefit-cost analysis, it was stated that there are a number of pitfalls that can affect the analysis of government projects. Actually, opportunities for error pervade benefit-cost analyses, from the very initial philosophy, through to the interpretation of a *B/C* ratio. It is important to talk about the more significant of these, both to help prevent analysts from erring, and to help those who may be reviewing a biased evaluation.

The major topics to be considered include the following:

1. Point of view (national, state, local, individual).
2. Selection of the interest rate.
3. Assessing benefit-cost factors.
4. Overcounting.
5. Unequal lives.
6. Tolls and fees.
7. Multiple-use projects.
8. Problems with the *B/C* ratio.

### 7.5.1 Point of View

The stance taken by the engineer in analyzing a public venture can have an extensive effect on the economic "facts." The analyst may take any of several viewpoints, including those of:

1. An individual who will benefit or lose.
2. A particular governmental organization.
3. A local area such as a city or county.
4. A regional area such as a state.
5. The entire nation.

The first of these viewpoints is not particularly interesting from the standpoint of economic analysis. Nonetheless, all too frequently, an isolated road is paved, a remote stretch of water or sewer line is extended under exceptional circumstances, or a seemingly ideal location for a public works facility is suddenly eliminated from consideration. In these cases, the "benefit-cost analysis," its review, and the implementation decision are usually made by a small, select group.

The other four viewpoints are, however, of considerable interest to those involved in public works evaluation. Analyzing projects or project components from viewpoint 2, that of a particular government agency, is analogous to economic comparisons in private enterprise. That is, only the gains and losses to the organization involved are considered. This viewpoint, which seems contrary to benefit-cost optimization to the public as a whole, may be appropriate under certain circumstances.

## Example 7.6

Consider a Corps of Engineers construction project in which the water table must be lowered in the immediate area so work can proceed. Any of several water cutoff or dewatering systems may be employed. Water cutoff techniques include driving a sheet pile diaphram or using a bentonite slurry trench to cut off the flow of water to the construction area. Dewatering methods include deep well turbines, an eductor system, or wellpoints for lowering the water level. It is sometimes appropriate for the Corps to evaluate these different techniques from an "organization" point of view, since each of the feasible methods provides the same service or outcome—a dry construction site. Therefore, the most economical decision from the Corps' point of view is also correct from the view of the public as a whole, since the benefits or contributions to the project are the same regardless of the method chosen.

The third point of view, that of a locality such as a city or county, is popular among local government employees and elected officials. Unfortunately, seemingly localized projects often impact a much wider range of the citizenry than is apparent.

## Example 7.7

County officials are to decide whether or not future refuse service should be county owned and operated, or whether a private contractor should be employed. The job requires front-end loader compaction trucks as well as roll on–off container capability. Primarily rural roads are traveled, and from 1 container (a roadside picnic area) to 50 containers (a large rurally located industrial plant) must be collected at each stop. Front-end loader containers range from 2 to 8 cubic yards, while roll on-off containers are sized from 15 to 45 cubic yards. Several trucks and drivers will be required, including a base for operations and maintenance.

The cost in dollars per ton of refuse collected, removed, and disposed of, is given below:

| | |
|---|---:|
| Personnel services | $ 6.08/ton |
| Materials, supplies, utilities | 5.13/ton |
| Maintenance and repair | 5.62/ton |
| Overhead | 3.41/ton |
| Depreciation | 3.61/ton |
| Five percent interest on half financed by bonds | 1.80/ton |
| Total county cost | $25.65/ton |
| Federal taxes foregone | $ 1.87/ton |
| State taxes foregone | 0.20/ton |
| Property taxes foregone | 1.54/ton |
| Eight percent return on half financed by tax money | 2.88/ton |
| Not necessarily paid by county | $ 6.49/ton |

County cost to provide refuse service will be $25.65/ton. The county is not, however, required to pay the additional $6.49/ton for federal and state taxes, *ad valorem* taxes, or a return on appropriated money, as would a private firm.

---

It is obvious from the example that a "local" county decision can affect a much wider public. Suppose, based on $25.65/ton, the county decided to own and operate the needed refuse service. Federal taxes of $1.87/ton that would have been paid by a private contractor will not be paid. Since the federal government will still have the same revenue requirements, the difference will be made up by passing on an infinitesimally small burden to the national public as a whole. Although this is easily rationalized at the local level—to spread a portion of the cost of county refuse service over the entire country—consider the result if every town, city, and county took this attitude.

An analogous argument follows for state income taxes foregone; however, now the burden is being spread over the people of the state. Even though this is much smaller than the national population, the burden per person to make up the lost state tax income is still very small. Again, providing refuse service at less cost to the local populace, at the expense of the state, is tempting from a parochial point of view.

If the county were to plan on not having to pay the *ad valorem* tax, the slack would be taken up by increasing the property tax rates in the county. Although this approach increases the burden on county property owners, that burden may be entirely disproportionate when compared to the refuse service each requires.

The last two points of view include a regional (e.g., a state) or national perspective. Ideally, a national outlook is preferable for local, regional, and national public works projects. Experience indicates, however, that the primary concern of public works officials and politicians is their particular constituency.

Perhaps the best advice for evaluators and decision makers in the public realm is to examine multiple viewpoints. That is, project evaluators are often not in a position to decide on a single specific point of view. They should instead present a thorough analysis clearly indicating any benefits or costs that depend on the perspective taken. Similarly, decision makers should require multiple points of view so they can be aware of the kind and degree of repercussions resulting from their actions.

## 7.5.2 Selection of the Interest Rate

The interest rate, discount rate, or minimum attractive rate of return is another premier factor to be decided upon when evaluating public works projects. In the route selection examples, the interest rate was taken at 8% with no question of the appropriateness of such a figure. Clearly, however, the interest rate has a significant effect on the net present worth or annual worth of cash flows in the private sector. Similarly, there is a significant effect on the net present worth of benefits minus costs or the benefit-cost ratio in public sector analyses.

## Example 7.8

Let us reconsider the highway route selection Examples 7.1 to 7.5 again, this time using a minimum attractive rate of return of 25%. Since route B was equivalent to route A at $i = 34\%$, the preference of route B to route A continues to hold. However, route C was equivalent to route B using a *MARR* of 22.5%. In the rate-of-return approach, it was determined that the incremental "rate of return" of route C over route B exceeded 20% but fell short of 25%. Now, however, using an interest rate of 25%, we find that spending the substantial additional funds for route C is not worthwhile.

---

One may argue with the above example, saying that $i = 25\%$ is higher than would ever be used. Such an argument is certainly not true for private industry. In public works, however, various knowledgeable individuals have explicitly proposed values on $i$ ranging from 0 to 13%, and rather vague support has been seen ranging from 0 to over 25%. Before considering various schools of thought on the appropriate interest rate, it is helpful to know how public activities are financed.

*Financing of Government Projects.* There are several different ways that units of government finance public sector projects. The most obvious way is, of course, through taxation such as income tax, property tax, sales tax, and road user tax. Another popular approach is through the issuance of bonds for either specific projects or general use. Other forms of borrowing, such as notes, may be thought of in the same category as bonds. A third type of fund raising includes income generating activities such as a municipally owned power plant, a toll road, or other activity where a charge is made to cover (or partially offset) the cost of the service performed. Although these are the primary sources of government funds, there are a number of ways in which this money may be passed from one government authority to another by way of direct payments, loans, subsidies, and grants.

Federal funds are raised through tax money and federal borrowing. Federal projects may then be financed through direct payment. In this case, no monetary return is expected by the government; however, the "return" is expressed through the benefits incurred by the public. Direct payment financing may be total as in the case of many corps of engineers projects, or partial, for example 90%, as in cost sharing with states for interstate highways.

Financing for projects of national interest and impact may also be available through no-interest or low-interest rate loans. Both are available for long periods of time, say up to 40 years, with terms obviously more favorable than could normally be expected from conventional sources of money. Such loans are available for financing large projects such as the Tennessee Valley Authority where revenues resulting from the projects are used to pay back the loans plus interest, if any. Certain university buildings such as dormitories may also be financed in this way. There are also occasions when principal payment deferment is permitted during the early years of the project.

Other forms of federal financing include subsidies and federal loan insurance. Subsidies are used to encourage projects or services believed to be in the public's best interest, such as in the area of transportation. Loan insurance is used to eliminate the private lending institution's risk, allowing lower interest loans over longer periods than conventionally available. Insured loans began with the Federal Housing Administration.

State and local public projects are financed from taxes or bonds. There are, however, constraints on bond financing. First, bond issues must be approved by the voters. Often, there must be a 60% or even two-thirds vote in favor of the bond. Second, in order to prevent excessive borrowing, states have limited the amount of bond debt that may be undertaken. This is often a fraction of the property valuation assessment in the local area. Finally, there are also restrictions that govern the payback requirements and lives of the bonds. Considering these restrictions, the temperament of the public regarding bond issues, and the future needs that are likely to require bond financing, care is obviously needed in selecting public works projects to be implemented or put before the people.

*Considerations in the Selection of the Interest Rate.* Many arguments over the correct philosophy to use in selecting the interest rate have surfaced over time. For practical purposes, most of these philosophies are somewhat aligned with one of the following positions:

1. A zero interest rate is appropriate when tax monies are used for financing.
2. The interest value need only reflect society's time preference rate.

3. The interest rate should match that paid by government for borrowed money.

4. The appropriate interest rate is dictated by the opportunity cost of those investments forgone by private investors who pay taxes or purchase bonds.

5. The appropriate interest rate is dictated by the opportunity cost of those investments foregone by government agencies due to budget constraints.

Advocates of a zero interest rate when tax money is used argue that current taxes require no principal or interest payment at all. Hence, current tax monies (e.g., highway user taxes) should be considered "free" money, and a no-interest or discount rate applied. Counterarguments to this stance point out that a zero (or even low) interest rate will allow very marginal projects or marginal "add-on" project enhancements to achieve a $B/C$ ratio greater than 1. This, in turn, takes money away from other projects that are truly deserving. If, in fact, it is not true that more deserving projects are precluded while very marginal projects (say 1 or 2% "return") are being approved, perhaps there is an excess of money available and taxes should be lowered. A final position against the zero interest rate advocates is that if government does not invest funds in high-benefit or "profitable" projects, the people should be able to retain and invest their own money to provide benefits more economically.

The "societal time preference rate" advocates contend that $i$ is merely "a planning parameter reflecting society's feelings about providing for the future as opposed to current consumption" says Howe [6]. Henderson [5] says $i$ is the rate that "reflects the government's judgment about the relative value which the community as a whole is believed to assign, or which the government feels it ought to assign, to present as opposed to future consumption at the margin." As such, Howe [6] makes clear that the societal time preference rate "need bear no relation to the rates of return in the private sector, interest rates, or any other measurable market phenomena." Estimated societal time preference rates tend to be quite low (e.g., 3 to 6%).

Many people back the use of an interest rate that matches that paid by government for borrowed money. This seems reasonable in that government bonds are in direct competition with other investment opportunities available in the private sector. Of course, many government bond coupon rates tend to be lower because of their tax exempt status (the bond interest is not taxable) than their private industry counterparts. One way of obtaining an interest figure is to determine the rate on "safe" long-term federal bonds [6]. Another possibility is to use the rate paid on borrowings by the particular government unit in question. That is, if a project will be financed by borrowing, the rate of interest to be paid is used. If nonborrowed funds are to be used, the appropriate rate is the average rate of interest being paid on long-term (over 15 years) borrowing.

There are also good arguments that the cost of government money may be too low an interest rate. For example, the opportunities forgone by other government agencies or investors in the private sector may have provided a far higher "return" than investments approved using the cost of government money. In addition, using a rate equal to the cost of government borrowings includes no provision for risk, nor does it include the subsidizing effect of the tax exemption. Finally, it is argued that plain good judgment dictates that projects worthy of approval under higher interest rates are more justifiable in cases where benefits and disbenefits received by the public are very non-uniform.

The next philosophy calls for an opportunity cost approach, taking into account many of the factors not considered in the pure cost of government borrowed money. For example, Howe [6] says this philosophy is that "no public project should be undertaken that would generate a rate of return less than the rate of return that would have been experienced on the private uses of funds that would be precluded by the financing of the public project (say, through taxes or bonds)." How are these private use rates determined? Consider a situation in which private investments would yield an average annual return of $i_1\%$. Also consider the rate of return on consumption, which is measured by the rate that consumers are willing to pay to consume now instead of later. This rate is, say, 10 to 13% for a house, 12 to 24% for a new or used car, and so on. Let this rate of return on consumption average out to $i_2\%$. Now, if a fraction $\alpha$ of government financing precludes private investments, while $1 - \alpha$ precludes consumption, the rate of return foregone to finance public projects is given as $i$.

$$i = \alpha i_1 + (1 - \alpha)i_2 \qquad 0 \le \alpha \le 1 \qquad (7.5)$$

An empirical study conducted by Haveman [4] in 1966 showed the appropriate weighted average to be 7.4% *at that time*. With the many economic changes since that time, a rate of perhaps 12 to 15% is more appropriate for the 1980s.

The last philosophy also requires an opportunity cost approach in which an artificial interest rate reflects the rates of return forgone on government projects by virtue of having insufficient funds. That interest rate is found by continuing to increase the value of $i$ until only the projects remain having $B/C > 1$ or $B - C > 0$ which can be afforded with monies available.

What conclusions can be reached from these philosophical arguments? What guidelines are available for evaluators of public works projects? Clearly, no one answer is universally applicable. But, as a general rule, we recommend using a rate that is at *least* as high as the average effective yield on tax exempt long-term government bonds. That is, projects or enhancements to projects not providing at least this rate of return should not be undertaken. This is not to say, however, just because a project is attractive at the *MARR,* that it should be implemented. Only the most worthy projects, in the environment of limited funds, should be constructed. These projects may be identified by conducting several analyses of prospective projects, continually raising the interest rate ($i = 6, 8, 10, 12, \ldots \%$) until only the projects remain that may be accomplished with available or legally borrowable funds.

### 7.5.3 Assessing Benefit-Cost Factors

The benefit-cost analyst knows, before starting a study, that placing a monetary figure on certain "societal benefits" may be difficult. Actually, there is even a more fundamental problem—what factors to assess. Some insight is available by considering the following four types of factors discussed by Cohn [2].

1. *"Internal"* effects are those which accrue directly or indirectly to the individual(s) or organization(s) with which the analyst is primarily concerned. These effects are always included in a benefit-cost analysis.

2. *"External"* technological (*or real*) effects are those which cause changes in the physical opportunities for consumption or production. For example, effects on navigation and water sport recreation due to a new hydroelectric plant would fall into this category. These effects should be included in an analysis.

3. *"External"* pecuniary effects relate to changes in the distribution of incomes through changes in the prices of goods, services, and production factors. For example, if a firm's expansion is sufficiently large to affect industry prices, an increase in its output is most likely to result in lower prices for output from other firms in the industry as a whole. Many authors agree that these effects can safely be ignored.

4. *Secondary effects* involve changes in the demand for and supply of goods, services, resources, and production factors which *arise from* a particular project. As an example, phosphate mining in Idaho on government lands will bring instant population increases to nearby small towns. Secondary effects include increasing the incomes of various producers such as vehicle repairmen and barbers who provide the vehicle repairmen with haircuts. McKean [9] contends that only incremental income arising from such effects should be included, if reasonable to do so.

### Example 7.9

A dam and reservoir are contemplated in an effort to reduce flood damage to homes and crops in a low area of northeastern Oklahoma. Annual damage to property varies from year to year, but averages approximately $230,000/year. The dam and reservoir contemplated should virtually eliminate damage to the area in question. No other benefits (e.g., irrigation, power generation, recreation) will be provided.

In performing a benefit-cost analysis of the flood control project, the engineer notes that the primary benefit to the public will be the $230,000/year damage prevented. However, the engineer argues that this will also cut back on money paid to contractors and servicemen for home and car repair, to health care units, for insurance premiums, and so forth. In other words, the building of a damn, it could be argued, will provide disbenefits to those who would normally receive part of their livelihood from helping flood-damaged families. Should the engineer include in the evaluation only the direct benefits to the flood damage victims, or should the other effects be included as well? That depends on how the analyst chooses to handle secondary effects, as indicated in the following discussion.

---

The disbenefits to those who would lose income if the dam and reservoir were built are considered *secondary effects*. That is, there would be a decrease in the demand for and supply of post-flood restoration goods and services. It is argued that only the *incremental* incomes or incremental profits (losses or lost profits in this case) should be considered when secondary effects are involved.

Another argument calls upon the "ripple" effect of the economy. That is, every secondary effect disbursement by one person or organization is a receipt to another person or organization. Each receipt then contributes to another disbursement, and so on. If the ripple philosophy were tracked or followed for the secondary effects of a particular alternative, the sum of receipts less disbursements would equal zero, and there would be no economic evaluation.

With either philosophy, the secondary effects of the example's flood control dam and reservoir would be small, if not negligible. This is intuitively reasonable, because the dam's main benefits represent the measure of the direct usefulness of the dam serving its intended purpose, whereas the diseconomies described are, in fact, secondary and diffuse.

Example 7.10 _____

Now, reconsider the previous flood control dam and reservoir of Example 7.9. Suppose the reservoir would cause a loss of agricultural land for grazing and crops. Should this loss be considered in the benefit-cost analysis? Yes, because it is an external real effect causing changes in the physical opportunities for consumption or production.

---

External technological effects are often well defined and, in practice, are usually included in benefit-cost analyses. External pecuniary effects, however, may not be vividly apparent, as illustrated in the following example.

Example 7.11 _____

A large irrigation project is being considered in the heart of cotton country. The irrigation will provide a significant effect on the quantity and quality of the cotton grown. This additional supply of cotton will, however, depress the price of cotton, lowering the profitability of other cotton growers. Also, the same effect will be felt throughout, say, the clothing industry and manufacturers of cotton substitutable products (products that may be used in place of cotton items) will likely have to reduce prices. At the same time, producers of cotton complementary goods (items that go well with or are used in conjunction with cotton products) will note increased demand, and may increase prices because of an insufficient supply.

Which of these effects would we include in an evaluation of the irrigation project? Of the factors mentioned above, none would be included in the analysis. Each of the effects described relates to changes in the distribution of incomes through changes in the prices of goods, services, and production factors. As such, they are considered external pecuniary effects, which are not "real" benefits or disbenefits and, hence, are not included.

In the above examples, internal and external technological effects were considered to be factors in an analysis, while external pecuniary and secondary effects were not included. There is, of course, no complete agreement on either these classifications or whether or not they should be evaluated. For the practitioner, good practical judgment is probably the best asset in deciding what factors to include.

As a guide, all identifiable effects of a project should be delineated. Some effects will clearly provide direct benefits or disbenefits to the public that should be counted. There may also be some factors that obviously should not be included. The third group—the controversial factors, if any—should be studied in depth and either included for good reason or to be used implicitly as supporting information. The reason for including or not considering these controversial factors should be stated in writing and become a part of the evaluation for the record.

### 7.5.4 Overcounting

A common dysfunction of trying to consider a large variety of effects in a benefit-cost analysis is to overcount, or unknowingly count some factors twice.

Example 7.12_____

Many years ago, Connecticut's Department of Health calculated the following loss from *preventable* disability, hoping to increase interest in health problems.
  Individual income lost

| | |
|---|---:|
| Number of wage earners affected | 350,000 |
| Average days per year lost due to preventable accidents | 4 |
| Average hourly wage | $2.50 |
| Total individual loss per year | $3,500,000 |

  Industry loss:

It is accepted that the loss to industry in disorganization, idle overhead, and lessened production is $2\frac{1}{2}$ times the wage loss, or
Total industry loss per year                              $8,750,000

These figures were totally inadequate because of a classic (but not overly apparent) case of double-counting. If we assume the loss to individuals is, correctly, their forgone earnings, then the loss to industry would be *its* forgone earnings, or its lost profits, and not the entire value of lessened production. As the figures stand above, lost wages are counted once from the viewpoint of the employee, and are again included as part of industry's losses. That is, Weisbrod [11] emphasizes that employee wages are double-counted.

Example 7.13

Reconsider Example 7.11 involving a cotton irrigation system. The increased quantity of cotton will require that additional gin and seed mill hands be employed, removing a significant number of persons from the welfare rolls. The amount of their new wage, equal to the sum of their old welfare payments plus some increase, represents an increase in real output and constitutes a legitimate national benefit of the project. To then add the reduction in welfare payments from the taxpayers to the unemployed would be to double-count welfare payments, once from the standpoint of the recipient, and once from the taxpayer's viewpoint.

## 7.5.5 Unequal Lives

When comparing one-shot public works projects having unequal lives, the planning horizon will commonly coincide with the longest lived alternative. When some projects are expected to have a long life while others have a shorter life, changes in the discount rate could change the attractiveness of some projects with respect to the alternatives.

Example 7.14

Two projects each have first costs of $200,000, with annual operating costs of $30,000. The life of Project A is 15 years, while that of Project B is 30 years and benefits accrued to the public are estimated at $60,000/year and $54,000/year, respectively. Each is a one-shot project; hence, Project A will have no benefits or costs after year 15. A 30-year planning horizon will be used. If the MARR is set at 8%, which project is more attractive?

$$(B - C)_A(8) = \$60,000(P|A\ 8,15) - \$30,000(P|A\ 8,15) - \$200,000$$
$$= \$60,000(8.5595) - \$30,000(8.5595) - \$200,000$$
$$= \$56,785$$

$$(B - C)_B(8) = \$54,000(P|A\ 8,30) - \$30,000(P|A\ 8,30) - \$200,000$$
$$= \$54,000(11.2578) - \$30,000(11.2578) - \$200,000$$
$$= \$70,187$$

Over a 30-year planning horizon, and using an interest rate of 8%, Project B is clearly better. Now suppose that insufficient money is available to implement all of the many projects, B included, that a government agency would like to do. A decision is made to consider all projects using an interest rate that reflects the opportunity cost of investments foregone by government agencies. That is, the interest rate is increased until only those projects that can be afforded remain. Assuming the interest rate is up to 12%, now which project is more desirable?

$$(B - C)_A(12) = \$60,000(P|A\ 12,15) - \$30,000(P|A\ 12,15) - \$200,000$$
$$= \$60,000(6.8109) - \$30,000(6.8109) - \$200,000$$
$$= \$4327$$

$$(B - C)_B(12) = \$54,000(P|A\ 12,30) - \$30,000(P|A\ 12,30) - \$200,000$$
$$= \$54,000(8.0552) - \$30,000(8.0552) - \$200,000$$
$$= -\$6675$$

At the higher interest rate, Project A is more favorable because of the increased emphasis on early year net benefits as opposed to the heavily discounted net benefits during the latter years of the planning horizon.

When a planning horizon shorter than some project durations is selected, a residual value must be estimated for those projects. The residual value is handled in the same way as a salvage value.

## Example 7.15

Reconsidering the previous example and selecting a 15-year planning horizon, let us now determine the more favorable alternative, assuming project B has a residual value of 40% of first cost. Let $i = 12\%$.

$$(B - C)_A(12) = \$60,000(P|A\ 12,15) - \$30,000(P|A\ 12,15) - \$200,000$$
$$= \$60,000(6.8109) - \$30,000(6.8109) - \$200,000$$
$$= \$4327$$

$$(B - C)_B(12) = \$54,000(P|A\ 12,15) + 0.4(\$200,000)(P|F\ 12,15)$$
$$-\$30,000(P|A\ 12,15) - \$200,000$$
$$= \$54,000(6.8109) + \$80,000(0.1827) - \$30,000(6.8109)$$
$$- \$200,000 = -\$21,922$$

## 7.5.6 Tolls, Fees, and User Charges

Tolls, fees, and user charges have an interesting effect on the fiscal aspects of public projects. If a toll, fee, or user charge is regarded as a payment or partial payment for benefits derived, it can be argued that net benefits received are reduced by the amount of the payment. Similarly, the amount of the payment decreases the cost of the project to the government. Thus, the $B/C$ ratio will change, but the $B - C$ measure of merit will remain constant so long as total user benefits remain constant.

## Example 7.16

Suppose 10,000 people/year attend a public facility that has an equivalent uniform annual cost of $20,000. The people, on the average, receive recreational benefits in the amount of $3 each. The $B/C$ ratio would be

$$B/C = \frac{3(\$10,000)}{\$20,000} = 1.5$$

and the $B - C$ measure of merit is

$$B - C = 3(\$10,000) - \$20,000 = \$10,000/\text{year}$$

Based on either criterion, the public facility appears worthwhile. Now suppose that a fee of $1.50/season is charged. The net benefits are now $3 - $1.50, or $1.50/person, and the government cost is reduced by $15,000/year. Thus, the $B/C$ and $B - C$ measures are as follows.

$$B/C = \frac{\$30,000 - \$15,000}{\$20,000 - \$15,000} = 3$$

$$B - C = \$30,000 - \$15,000 - (\$20,000 - \$15,000) = \$10,000/\text{year}$$

Note that in the example $B/C$ changed while $B - C$ did not. This phenomenon will be discussed in a subsequent section on "Problems With the $B/C$ Ratio."

It might be concluded that tolls, fees, and user charges are irrelevant, at least with respect to the $B - C$ measure of merit. This, however, *is not* true if the number of users or degree of use is linked to the fee charged, as it almost always will be.

## Example 7.17

As an extreme, suppose that the 10,000 users of the public facility in the previous example receive different levels of benefits, but they average out to $3/person. The actual breakout is that 8000 persons perceive $1.45 worth of enjoyment, 1000 persons perceive $3 worth, and 1000 persons expect to derive $15.40 in recreational benefits. With a user fee of $1.50/person, only 2000 will patronize the facility. Thus, the $B/C$ and $B - C$ measures would be:

$$B/C = \frac{1000(\$3) + 1000(\$15.40) - 2000(\$1.50)}{\$20,000 - 2000(\$1.50)} = \frac{\$15,400}{\$17,000} = 0.91$$

$$B - C = [1000(\$3) + 1000(\$15.40) - 2000(\$1.50)] - [\$20,000 - 2000(\$1.50)]$$
$$= -\$1600$$

In this case, the reaction of demand to a fee would cause the costs to exceed realized benefits. Thus, when tolls, fees, and user charges are expected, their effect on user demand, and hence total user benefits, must be determined and accounted for.

### 7.5.7 Multiple-Use Projects

Multiple-use projects receive a great deal of attention, both pro and con. Multiple uses, and hence multiple benefits, are often available at slight incremental costs over single-use projects. Of course, the incremental capital and net operating costs required for an additional use must provide at least a like worth of benefits.

Example 7.18 _____

A dam and reservoir for irrigation will provide present worth benefits of $25 million over the next 50 years. The present worth cost of construction, operation, and maintenance of the irrigation facility will be $14,500,000. A single purpose flood control dam providing present worth benefits of $6 million would cost a present worth of $9 million. Suitable design modifications can be made to the irrigation dam and reservoir to provide the flood control benefits, too, at a total package present worth cost of $18,500,000. Funds permitting, what should be done?

First, it must be understood that the benefits and costs discounted to the present were discounted at the *MARR* deemed suitable by the decision maker. Assuming this to be true, it is clear that the irrigation project is worthwhile, providing a benefit-cost ratio of

$$B/C_{\text{irrigation}} = \frac{\$25,000,000}{\$14,500,000} = 1.72^1$$

As a single-purpose facility, a flood control dam would not provide benefits commensurate with its costs, yielding a *B/C* ratio of

$$B/C_{\text{flood control}} = \frac{\$6,000,000}{\$9,000,000} = 0.67^1$$

As a multiple-use facility, however, the flood control benefits may be provided at a sufficiently low incremental cost to be justifiable. The incremental *B/C* ratio is

$$\Delta B/\Delta C_{\substack{\text{irrigation plus} \\ \text{flood control}}} = \frac{\$6,000,000}{\$18,500,000 - \$14,500,000} = 1.5$$

Thus, a multiple-use facility should be built.

_____

[1] This is also the incremental *B/C* ratio of irrigation over doing nothing.

The example illustrates how multiple uses can draw on each other, providing benefits economically that could never have been provided using a single-purpose facility (e.g., the *B/C* flood control ratio of 0.67). Multiple-purpose projects also have their problems. For example, it is frequently desirable to "allocate" the costs of a project to its various uses.

## Example 7.19

A city's refuse is used to fire a power generation facility owned by a municipality. Not only is electrical power supplied to a segment of the city, but burning of the refuse after processing has virtually eliminated the need for an expensive landfill operation. Since this project is self-supporting, construction, operation, and maintenance costs must be allocated between the disposal and power benefits provided in order to determine the user charge for refuse disposal and electrical energy. Arguments for cost allocation range from (1) no costs should be allocated to refuse disposal because the refuse is being used in place of fuel oil or coal and, in fact, a credit should be issued, to (2) refuse disposal should receive sufficient cost allocation to raise rates above those for conventional disposal to include the aesthetic benefits of no unsightly public landfill.[2]

Another problem with multiple-purpose projects includes proponents from and opposition by various interest groups, on the basis of politics, environment, and the like. Yet another problem has to do with coordinating project finances, where some aspects may be federally financed with no payback required, other aspects may be self-supporting, and yet others may require state funds. Finally, there may be conflicts between the various purposes involved. That is, an enhancement of one use may detract from another use. Nevertheless, multiple-use projects are here to stay and will be confronted frequently by engineers in the public sector.

### 7.5.8 Problems with the *B/C* Ratio

There are two frequent problems with the *B/C* ratio that require an explanation and warning. Either can give misleading results that may cause an otherwise perfect analysis to point toward the wrong project.

First, it is sometimes difficult to decide whether an item is a benefit to the public or a cost savings to the government. Similarly, there is often uncertainty between disbenefits and costs.

## Example 7.20

A project provides annual equivalent benefits of $100,000, disbenefits of $60,000/year, and annual costs of $5000. What is the benefit-cost ratio?

---

[2] These extremes in arguments have actually been used by public officials of one major city.

Let us first calculate the $B/C$ ratio for the problem as stated.

$$B/C = \frac{\$100,000 - \$60,000}{\$5000} = 8$$

Such a high ratio leads one to believe that the project is outstanding.

Now another analyst notes that the government will reimburse those incurring damages from the project in an amount equivalent to $60,000/year. Thus, the analyst concludes that the public disbenefits have been compensated for by the government and calculates a $B/C$ ratio of

$$B/C = \frac{\$100,000}{\$5000 + \$60,000} = 1.54$$

which is considerably lower.

Example 7.20 shows that a wide range of $B/C$ ratios may reasonably be obtained on a single project simply by interpreting certain elements of the problem differently. The resolution of this problem is not difficult. The analyst should simply calculate the net benefits less the net costs.

## Example 7.21

Let us reconsider Example 7.20 and calculate the annual net benefits less net costs.

The first analyst would have calculated

$$B - C = (\$100,000 - \$60,000) - (\$5000) = \$35,000/\text{year}$$

and the second analyst would have calculated

$$B - C = (\$100,000) - (\$5000 + \$60,000) = \$35,000/\text{year}$$

Calculating $B - C$ eliminates the inherent bias in the $B/C$ ratio and does not require an incremental approach between alternatives where benefits and costs are known directly for each alternative. That is, if mutually exclusive alternatives are involved, over the same time horizon, the one having the highest $B - C$ value should be selected. Unfortunately, the $B/C$ ratio is by far the more popular criterion of the two.

The potentiality for the other $B/C$ problem was illustrated in the dam and reservoir irrigation example. That is, when the $B/C$ ratio is used, it should be based on *incremental benefits* and *incremental costs*.[3] Simply to calculate the $B/C$ ratio of each alternative and take the one with the largest ratio is incorrect and will frequently lead to errors in project selection.

---

[3] Actually, the $B/C$ irrigation ratio in the referenced example is a ratio of incremental benefits to incremental costs where the pairwise comparison is between irrigation and doing nothing.

Example 7.22

In the dam and reservoir of Example 7.18 we calculated the $B/C$ irrigation ratio to be 1.72. To compare this value against a total project $B/C_{\text{irrigation+flood control}}$ ratio of

$$\frac{\$25,000,000 + \$6,000,000}{\$18,500,000} = 1.68$$

would cause us to select irrigation only, in error.

A related error is to require the incremental $B/C$ ratio to be above that for the previous incremental $B/C$ ratio. In the dam and reservoir example, had the incremental $B/C$ ratio, $\Delta B/\Delta C_{\text{irrigation plus flood control}} = 1.5$ been compared against the $B/C_{\text{irrigation}} = 1.72$ ratio, again an incorrect conclusion would have resulted. As long as the incremental $B/C$ ratio exceeds 1, the incremental benefits justify the incremental costs. In this regard, the $B/C$ ratio criterion is closely akin to the rate-of-return criterion discussed in Chapter Four.

## 7.6 COST-EFFECTIVENESS ANALYSIS

To this point it has been assumed that public project effects were measurable, either directly or indirectly, in monetary terms. There are circumstances, however, when project outputs are not measurable monetarily and must be expressed in physical units appropriate to the project. In these cases, *cost-effectiveness analysis* has proven to be a useful technique for deciding between projects or systems for the accomplishment of certain goals. Although cost-effectiveness is most often associated with the economic evaluation of complex defense and space systems, it has also proven useful in the social and economic sectors. In fact, it has roots dating back to Arthur M. Wellington's *The Economic Theory of Railway Location* in 1887 [12].

Cost-effectiveness analyses require that three conditions be met, according to Kazanowski [7]. They are:

1. Common goals or purposes must be identifiable and attainable.
2. There must be alternative means of meeting the goals.
3. There must be perceptible constraints for bounding the problem.

Common goals are required in order to have a basis for comparison. For example, it would not make sense to compare a submarine with a sophisticated single sideband communication network. Obviously, alternative methods of accomplishing the goals must be available in order to have a comparison. Finally, reasonable bounds for constraining the problem by time, cost, and or effectiveness are necessary to limit and better define the alternatives to be considered.

### 7.6.1 The Standardized Approach

Kazanowski [7] presents 10 standardized steps that constitute a correct approach to cost-effectiveness analyses. They are, in their usual order, presented below.

**Step 1.** *Define the goals,* purpose, missions, and so forth, that are to be met. Cost effectiveness analysis will identify the best alternative way of meeting these goals.

**Step 2.** *State* the *requirements* necessary for attainment of the goals. That is, state any requirements that are essential if the goals are to be attained.

**Step 3.** *Establish* evaluation *measures* which relate capabilities of alternatives to requirements. An excellent list of evaluation criteria is contained in Kazanowski [8]. Typical measures are performance, availability, reliability, maintainability, and so forth.

**Step 4.** *Select* the fixed-effectiveness or fixed-cost *approach.* The fixed effectiveness criterion is minimum alternative cost to achieve the specified goals or effectiveness levels. Alternatives failing to achieve these levels may either be eliminated or assessed penalty costs. The fixed cost criterion is the amount of effectiveness achieved at a given cost. "Cost" is usually taken to mean a present worth or annual equivalent of "life cycle cost" which includes research and development, engineering, construction, operation, maintenance, salvage, and other costs incurred throughout the life cycle of the alternative.

**Step 6.** *Determine capabilities* of the alternatives in terms of the evaluation measures.

**Step 7.** *Express* the alternatives and their *capabilities* in a suitable manner.

**Step 8.** *Analyze* the various *alternatives* based upon the effectiveness criteria and cost considerations. Often, some alternatives are clearly dominated by others and should be removed from consideration.

**Step 9.** *Conduct* a *sensitivity analysis* to see if minor changes in assumptions or conditions cause significant changes in alternative preferences.

**Step 10.** *Document all* considerations, analyses, and decisions from the above nine steps.

Clearly, one cannot become a cost-effectiveness expert by reading these 10 steps. They do, however give an indication of what is involved in a cost-effectiveness study. Excellent detailed presentations on cost effectiveness are available in English [3] and Blanchard [1].

---

Example 7.23———————————————————————————————

Three different propulsion systems are under consideration. The life cycle cost is not to exceed $2.4 million. A single effectiveness measure is decided on, that being reliability. Reliability would be defined as "the probability that the propulsion system will perform without failure under given conditions for a given period of time." Four contractors submit candidate systems, which are evalu-

TABLE 7.2. Evaluation of Candidate
Propulsion Systems for Example 7.23

| Propulsion System | Life Cycle Cost in Millions | Reliability |
|---|---|---|
| 1 | 2.4 | 0.99 |
| 2 | 2.4 | 0.98 |
| 3 | 2.0 | 0.98 |
| 4 | 2.0 | 0.97 |

ated as shown in Table 7.2. Since there is only one effectiveness measure, the results given in Table 7.2 may be expressed graphically as shown in Figure 7.1.

If we look at these systems in pairs, comparing just 1 and 2 is equivalent to a fixed-cost comparison in which we prefer system 1 because of its higher reliability. Similar reasoning leads us to prefer system 3 over system 4. If we were to compare only systems 2 and 3, this would be a fixed-effectiveness comparison, in which case we prefer system 3 due to its lower cost. Clearly, system 1 dominates 2, 3 dominates 2, and 3 dominates 4. We are left with making the decision between systems 1 and 3. This choice will depend on whether or not an additional percent reliability justifies the expenditure of an additional $400,000.

In this example, it is tempting to go the next step and say that based on the ratios of reliability to cost 0.99/$2.4 million and 0.98/$2.0 million, system 3 should be selected. Unfortunately, apparent as such a decision may seem, the correct decision may depend on many other considerations, not the least of

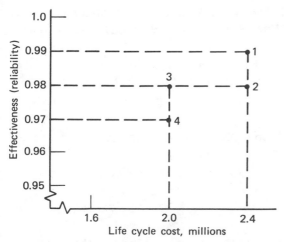

FIGURE 7.1. Effectiveness (reliability versus life cycle cost.

which are the payload (perhaps human life) and the consequences of system failure (perhaps total destruction with no payroll recovery).

## 7.7 SUMMARY

Benefit-cost and cost-effectiveness analyses are accepted facts of life in the public sector. Because of the rapid shifting of the majority of employees from private industry to the public and service sectors, no doubt tomorrow's economic evaluators will be more and more closely involved with these techniques over time.

## BIBLIOGRAPHY

1. Blanchard, Ben S., "Cost Effectiveness/Life-Cycle Cost Analysis," unpublished report, Virginia Polytechnic Institute and State University, Blacksburg, Virginia, 24061.
2. Cohn, Elchanan, *Public Expenditure Analysis,* D. C. Heath, 1972.
3. English, J. Morley (Editor), *Cost Effectiveness,* Wiley, 1968.
4. Haveman, R. H., "The opportunity cost of displaced private spending and the social discount rate." *Water Resources Research, 5* (5), 1969, pp. 947–957.
5. Henderson, P. D., "The Investment Criteria for Public Enterprises," in *Public Enterprises,* edited by Ralph Turvey, Penguin Books, 1968.
6. Howe, Charles W., *Benefit-Cost Analysis for Water System Planning,* Water Resources Monograph 2, American Geophysical Union, Washington, D.C., 1971.
7. Kazanowski, A. D., "A Standardized Approach to Cost-Effectiveness Evaluations," in J. Morley English (Editor), *Cost Effectiveness,* Wiley, 1968.
8. Kazanowski, A. D., "Some Cost-Effectiveness Evaluation Criteria," Appendix B in J. Morley English (Editor), *Cost Effectiveness,* Wiley, 1968.
9. McKean, Roland N., *Efficiency in Government Through Systems Analysis,* Wiley, 1958.
10. Prest, A. R., and Turvey, Ralph, "Cost-Benefit Analysis: A Survey," *Economic Journal, 75,* December 1965, pp. 683–735.
11. Weisbrod, Burton A., *Economics of Public Health,* University of Pennsylvania Press, Philadelphia, 1961, pp. 7–8.
12. Wellington, A. M., *The Economic Theory of Railway Location,* Wiley, 1887.
13. United States Code, 1940 edition, U.S. Government Printing Office, Washington, D.C., p. 2964.

# PROBLEMS

1. Identify the benefits, both monetary and intangible, that would accrue to the public for the following projects: (a) a museum, (b) a fence enclosed walkway over a busy four-lane road connecting two city subdivisions, one having a playground and park, (c) a city sanitary system, and (d) a visiting troupe of performers to appear in a downtown city mall for several days. (7.3)

2. Three alternatives are available, A, B, and C. Their respective annual benefits, disbenefits, costs, and savings are as follows:

|  | A | B | C |
|---|---|---|---|
| Benefits | $200,000 | $300,000 | $400,000 |
| Disbenefits | 42,000 | 75,000 | 118,000 |
| Costs | 150,000 | 250,000 | 325,000 |
| Savings | 15,000 | 40,000 | 55,000 |

(a) Calculate the $B/C$ ratios for each project. Can you tell from these calculations which should be selected?
(b) Determine which should be selected using the incremental $B/C$ ratio.
(c) Calculate $B - C$ for each alternative. (7.4)

3. Two four-lane roads intersect and traffic is controlled by a standard green, yellow, red stoplight. From each of the four directions a left turn is permitted from the inner lane; however, this impedes the flow of traffic while the person desiring to turn left waits until a safe turn may be accomplished. The light operates on a cycle allowing 30 seconds of green–yellow, followed by 30 seconds of red light for each direction. Approximately an eighth of the 8000 vehicles using the intersection daily are held up for one full minute cycle of the light, and made to perform one extra stop-start operation, solely because of the left turn bottleneck. These delays are only during the 260 working days/year. A stop-start costs 1.0¢/vehicle. A widening of the inter-section to accommodate a left-turn lane plus a new lighting system to provide for a left-turn signal will cost $100,000. The cost of time for commercial traffic is three times that of private traffic, and private traffic accounts for 80% of the vehicles. How much must commercial traffic time be worth to justify the intersection changes if the interest rate is 8% and a lifetime of 10 years is expected? (7.4)

4. A city is trying to decide between coal, fuel oil 3, low-sulphur fuel oil, and natural gas to power their electrical generators. Fuel forecasts indicate the following needs for the upcoming year, and the gradient increase during each subsequent year.

| Fuel | First Year | Gradient Each Subsequent Year |
|---|---|---|
| Coal | 460,000 tons | 23,000 tons |
| Fuel oil 3 | 1,760,000 barrels | 88,000 barrels |
| Low-sulphur fuel oil | 1,820,000 barrels | 91,000 barrels |
| Natural gas | 11,000 $10^6$ cubic feet | 550 $10^6$ cubic feet |

The cost of the fuel, transportation, and various pollution effects have been estimated as follows,

| Fuel | Cost | Transportation and Storage | Health | Crops | Uncleanliness |
|------|------|---------------------------|--------|-------|---------------|
| Coal | $5.80/ton | $2.25/ton | $0.70/ton | $1.75/ton | $1.05/ton |
| Fuel oil 3 | 2.15/barrel | 0.20/barrel | 0.10/barrel | 0.25/barrel | 0.15/barrel |
| Low-sulphur fuel oil | 2.75/barrel | 0.20/barrel | 0.03/barrel | 0.075/barrel | 0.045/barrel |
| Natural gas | 0.00058/cubic foot | 0.00002/cubic foot | Negligible | Negligible | Negligible |

Calculate the annual equivalent benefits and costs of these fuels considering a life of 30 years and (a) $i = 5\%$ and (b) $i = 10\%$. Use an incremental $B/C$ analysis to determine the best fuel to use. (7.4)

5. The following costs and benefits have been listed for a dam and reservoir which will be used for both flood control and electrical power generation. Investment

| | |
|---|---|
| Dam, including access roads, clearing, and foundation treatment | $31,330,000 |
| Generation equipment and transmission apparatus | 16,176,000 |
| Land | 2,200,000 |
| Highway relocation | 2,770,000 |
| Miscellaneous | 180,000 |

Operating and maintenance costs
   Two percent of investment during the first year and increasing by 5% of itself each subsequent year.
Annual benefits

| | |
|---|---|
| Flood losses prevented | $ 835,000 |
| Property value enhancement | 149,000 |
| Power value | 3,190,000 |

For a planning horizon of 50 years and $i = 5\%$, what is the $B/C$ ratio? (7.4)

6. A municipal zoo is to be enlarged, and the initial cost of the enlargement for physical facilities will be $400,000. Animals for the addition will cost another $100,000. Maintenance, food, and animal care will run $62,000/year. The zoo is expected to be in operation for an indefinite period; however, a study period of 20 years is to be assumed, with a residual (salvage) value of 50% for all physical facilities. Interest is 8%. An estimated 300,000 persons will visit the zoo each year, and they will receive, on the average, an additional $0.50/person in enjoyment when the new area is complete. Should the new area be built? Use a $B - C$ measure of merit. (7.4)

7. A highway is to be built connecting Baldhill with Broken Arrow (Baldhill is south of Bixby). Route A follows the old road and costs $4 million initially and $210,000/year thereafter. A new route, B, will cost $7 million initially and $180,000/year thereaf-

ter. Route C is simply an enhanced version of route B with wider lanes, shoulders, and so on. It will cost $9 million at first, plus $240,000/year to maintain. Relevant annual user costs considering time, operation, and safety are $1 million for A, $700,000 for B, and $500,000 for C. Using a *MARR* of 8%, a 15-year study period, and a residual (salvage) value of 50% of first cost, which should be constructed? Use a *B/C* analysis. Which route is preferred if $i = 0$? (7.4)

8. Solve problem 7 using the $B - C$ criterion. (7.4)

9. Seven projects are available as summarized below. It is desired to select only the projects that government can afford on an available budget of $75,000 for first cost. Operating and maintenance costs are no worry. All have a favorable *B/C* ratio at the cost of money, $i = 6\%$. Raise the interest rate until only those projects remain that continue to have a $B/C > 1$ and that government can afford. Use a 10-year planning horizon. Which projects are selected? What is the opportunity cost of those investments foregone by government? (7.5)

| Project | First Cost | Operating and Maintenance | Residual Value After 10 Years | Benefits/Year |
|---|---|---|---|---|
| A | $33,000 | $3,000 | $16,000 | $14,400 |
| B | 27,000 | 1,500 | 8,000 | 7,630 |
| C | 41,000 | 2,500 | 12,000 | 10,800 |
| D | 38,000 | 1,600 | 14,500 | 8,460 |
| E | 30,000 | 4,200 | 16,000 | 8,600 |
| F | 34,000 | 600 | 8,000 | 5,400 |
| G | 25,000 | 5,100 | 3,000 | 8,980 |

10. A proposed expressway is under study and is found to have an unfortunately high annual cost of $2,200,000 as compared to benefits of only $1,300,000. However, with only inconsequential changes in design, the highway may be eligible for incorporation into the interstate highway system, in which case 90% of the cost would be paid by the federal government. If so, it is argued that the expressway would cost only $220,000/year, yielding a handsome *B/C* ratio of 5.91. Is this reasoning sound? Discuss the reasoning from different points of view. (7.5)

11. Many benefits resulting from public sector projects may be argued to be at the expense of someone else, thus counterbalancing the supposed direct benefits. For example, in Example 7.1, operating costs saved include the cost of fuel, oil, tires, wear and tear, and the like. This is money that will *not* be spent at service stations, tire stores, and new car dealers. The safety costs saved include fees that otherwise would go to lawyers, hospitals, doctors, auto repair shops, and so forth. Do you think the fact that the direct savings represent lost revenues to others should void or nullify these types of benefits in a benefit-cost analysis? Why or why not? (7.5)

12. Three projects, each having a first cost of $1 million and annual operating costs of $100,000 are proposed. The lives of Projects A, B, and C are 20, 30, and 40 years, respectively, after which the project will be over, providing no benefits and requir-

ing no costs. The annual benefits provided over the lives of Projects A, B, and C will be $268,000, $250,000, and $243,000, respectively.

Select only one project to be implemented. Use a present worth basis and a $B - C$ measure of merit. First, use an *MARR* of 5%, and a study period of 40 years. Projects A and B will be void during the last 20 and 10 years of the study period, respectively. Second, continually increase the interest rate, again using a study period of 40 years, until only one project remains with a favorable $B - C$ value. Do the two methods point to the same project? Why or why not? (7.5)

13. A recreation area suitable for camping, picnicking, hiking, and water sports is to be developed at an annual equivalent cost of $112,500, including initial cost of construction (clearing, water, restrooms, road, etc.), operation, upkeep, and security. An average of 50 families will camp each night throughout the year. In addition, another 100 persons will be admitted for day use of the recreational facilities. The perceived benefits provided to overnight and day users will vary, due to the subjective nature of people. However, the average family camping at the area is estimated to be willing to pay $5 for the privilege, and the average day user $1. What is the $B/C$ ratio of this recreational area? $B - C$? Should it be built? (7.5)

14. In problem 13, assume that 50% of the potential camping families perceive $5 in benefits, 25% perceive $7.50 in benefits, and 25% expect to receive $10 worth of recreation. If a charge of $7.50 is imposed on campers only, and day users are permitted free, recalculate $B/C$ and $B - C$. Should the facility be built? (7.5)

15. Reconsider problem 6 and assume that half of the people will derive $0.50 of enjoyment per person, while the other half anticipate only $0.25 in benefits. With this information available, and assuming that an incremental $0.25 entrance fee/ person will be charged separately for attendance to the addition, do you recommend building the addition? (7.5)

16. Comment on the following analysis if you think it is incorrect. Benefits of $1.75 each are now received by 9000 persons and benefits are perceived to be $3 each by another 9000 users. The annual cost of the recreational facility is $36,000. Commissioners have argued that an entrance fee of $2 should be charged to make the facility self-supporting. They argue that $2 is below the average benefit received per person. They do point out, though, that 9000 persons do not perceive but a $1.75 recreational value, and hence a $0.25 disbenefit per person should be noted. The $B/C$ ratio is then

$$B/C = \frac{\$3(9,000) + \$1.75(9,000) - \$0.25(9,000)}{36,000 - \$2(18,000)} = \infty$$

and the user fee of $2 should definitely be implemented immediately. What do you think? (7.5)

17. A government has the following estimates:

| | |
|---|---|
| Annual benefits | $1,150,000 |
| Annual disbenefits | 850,000 |
| Annual costs | 765,000 |
| Annual savings | 750,000 |

(a) Calculate the $B/C$ ratio.

(b) Mistakenly treating disbenefits as costs and savings as benefits, calculate the $B/C$ ratio.

(c) Calculate $B - C$.

(d) Which do you prefer, $B - C$ or $B/C$? Why? (7.5)

18. Analysts recognize that with a $2 entrance fee as in problem 16, only 9000 persons will utilize the facility, and the benefits derived by 9000 more persons at $1.75 each will be lost. They argue that the *true* benefits are those received only by 9000 persons at $3 each, less the $2 fee, less a disbenefit of $1.75(9000) for the persons who chose not to pay $2 to enter and lose the previous recreational enjoyment from the facility. As such, the $B/C$ ratio is as follows:

$$B/C = \frac{\$3(9,000) - \$2(9,000) - \$1.75(9,000)}{36,000 - \$2(9,000)} = -.38$$

Do you agree with this analysis? Why or why not? (7.5)

19. Four designs are under consideration for a space shuttle designed to carry a particular type of payload. It is estimated that each will have equal lives; however, the first cost, operating cost, and renovation cost, as well as the salvage value, will differ, as will the size of the payload that may be carried. These characteristics are summarized below for shuttles having lives of three years, and fired once per year at the end of the year.

| Shuttle Design | First Cost | Operating Cost/Firing | Renovation Cost/Firing | Salvage Value | Payload in Units |
|---|---|---|---|---|---|
| 1 | $28,000,000 | $8,000,000 | $1,400,000 | $2,800,000 | 180 |
| 2 | 32,000,000 | 6,400,000 | 2,400,000 | 3,200,000 | 155 |
| 3 | 34,000,000 | 7,200,000 | 1,900,000 | 3,400,000 | 140 |
| 4 | 40,000,000 | 5,600,000 | 1,200,000 | 4,000,000 | 175 |

Based on $i = 9\%$, and using a cost-effectiveness analysis, which designs may be eliminated and which should be considered further? (7.6)

20. Six subsystem designs have been proposed for a critical part of a communications network. Each will be "burned in" to eliminate the infant mortality problem. Hence, reliability of the subsystem will have a constant hazard rate. That is, the failure density function will be exponential and reliability may be expressed as

$$R(t) = e^{-\lambda t}$$

where $t$ is the mission time. Mission time for the subsystem will be one year, as a thorough annual preventative maintenance and recalibration procedure is standard, returning the subsystem to "new" condition. The unit will be on call 24 hours/day, seven days/week. The design data are as follows:

| System | Annual Equivalent Initial Cost | Annual Maintenance Cost | $\lambda$ |
|--------|-------------------------------|-------------------------|-----------|
| A | $10,000 | $2,000 | $1.1446886 \times 10^{-5}$ |
| B | 10,800 | 1,700 | $1.0989011 \times 10^{-5}$ |
| C | 11,200 | 2,100 | $1.175824 \times 10^{-5}$ |
| D | 12,000 | 1,600 | $1.070000 \times 10^{-5}$ |
| E | 12,000 | 1,800 | $1.081250 \times 10^{-5}$ |
| F | 14,100 | 1,200 | $0.998000 \times 10^{-5}$ |

Using a cost-effectiveness analysis, plot reliability versus annual cost. Which subsystems can be eliminated from consideration? How might you decide among the rest? (7.6)

## CHAPTER EIGHT

# BREAK-EVEN, SENSITIVITY, AND RISK ANALYSES

## 8.1 INTRODUCTION

In the previous chapters we have assumed that all of the values of the parameters of the economic models were known with certainty. In particular, correct estimates of the values for the length of the planning horizon, the minimum attractive rate of return, and each of the individual cash flows were assumed to be available. In this chapter we consider the consequences of estimating the parameters incorrectly.

The discussion will concentrate on answering a number of "what if . . ." questions concerning the effects of different parameter values on the measure of economic effectiveness of interest. For example, when we are completely *uncertain* of the possible values a parameter can take on, we will be interested in determining the set of values for which an investment alternative is justified economically and the set of values for which an alternative is not justified; this process is called *break-even analysis*. Alternatively, when we are reasonably sure of the possible values a parameter can take on, but *uncertain* of their chances of occurrence, we will be interested in the sensitivity of the measure of merit to various parameter values; this process is referred to as *sensitivity analysis*. Finally, when probabilities can be assigned to the occurrence of the various values of the parameters, we can make probability statements concerning the values of the measure of merit for the various alternatives; this process is referred to as *risk analysis*.

The conditions that lead to break-even and sensitivity analyses are described in the economic analysis literature as conditions under *uncertainty*, since one is completely uncertain of the chances of a parameter taking on a given value. When the conditions are such that probabilities can be assigned to the various parameter values, the decision environment that results is said to be a decision under *risk*; hence, the term risk analysis is used.

One of the most perplexing (and frustrating) experiences for the engineering economic analyst is to encounter an individual who refuses to analyze systematically the economic performances of investment alternatives because "it's not possible to estimate future cash flows exactly." Two responses come to mind. First, failure to analyze the alternatives systematically implies a qualitative judgment will be made, based on even more imperfect information than would be available using a systematic approach. Second, the selection decision might not require precise estimates of the cash flows. In the latter case, the selection decision might be relatively insensitive to a broad range of possible values for the cash flows.

The purpose of break-even and sensitivity analyses is to *reduce* the amount of information needed to make a good decision. Instead of needing a "point" estimate for, say, the salvage value of a piece of equipment, an "interval" estimate might be sufficient. Break-even analysis can reduce the forecasting requirement to determining if the magnitude of a cash flow will exceed a specific value.

It is important to recognize that break-even, sensitivity, and risk analyses are intended to make more realistic the economic comparison of investment alternatives. Without using such approaches, one must make decisions under the assumption of perfect information regarding the future.

In Chapter Nine we will present *prescriptive* or *normative* models under uncertainty and risk conditions, models will be presented that allow us to *prescribe* the best or optimal course of action using a number of different decision criteria. In this chapter we present *descriptive* models under uncertainty and risk conditions. Instead of attempting to *prescribe* the action to be taken, we attempt to *describe* the behavior of the measure of merit under conditions of uncertainty and risk. Thus, we view break-even, sensitivity, and risk analyses as descriptive processes, not normative processes. Our objective in this chapter will be to gain insight into the behavior of the measure of merit under conditions of uncertainty and risk; the decision of how to respond, given the information from the sensitivity analysis depends on the decision process employed by the decision maker. Hence, the decision models from Chapter Nine may be used to assist the decision maker in making the final selection.

## 8.2 BREAK-EVEN ANALYSIS

Although we did not label the process as such, in Chapter Three we performed a number of break-even analyses; there we referred to the process as *equiva-*

*lence*. In particular, in presenting the concept of equivalence, a situation was posed and you were asked to determine the value of a particular parameter in order for two cash flow profiles to be equivalent. Another way of stating the problem could have been "Determine the value of $X$ that will yield a break-even situation between the two alternatives." In this case, $X$ denotes the parameter whose value is to be determined. Additionally, when the cash flow profiles for two alternatives are equivalent, a break-even situation can be said to exist between the two alternatives.

Another application of break-even analysis occurs in the use of the rate-of-return method. Specifically, the internal rate of return can be interpreted as the break-even value of the reinvestment rate, since such a reinvestment rate will yield a zero future worth for either an individual alternative or the differences in two alternatives.

Break-even analysis is certainly not an unfamiliar concept. Furthermore, the information obtained from a break-even analysis can be of considerable aid to anyone faced with an investment alternative involving a degree of uncertainty concerning the value of some parameter. The term "break-even" is derived from the desire to determine the value of a given parameter that will result in neither a profit nor a loss.

## Example 8.1

Suppose a firm is considering manufacturing a new product and the following data have been provided:

| | |
|---|---|
| Sales price | $12.50/unit |
| Equipment cost | $200,000 |
| Overhead cost | $50,000/year |
| Operating and maintenance cost | $25/operating hour |
| Production time/1000 units | 100 hours |
| Planning horizon | 5 years |
| Minimum attractive rate of return | 15% |

Assuming a zero salvage value for all equipment at the end of 5 years, and letting $X$ denote the annual sales for the product, the annual worth for the investment alternative can be determined as follows:

$$AW(15\%) = -\$200,000(A|P\ 15,5) - \$50,000 - 0.100(\$25)X + \$12.50X$$
$$= -\$109,660 + \$10.00X$$

Solving for $X$ yields a *break-even* sales value of 10,966/year. If it is felt that annual sales of at least 10,966 units can be achieved each year, then the alternative appears to be worthwhile economically. Even though one does not know with certainty how many units of the new product will be sold annually, it is felt that the information provided by the break-even analysis will assist management in deciding whether or not to undertake the new venture.

A graphical representation of the example is given in Figure 8.1. The chart is referred to as a *break-even* chart, since one can determine graphically the

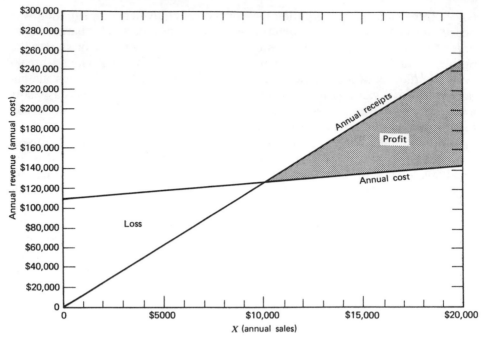

FIGURE 8.1. Break-even chart.

*break-even point* by observing the value of $X$ when annual revenue equals annual cost.

Example 8.2 _____

Consider a contractor who experiences a seasonal pattern of activity for compressors. The manager currently owns eight compressors and suspects that this number will not be adequate to meet the demand. The contractor realizes that there will arise situations when more than eight compressors will be required, and is considering purchasing an additional compressor for use during heavy demand periods.

A local equipment rental firm will rent compressors at a cost of $50/day. Compressors can be purchased for $6000. The difference in operating and maintenance costs between owned and rented compressors is estimated to be $3000/year.

Letting $X$ denote the number of days a year that more than eight compressors are required, the following break-even analysis is performed. A planning horizon of 5 years, zero salvage values, and 20% minimum attractive rate of return are assumed.

*Annual worth (purchasing compressor)*

$$AW_1(20\%) = -\$6000(A|P\ 20,5) - \$3000$$
$$= -\$5006.40$$

*Annual worth (renting compressor)*

$$AW_2(20\%) = -\$50X$$

Setting the annual worths equal for the two alternatives yields a *break-even value* of $X = 100.128$ days/year.

Hence, if the contractor anticipates that a demand will exist for an additional compressor more than 100 days/year over the next five years, then an additional compressor should be purchased.

---

Example 8.3

Let us extend the previous example problem to the question of how many compressors should be purchased. It is interesting to note that compressors should continue to be purchased so long as each one reduces the number of compressor-rental-days by at least 100/year over the 5-year period. To illustrate, if the contractor provides a forecast of daily demand for compressors as given in Table 8.1, it is seen that one additional compressor should be purchased.

To show that one more compressor should be purchased, let us compute the annual worths for the alternatives of purchasing 0, 1, 2, and 3 additional compressors. From Table 8.1,

$$AW_0(20\%) = -\$50(290)$$
$$= -\$14,500$$

$$AW_1(20\%) = -\$6000(A|P\ 20,5) - \$3000 - \$50(170)$$
$$= -\$13,506.40$$

$$AW_2(20\%) = -\$12,000(A|P\ 20,5) - \$6000 - \$50(90)$$
$$= -\$14,512.80$$

$$AW_3(20\%) = -\$18,000(A|P\ 20,5) - \$9000 - \$50(40)$$
$$= -\$17,019.20$$

Thus, one additional compressor should be purchased in order to provide a total of nine compressors to meet annual demand.

---

Example 8.4

As a fourth illustration of break-even analysis, consider an office building that is being converted from a coal burning furnace to a furnace that burns either

TABLE 8.1. Forecasts of Demand for Compressors

| | $X$<br>Number of Compressors<br>Demanded On a Given Day | $f(X)$<br>Number of Days per Year<br>Demand Equals $X$ |
|---|---|---|
| | ≤ 8 | 140 |
| | 9 | 40 |
| | 10 | 30 |
| | 11 | 20 |
| | 12 | 20 |
| | 13 | 10 |
| | ≥14 | 0 |

| $N$<br>Number of<br>Compressors<br>Owned by<br>the Contractor | Number of<br>Compressor-Rental-Days<br>if $N$ are Owned | Difference in Number<br>of Compressor-Rental-<br>Days if $N$ versus $N + 1$<br>Are Owned |
|---|---|---|
| 8 | 40(1) + 30(2) + 20(3) + 20(4) + 10(5) = 290 | — |
| 9 | 30(1) + 20(2) + 20(3) + 10(4) = 170 | 120 |
| 10 | 20(1) + 20(2) + 10(3) = 90 | 80 |
| 11 | 20(1) + 10(2) = 40 | 50 |
| 12 | 10(1) = 10 | 30 |
| 13 | 0 | 10 |

natural gas or number two fuel oil. The cost of converting to natural gas is estimated to be $50,000 initially; additionally, annual operating and maintenance costs are estimated to be $6000 less than that experienced currently. Approximately 1000 Btu's are produced per cubic foot of natural gas; it is estimated natural gas will cost $0.001/cubic foot. The cost of converting to number two fuel oil is estimated to be $110,000 initially; annual operating costs are estimated to be $4000 less than that experienced using the coal furnace. Approximately 140,000 Btu's are produced per gallon of fuel oil; number two fuel oil is anticipated to cost $0.12/gallon. A planning horizon of 20 years is used; zero salvage values and a 20% minimum attractive rate of return are assumed. It is desired to determine the break-even value for annual Btu requirements for the heating system.

Letting $X$ denote the Btu requirement per year, the equivalent uniform annual cost for the natural gas alternative is found to be

$$EUAC_1(10\%) = \$50{,}000(A|P\ 20{,}20) - \$6000 + \frac{\$0.001}{1000}\,X$$

$$= \$4270 + \$1.0 \times 10^{-6}X$$

Similarly, for the number two fuel oil, the equivalent uniform annual cost will be

$$EUAC_2(10\%) = \$110,000(A|P\ 20,20) - \$4000 + \frac{\$0.12}{140,000}X$$

$$= \$18,594 + \$0.8571 \times 10^{-6}X$$

Equating the annual costs and solving for $X$ yields a break-even value of approximately $10.035 \times 10^9$ Btu's/year. Since a consulting engineer has estimated that Btu requirements will be considerably less than the break-even value, it is decided to convert to natural gas.

Example 8.5_____

A firm is contemplating purchasing versus leasing a computer. If the computer is purchased, a service contract can be entered into with the computer supplier for maintenance and software support. Alternatively, the company can perform its own maintenance and software support. Finally, if the computer is leased, the lease cost will include maintenance and software support.

The following cost estimates are available for the 5-year study period. Outright purchase of the computer will cost $500,000 initially. Purchase of the computer with a 5-year service contract will be $600,000 initially, followed by four annual payments of $55,000. Alternately, the computer can be leased for $165,000/year payable at the beginning of the year.

The firm is uncertain as to the cost of doing its own maintenance and software support. However, they do believe the first year's cost will be approximately $40,000 and that it will increase at an unknown annual compound rate. Additionally, the firm is unsure of the $MARR$ to be used; however, they have agreed to use either 10% or 15% in their calculations.

Using break-even analysis to provide a recommendation to the firm reveals the following:

*Present Worth of Purchase with Service Contract Alternative*

$$PW(10\%) = -\$600,000 - \$55,000(P|A\ 10,4)$$
$$= -\$774,344.50$$

$$PW(15\%) = -\$600,000 - \$55,000(P|A\ 15,4)$$
$$= -\$757,025$$

*Present Worth of Lease Alternative*

$$PW(10\%) = -\$165,000 - \$165,000(P|A\ 10,4)$$
$$= -\$688,033.50$$

$$PW(15\%) = -\$165,000 - \$165,000(P|A\ 15,4)$$
$$= -\$636,075$$

The lease alternative is preferred to the alternative of purchasing with a service contract regardless of the *MARR* selection. Thus, the problem reduces to a comparison of outright purchasing the computer versus leasing the computer.

*Present Worth of Purchase Alternative*

$$PW(10\%) = -\$500{,}000 - \$40{,}000(P|A\ 10,j,5)$$

$$PW(15\%) = -\$500{,}000 - \$40{,}000(P|A\ 15,j,5)$$

Equating the present worths of leasing and purchasing and reducing yields the following

$$PW(10\%) = -\$688{,}033.50 = -\$500{,}000 - \$40{,}000(P|A\ 10,j,5)$$

or

$$4.7008 = (P|A\ 10,j,5)$$

or

$$j \approx 11.9\%$$

$$PW(15\%) = -\$636{,}075 = -\$500{,}000 - \$40{,}000(P|A\ 15,j,5)$$

or

$$3.4019 = (P|A\ 15,j,5)$$

or

$$j \approx 1.0\%$$

From the break-even analysis is is concluded that the computer should be leased if *MARR* is 15%, since at least a 1% annual increase in maintenance and software support is anticipated. If the *MARR* is 10% then an annual increase of at least 11.5% in maintenance and software support costs is needed before leasing is preferred. Based on the analysis the firm decides to lease the computer. Do you agree with their decision?

---

Example 8.6 ————————————————————————————————

A manufacturing firm has the following annual revenue and cost functions:

$$\text{Annual revenue} = AR(x) = \$1500x - \$0.05x^2$$

$$\text{Annual cost} = AC(x) = \$1{,}200{,}000 + \$1000x - \$0.01x^2$$

where $x$ is the annual production volume. It is desired to determine the break-even production volume.

The net profit before taxes is equal to the difference in the revenue and cost functions, or

$$\text{Net profit} = NP(x) = \$500x - \$0.04x^2 - \$1{,}200{,}000$$

Setting the net profit to zero and solving the quadratic equation for $x$ yields values of $x = 3240$ and $x = 9260$. Thus, the operation will be profitable so long as annual production is between 3240 and 9260 units.

Example 8.7_____

A manufacturing firm is performing a make-or-buy analysis for a particular component. They find that the cost to manufacture $x$ units in-house is

$$TC(x) = \$1000 + \$150x - \$0.01x^2$$

A quotation has been received from an outside supplier to provide $x$ units at a cost of \$3000 + \$65x. For what range of production volumes should the component be purchased from the outside supplier?

Equating the two costs and solving for $x$ gives the following:

$$\$1000 + \$150x - \$0.01x^2 = \$3000 + \$65x$$

$$\$0.01x^2 - \$85x + \$2000 = 0$$

$$x = 23.6 \text{ and } 8476$$

Thus, for a production volume between 23 and 8476 the firm should purchase the components. If more than 8476 units (or less than 23.6 units) are to be produced, then the firm should manufacture the component.

## 8.3 SENSITIVITY ANALYSIS

Break-even analysis is normally used when an accurate estimate of a parameter cannot be provided, but intelligent judgments can be made as to whether or not the parameter's value is less than or greater than some break-even value. Sensitivity analysis is used to analyze the effects of making errors in estimating parameter values.

Although the analysis techniques employed in break-even analysis and sensitivity analysis are quite similar, there are some subtle differences in the objectives of each. Because of the similarities in the two, it is not uncommon to see the terms used interchangeably.

Example 8.8_____

To illustrate what we mean by a sensitivity analysis, consider the investment alternative depicted in Figure 8.2. The alternative is to be compared against the "do nothing" alternative, which has a zero present worth. If errors are made in estimating the size of the required investment (\$10,000), the magnitude of the annual receipt (\$3000), the duration of the project (5 years), the minimum attractive rate of return (12%), and/or the form of the series of receipts (uniform versus nonuniform), then the economic desirability of the alternative might be

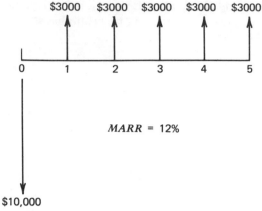

$10,000

FIGURE 8.2. Cash flow diagram.

affected. It is anticipated that the future states (possible values) for each parameter will be contained within an interval having a range from −40 to +40% of the initial estimate. Hence, in the uncertain environment we have defined an infinite number of future states are to be considered in the continuum from −40 to +40% of the initial estimate.

If it is assumed that all estimates are correct except the estimate of annual receipts, the annual worth for the alternative can be given as

$$AW(12\%) = - \$10,000(A|P\ 12,5) + \$3000(1 + X)$$

where $X$ denotes the percent error in estimating the value for annual receipts. Plotting annual worth as a function of the percent error in estimating the value of annual receipts yields the straight line having positive slope shown in Figure 8.3. Performing similar analyses for the initial investment required, the duration of the investment (planning horizon) and the minimum attractive rate of return yields the results given in Figure 8.3.

As shown in Figure 8.3, the net annual worth for the investment is affected differently by errors in estimating the values of the various parameters. The net annual worth is relatively insensitive to changes in the minimum attractive rate of return; in fact, as long as the *MARR* is less than approximately 15.25%, the investment will be recommended. A *break-even* situation exists if either the annual receipts decrease by approximately 7.47% to $2774/year or the required investment increases by approximately 8.14% or $10,814. If the project life is 4 years or less, then the investment will not be profitable.

The analysis depicted in Figure 8.3 examines the sensitivity of individual parameters one at a time. In practice, estimation errors can occur for more than one parameter. In such a situation, instead of a sensitivity curve, a sensitivity surface is needed.

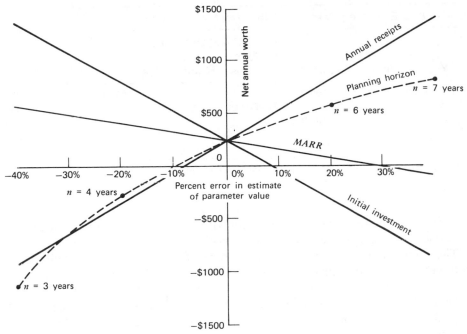

FIGURE 8.3. Deterministic sensitivity analysis.

Example 8.9

Consider an investment alternative involving the modernization of a warehousing operation in which automated storage and retrieval equipment is to be installed in a new warehouse facility. The building has a projected life of 30 years and the equipment has a projected life of 15 years. A minimum attractive rate of return of 15% is to be used in the economic analysis. It is anticipated that the new warehouse system will require 70 fewer employees than the present system. Each employee costs approximately $18,000/year. The building is estimated to cost $2,500,000 and the equipment is estimated to cost $3,500,000. Annual operating and maintenance costs for the building and equipment are estimated to be $150,000/year more than the current operation. Existing equipment and buildings not included in the new warehouse have terminal salvage values totaling $600,000. Investing in the new warehouse will negate the need to replace existing equipment in the future; the present worth savings in replacement cost is estimated to be $200,000. The estimated cash flow profile for the investment alternative is given in Table 8.2.

A 30-year planning horizon is to be employed in the analysis. Since equipment life is estimated to be 15 years, it is assumed that identical replacement equipment will be purchased after 15 years. Furthermore, since constant worth dollar estimates are being used in contending with inflationary effects, it is

TABLE 8.2. Estimated Cash Flows for Warehouse System

| EOY | Building | Equipment | Labor Savings | Operating and Maintenance | Salvage* | Total |
|---|---|---|---|---|---|---|
| 0 | -$2,500,000 | -$3,500,000 | | | $800,000 | -$5,200,000 |
| 1-15 | | | $1,260,000 | -$150,000 | | 1,110,000 |
| 15 | | - 3,500,000 | | | | - 3,500,000 |
| 16-30 | | | 1,260,000 | - 150,000 | | 1,110,000 |
| 30 | 0 | 0 | | | | 0 |

* Includes present worth of savings in equipment replacement.

assumed that the replacement equipment will have cash flows that are identical to those that occur during the first 15 years.

The architectural and engineering estimate of $2,500,000 for the building is believed to be quite accurate, as are the estimates of $150,000 for annual operating and maintenance costs, terminal salvage values totaling $600,000, and savings of $200,000 in replacement costs. However, it is felt that the estimate of $3,500,000 for equipment and the labor savings estimate of 70 employees are subject to error. A sensitivity analysis for these two parameters is to be performed on a before tax basis.

Letting $x$ denote the percent error in the estimate of equipment cost and $y$ denote the percent error in estimating the annual labor savings, it can be seen that the warehouse modernization will be justified economically, if

$$PW = -\$2,500,000 - \$3,500,000(1 + x)[1 + (P|F\ 15,15)]$$
$$- \$150,000(P|A\ 15,30) + \$18,000(70)(1 + y)(P|A\ 15,30)$$
$$+ \$800,000 \geq 0$$

or if

$$PW = \$1,658,110 - \$3,930,150x + \$8,273,160y \geq 0$$

Solving for $y$ gives

$$y \geq - 0.2004 + 0.47505x$$

Plotting the equation, as shown in Figure 8.4, indicates the favorable region $(PW > 0)$ lies above the breakeven line and the unfavorable region $(PW < 0)$ lies below the breakeven line.

If no errors are made in estimating the equipment cost, that is, $x = 0$, then up to a 20.04% reduction in annual labor savings can be tolerated. Likewise, if no errors are made in estimating the annual labor savings, that is, $y = 0$, then up to a 42.185% increase in equipment cost could occur and the warehouse modernization continue to be justified economically.

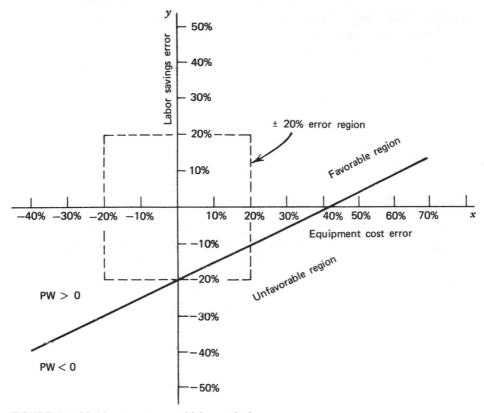

FIGURE 8.4. Multiparameter sensitivity analysis.

Since very little of the ±20% estimation error zone results in a negative present worth, it appears that the recommendation to modernize the warehouse is insensitive to errors in estimating either equipment cost, labor savings, or both. However, the decision is more sensitive to errors in estimating labor savings than in estimating equipment costs.

(This particular example illustrates the difficulties in distinguishing between break-even and sensitivity analyses. However, no matter what you call it, the analysis can be quite beneficial in gaining added understanding of the possible outcomes associated with a new venture involving a large capital investment.)

Example 8.10

To illustrate the use of sensitivity analysis in comparing investment alternatives, consider three mutually exclusive alternatives having cash flow profiles as given in Table 8.3.

TABLE 8.3. Cash Flow Profiles for Three
Mutually Exclusive Investment
Alternatives

| $t$ | $A_{1t}$ | $A_{2t}$ | $A_{3t}$ |
|---|---|---|---|
| 0 | $0 | −$10,000 | −$15,000 |
| 1 | −12,000 | − 9,000 | − 9,000 |
| 2 | −12,000 | − 9,000 | − 8,000 |
| 3 | −12,000 | − 9,000 | − 7,000 |
| 4 | −12,000 | − 9,000 | − 6,000 |
| 5 | −12,000 | − 9,000 | − 5,000 |

Suppose some uncertainty exists concerning the appropriate *MARR* to use in the analysis. As depicted in Figure 8.5, Alternative 3 is preferred if the *MARR* is less than approximately 17%; Alternative 1 is preferred if the *MARR* is greater than 17%; and Alternative 2 is never preferred.

If the internal rate of return method were used to compare the alternatives, the differences in Alternatives 1 and 2 would be analyzed and a rate of return (break-even value) of 15.02% would be obtained; next, Alternatives 2 and 3 would be compared and a break-even value of 19.48% would be obtained. The combination of Alternatives 1 and 3 would not be considered when the rate of return method is used.

Example 8.10 illustrates two points: the rates of return obtained from the internal rate of return method are actually break-even values for the minimum attractive rate of return for the combinations of alternatives considered; and sensitivity analysis and break-even analysis are closely related.

The final approach we consider in performing sensitivity analyses assumes the possible number of values for the parameters can be reasonably represented by three values for each: pessimistic, most likely, and optimistic estimates will be provided. Given the three estimates of the value of each parameter, the value of the measure of merit is determined for each combination of estimates.

Example 8.11 _____

To illustrate the suggested approach, consider once more the example problem depicted in Figure 8.2 and involving an initial investment of $10,000 followed by annual receipts of $3000 for five years. Realizing that any or all of the parameter estimates could be in error, pessimistic, most likely, and optimistic estimates are provided in Table 8.4 for the parameters. As can be seen, there are $3^m$ combinations to consider, where $m$ is the number of parameters to be

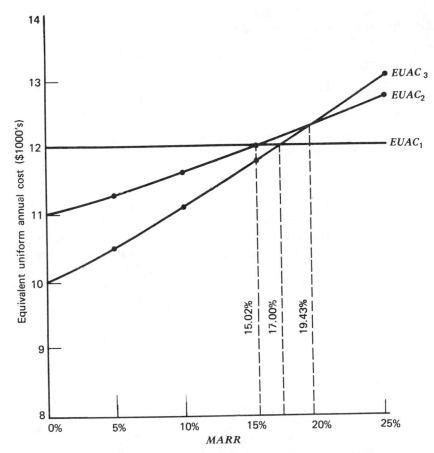

FIGURE 8.5. Sensitivity of equivalent uniform annual cost to the MARR.

included in the sensitivity analysis. Thus, in this case, 81 combinations are to be considered. Four representative combinations are given below.

$$PW = -\$12,000 + \$4000(P|A\ 15,4) = -\$580$$

$$PW = -\$8000 + \$2500(P|A\ 20,6)9 = \$313.75$$

$$PW = -\$10,000 + \$3000(P|A\ 15,4) = -\$1435$$

$$PW = -\$12,000 + \$3000(P|A\ 12,5) = -\$1185.60$$

The range on present worth can be obtained by evaluating the totally pessimistic case and the totally optimistic case.

$$PW(\text{pessimistic}) = -\$12,000 + \$2500(P|A\ 20,4) = -\$5528.25$$

$$PW(\text{optimistic}) = -\$8000 + \$4000(P|A\ 12,6) = \$8445.60$$

TABLE 8.4. Pessimistic, Most Likely, and Optimistic Estimates

|  | Pessimistic* | Most Likely | Optimistic |
|---|---|---|---|
| Investment | −$12,000 | −$10,000 | −$8,000 |
| Annual receipts | $ 2,500 | $ 3,000 | $4,000 |
| Discount rate | 20% | 15% | 12% |
| Planning horizon | 4 years | 5 years | 6 years |

* Here we interpret the pessimistic value as a value which will lower the present worth; hence, a high discount rate is viewed as a pessimistic value.

A tabulation of the present worths for the 81 possible combinations of pessimistic (P), most likely (M), and optimistic (O) estimates is provided in Table 8.5. The legend (PMOM) indicates a pessimistic estimate for the investment required, a most likely estimate for annual receipts, an optimistic estimate for the discount rate, and a most likely estimate for the planning horizon were used to determine the present worth of −$1185.60. Plotting the results in the form of a frequency histogram yields the frequencies provided in Table 8.6.

TABLE 8.5. Present Worth Values for Combination of Estimates

| PPPP | −$5,528.25 | MPPP | −$3,528.25 | OPPP | −$1,528.25 |
|---|---|---|---|---|---|
| PPPM | − 4,523.50 | MPPM | − 2,523.50 | OPPM | − 523.50 |
| PPPO | − 3,686.25 | MPPO | − 1,686.25 | OPPO | 313.75 |
| PPMP | − 4,862.50 | MPMP | − 2,862.50 | OPMP | − 862.50 |
| PPMM | − 3,619.50 | MPMM | − 1,619.50 | OPMM | 380.80 |
| PPMO | − 2,538.75 | MPMO | − 538.75 | OPMO | 1,461.25 |
| PPOP | − 4,406.75 | MPOP | − 2,406.75 | OPOP | − 406.75 |
| PPOM | − 2,988.00 | MPOM | − 988.00 | OPOM | 1,012.00 |
| PPOO | − 1,721.50 | MPOO | 278.50 | OPOO | 2,278.50 |
| PMPP | − 4,233.90 | MMPP | − 2,233.90 | OMPP | − 233.90 |
| PMPM | − 3,028.20 | MMPM | − 1,028.20 | OMPM | 971.80 |
| PMPO | − 2,023.50 | MMPO | − 23.50 | OMPO | 1,976.50 |
| PMMP | − 3,435.00 | MMMP | − 1,435.00 | OMMP | 565.00 |
| PMMM | − 1,943.40 | MMMM | 56.60 | OMMM | 2,056.60 |
| PMMO | − 646.50 | MMMO | 1,353.50 | OMMO | 3,353.50 |
| PMOP | − 2,888.10 | MMOP | − 888.10 | OMOP | 1,111.90 |
| PMOM | − 1,185.60 | MMOM | 814.40 | OMOM | 2,814.40 |
| PMOO | 334.20 | MMOO | 2,334.20 | OMOO | 4,334.20 |
| POPP | − 1,645.20 | MOPP | 354.80 | OOPP | 2,354.80 |
| POPM | − 37.60 | MOPM | 1,962.40 | OOPM | 3,962.40 |
| POPO | 1,302.00 | MOPO | 3,302.00 | OOPO | 5,302.00 |
| POMP | − 580,00 | MOMP | 1,420.00 | OOMP | 3,420.00 |
| POMM | 1,408.80 | MOMM | 3,408.80 | OOMM | 5,408.80 |
| POMO | 3,138.00 | MOMO | 5,138.00 | OOMO | 7,138.00 |
| POOP | 149.20 | MOOP | 2,149.20 | OOOP | 4,149.20 |
| POOM | 2,419.20 | MOOM | 4,419.20 | OOOM | 6,419.20 |
| POOO | 4,445.60 | MOOO | 6,445.60 | OOOO | 8,445.60 |

TABLE 8.6. Frequency Tabulation
of Present Worths

| Present Worth | Frequency |
|---|---|
| -$6,000 to -$5,000.01 | 1 |
| - 5,000 to - 4,000.01 | 4 |
| - 4,000 to - 3,000.01 | 5 |
| - 3,000 to - 2,000.01 | 8 |
| - 2,000 to - 1,000.01 | 9 |
| - 1,000 to - 0.01 | 11 |
| 0 to 999.99 | 10 |
| 1,000 to 1,999.99 | 9 |
| 2,000 to 2,999.99 | 7 |
| 3,000 to 3,999.99 | 6 |
| 4,000 to 4,999.99 | 4 |
| 5,000 to 5,999.99 | 3 |
| 6,000 to 6,999.99 | 2 |
| 7,000 to 7,999.99 | 1 |
| 8,000 to 8,999.99 | 1 |

Unfortunately, it is difficult to draw any significant conclusions from the data given in Tables 8.5 and 8.6 concerning the parameter that has the greatest effect on present worth. Instead, the use of the three estimates for each parameter provides us with a feel for the possible values of present worth and their relative frequencies. However, we should not place too much weight on the relative frequencies, since we have not verified that each combination of estimates is equally likely to occur. The assignment of probabilities to combinations of parameter estimates lies in the domain of risk analysis and will be treated next.

## 8.4 RISK ANALYSIS

Risk analysis will be defined as the process of developing probability distributions for some measure of merit for an investment proposal. Typically, probability distributions are developed for either present worth, annual worth, or the rate of return for an individual investment alternative. Consequently, probability distributions are required for random variables such as the cash flows, planning horizon, and the discount rate. The probability distributions are then aggregated analytically or through simulation to obtain the desired probability distribution for the measure of merit.

The cash flow occurring in a given year is often a function of a number of other variables such as selling prices, size of the market, share of the market, market growth rate, investment required, inflation rate, tax rates, operating costs, fixed costs, and salvage values of all assets. The values of a number of these random variables can be correlated with each other, as well as autocorrelated.[1] Consequently, an analytical development of the probability distribution for the measure of merit is not easily achieved in most real-world situations. Thus, simulation is widely used in performing risk analyses.

Risk analysis has gained acceptance in a large number of industries. One industry survey indicated that over 50% of the companies responding to the survey were using risk analysis for operational and/or strategic planning.

By incorporating the concept of utility theory, it is possible to extend risk analysis and obtain a normative or prescriptive model. To be more specific, by combining a manager's utility function and the probability distribution for, say, present worth, it is possible to specify the alternative that maximizes expected utility. Lifson [12] has recommended this approach previously. Hillier [7, 8, 9] incorporates utility theory in applying risk analysis to interrelated investment alternatives.

Our discussion will concentrate on the contributions of risk analysis as a descriptive technique instead as a normative technique. For those interested in extending our discussion to the normative mode as well as the incorporation of utility theory with risk analysis, see Hillier [7, 8, 9].

### 8.4.1 Distributions

The risk analysis procedure was developed to take into consideration the imprecision in estimating the values of the inputs required in making economic evaluations. The imprecision is represented in the form of a probability distribution.

Probability distributions for the random variables are usually developed on the basis of subjective probabilities. Typically, the further an event is into the future, the less precise is our estimate of the value of the outcome of the event. Hence, by letting the variance reflect our degree of precision, we would expect the variance of the probability distributions to increase with time.

Among the theoretical probability distributions commonly used in risk analysis are the normal distribution and the beta distribution. Examples of these distributions are depicted in Figures 8.6 and 8.7. For a discussion of a number of distributions and their process generators in the context of simulation, see Schmidt and Taylor [14] and Shannon [15], among others.

In some situations the subjective probability distribution cannot be represented accurately using a well-known theoretical distribution. Instead, one must estimate directly the probability distribution for the random variable.

One approach that can be used to estimate the subjective probability distribution is to provide optimistic, pessimistic, and most likely estimates for the random variable. The optimistic and pessimistic values should be ones that you do not anticipate will be exceeded with a significant probability (e.g., 1% chance). Given the practical limits on the range of values anticipated for the

---

[1] The term autocorrelation means correlated with itself over time.

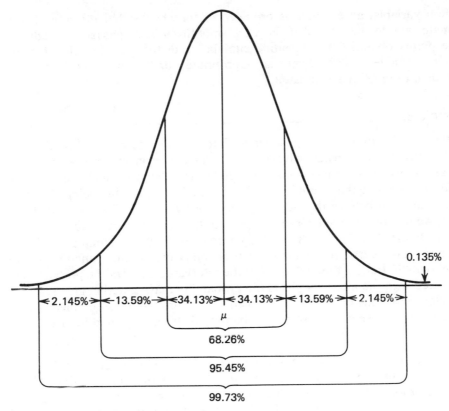

FIGURE 8.6. Normal distribution.

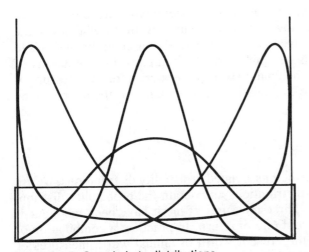

FIGURE 8.7. Sample beta distributions.

random variable, an estimate is provided of the chance that the most likely estimate will not be exceeded. Next, a smooth curve is passed through the three points obtained; the resulting cumulative distribution function for the random variable is divided into an appropriate number of intervals and the individual probabilities estimated.

Example 8.12 _____

To illustrate the process, suppose we wish to develop the probability distribution for, say, the anticipated salvage value of a machine in 5 years. We estimate the salvage value will range from $0 to $3000, with the most likely value being $1250. We estimate that there is a 40% chance of salvage value being less than $1250. Since we believe the extreme values of $0 and $3000 are not likely to occur, we use an S-shaped curve to represent the cumulative distribution function, as depicted in Figure 8.8a. Using intervals of $500, the cumulative distribution function is transformed into the probability distribution function shown in Figure 8.8b. Letting the midpoints of the intervals represent the probabilities associated with the intervals yields the probability mass function given in Figure 8.8c. Depending on the use to be made of the probabilities obtained, either of the three representations of the probability distribution could be used.

_____

The process described above is quite subjective, since a number of different curves could be used to describe the cumulative distribution function. However, the probability distribution itself is subjectively based. What is sought is a probability distribution that best describes one's beliefs about the outcomes of the random variable. If the probability distribution obtained above does not reflect your beliefs, the process should be repeated until a satisfactory probability distribution is obtained.

The critics of risk analysis cite the degree of subjectivity involved in developing probability distributions. However, it is argued by those who favor the technique that the only alternative to using subjectively based probabilities is to use the traditional single estimate approach, which implies that conditions of certainty exist. Interestingly, the final distribution obtained for the measure of merit is often quite insensitive to deviations in the shapes of the distributions for the parameters.

## 8.4.2 Risk Aggregation

Given the essential factors and their associated probability distributions, we are in a position to aggregate the distributions and obtain the probability distribution for the measure of merit. Three measures of merit have been mentioned: present worth, annual worth, and rate of return. In practice, a combination of the rate of return and either the present worth or annual worth measures of merit are often used. Each method of comparing investment alternatives has its

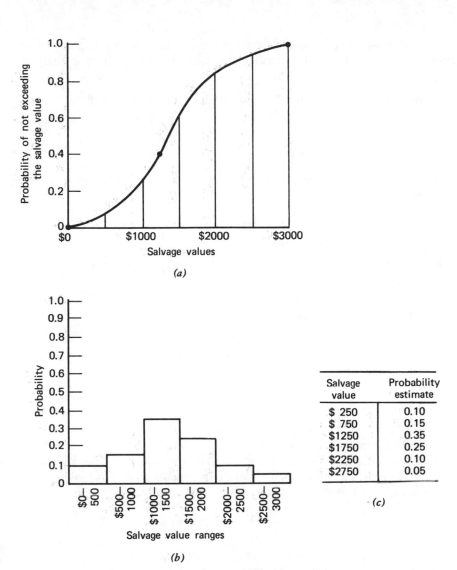

FIGURE 8.8. Developing the subjective probability distribution.

| Salvage value | Probability estimate |
|---|---|
| $ 250 | 0.10 |
| $ 750 | 0.15 |
| $1250 | 0.35 |
| $1750 | 0.25 |
| $2250 | 0.10 |
| $2750 | 0.05 |

*(c)*

limitations. For example, if the rate of return is being determined, the explicit reinvestment rate of return method is recommended, since multiple rates of return can arise when using the internal rate of return method.

Risk aggregation is achieved in basically two ways: by using simulation and analytically. Analytic approaches can be used in a number of simple cases. For more complex situations involving a large number of variables, simulation is used.

*Simulation Approaches.* Simulation, in the general sense, may be thought of as performing experiments on a model. Basically, simulation is an 'if . . . , then . . ." device (i.e., *if* a certain input is specified, *then* the output can be determined). Some of the major reasons for using simulation in risk analysis are:

1. Analytic solutions are impossible to obtain without great difficulty.
2. Simulation is useful in selling a system modification to management.
3. Simulation can be used as a verification of analytical solutions.
4. Simulation is very versatile.
5. Less background in mathematical analysis and probability theory is generally required.

Some of the major disadvantages of simulation are:

1. Simulations can be quite expensive.
2. Simulations introduce a source of randomness not present in analytic solutions.
3. Simulations do not reproduce the input distributions exactly (especially the tails of the distribution).
4. Validation is easily overlooked in using simulation.
5. Simulation is so easily applied it is often used when analytic solutions can be easily obtained at considerably less cost.

## Example 8.13

To illustrate the simulation approach in performing a risk analysis, consider an investment of $10,000 over a 4-year period that returns $R_t$ at the end of year $t$, with $R_t$ being a statistically independent random variable. Using a *MARR* of 20% it is desired to use simulation to estimate the probability of the investment being unprofitable (or risky!). The following probability distribution is assumed for $R_t$.

| $R_t$ | Probability | Random Number |
|-------|-------------|---------------|
| $2,000 | 0.10 | 0 |
| 3,000 | 0.20 | 1, 2 |
| 4,000 | 0.30 | 3, 4, 5 |
| 5,000 | 0.40 | 6, 7, 8, 9 |

(One-digit random numbers are assigned to the values of the random variables in proportion to their probability of occurrence.)

Consulting the random numbers given in Table 8.6 and beginning with column 1, row 1 the following sequence of 4 random digits is obtained: 9, 7, 5, and 9. Letting the first random digit (9) represent the return in the first year ($5000), the second random digit (7) represent the return in the second year

TABLE 8.6. Two-Digit Random Numbers

| | | | | | | | |
|---|---|---|---|---|---|---|---|
| 90 | 43 | 78 | 83 | 82 | 99 | 54 | 02 |
| 78 | 31 | 58 | 98 | 68 | 09 | 87 | 80 |
| 51 | 81 | 42 | 35 | 21 | 42 | 03 | 62 |
| 93 | 97 | 15 | 95 | 07 | 56 | 60 | 39 |
| 27 | 37 | 12 | 63 | 31 | 35 | 66 | 93 |
| 79 | 39 | 44 | 22 | 83 | 96 | 51 | 00 |
| 89 | 61 | 73 | 29 | 43 | 84 | 91 | 34 |
| 29 | 38 | 30 | 84 | 90 | 18 | 00 | 10 |
| 97 | 64 | 33 | 29 | 17 | 48 | 26 | 04 |
| 07 | 64 | 15 | 02 | 44 | 32 | 92 | 99 |
| 82 | 13 | 50 | 83 | 35 | 39 | 50 | 51 |
| 59 | 83 | 21 | 30 | 86 | 90 | 16 | 09 |
| 04 | 46 | 19 | 63 | 60 | 53 | 33 | 97 |
| 96 | 54 | 91 | 43 | 44 | 40 | 09 | 02 |
| 31 | 27 | 71 | 78 | 03 | 65 | 53 | 62 |
| 03 | 45 | 70 | 42 | 22 | 16 | 67 | 13 |
| 08 | 35 | 45 | 92 | 79 | 97 | 46 | 02 |
| 37 | 60 | 80 | 55 | 05 | 35 | 75 | 57 |
| 90 | 43 | 63 | 17 | 56 | 21 | 69 | 09 |
| 22 | 07 | 69 | 85 | 38 | 74 | 02 | 58 |
| 05 | 33 | 79 | 00 | 69 | 29 | 67 | 08 |
| 48 | 97 | 91 | 14 | 53 | 00 | 03 | 42 |
| 94 | 68 | 64 | 58 | 97 | 32 | 27 | 80 |
| 15 | 39 | 85 | 87 | 82 | 38 | 52 | 16 |
| 09 | 37 | 81 | 73 | 37 | 01 | 66 | 84 |

($5000), and so on yields simulated returns of $5000, $5000, $4000, and $5000 over the 4-year period for a present worth of $2364.80. Obviously, one simulated investment is not sufficient to estimate the probability of the investment being profitable. Continued simulation yields the results given in Table 8.7. For the 20 simulated investments, 6 had negative present worths; thus, the probability of the investment being unprofitable might be estimated to be 0.30. However, one should be cautioned against drawing conclusions from only 20 simulated investments. In practice, it would be desirable to develop a computer program to perform the simulation and obtain perhaps 1000 simulated investments before estimating the probability of a negative present worth.

Example 8.14

As a second illustration of risk analysis, consider the following situation. An individual is planning on purchasing a used microcomputer for $1000 and performing certain billing and accounting functions for several retail businesses in the neighborhood. It is anticipated that the business will last only 4 years

TABLE 8.7. Simulation Results for Example 8.13

| Investment Trial | Year 1 | | Year 2 | | Year 3 | | Year 4 | | Present Worth, PW(20%) |
|---|---|---|---|---|---|---|---|---|---|
| | RN | Value | RN | Value | RN | Value | RN | Value | |
| 1 | 9 | $5,000 | 7 | $5,000 | 5 | $4,000 | 9 | $5,000 | $2,364.80 |
| 2 | 2 | 3,000 | 7 | 5,000 | 8 | 5,000 | 2 | 3,000 | 312.30 |
| 3 | 9 | 5,000 | 0 | 2,000 | 8 | 5,000 | 5 | 4,000 | 378.00 |
| 4 | 0 | 2,000 | 9 | 5,000 | 3 | 4,000 | 0 | 2,000 | − 1,582.00 |
| 5 | 0 | 2,000 | 3 | 4,000 | 9 | 5,000 | 2 | 3,000 | − 1,215.40 |
| 6 | 0 | 2,000 | 4 | 4,000 | 9 | 5,000 | 1 | 3,000 | − 1,215.40 |
| 7 | 0 | 2,000 | 0 | 2,000 | 8 | 5,000 | 1 | 3,000 | − 2,604.20 |
| 8 | 3 | 4,000 | 7 | 5,000 | 9 | 5,000 | 9 | 5,000 | 2,110.20 |
| 9 | 9 | 5,000 | 7 | 5,000 | 7 | 5,000 | 2 | 3,000 | 1,978.90 |
| 10 | 9 | 5,000 | 4 | 4,000 | 6 | 5,000 | 1 | 3,000 | 1,284.50 |
| 11 | 3 | 4,000 | 8 | 5,000 | 7 | 5,000 | 0 | 2,000 | 663.30 |
| 12 | 2 | 3,000 | 5 | 4,000 | 8 | 5,000 | 4 | 4,000 | 100.20 |
| 13 | 5 | 4,000 | 9 | 5,000 | 4 | 4,000 | 3 | 4,000 | 1,049.20 |
| 14 | 8 | 5,000 | 9 | 5,000 | 3 | 4,000 | 3 | 4,000 | 1,882.50 |
| 15 | 6 | 5,000 | 3 | 4,000 | 6 | 5,000 | 6 | 5,000 | 2,249.10 |
| 16 | 1 | 3,000 | 8 | 5,000 | 4 | 4,000 | 5 | 4,000 | 215.90 |
| 17 | 2 | 3,000 | 4 | 4,000 | 3 | 4,000 | 6 | 5,000 | 3.80 |
| 18 | 4 | 4,000 | 0 | 2,000 | 3 | 4,000 | 9 | 5,000 | − 551.70 |
| 19 | 6 | 5,000 | 3 | 4,000 | 3 | 4,000 | 3 | 4,000 | 1,188.10 |
| 20 | 1 | 3,000 | 1 | 3,000 | 7 | 5,000 | 7 | 5,000 | − 111.90 |

because of the growth of competition. If income in the third year exceeds expenses by more than $400, operations will continue the fourth year. However, if income is less than or equal to $400 more than expenses in the third year, the computer and software will probably be sold at the end of the third year. For simplicity a probability of 0.70 is assigned to the possibility of selling out at the end of the third year, given income is not greater than $400 above expenses in the third year. Let

$E_j$ = expense for year $j$
$I_j$ = income for year $j$
$n$ = life of the investment
$S_n$ = salvage value based on an $n$-year life

Assuming a zero discount rate for simplicity of calculations, the present worth of the investment is given as

$$PW = - \$1000 + \sum_{j=1}^{n} (I_j - E_j) + S_n$$

The probability distributions assumed to hold for this example are provided in Table 8.8. As in the previous example, a computer should be used to perform the simulation. However, to illustrate the technique, we will manually perform

TABLE 8.8. Data for the Risk Analysis Example Problem

| $E_1$ | $p(E_1)$ | RN | $I_1$ | $p(I_1)$ | RN |
|---|---|---|---|---|---|
| $200 | 0.25 | 00–24 | $ 400 | 0.50 | 00–49 |
| 300 | 0.50 | 25–74 | 600 | 0.50 | 50–99 |
| 400 | 0.25 | 75–99 | | | |

| $E_2$ | $p(E_2)$ | RN | $I_2$ | $p(I_2)$ | RN |
|---|---|---|---|---|---|
| $300 | 0.10 | 00–09 | $ 500 | 0.20 | 00–19 |
| 400 | 0.40 | 10–49 | 750 | 0.40 | 20–59 |
| 500 | 0.40 | 50–89 | 1,000 | 0.40 | 60–99 |
| 600 | 0.10 | 90–99 | | | |

| $E_3$ | $p(E_3)$ | RN | $I_3$ | $p(I_3)$ | RN |
|---|---|---|---|---|---|
| $400 | 0.20 | 00–19 | $ 800 | 0.30 | 00–29 |
| 500 | 0.30 | 20–49 | 1,000 | 0.50 | 30–79 |
| 600 | 0.30 | 50–79 | 1,200 | 0.20 | 80–99 |
| 700 | 0.20 | 80–99 | | | |

| $E_4$ | $p(E_4)$ | RN | $I_4$ | $p(I_4)$ | RN |
|---|---|---|---|---|---|
| $500 | 0.25 | 00–24 | $ 600 | 0.25 | 00–24 |
| 600 | 0.25 | 25–49 | 800 | 0.25 | 25–49 |
| 700 | 0.25 | 50–74 | 1,000 | 0.25 | 50–74 |
| 800 | 0.25 | 75–99 | 1,200 | 0.25 | 75–99 |

| $N$ | $p(N|I_3 - E_3 \leq \$400)$ | RN |
|---|---|---|
| 3 | 0.70 | 00–69 |
| 4 | 0.30 | 70–99 |

| $N$ | $p(N|I_3 - E_3 > \$400)$ |
|---|---|
| 4 | 1.00 |

| $S_3$ | $p(S_3)$ | RN | $S_4$ | $p(S_4)$ | RN |
|---|---|---|---|---|---|
| $400 | 0.50 | 00–49 | $ 300 | 0.60 | 00–59 |
| 500 | 0.50 | 50–99 | 400 | 0.40 | 60–99 |

10 simulations of the investment. The table of two-digit random numbers given in Table 8.6 will be used. Using the worksheet given in Table 8.9, 10 simulations of the investment yielded an average present worth of $805 for the investment. Of course, 10 simulations is not an adequate number of trials to draw strong conclusions concerning the investment. However, the example does demonstrate the simulation approach.

To illustrate the approach taken, the first random number selected will provide the simulated value for expenses in the first year. A random number of 90

TABLE 8.9. Simulation of the Sample Investment Problem in Example 8.14

| Trial | Year | RN(E) | E | RN(I) | I | I − E | RN(N) | RN(S) | S | PW |
|-------|------|-------|-----|-------|-------|-------|-------|-------|-------|-------|
| 1 | 1 | 90 | $400 | 78 | $ 600 | $200 | | | | |
| | 2 | 51 | 500 | 93 | 1,000 | 500 | | | | |
| | 3 | 27 | 500 | 79 | 1,000 | 500 | — | | | |
| | 4 | 89 | 800 | 29 | 800 | 0 | | 97 | $400 | $ 600 |
| 2 | 1 | 07 | 200 | 82 | 600 | 400 | | | | |
| | 2 | 59 | 500 | 04 | 500 | 0 | | | | |
| | 3 | 96 | 700 | 31 | 1,000 | 300 | 03 | 08 | 400 | 100 |
| | 4 | — | — | — | — | — | | | | |
| 3 | 1 | 37 | 300 | 90 | 600 | 300 | | | | |
| | 2 | 22 | 400 | 05 | 500 | 100 | | | | |
| | 3 | 48 | 500 | 94 | 1,200 | 700 | — | | | |
| | 4 | 15 | 500 | 09 | 600 | 100 | | 43 | 300 | 500 |
| 4 | 1 | 31 | 300 | 81 | 600 | 300 | | | | |
| | 2 | 97 | 600 | 37 | 750 | 150 | | | | |
| | 3 | 39 | 500 | 61 | 1,000 | 500 | — | | | |
| | 4 | 38 | 600 | 64 | 1,000 | 400 | | 64 | 400 | 750 |
| 5 | 1 | 13 | 200 | 83 | 600 | 400 | | | | |
| | 2 | 46 | 400 | 54 | 750 | 350 | | | | |
| | 3 | 27 | 500 | 45 | 1,000 | 500 | — | | | |
| | 4 | 35 | 600 | 60 | 1,000 | 400 | | 43 | 300 | 950 |
| 6 | 1 | 07 | 200 | 33 | 400 | 200 | | | | |
| | 2 | 97 | 600 | 68 | 1,000 | 400 | | | | |
| | 3 | 39 | 500 | 37 | 1,000 | 500 | — | | | |
| | 4 | 78 | 800 | 58 | 1,000 | 200 | | 42 | 300 | 600 |
| 7 | 1 | 15 | 200 | 12 | 400 | 200 | | | | |
| | 2 | 44 | 400 | 73 | 1,000 | 600 | | | | |
| | 3 | 30 | 500 | 33 | 1,000 | 500 | — | | | |
| | 4 | 15 | 500 | 50 | 1,000 | 500 | | 21 | 300 | 1,100 |
| 8 | 1 | 19 | 200 | 91 | 600 | 400 | | | | |
| | 2 | 71 | 500 | 70 | 1,000 | 500 | | | | |
| | 3 | 45 | 500 | 80 | 1,200 | 700 | — | | | |
| | 4 | 63 | 700 | 69 | 1,000 | 300 | | 79 | 400 | 1,300 |
| 9 | 1 | 91 | 400 | 64 | 600 | 200 | | | | |
| | 2 | 85 | 500 | 81 | 1,000 | 500 | | | | |
| | 3 | 83 | 700 | 98 | 1,200 | 500 | — | | | |
| | 4 | 35 | 600 | 95 | 1,200 | 600 | | 63 | 400 | 1,200 |
| 10 | 1 | 22 | 200 | 29 | 400 | 200 | | | | |
| | 2 | 84 | 500 | 29 | 750 | 250 | | | | |
| | 3 | 02 | 400 | 83 | 1,200 | 800 | — | | | |
| | 4 | 30 | 600 | 63 | 1,000 | 400 | | 43 | 300 | 950 |
| | | | | | | | | | | 8,050 |

was obtained from row 1, column 1 in Table 8.6. Consulting Table 8.8, it is seen that a random number of 90 represents an expense of $400; hence $400 is entered appropriately on the worksheet given in Table 8.9. The second random number is selected from row 2, column 1 of Table 8.6 to generate the income for year 1. A random number of 78 is obtained and, from Table 8.8, a simulated income of $600 is obtained for the first year. Continuing through the third year, it is found that income exceeds expenses by $500; hence the business will continue through the fourth year. A random number of 97 is drawn to generate the salvage value of $400 for the investment.

Note that in the second trial it was decided that the business should be discontinued after 3 years. Furthermore, income was never less than expenses in any year; this illustrates insufficient observations (trials) have been obtained since, in year 2, expense can exceed income with probability

$$Pr(E_2 = 600 \text{ and } I_2 = 500) = 0.10(0.20) = 0.02$$

and in year 4 expense can exceed income with probability

$$Pr(E_4 = 700 \text{ and } I_4 = 600) + Pr(E_4 = 800 \text{ and } I_4 = 600) = 0.125$$

---

As an alternative to using the discrete probability distributions given in Table 8.8, we might employ continuous distributions such as the normal, gamma, or beta distributions to represent income and expenses. If such distributions are to be used, appropriate techniques for generating simulated values of the random variables are required. Since a detailed treatment of simulation is beyond the scope of this text, we refer you to texts devoted to simulation for additional discussion [14], [15].

*Analytic Approaches.* As an illustration of the use of analytic approaches in developing the probability distribution for present worth, consider the following present worth relation.

$$PW = \sum_{j=0}^{n} C_j(1 + i)^{-j} \qquad (8.1)$$

Suppose the cash flows, $C_j$, are *random variables* with expected values $E(C_j)$ and variances $V(C_j)$. Since the *expected value* of a sum of random variables is given by the sum of the expected values of the random variables, the expected present worth is given by

$$E(PW) = \sum_{j=0}^{n} E[C_j(1 + i)^{-j}] \qquad (8.2)$$

Furthermore, since the expected value of the product of a constant and a random variable is given by the product of the constant and the expected value of the random variable,

$$E[C_j(1 + i)^{-j}] = E(C_j)(1 + i)^{-j} \qquad (8.3)$$

Substituting Equation 8.3 into Equation 8.2 yields

$$E(PW) = \sum_{j=0}^{n} E(C_j)(1 + i)^{-j}$$

Hence, we see that the expected present worth of a series of cash flows is found by summing the present worths of the expected values of the inidividual cash flows.

Example 8.15_____

Recall Example 8.13 involving an investment of $10,000 of a 4-year period. The expected value of the return in a given year is given by

$$E[\text{Return}] = \$2000(0.10) + \$3000(0.20)$$
$$+ \$4000(0.30) + \$5000(0.40)$$
$$= \$4000$$

Thus, the expected present worth for the investment situation is

$$E(PW) = \$4000(P|A\ 20,4) - \$10,000$$
$$= \$354.80$$

(From Table 8.7, a computation of the average simulated present worth yields $425.01. This serves to illustrate the need for many simulated trials.)

_____

To determine the variance of the present worth, we first recall that if $X_1$, $X_2$, $X_3$, and $Y$ are random variables related as follows,

$$Y = X_1 + X_2 + X_3 \qquad (8.4)$$

then the variance of $Y$ is given by

$$\text{Var}(Y) = E(Y^2) - E(Y)^2 \qquad (8.5)$$

or

$$\text{Var}(Y) = E[(X_1 + X_2 + X_3)^2] - E(X_1 + X_2 + X_3)^2$$

Expanding and collecting terms yields

$$
\begin{aligned}
\mathrm{Var}(Y) &= E(X_1^2 + X_2^2 + X_3^2 + 2X_1X_2 + 2X_1X_3 + 2X_2X_3) \\
&\quad - [E(X_1) + E(X_2) + E(X_3)]^2 \\
&= E(X_1^2) + E(X_2^2) + E(X_3^2) + 2E(X_1X_2) + 2E(X_1X_3) + 2E(X_2X_3) \\
&\quad - E(X_1)^2 - E(X_2)^2 - E(X_3)^2 - 2E(X_1)E(X_2) - 2E(X_1)E(X_2) \\
&\quad - 2E(X_2)E(X_3) \\
&= E(X_1^2) - E(X_1)^2 + E(X_2^2) - E(X_2)^2 + E(X_3^2) - E(X_3)^2 \\
&\quad + 2[E(X_1X_2) - E(X_1)E(X_2)] + 2[E(X_1X_3) - E(X_1)E(X_3)] \\
&\quad + 2[E(X_2X_3) - E(X_2)E(X_3)]
\end{aligned}
\tag{8.6}
$$

Since $E(X_k^2) - E(X_k)^2$ defines the variance of the random variable $X_k$ and $E(X_pX_k) - E(X_p)E(X_k)$ defines the covariance of the random variables $X_p$ and $X_k$, we see that

$$
\boxed{
\begin{aligned}
\mathrm{Var}(Y) &= \mathrm{Var}(X_1) + \mathrm{Var}(X_2) + \mathrm{Var}(X_3) + 2\,\mathrm{Cov}(X_1X_2) \\
&\quad + 2\,\mathrm{Cov}(X_1X_3) + 2\,\mathrm{Cov}(X_2X_3)
\end{aligned}
}
\tag{8.7}
$$

where $\mathrm{Cov}(X_pX_k)$ denotes the covariance of $X_p$ and $X_k$. If the random variables $X_p$ and $X_k$ are statistically independent, then $\mathrm{Cov}(X_pX_k) = 0$.

Generalizing to the sum of $(n + 1)$ random variables, if

$$
\boxed{ Y = \sum_{j=0}^{n} X_j }
\tag{8.8}
$$

then the expected value and variance of $Y$ are given by

$$
\boxed{ E(Y) = \sum_{j=0}^{n} E(X_j) }
$$

$$
\boxed{ \mathrm{Var}(Y) = \sum_{j=0}^{n} \mathrm{Var}(X_j) + 2 \sum_{j=0}^{n-1} \sum_{k=j+1}^{n} \mathrm{Cov}(X_jX_k) }
\tag{8.9}
$$

Additionally, if $a_1$ and $a_2$ are constants and $X_1$, $X_2$, and $Y$ are random variables related by

$$
\boxed{ Y = a_1X_1 + a_2X_2 }
\tag{8.10}
$$

recall that

$$\mathrm{Var}(Y) = a_1{}^2\,\mathrm{Var}(X_1) + a_2{}^2\,\mathrm{Var}(X_2) + 2a_1a_2\,\mathrm{Cov}(X_1X_2) \qquad (8.11)$$

Combining the relationships given in Equations 8.9 and 8.11, the variance of present worth is found to be

$$\mathrm{Var}(PW) = \sum_{j=0}^{n} \mathrm{Var}(C_j)(1 + i)^{-2j}$$
$$+ 2\sum_{j=0}^{n-1}\sum_{k=j+1}^{n} \mathrm{Cov}(C_jC_k)(1 + i)^{-(j+k)} \qquad (8.12)$$

## Example 8.16

Continuing Example 8.15 to compute the variance of present worth, recall from Example 8.13 the random variables are statistically independent. Thus, $\mathrm{Cov}(C_jC_k) = 0$ for all $j$ and $k$. The variance of an annual return is determined as follows:

$$\mathrm{Var}[\mathrm{Return}] = E[\mathrm{Return}^2] - E[\mathrm{Return}]^2$$

and

$$\begin{aligned}
E[\mathrm{Return}^2] &= (\$2000)^2(0.10) + (\$3000)^2(0.20) \\
&\quad + (\$4000)^2(0.30) + (\$5000)^2(0.40) \\
&= 17 \times 10^6
\end{aligned}$$

Thus

$$\begin{aligned}
\mathrm{Var}[\mathrm{Return}] &= (17 \times 10^6) - (4000)^2 \\
&= 1 \times 10^6
\end{aligned}$$

The standard deviation is the square root of the variance and equals

$$SD[\mathrm{Return}] = \$1000$$

To compute the variance of present worth, from Equation 8.12

$$\begin{aligned}
\mathrm{Var}(PW) &= \mathrm{Var}[\mathrm{Investment}] + \sum_{t=1}^{4} \mathrm{Var}(R_t)(1.20)^{-2t} \\
&= 0 + \sum_{t=1}^{4} (1 \times 10^6)(1.20)^{-2t} \\
&= (1 \times 10^6) \sum_{t=1}^{4} (P|F\ 20, 2t) \\
&= (1 \times 10^6)(0.6944 + 0.4823 + 0.3349 + 0.2326) \\
&= 1.7442 \times 10^6
\end{aligned}$$

and

$$SD(PW) = \$1320.68$$

---

Hillier [5] argues that it is probably unrealistic to expect investment analysts to develop accurate estimates for covariances in Equation 8.12. Consequently, he suggests that the net cash flow in any year be divided into those components of cash flow that are reasonably independent from year to year and those that are correlated over time. Specifically, it is assumed that

$$C_j = X_j + Y_{j1} + Y_{j2} + \cdots + Y_{jm} \tag{8.13}$$

where the $X_j$ values are mutually independent over $j$ but, for a given value of $h$, $Y_{0h}, Y_{1h}, \ldots, Y_{nh}$ are *perfectly* correlated.

When two random variables $X$ and $Y$ are perfectly correlated, one can be expressed as a linear function of the other. Hence, if

$$Y = a + bX \tag{8.14}$$

where $a$ and $b$ are constants, then the covariance of $X$ and $Y$ is given by

$$\text{Cov}(XY) = E(XY) - E(X)E(Y) \tag{8.15}$$

However, from Equation 8.14, we note that

$$E(Y) = a + bE(X) \tag{8.16}$$

and

$$\text{Var}(Y) = b^2 \, \text{Var}(X) \tag{8.17}$$

Substituting Equations 8.14 and 8.16 into Equation 8.15 gives

$$\begin{aligned} \text{Cov}(XY) &= E(aX + bX^2) - E(X)[a + bE(X)] \\ &= aE(X) + bE(X^2) - aE(X) - bE(X)^2 \\ &= b \, \text{Var}(X) \end{aligned} \tag{8.18}$$

Recalling that the *standard deviation* of a random variable is defined as the square root of the variance of the random variable, we see that Equation 8.18 can be expressed as

$$\text{Cov}(XY) = bSD(X)SD(X) \tag{8.19}$$

where $SD(X)$ denotes the standard deviation of $X$. However, from Equation 8.17, we see that $SD(Y)$ equals $bSD(X)$. Hence,

$$\boxed{\mathrm{Cov}(XY) = SD(X)SD(Y)} \qquad (8.20)$$

when $X$ and $Y$ are perfectly correlated.

The model suggested by Hillier yields the following expressions for the expected present worth and variance of present worth:

$$\boxed{E(PW) = \sum_{j=0}^{n} E(X_j)(1 + i)^{-j} + \sum_{j=0}^{n} \sum_{h=1}^{m} E(Y_{jh})(1 + i)^{-j}} \qquad (8.21)$$

and

$$\boxed{\mathrm{Var}(PW) = \sum_{j=0}^{n} \mathrm{Var}(X_j)(1 + i)^{-2j} + \sum_{h=1}^{m} \left[ \sum_{j=0}^{n} SD(Y_{jh})(1 + i)^{-j} \right]^2} \qquad (8.22)$$

where $SD(Y_{jh})$ denotes the standard deviation of the random variable $Y_{jh}$.

A very important theorem from probability theory, the *central limit theorem*, is usually invoked at this point. The central limit theorem establishes under very general conditions that the sum of independently distributed, random variables tends to be distributed normally as the number of terms in the summation approaches infinity. Hence, it is argued that present worth, as defined by Equation 8.1, is normally distributed with mean and variance, as given by Equations 8.21 an 8.22. Of course, this assumes that the $C_j$ are statistically independent. However, the view is usually taken that the normal distribution is a reasonable approximation to the distribution of present worth.

Given that present worth is assumed to be normally distributed, one can compute for each investment alternative the probability of achieving a given aspiration level. For example, one is usually interested in knowing the probability that present worth is less than zero. Some analysts interpret this to be a measure of the risk associated with an investment alternative.

In performing an analysis of the risk associated with an investment alternative, we have treated the simplest situation; only the cash flows were considered to be random variables and the measure of effectiveness employed was present worth, not rate of return. When either the discount rate or the planning horizon are treated as random variables and when the probability distribution for rate of return is desired, simulation approaches are usually employed.

## Example 8.17

As an illustration of the analytic approach in developing the probability distribution for $PW$, consider the following example problem based on one given by Giffin [3].

A flight school operator is considering the alternatives of purchasing a utility category training aircraft versus purchasing an acrobatic version of the same aircraft. Having been in the flight training business for a number of years, the

TABLE 8.10. Estimated Net Cash Flows for a Utility Model Aircraft*

| Year | Source | Symbol | Expected Value | Range | Standard Deviation |
|------|--------|--------|---------------|-------|--------------------|
| 0 | Purchase | $C_0$ | −$11,000 | $9,500–12,500 | $500 |
| 1 | Income-expense | $C_1$ | 2,200 | 2,050– 2,350 | 50 |
| 2 | Income-expense | $C_2$ | 2,200 | 1,900– 2,500 | 100 |
| 3 | Income-expense | $C_3$ | 2,200 | 1,900– 2,500 | 100 |
| 4 | Income-expense | $C_4$ | 2,000 | 1,700– 2,300 | 100 |
| 5 | Income-expense | $C_{51}$ | 1,000 | 700– 1,300 | 100 |
| 5 | Salvage | $C_{52}$ | 6,000 | 4,800– 7,200 | 400 |

* The data have been modified slightly from those given by Giffin [3].

operator has reason to believe that income and expense from a utility category aircraft will be nearly independent from year to year. The pertinent data are summarized in Table 8.10.

Investment in the acrobatic aircraft is a more risky but promising investment. The acrobatic aircraft is also expected to have a life of 5 years. The operator feels that maintenance costs will be nearly independent year to year with this aircraft. However, since the flight school has never offered an acrobatic course before, there is some uncertainty regarding the demand for time in such an aircraft. It is felt that the net cash flow from the sale of flight time for each of the five years will be perfectly correlated. The pertinent data are summarized in Table 8.11. An interest rate of 10% is to be used in the analysis.

From Equations 8.21 and 8.22, we obtain the following:

$$E(PW_1) = \$190 \qquad E(PW_2) = \$355$$

$$\mathrm{Var}(PW_1) = 334{,}741 \qquad \mathrm{Var}(PW_2) = 11{,}631{,}930$$

TABLE 8.11. Estimated Net Cash Flows for an Acrobatic Model Aircraft*

| Year | Source | Symbol | Expected Value | Range | Standard Deviation |
|------|--------|--------|---------------|-------|--------------------|
| 0 | Purchase | $X_0$ | −$14,000 | $13,100–14,900 | $ 300 |
| 1 | Expense | $X_1$ | − 10,000 | 9,100–10,900 | 300 |
| 2 | Expense | $X_2$ | − 10,000 | 9,100–10,900 | 300 |
| 3 | Expense | $X_3$ | − 11,000 | 10,100–11,900 | 300 |
| 4 | Expense | $X_4$ | − 12,000 | 10,800–13,200 | 400 |
| 5 | Expense | $X_{51}$ | − 12,000 | 10,800–13,200 | 400 |
| 1 | Income | $Y_{11}$ | 12,500 | 9,500–15,500 | 1,000 |
| 2 | Income | $Y_{21}$ | 13,500 | 10,500–16,500 | 1,000 |
| 3 | Income | $Y_{31}$ | 13,500 | 10,800–16,200 | 900 |
| 4 | Income | $Y_{41}$ | 14,500 | 12,100–16,900 | 800 |
| 5 | Income | $Y_{51}$ | 13,500 | 11,400–15,600 | 700 |
| 5 | Salvage | $X_{52}$ | 7,500 | 6,000– 9,000 | 500 |

* The data have been modified slightly from those given by Giffin [3].

Assuming normally distributed $PW$, the probability of an equivalent present worth less than zero is found to be

$$Pr(PW_1 < 0 | i = 10\%) = 0.37$$

for the utility model aircraft and

$$Pr(PW_2 < 0 | i = 10\%) = 0.46$$

for the acrobatic model aircraft. Here we are faced with a choice between the alternative having the greatest expected value and the alternative having the smallest variance. The choice will depend on the owner's attitudes toward risk. At this point we leave the flight school operator "up in the air" relative to a choice between the investment alternatives. If the decision were yours to make, which model aircraft would you choose? Why?

---

Risk analysis offers a number of important advantages over traditional deterministic approaches. Klausner [11] summarizes some of the most significant advantages as follows:

1. Uncertainty Made Explicit: The uncertainty that an estimator feels about his estimate of an element value is brought out into the open and incorporated into the investment analysis. The analysis technique permits maximum information utilization by providing a vehicle for the inclusion of "less likely" estimates in the analysis.

2. More Comprehensive Analysis. This technique permits a determination of the effect of simultaneous variation of all the element values on the outcome of an investment. This approximates the "real world" conditions under which an actual investment's outcome will be determined. The Probabilistic Cash Flow Simulation generates an overall indication of potential variation in outcome and project risk. This indicator, in the form of a probability distribution, accounts statistically for element interaction.

3. Variability of Outcome Measured. One of the most significant advantages of this analysis technique is that it gives a measure of the dispersion around the investment outcome based on the expected cash flow. This dispersion, or variability, is an important consideration in the comparison of alternative investments. Other things being equal, lower variability for the same return is usually desirable. The probability distribution associated with each investment's outcome gives a clear picture of this important evaluation consideration.

4. Promotes More Reasoned Estimating Procedures. By requiring that element values be given as probability distributions rather than as single values, more reasoned consideration is given to the estimating procedure. Judgment is applied to the individual element values rather than to the investment's outcome which is jointly determined by all of the elements. Thinking through the uncertainties in a project and recognizing what is known and unknown will go far toward ensuring the best investment decision. Understanding and dealing effectively with uncertainty and risk is the key to rational decision making.

Risk analysis is a technique that has been used by a number of firms to improve the decision-making process in a risk environment. When applied properly, risk analysis can enhance significantly the manager's understanding of the risks associated with an investment alternative. The major premise underlying risk analysis is the belief that a manager can make better decisions when he or she has a fuller understanding of the implications of the decision.

## 8.5 SUMMARY

Descriptive approaches were presented in this chapter for coping with risk and uncertainty conditions in performing economic analyses. In particular, break-even, sensitivity, and risk analyses were discussed. In Chapter Nine, we employ normative or prescriptive approaches for dealing with risk and uncertainty.

The extent to which supplementary analyses are justified depends largely on the magnitude of the initial investment. In particular, if an analysis has been performed using the techniques described in Chapters Four through Seven and one of the alternatives is not clearly the superior alternative, a risk analysis will probably yield additional insight concerning the decision to be made. If either small amounts of money are involved or if the choice is not close using the techniques described in Chapters Four through Seven under assumed certainty, then supplementary analyses are not justified. In those cases where supplementary analysis is justified and a sensitivity analysis is not sufficient to allow a decision, a risk analysis is performed.

## BIBLIOGRAPHY

1. Canada, J. R., and White, J. A., Jr., *Capital Investment Decision Analysis for Management and Engineering,* Prentice-Hall, 1980.

2. Fleischer, G. A., *Risk and Uncertainty: Non-Deterministic Decision Making in Engineering Economy,* AIIE Monograph Series, American Institute of Industrial Engineers, Inc., 1975.

3. Giffin, W. C., *Introduction to Operations Engineering,* R. D. Irwin, 1971.

4. Hertz, D. B., "Risk Analysis in Capital Investments," *Harvard Business Review,* 42 (1), January–February 1964.

5. Hillier, F. S., "The Derivation of Probabilistic Information for the Evaluation of Risky Investments," *Management Science, 9* (3), April 1963.

6. Hillier, F. S., "Supplement to 'The Derivation of Probabilistic Information Evaluation of Risky Investments,'" *Management Science, 11* (3), January 1965.

7. Hillier, F. S., *The Evaluation of Risky Interrelated Investments,* North-Holland Publishing, 1969.

8. Hillier, F. S., "A Basic Approach to the Evaluation of Risky Interrelated Investments," Chapter 1 in *Studies in Budgeting,* R. F. Byrne, W. W. Cooper, A. Charnes, O. A. Davis, and D. A. Gilford, Editors, North-Holland Publishing, 1971.

9. Hillier, F. S., "A Basic Model for Capital Budgeting of Risky Interrelated Projects," *The Engineering Economist, 17* (1), 1971.

10. Kaplan, S., and Barish, N., "Decision Making Allowing for Uncertainty of Future Investment Opportunities," *Management Science, 13* (10), June 1967.

11. Klausner, R. F., "The Evaluation of Risk in Marine Capital Investments," *The Engineering Economist, 14* (4), Summer 1969.

12. Lifson, M., "Evaluation Modeling in the Context of the Systems Decision Process," *Decision and Risk Analysis: Powerful New Tools for Management,* Proceedings of the Sixth Triennial Symposium, published by *The Engineering Economist,* 1971.

13. Miller, E. C., *Advanced Techniques for Strategic Planning,* AMA Research Study 104, American Management Association, Inc., 1971.

14. Schmidt, J. W., and Taylor, R. E., *Simulation and Analysis of Industrial Systems,* R. D. Irwin, 1970.

15. Shannon, R. E., *Systems Simulation: the Art and Science,* Prentice-Hall, 1975.

## PROBLEMS

1. An aluminum extrusion plant manufactures a particular product at a variable cost of $0.04 per unit, including material cost. The fixed costs associated with manufacturing the product equal $30,000/year. Determine the break-even value for annual sales if the selling price per unit is (a) $0.40, (b) $0.30, and (c) $0.20. (8.2)

2. Two 100-horsepower motors are under consideration by the Mighty Machinery Company. Motor Q costs $5000 and operates at 90% efficiency. Motor R costs $3500 and is 88% efficient. Annual operating and maintenance costs are estimated to be 15% of the initial purchase price. Power costs 3.2¢/kilowatt-hour. How many hours of full-load operation are necessary each year in order to justify the purchase of motor Q? Use a 15-year planning horizon; assume that salvage values will equal 20% of the initial purchase price; and let the *MARR* be 20%. (*Note.* 0.746 kilowatts = 1 horsepower.) (8.2)

3. A consulting engineer is considering two pumps to meet a demand of 15,000 gallons/minute at 12 feet total dynamic head. The specific gravity of the liquid being pumped is 1.50. Pump A operates at 70% efficiency and costs $12,000; pump B operates at 75% efficiency and costs $18,000. Power costs $0.015/kilowatt-hour. Continuous pumping for 365 days/year is required (i.e., 24 hours/day). Using a *MARR* of 20% and assuming equal salvage values for both pumps, how many years of service are required for pump B to be justified economically? (*Note.* Dynamic head times gallon/minute times specific gravity divided by 3960 equals horsepower required. Horsepower times 0.746 equals kilowatts required.) (8.2)

4. Owners of a nationwide motel chain are considering locating a new motel in Duluth, Georgia. The complete cost of building a 150-unit motel (excluding furnishings) is $5

million; the firm estimates that the furnishings in the motel must be replaced at a cost of $1,875,000 every 5 years. Annual operating and maintenance cost for the facility is estimated to be $125,000. The average rate for a unit is anticipated to be $45/day. A 15-year planning horizon is used by the firm in evaluating new ventures of this type; a terminal salvage value of 20% of the original building cost is anticipated; furnishings are estimated to have no salvage value at the end of each 5 year replacement interval; land cost is not to be included. Determine the break-even value for the daily occupancy percentage based on a *MARR* of (a) 0%, (b) 10%, (c) 20%, and (d) 30%. (Assume that the motel will operate 365 day/year.) (8.2)

5. A manufacturing plant in Michigan has been contracting snow removal at a cost of $325/day. The past three years have produced heavy snowfalls, resulting in the cost of snow removal being of concern to the plant manager. The plant engineer has found that a snow-removal machine can be purchased for $65,000; it is estimated to have a useful life of 10 years, and a zero salvage value at that time. Annual costs for operating and maintaining the equipment are estimated to be $15,000. Determine the break-even value for the number of days per year that snow removal is required in order to justify the equipment, based on a *MARR* of (a) 0%, (b) 10%, and (c) 30%. (8.2)

6. A business firm is contemplating the installation of an improved material handling system between the packaging department and the finished goods warehouse. Two designs are being considered. The first consists of a driverless tractor system involving three tractors on the loop, with four trailers pulled by each tractor. The second design consists of a pallet conveyor installed between packaging and the warehouse. The driverless tractor system will have an initial equipment cost of $280,000 and annual operating and maintenance costs of $25,000. The pallet conveyor has an initial cost of $360,000 and annual operating and maintenance costs of $2000. The firm is not sure what planning horizon to use in the analysis; however, the salvage value estimates given in the following table have been developed for various planning horizons. Using a *MARR* of 20%, determine the break-even value for *N*, the planning horizon. (8.2)

Salvage Value Estimates

| N | Driverless Tractor | Pallet Conveyor |
|---|---|---|
| 1 | $230,000 | $300,000 |
| 2 | 185,000 | 245,000 |
| 3 | 145,000 | 200,000 |
| 4 | 110,000 | 160,000 |
| 5 | 80,000 | 125,000 |
| 6 | 55,000 | 95,000 |
| 7 | 35,000 | 70,000 |
| 8 | 20,000 | 50,000 |
| 9 | 10,000 | 35,000 |
| 10 | 5,000 | 25,000 |
| 11 | — | 20,000 |
| 12 | — | 20,000 |

**7.** A machine can be purchased at $t = 0$ for $20,000. The estimated life is 5 years, with an estimated salvage value of zero at that time. The average annual operating and maintenance expenses are expected to be $4000. If $MARR = 20\%$, what must the average annual revenues be in order to be indifferent between (a) purchasing the machine, or (b) "do nothing"? (8.2)

**8.** The motor on a gas-fired furnace in a small foundry is to be replaced. Three different 15-horsepower electric motors are being considered. Motor X sells for $1500 and has an efficiency rating of 90%; motor Y sells for $1000 and has a rating of 85%; motor Z sells for $750 and is rated to be 80% efficient. The cost of electricity is $0.05/kilowatt-hour. An 8-year planning horizon is used, and zero salvage values are assumed for all three motors. A $MARR$ of 25% is to be used. Assume that the motor selected will be loaded to capacity. Determine the range of values for annual usage of the motor (in hours) that will lead to the preference of each motor. (*Note.* 0.746 kilowatts = 1 horsepower.) (8.2)

**9.** Two condensers are being considered by the Ajax Company. A copper condenser can be purchased for $3000; annual operating and maintenance costs are estimated to be $500. Alternatively, a ferrous condenser can be purchased for $2500; since the Ajax Company has not had previous experience with ferrous condensers, they are not sure what annual operating and maintenance cost estimate is appropriate. A 5-year planning horizon is to be used, salvage values are estimated to be 25% of the original purchase price, and a $MARR$ of 20% is to be used. Determine the break-even value for the annual operating and maintenance cost for the ferrous condenser. (8.2)

**10.** A firm has decided to manufacture widgets. There are two production processes available for consideration. Process A involves the purchase of a $22,000 machine that will last for 10 years and have a $2000 salvage value at that time. Annual operating and maintenance costs amount to $15,000/year for the machine. In addition, using Process A requires additional costs amounting to $0.15/widget produced. The second process, Process B, requires an investment of $15,000 in a machine that will last for 10 years and have a $2000 salvage value at that time. Annual operating and maintenance costs amount to $13,000/year for the machine. Additional costs of $0.20/widget produced result when Process B is used.
With a 15% interest rate, for what annual production volume is Process A preferred? (8.2)

**11.** Two manufacturing methods are being considered. Method A has a fixed cost of $5000 and a variable cost of $50. Method B has a fixed cost of $2500 and a variable cost of $150. For what production volume would one prefer (a) Method A, and (b) Method B? (8.2)

**12.** The ABC Company is faced with three proposed methods for making one of their products. Method A involves the purchase of a machine for $25,000. It will have a 7-year life, with a zero salvage value at that time. Using Method A involves additional costs of $0.20/unit of product produced per year. Method B involves the purchase of a machine for $35,000. It will have a $2000 salvage value when disposed of in 7 years. Using Method B involves additional costs of $0.15/unit of product produced per year. Method C involves buying a machine for $30,000. It will have a $2000 salvage value when disposed of in 7 years. Additional costs of $0.25/unit of product per year arise when Method C is used. A 25% interest rate is used by the ABC Company in evaluating investment alternatives. For what range of values for annual production volume is each method preferred? (8.2)

13. Consider the following annual revenue and cost functions for a firm.

$$AR(x) = \$2250x - \$0.075x^2$$

$$AC(x) = \$1,500,000 + \$1500x - \$0.025x^2$$

Where $x$ is the annual production volume. Determine the break-even production volume(s). (8.2)

14. In performing a make-or-buy analysis the following cost functions were obtained

$$M(x) = \$2500 + \$125x - \$0.025x^2$$

$$B_1(x) = \$5000 + \$75x$$

For what range of production volumes $(x)$ should the component be purchased rather than made? (8.2)

15. In problem 14 suppose the firm is approached by a second supplier with the following quotation,

$$B_2(x) = \$2000 + \$100x + 0.005x^2$$

What would be your recommendation to the firm for various values of production volumes? (8.2)

16. In problem 4 suppose the following pessimistic, most likely, and optimistic estimates are given for building cost, furnishings cost, annual operating and maintenance costs and the average rate per occupied unit.

| | Pessimistic | Most Likely | Optimistic |
|---|---|---|---|
| Building cost | $7,500,00 | $5,000,000 | $4,000,000 |
| Furnishings cost | 3,000,000 | 1,875,000 | 1,000,000 |
| Annual operating and maintenance costs | 200,000 | 125,000 | 75,000 |
| Average rate | 25/day | 45/day | 55/day |

Determine the pessimistic and optimistic limits on the break-even value for the daily occupancy percentage based on a MARR of 20%. Assume the motel will operate 365 days/year. (8.3)

17. In problem 6 suppose a 10 year planning horizon is specified. Perform a sensitivity analysis comparable to that given in Figure 8.3 to determine the effect of errors in estimating the *differences* in initial investment, the *differences* in annual operating and maintenance costs, and the *differences* in salvage values for the two investment alternatives. (8.3)

18. A warehouse modernization plan requires an investment of $3 million in equipment; at the end of the 10-year planning horizon, it is anticipated the equipment will have a

salvage value of $600,000. Annual savings in operating and maintenance costs due to the modernization are anticipated to total $1,400,000/year. A *MARR* of 20% is used by the firm. Perform a sensitivity analysis to determine the effects on the economic feasibility of the plan due to errors in estimating the initial investment required, and the annual savings. (8.3)

19. An investment of $10,000 is to be made into a savings account. The interest rate to be paid each year is uncertain; however, it is estimated that it is twice as likely to be 7% as it is to be 6% and it is equally likely to be either 6% or 8%. Determine the probability distribution for the amount in the fund after 3 years, assuming the interest rate is not autocorrelated. (7.4)

20. Given an initial investment of $10,000, annual receipts of $3500, and an uncertain life for the investment. Use a 25% *MARR*. Let the probability distribution for the life of the investment be given as follows:

| N | p(N) |
|---|------|
| 1 | 0.10 |
| 2 | 0.15 |
| 3 | 0.20 |
| 4 | 0.25 |
| 5 | 0.15 |
| 6 | 0.10 |
| 7 | 0.05 |

   (a) Perform 20 simulations and estimate the probability of the investment being profitable.
   (b) Analytically determine the probability of the investment being profitable. (8.4)

21. In problem 20 suppose the *MARR* is not known with certainty and the following probability distribution is anticipated to hold.

| i | p(i) |
|------|------|
| 0.20 | 0.20 |
| 0.25 | 0.60 |
| 0.30 | 0.20 |

   Use simulation and analytical methods to estimate or determine the probability of the investment being profitable. (8.4)

22. In problem 20 suppose the magnitude of the annual receipts (R) is subject to random variation. Assume that each annual receipt will be identical in value and the annual receipt has the following probability distribution:

| R | p(R) |
|--------|------|
| $3,000 | 0.20 |
| 3,500 | 0.60 |
| 4,000 | 0.20 |

Use simulation and analytical methods to estimate or determine the probability of the investment being profitable. (8.4)

23. In problem 20 suppose the minimum attractive rate of return is distributed as given in problem 21 and suppose the annual receipts are distributed as given in problem 22. Analytically determine the probability of the investment being profitable.

24. In problem 23 suppose the initial investment is equally likely to be either $9,000 or $11,000. Analytically determine the probability of the investment being profitable. (8.4)

25. Suppose $n = 4$, $i = 0\%$, a $11,000 investment is made, and the receipt in year $j$, $j = 1, \ldots, 4$, is statistically independent and distributed as in problem 22. Determine the probability distribution for present worth. (8.4)

26. Recall Example 8.14. Continue the simulation for 10 additional trials and compute the cumulative average present worth. (8.4)

27. In Example 8.14, suppose the individual had used a minimum attractive rate of return of 10% in the calculations. Would the first 10 simulation trials have yielded a positive average present worth? (8.4)

28. Recall Example 8.13. Continue the simulation for 10 additional investment trials and estimate the probability of a negative present worth and compute the average present worth over the total of 30 trials. (8.4)

29. Company W is considering investing $10,000 in a machine. The machine will last $N$ years at which time it will be sold for $L$. Maintenance costs for this machine are estimated to increase by 10%/year over its life. The maintenance cost for the first year is estimated to be $1000. The company has a 20% *MARR*. Based on the probability distributions given below for $N$ and $L$, what is the expected equivalent uniform annual cost for the machine? (8.4)

| N | L | p(N) |
|---|---|---|
| 6 | $3,000 | 0.2 |
| 8 | 2,000 | 0.4 |
| 10 | 1,000 | 0.4 |

30. Consider an investment alternative having a 6-year planning horizon and expected values and variances for statistically independent cash flows as given below:

| j | E(C_j) | V(C_j) |
|---|---|---|
| 0 | -$25,000 | $625 \times 10^4$ |
| 1 | 4,000 | $16 \times 10^4$ |
| 2 | 5,000 | $25 \times 10^4$ |
| 3 | 6,000 | $36 \times 10^4$ |
| 4 | 7,000 | $49 \times 10^4$ |
| 5 | 8,000 | $64 \times 10^4$ |
| 6 | 9,000 | $81 \times 10^4$ |

Using a discount rate of 30%, determine the expected values and variances for both present worth and annual worth. Based on the central limit theorem, compute the probability of a positive present worth; compute the probability of a positive annual worth. (8.4)

31. Solve problem 30 using a discount rate of (a) 0%, (b) 15%. (8.4)

32. Solve problem 30 when the $C_j$, $j = 1, \ldots, 6$, are perfectly correlated. (8.4)

33. Two investment alternatives are being considered. Alternative A requires an initial investment of $13,000 in equipment; annual operating and maintenance costs are anticipated to be normally distributed, with a mean of $5000 and a standard deviation of $500; the terminal salvage value at the end of the 8-year planning horizon is anticipated to be normally distributed with a mean of $2000 and a standard deviation of $800. Alternative B requires end of year annual expenditures over the planning horizon, with the annual expenditure being normally distributed with a mean of $7500 and a standard deviation of $750. Using a *MARR* of 25%, what is the probability that Alternate A is the most economic alternative? (8.4)

34. In problem 33, suppose the *MARR* were 20% with probability 0.25, 25% with probability 0.50, and 30% with probability 0.25, what is the probability that Alternative A is the most economic alternative? (8.4)

35. Two investment alternatives are under consideration. The data for the alternatives are given below. It is assumed the cash flow are not autocorrelated.

| EOY | E[CF(A)] | SD[CF(A)] | E[CF(B)] | SD[CF(B)] |
|-----|----------|-----------|----------|-----------|
| 0 | -$10,000 | $1,000 | -$15,000 | $1,500 |
| 1 | 4,000 | 400 | 8,000 | 800 |
| 2 | 5,000 | 500 | 8,000 | 800 |
| 3 | 6,000 | 600 | 8,000 | 800 |
| 4 | 7,000 | 700 | 8,000 | 800 |
| 5 | 8,000 | 800 | 8,000 | 800 |
| 6 | 9,000 | 900 | 8,000 | 800 |

(a) For each alternative, determine the mean and standard deviation for present worth and annual worth using a *MARR* of 20%.
(b) For each alternative, based on the central limit theorem, compute the probability of a positive present worth using a *MARR* of 20%.
(c) Develop the mean and standard deviation for the *incremental* present worth using a *MARR* of 20%.
(d) Based on the central limit theorem, compute the probability of a positive incremental present worth using a *MARR* of 20%. (8.4)

## CHAPTER NINE
# DECISION MODELS

## 9.1 INTRODUCTION

This chapter presents basic models for displaying decision situations involving risk and uncertainty and suggests criteria for choosing among alternatives. As such, these models will be *prescriptive* or *normative* in that they prescribe a decision action to be taken. The risk-analysis discussion and models presented in Chapter Eight were descriptive; they described the behavior of the measure of effectiveness under conditions of uncertainty and risk. This chapter is concerned with specifying the preferred choice from a set of feasible alternatives and represents the final step in the six step procedure for comparing alternatives presented in Chapter Five.

The first section of this chapter discusses a matrix model for decisions under risk and uncertainty and suggests criteria for choosing among alternatives. The second section presents a *network* or *tree model* for decision situations involving a sequence of decisions over a time horizon where the outcomes of alternatives are uncertain. The final section of the chapter offers an introductory discussion on methodology for choosing among alternatives when multiple measures of effectiveness are involved. The matrix model for decisions under risk and uncertainty is now introduced by an example concerning material handling.

## Example 9.1

For purposes of introducing a matrix model for decision making, consider the following manufacturing situation. A new product group is to be added by the ABC Company with initial production scheduled 2 years hence. The new product group, which will consist of a single product in varying sizes and finishes, requires that a new and separate facility be added to the existing plant. The company is presently completing plans for the new facility, and plant construction will begin in the near future. Excess floor space is planned in order to increase production if demand increases. This increase is expected, with demand reaching stability after 5 years of production.

The manufacturing system consists, essentially, of several machining centers that operate on a job-shop basis instead of on a production-line basis. The manufacturing sequence is fixed except for slight variations, depending on product size and finish required. Thus, a subproblem of the total manufacturing system is to determine the material handling requirements. After considerable study and discussion, it has been decided that industrial lift trucks provide the flexibility of handling required in the job-shop environment. Now, questions arise as to the type of lift truck, the number required, the attachments needed, whether they should be leased or purchased and, if leased, which of several leasing plans should be chosen, and so forth. The answer to these and other relevant questions depends primarily on the demand for the new product during the first 5-year production period. This demand is uncertain, and ABC Company management can influence demand through advertising only to a limited degree.

Management could make a decision on material handling requirements based on the demand forecast for the first year of production, then evaluate the situation at the end of the first year, make a new decision for the next year, reevaluate, and so on, in sequential fashion. However, let us assume that it has been decided to choose a material handling system now for the first 3-year production period and then reevaluate at the end of that period.

Given the above managerial decision, the analysts investigating the material handling requirements make the judgment that demand for the product group during the 3-year planning horizon will be such that either four, five, or six lift trucks are required. They further reason that the feasible alternatives are:

$A_1$, $A_2$, $A_3$—lease four, five, or six trucks, respectively

$A_4$, $A_5$, $A_6$—purchase four, five, or six trucks, respectively

These lease plans do not include a maintenance contract, since the ABC Company presently employs skilled maintenance personnel. Furthermore, the analysts have not considered any lease plan with an "option to purchase" clause in order to avoid "contractual sales" income tax complexities.

The analysts select before-tax, equivalent uniform annual cost as the single measure of effectiveness for a given "$A_j$ versus number of trucks required" combination. It is necessary to make gross estimates of these annual costs, but essentially, for leasing, there is a fixed monthly charge per truck plus a judgmental penalty cost associated with production delays when the number of trucks required exceeds the number of trucks available. On the other hand, the company can obtain a more favorable leasing charge as the number of trucks leased increases. For the purchase alternatives, similar fixed (actually fixed and variable) annual operating and maintenance costs, penalty costs, and quantity discounts are involved. After much study, somewhat gross estimations, and a liberal application of judgment, the analysts present the model of Table 9.1 to management for a final decision.

On the basis of the single, equivalent uniform annual cost measure of effectiveness, any *purchase* alternative is preferable to any *lease* alternative. For example, the purchase alternative $A_4$ has a lower equivalent annual cost than any lease alternative whether four, five, or six trucks are required. Thus, Alternative $A_4$ dominates the $A_1$, $A_2$, and $A_3$ alternatives, and they may therefore be deleted from the decision matrix. (It is noted that this might not have been the case if after-tax annual cost values had been calculated. Furthermore, a consideration of intangible factors by management could result in a lease alternative being chosen over a purchase alternative even though annual costs are higher.) The purchase alternatives $A_5$ and $A_6$ also dominate the three lease alternatives, but no one purchase alternative dominates any other purchase alternative. That is, no single purchase alternative has lower annual costs than any other purchase alternative when viewed across all "number of trucks required" conditions, anyone of which it is assumed can occur. Dominance will be discussed in greater detail subsequently.

A solution to the lift truck problem will not be presented here. The example has been cited to illustrate the matrix format of Table 9.1 and to suggest the model as an effective portrayal of many decision situations.

TABLE 9.1. Decision Model for the Lift Truck Example

| Number of Trucks Required<br>Alternatives | 4<br>Equivalent<br>Costs | 5<br>Uniform | 6<br>Annual |
|---|---|---|---|
| $A_1$—Lease four | $18,000 | $20,000 | $24,000 |
| $A_2$—Lease five | 20,000 | 20,000 | 22,000 |
| $A_3$—Lease six | 21,000 | 21,000 | 21,000 |
| $A_4$—Purchase four | 12,500 | 14,500 | 18,500 |
| $A_5$—Purchase five | 14,000 | 14,000 | 16,000 |
| $A_6$—Purchase six | 15,000 | 15,000 | 15,000 |

The elements in the decision model of Table 9.1 are now summarized as:

**States** = the number of lift trucks required in the 3-year planning horizon.

**Feasible alternatives** = the number of lift trucks available, whether leased or purchased.

**Outcomes** = the results of particular "state versus alternative" combinations. In this case, the general outcomes are that the number of trucks required is less than, equal to, or greater than the number of trucks available.

**Value of the outcomes** = the measure of effectiveness for a particular "state versus alternative" combination. In this case, equivalent uniform annual costs.

A matrix decision model with general symbolism, adapted from Morris [15], is given in Figure 9.1 and further discussed in the next section.

## 9.2 THE MATRIX DECISION MODEL

The symbolism employed is defined as follows:

$A_j$ = an *alternative* or *strategy* under the decision maker's control, where $j = 1, 2, \ldots , n$.

| $A_j$ $\diagdown$ $S_k$ | $p_1$ $S_1$ | $p_2$ $S_2$ | $\cdots$ | $\cdots$ | $p_k$ $S_k$ | $\cdots$ | $\cdots$ | $p_m$ $S_m$ |
|---|---|---|---|---|---|---|---|---|
| $A_1$ | $V(\theta_{11})$ | $V(\theta_{12})$ | $\cdots$ | $\cdots$ | $V(\theta_{1k})$ | $\cdots$ | $\cdots$ | $V(\theta_{1m})$ |
| $A_2$ | $V(\theta_{21})$ | $V(\theta_{22})$ | $\cdots$ | $\cdots$ | $V(\theta_{2k})$ | $\cdots$ | $\cdots$ | $V(\theta_{2m})$ |
| $\vdots$ | $\vdots$ | $\vdots$ | | | $\vdots$ | | | $\vdots$ |
| $\vdots$ | $\vdots$ | $\vdots$ | | | $\vdots$ | | | $\vdots$ |
| $A_j$ | $V(\theta_{j1})$ | $V(\theta_{j2})$ | $\cdots$ | $\cdots$ | $V(\theta_{jk})$ | $\cdots$ | $\cdots$ | $V(\theta_{jm})$ |
| $\vdots$ | $\vdots$ | $\vdots$ | | | $\vdots$ | | | $\vdots$ |
| $\vdots$ | $\vdots$ | $\vdots$ | | | $\vdots$ | | | $\vdots$ |
| $A_n$ | $V(\theta_{n1})$ | $V(\theta_{n2})$ | $\cdots$ | $\cdots$ | $V(\theta_{nk})$ | $\cdots$ | $\cdots$ | $V(\theta_{nm})$ |

FIGURE 9.1. A matrix decision model with general symbolism.

$S_k$ = a *state* or *possible future* that can occur given that alternative $A_j$ is chosen, where $k = 1, 2, \ldots, m$.

$\theta_{jk}$ = the *outcome* of choosing alternative $A_j$ and having state $S_k$ occur.

$V(\theta_{jk})$ = the *value* of outcome $\theta_{jk}$, which may be in terms of dollars, time, distance, or utility.

$p_k$ = the *probability* that state $S_k$ will occur.[1]

The term $p_k$ was not previously introduced in the lift truck decision problem but will subsequently serve as a means by which decisions may be categorized into decisions under assumed certainty, decisions under risk, and decisions under uncertainty. This categorization has essentially been made in Chapter Eight but will be repeated in order to present decision rules for the normative models of this chapter.

The matrix model, in general, describes a set of alternatives available where a single alternative is to be selected at the present time. It is implied that the outcomes possible for a given alternative do not result in the necessity for subsequent decisions at future times. If a sequence of decisions were involved, the decision tree model presented later would probably be more appropriate.

It is also assumed in the matrix model that the alternatives are mutually exclusive, feasible, and represent the total set of alternatives the decision maker may wish to consider. Recall from the discussion in Chapter Five that mutually exclusive alternatives cannot occur together, and thus a single alternative will be chosen from the feasible set. The question now arises, "What constitutes feasibility?" and the answer to the question lies in the original statement of objectives that the alternative is to accomplish. For example, if one were considering the replacement of a production machine because of increasing maintenance costs and inadequate capacity, an objective might be that the new machine must be capable of producing 1000 units of Product A per day. Those machines incapable of satisfying this constraint of 1000 units/day are infeasible and rejected from further consideration. Those machines capable of meeting or exceeding the requirement of 1000 units/day are feasible, and the "best" replacement machine is then selected from the feasible set according to a more discriminating criterion, such as minimum estimated equivalent annual cost or maximum estimated annual profit.

Determining the "total set" of feasible alternatives is also a difficulty deserving additional comment. Rarely does the analyst have the time or even the insight to define all the possible alternatives available in a given decision situation. How to execute a search procedure and when to terminate the search are interesting and important questions, but will not be pursued here. Suffice it to say that experience plays a major role in the search procedure, and when to terminate the search is primarily a matter of judgment in the practical world of business decisions. The decision to terminate the search for alternatives is a trade-off of the cost of finding an additional alternative versus the benefits

---

[1] It is assumed that the probability of a particular state occurring does not depend on the alternative chosen by the decision maker.

resulting from the discovery of the additional alternative. Admittedly, the benefits cannot actually be evaluated until after the alternative is discovered and thus, this guideline for terminating search is largely conceptual in nature.

Experience and judgment also play an important role in defining the relevant states for a given decision situation and the set of feasible alternatives involved. To convey further the meaning of a "state," other terms in the literature are *possible futures, states of nature, external conditions,* and *future events.*

In the discussion on risk analysis in Chapter Eight, states were values for the random variables of purchase price, annual operating expenses, annual revenues, service lives, and salvage values. However, in other decision situations, the states may be defined more grossly; for instance, the market demand for product A will be low, medium, or high. Furthermore, the number of states may be finite or infinite. For the examples of this text, the number of states will be assumed finite and few in number. This seems a realistic assumption for most business and engineering decisions because, although the states may actually be a very large finite number or even infinite, the decision maker wishes to limit the scope of the problem and uses a sample of states. Using the industrial lift truck example presented earlier, knowledge of the machining processes involved and a recorded history of machining times can provide the basis for estimating values for the states in terms of the number of trucks required to service different levels of production output. This same experience can also provide the range of state values to consider. In any case, it will be assumed here that once the states are defined, they are collectively exhaustive (i.e., the total set of possible states). Furthermore,

1. the occurrence of one state precludes the occurrence of the others (i.e., they are mutually exclusive),
2. the occurrence of a state is not influenced by the alternative selected by the decision maker, and
3. the occurrence of a state is not known with certainty by the decision maker.

For each alternative-versus-state combination, there is an outcome and a value associated with the outcome. In the previous industrial lift truck example, a given "number of trucks available versus number of trucks required" combination resulted in the trucks available being short, over, or equal to the trucks required. For each outcome, it was assumed that an annual equivalent dollar cost could be calculated and was the preferred decision criterion. Other value measurements may, of course, be appropriate in other decision situations. The matrix values may be one of two general kinds—objective or subjective. Objective values represent physical quantities such as dollars, hours, liters, kilograms, and so forth. With the exception of the section on multiple goals in this chapter, the subsequent examples are concerned with dollars or objective values.[2] Subjective values, on the other hand, are numbers that repre-

---

[2] A linear utility function for the decision maker is also assumed. That is, minimizing costs or maximizing profits are both the same as maximizing the decision maker's utility.

sent the decision maker's relative preferences; for example, a score of 90 on an examination is preferred to a score of 60. The values of 90 and 60 are subjective values assigned on an arbitrary scale of, say, 0 to 100. An example of a subjective value matrix will be given in the multiple-goals section of this chapter. In any case, as a general statement, the problems of measurement and value determination are often very substantial in real-world situations.

Let the matrix model of Figure 9.1 now be recalled in order to classify decisions on the basis of the amount of information available to the decision maker in regard to the probability of occurrence for a given state.

## 9.3 DECISIONS UNDER ASSUMED CERTAINTY

It is reasonable to assume in many decision situations that only one state is relevant. Then, the decision maker assumes this single state will occur with certainty (i.e., with probability $= 1.0$). This kind of case is termed a *decision under assumed certainty*. For example, it may be a matter of company policy to depreciate new equipment purchases fully in 5 years. If, in an economic analysis of several new equipment candidates, the measure of effectiveness is equivalent annual cost, the determination of these annual costs would be based on a 5-year life for each candidate. The outcome of a 5-year life and the associated annual operating costs for each candidate are assumed certain. The actual functional life of an equipment candidate might be 4, 5, 6, 7, 8, . . . , years. However, a judgment is made by the decision maker to suppress uncertainty because of convenience or practicality and assume only the single state of a 5-year life. All of the analysis in Chapters Three to Seven was based on such judgments, and single-value estimates for operating costs, service life, and the like were examples of assumed certainty.

In terms of the matrix decision model, a decision under assumed certainty would appear as follows:

|         | $S$              |
|---------|------------------|
| $A_1$   | $V(\theta_1)$    |
| $A_2$   | $V(\theta_2)$    |
| $\vdots$ | $\vdots$        |
| $A_j$   | $V(\theta_j)$    |
| $\vdots$ | $\vdots$        |
| $A_{n-1}$ | $V(\theta_{n-1})$ |
| $A_n$   | $V(\theta_n)$    |

If, in the above model, the values for each alternative were profits (or gains), the principle for choice might be "choose the alternative that maximizes profit." Conversely, if the values were costs (or losses), one might "choose the alternative that minimizes costs."

Example 9.2_____

Suppose that an investor is considering a $10,000 purchase of United States government securities. Different government agencies offer such securities, and they are available on the open market. Assuming that the various securities have common maturity dates, the investor judges that each agency is financially secure and payoffs are certain. A logical criterion for choice among the securities would be the effective annual yield on the $10,000 purchase. (It is assumed there is no reason for allocating the $10,000 to different securities.) Suppose the investor considers five government securities and, for a maturity date of 5 years hence, has calculated the effective annual yield on each security to be

$$
\begin{array}{ll}
A_1 & 8.0\% \\
A_2 & 7.3\% \\
A_3 & 8.7\% \\
A_4 & 6.0\% \\
A_5 & 6.5\%
\end{array}
$$

Since the investor would logically desire to maximize the yield on the investment, $A_3$ would be chosen. Recall that equal risk and maturity dates were assumed for each security.

Example 9.3_____

The cash flow profiles for three mutually exclusive investment alternatives are given in Table 9.2. Considering the minimum attractive rate of return of 10% as

TABLE 9.2. Cash Flow Profiles for Three Mutually Exclusive Investment Alternatives ($MARR$ = 10%)

| $t$ | $A_1$ | $A_2$ | $A_3$ |
|---|---|---|---|
| 0 | $ 0 | -$10,000 | -$15,000 |
| 1 | 4,000 | 7,000 | 7,000 |
| 2 | 4,000 | 7,000 | 8,000 |
| 3 | 4,000 | 7,000 | 9,000 |
| 4 | 4,000 | 7,000 | 10,000 |
| 5 | 4,000 | 7,000 | 11,000 |
| Present worth | $15,163.20 | $16,535.60 | $18,397.33 |

TABLE 9.3.
Present Worths of
Three Mutually
Exclusive
Investment
Alternatives
($MARR = 10\%$)

| $A_j$ | $S_1$ |
|---|---|
|  | $i = 10\%$ |
| $A_1$ | $15,163.20 |
| $A_2$ | 16,535.60 |
| $A_3$ | 18,397.33 |

the single state that will occur with certainty and the present worth of each alternative as the value of interest, the matrix model of this decision appears in Table 9.3. If the principle for choice is "select the alternative that maximizes the present worth," $A_3$ should be chosen.

## 9.4 DECISIONS UNDER RISK

A decision situation is called a *decision under risk* when the decision maker elects to consider several states and, the probabilities of their occurrence are explicitly stated. In some decision problems, the probability values may be objectively known from historical records or objectively determined from analytical calculations. For example, if an unbiased die is rolled once, the six possible states are that one, two, three, four, five, or six dots will occur on the top face exposed. The objective probability that each state will occur prior to the roll is 1/6. The sum of the individual probabilities over all states equals the value 1.0. This property of $\Sigma p_j = 1.0$ agrees with an earlier statement that, once the set of possible states is decided on, the states are treated as if they constituted a set of mutually exclusive and collectively exhaustive events. Indeed, the property of $\Sigma p_j = 1.0$ is a necessary restriction if probability theory is to be used as a guide to decision making.

The decision maker may not have past records available to arrive at objective probability values. This was the case in the industrial truck example earlier. However, if the decision maker feels that experience and judgment are sufficient to assign probability values subjectively to the occurrence of each state, the decision is still treated as one under risk. The restriction of $\Sigma p_k = 1.0$ still applies in this case. Principles of choice for selecting an alternative in a decision under risk, which commonly appear in the literature on decision theory, are illustrated in Example 9.4.

# Example 9.4

The three mutually exclusive investment alternatives presented in Table 9.2 will again be used to illustrate guidelines for a decision under risk. The single state previously defined was a minimum attractive rate of return equal to 10%. For the purpose of this example, the two additional states of $MARR = 20\%$ and $MARR = 30\%$ will be assumed. The realism of considering these states as probabilistic in nature may be questionable, but an argument supporting this claim can be posed. In periods of an unstable economy, the analyst may perceive the $MARR$ as uncertain and be willing to define different values with associated probabilities. The matrix model of Table 9.4 assumes that this is the case, and cell values are present worths (rounded) for a 5-year planning horizon.

The states of $S_1 = 10\%$, $S_2 = 20\%$, and $S_3 = 30\%$ may have been defined by judging these values to be optimistic, most likely, and pessimistic estimates, as discussed in Chapter Eight. A $MARR$ that lowers the present worth of the investment alternative is considered a pessimistic $MARR$. The chance of occurrence for the states may have been reasoned as follows: $S_3$ is twice as likely to occur as $S_2$, which, in turn is three times as likely to occur as $S_1$. This reasoning can be stated as the equation $p_1 + 3p_1 + 6p_1 = 1.0$, and $p_1 = 0.10$, $p_2 = 0.30$, and $p_3 = 0.60$ determined.

## 9.4.1 Dominance

The *dominance* principle is described as follows. Given two alternatives, if one would always be preferred no matter which state occurs, this preferred alternative is said to dominate the other, and the dominated alternative can be deleted from further consideration. Applying this principle will not necessarily solve the decision problem by selecting a unique alternative, but it may reduce the number of alternatives for further consideration.

The dominance principle is a way of deciding which alternatives not to select. If the values are in terms of *costs,* a more formal statement of the domi-

TABLE 9.4. Matrix Model with States of $MARR = 10\%$, 20%, and 30%

| $A_j$ | MARR | $p_1 = 0.10$ $S_1$ 10% | $p_2 = 0.30$ $S_2$ 20% | $p_3 = 0.60$ $S_3$ 30% |
|---|---|---|---|---|
| $A_1$ | | $15,163 | $11,962 | $9,742 |
| $A_2$ | | $16,536 | $10.934 | $7,049 |
| $A_3$ | | $18,397 | $10,840 | $5,679 |

398 • DECISION MODELS

nance relation is: *If there exists a pair of alternatives $A_j$ and $A_l$ such that V* $(\theta_{jk}) \leq V(\theta_{lk})$ *for all k, $A_j$ is said to dominate $A_l$. Alternative $A_l$ may then be discarded from the decision problem.* If the values of a decision matrix are in terms of *gains*, the condition for $A_j$ to dominate $A_l$ is that $V(\theta_{jk}) \geq V(\theta_{lk})$ for all $k$.

In Table 9.4, no alternative has a present worth value greater than the present worth value of another alternative for all states. Thus, there is no dominant alternative for this example, but recall that there were dominated alternatives in Table 9.1.

## 9.4.2 Expectation-Variance Principle

If $X$ is a discrete random variable that is defined for a finite number of values and $p(x)$ denotes the probability of a particular value occurring, then the expected or mean value of the random variable is defined as

$$E(X) = \sum_{\text{all } x} xp(x) \tag{9.1}$$

The variance of the random variable is defined as $Var(X)$, where

$$Var(X) = \sum_{\text{all } x} [x - E(X)]^2 p(x) \tag{9.2}$$

For the matrix of Figure 9.1, the random variable is $V(\theta_{jk})$, and the expected value of an alternative $A_j$ is

$$E(A_j) = \sum_{\text{all } k} V(\theta_{jk}) p_k \tag{9.3}$$

where $p_k$ denotes the probability of state $k$ occurring.

The expected values for each of the alternatives of Table 9.4 are calculated as

$$E(A_1) = \$15,163(0.10) + \$11,962(0.30) + \$9742(0.60)$$
$$= \$10,950.10$$

$$E(A_2) = \$16,536(0.10) + \$10,934(0.30) + \$7049(0.60)$$
$$= \$9163.20$$

$$E(A_3) = \$18,397(0.10) + \$10,840(0.30) + \$5679(0.60)$$
$$= \$8499.10$$

If the principle of maximizing (minimizing) the expected gain (loss) is followed, alternative $A_1$ would be chosen, since the expected values are positive present worths. A corollary of the expected-value criterion is that, in the event of a tie in expected value for two or more alternatives, the alternative having minimum variance should be chosen. The argument for this secondary criterion is, essentially, that a larger variance is indicative of greater uncertainty, and since most decision makers have an aversion to uncertainty, the alternative with smallest variance should be chosen. Conrath [4], in an article on the implementation of statistical decision theory in practical decisions under risk, states that ". . . as yet we have little guidance as to what the trade-offs between mean and variance might be or ought to be." That the expected value and the variance are two independent decision criteria instead of dependent criteria is an interesting speculation (or perhaps observation) that deserves further research.

### 9.4.3 Most Probable Future Principle

If, in a decision under risk, one state has a probability of occurrence considerably greater than any other, the *most probable future principle* is to consider this state as certain and all other states as having a zero chance of occurrence. The decision is thereby reduced to a decision under assumed certainty. Then an alternative is chosen that maximizes (or minimizes) the measure of effectiveness being used. For Table 9.4, $S_3$ has a probability of 0.60 assigned and is therefore the most probable state. Assuming certainty for $S_3$ or $p_3 = 1.0$, the reduced matrix is

$$\begin{array}{cc} & S_3 \\ A_1 & \$9742 \\ A_2 & \$7049 \\ A_3 & \$5679 \end{array}$$

To obtain the maximum present worth, $A_1$ is selected as the preferred alternative.

Another example of the most probable future principle is the case of a person commuting to work. Each morning that the person leaves for work there is a nonzero probability that an accident could occur and serious injury result. However, the most probable future is that the commuter will reach work safely. The commuter considers this future certain, or else he (she) would probably not go to work!

As a final remark on this principle of choice, it seems appealing only if one state has a significantly higher probability of occurring than any other and the values in the total matrix do not differ significantly. One would not wish to suppress the consideration of a state that, albeit with low probability, would inflict a severe loss on the decision maker. An example of this is the matrix below where the positive values are profits.

|       | $p_1 = 0.95$ | $p_2 = 0.05$ |
|-------|--------------|--------------|
| $A_1$ | $10          | $40          |
| $A_2$ | $50          | -$5,000      |

Following the most probable future principle, $A_2$ would be chosen. However, the suppression of $S_2$ does not alter the fact that $S_2$ could occur, by virtue of $p_2 = 0.05$; in that event the decision maker would suffer a loss of $5000.

### 9.4.4 Aspiration-Level Principle

In most real-world decisions, the complexity of the decision prevents the discovery and selection of an alternative that will yield the single "best" result, and decision makers set their goals in terms of outcomes that are good enough. That is, decision makers set aspiration levels and then evaluate alternatives against them. An interpretation of this philosophy in terms of a decision under risk is to select an alternative that maximizes the probability of achieving the desired aspiration level. Typical aspiration-level objectives might be to choose the alternative that maximizes the probability of either (1) a 25% internal rate of return, (2) annual costs less than $8000, or (3) annual profits greater than $9500. In terms of the example of Table 9.4, let the objective be to choose the alternative that maximizes the probability that the present worth value will be equal to or greater than $8000. This objective can be symbolized as $P(PW \geq \$8000)$. From Table 9.4 it can be seen that, for $A_1$, the present worth will be greater than $8000 when $S_1$, $S_2$, or $S_3$ occurs. Thus, a present worth greater than $8000 is certain if $A_1$ is chosen. Because the states are mutually exclusive, the additive-probability law permits the sum of the individual state probabilities to be the probability of the joint event, "$PW \geq \$8000$" for $A_1$. The calculations for each alternative are

$$\text{For } A_1, \ P(PW \geq \$8000) = P(S_1) + P(S_2) + P(S_3)$$
$$= 0.10 + 0.30 + 0.60 = 1.0$$

$$\text{For } A_2, \ P(PW \geq \$8000) = P(S_1) + P(S_2)$$
$$= 0.10 + 0.30 = 0.40$$

$$\text{For } A_3, \ P(PW \geq \$8000) = P(S_1) + P(S_2)$$
$$= 0.10 + 0.30 = 0.40$$

and $A_1$ maximizes $P(PW \geq \$8000)$.

A variety of aspiration level objectives are, of course, possible for even a *given* decision under risk. The relevant aspiration-level objective is a decision maker's preference.

Do not interpret any of the principles for choice above or those to follow in the next section as the way an alternative *should* be selected. Instead, the principles for choice presented result from observing how people *seem* to decide on alternatives; therefore, they are offered only as possible guidelines to the decision maker.

## 9.5 DECISIONS UNDER UNCERTAINTY

A decision situation where several states are possible and sufficient information is not available to assign probability values to their occurrence is termed a *decision under uncertainty.* It can be argued that any decision maker, if able to define the states, should have sufficient knowledge to make at least gross subjective probability estimates for these states. On the other hand, in situations such as (1) installing new safety devices on machinery where the states are the number of injuries expected per year, or (2) the introduction of a new product where the states are units demanded per year, one may feel quite inadequate to assign probabilities to these states. This may be the case even though the analyst is not totally lacking in knowledge concerning the number of injuries expected per year or the units of new product demanded per year. The principles for choice subsequently presented are based on the premise that probabilities cannot be assigned.

### 9.5.1 The Laplace Principle

The philosophy of the *Laplace Principle,* named after an early nineteenth-century mathematician [6], is simply that if one cannot assign probabilities to the states, the states should be considered as equally probable. Then, consider the decision as one under risk. Applying this principle to the previous example of Table 9.4, the probability value assigned to each of the three states is 1/3. Then,

$$E(A_1) = \$15{,}163(1/3) + \$11{,}962(1/3) + \$9742(1/3)$$
$$= \$12{,}289$$

$$E(A_2) = \$16{,}536(1/3) + \$10{,}934(1/3) + \$7049(1/3)$$
$$= \$11{,}506.33$$

$$E(A_3) = \$18{,}397(1/3) + \$10{,}840(1/3) + \$5679(1/3)$$
$$= \$11{,}638.67$$

and $A_1$ would be chosen to maximize expected present worth.

### 9.5.2 Maximin and Minimax Principles

The *maximin* and *minimax* principles for choice represent a single philosophy, depending on whether the matrix values are *gains* or *losses,* respectively. These principles hold considerable appeal for the conservative or pessimistic decision maker. In applying the principles, if the matrix values are *gains,* the minimum gain associated with each alternative (when viewed over all states) is determined, and the maximum value in the set of minimum values designates the alternative to be chosen. In other words, determine the worst value possible for *each* alternative and then choose the best value from the set of worst values. Applying the maximin principle to the example problem of Table 9.4 gives

| | Minimum Present Worth Value | Maximum of These |
|---|---|---|
| $A_1$ | $9742 | $9742 |
| $A_2$ | 7049 | |
| $A_3$ | 5679 | |

Thus, alternative $A_1$ is selected as the alternative that will maximize the minimum present worth value that could occur. More formally stated, the maximin principle is to

*Select the alternative, j, associated with the* $\underset{j}{max}\ \underset{k}{min}\ V(\theta_{jk})$.

In the case of a decision dealing with *losses* or *costs,* the minimax philosophy is also one of extreme pessimism. In applying the minimax principle to a matrix of costs, the maximum cost associated with each alternative (when viewed over all states) is determined, and the minimum value in the set of maximum values designates the alternative to be chosen. Again, this is choosing the best (or minimum) cost from the set of worst (or maximum) costs. An example to illustrate the application of the minimax principle is given in Table 9.5, where the cell values are equivalent uniform annual costs for the alternatives. Thus, from Table 9.5, alternative $A_1$ would be chosen. Formally stated, the minimax principle is to

*Select the alternative, j, associated with the* $\underset{j}{min}\ \underset{k}{max}\ V(\theta_{jk})$.

### 9.5.3 Maximax and Minimin Principles

As illustrated in the previous section, the maximin principle for gains and the minimax principle for losses represent a philosophy of extreme pessimism on behalf of the decision maker. For the person who is an extreme optimist in a given decision situation, an *optimistic* rule for choice among alternatives that involve *gains* is the *maximax principle.* That is, the decision maker desires to select the alternative that affords the opportunity to obtain the largest value given in the matrix. A formal statement for the maximax principle is to

*Select the alternative, j, associated with the* $\underset{j}{max}\ \underset{k}{max}\ V(\theta_{jk})$.

Then, if the matrix cell values were losses, an optimistic philosophy of choice is to select the alternative that affords the opportunity to obtain the minimum loss value given in the matrix. A formal statement for the *minimin principle* involving *loss* values is

*Select the alternative, j, associated with the* $\underset{j}{min}\ \underset{k}{min}\ V(\theta_{jk})$.

TABLE 9.5. Matrix of Equivalent Uniform Annual Costs

| Alternatives | States $S_1$ | $S_2$ | $S_3$ | $S_4$ | Maximum Cost | Minimum of These |
|---|---|---|---|---|---|---|
| $A_1$ | $7,000 | $10,000 | $ 6,500 | $ 5,000 | $10,000 | $10,000 |
| $A_2$ | 9,000 | 4,000 | 9,000 | 15,000 | 15,000 | |
| $A_3$ | 4,000 | 6,000 | 12,000 | 8,500 | 12,000 | |
| $A_4$ | 2,000 | 8,000 | 3,000 | 11,000 | 11,000 | |

Applying the maximax principle to the gains example of Table 9.4 selects alternative $A_3$, and applying the minimin principle to the costs example of Table 9.5 selects alternative $A_4$.

### 9.5.4 Hurwicz Principle

The *Hurwicz principle* considers that a decision maker's view may, in the case of gains, fall between the extreme pessimism of the maximin principle and the extreme optimism of the maximax principle and offers a method by which various levels of optimism–pessimism may be incorporated into the decision. The Hurwicz principle defines an index of optimism, $\alpha$, on a scale from 0 to 1. A value of $\alpha = 0$ thus indicates zero optimism or extreme pessimism and, conversely, a value of $\alpha = 1$ indicates extreme optimism.

Assuming that a decision maker is able to reflect a degree of optimism by assigning a specific value to $\alpha$, and again emphasizing that the decision is in terms of *gains,* the maximum gain value for each alternative is multiplied by $\alpha$ and the minimum gain value for each alternative is multiplied by $(1 - \alpha)$. The linear sum of these two products for each alternative $j$ is called the Hurwicz value, $H_j$, and the alternative that maximizes this value is selected. That is,

*Select an index of optimism, $\alpha$, such that $0 \le \alpha \le 1$. For each alternative j, compute $H_j = \alpha [max_k V(\theta_{jk})] + (1 - \alpha) [min_k V(\theta_{jk})]$ and select the alternative that maximizes this quantity.*

Note that if $\alpha = 0$ (extreme pessimism), only the minimum gain for each alternative will be examined, and the maximum of these values would then be selected—the maximin principle of choice. If $\alpha = 1$ (extreme optimism), the maximax principle would be executed.

Suppose the decision maker concerned with the example problem of Table 9.4 was a middle-of-the-road type of person and assigns $\alpha = 0.5$. Then

$$H_1 \text{ for } A_1 = (0.5)(\$15,163) + (0.5)(\$9742) = \$12,452.50$$

$$H_2 \text{ for } A_2 = (0.5)(\$16,536) + (0.5)(\$7049) = \$11,792.50$$

$$H_3 \text{ for } A_3 = (0.5)(\$18,397) + (0.5)(\$5679) = \$12,038.00$$

Choosing the maximum of these values is to select alternative $A_1$.

Shortcomings of the Hurwicz principle include (1) ignoring intermediate values for each alternative, (2) the inability to select a particular alternative when two or more alternatives have the same Hurwicz value, and (3) the practical difficulty of assigning a specific value to $\alpha$. This latter objection can be circumvented by the graphic solution discussed below.

It will be noted from the expression for calculating the Hurwicz value that the equation is linear in $\alpha$, which ranges in value from 0 to 1.0. Any alternative thus yields a linear function that can be plotted and a preferred alternative determined for various ranges of $\alpha$. An example to illustrate the procedure is given in Table 9.6, where the cell values are gains.

TABLE 9.6. Data for the
Hurwicz Example

|       | $S_1$ | $S_2$ | $S_3$ |
|-------|-------|-------|-------|
| $A_1$ | $ 2  | $10   | $30   |
| $A_2$ | 18    | 18    | 18    |
| $A_3$ | 24    | 12    | 10    |

Applying the principle for the data of Table 9.6, the Hurwicz values are

$$H_1 \text{ for } A_1 = \alpha(\$30) + (1 - \alpha)(\$2) = 28\alpha + 2$$
$$H_2 \text{ for } A_2 = \alpha(\$18) + (1 - \alpha)(\$18) = 18$$
$$H_3 \text{ for } A_3 = \alpha(\$24) + (1 - \alpha)(\$10) = 14\alpha + 10$$

These equations are plotted in Figure 9.2.

From Figure 9.2 it can be seen that if a decision maker's index of optimism is less than $\alpha_1$, alternative $A_2$ would be preferred; if it is greater than $\alpha_1$, alternative $A_1$ would be preferred. A specific value for $\alpha_1$ can be determined graphically or analytically by, in this case, equating the Hurwicz value functions for any pair of alternatives (since there is a single intersection for the three linear functions). For example,

$$H_1 = H_2$$
$$28\alpha_1 + 2 = 18 \quad \text{and} \quad \alpha_1 = 4/7$$

If the decision problem involves costs or losses, a reinterpretation of the Hurwicz principle is necessary:

*Select an index of optimism, $\alpha$, such that $0 \leq \alpha \leq 1$. For each alternative, $A_j$, compute $H_j = \alpha[\min_k V(\theta_{jk})] + (1 - \alpha) [\max_k V(\theta_{jk})]$ and select the alternative that minimizes this quantity.*

Note that if $\alpha = 0$, only the maximum costs would be examined for each

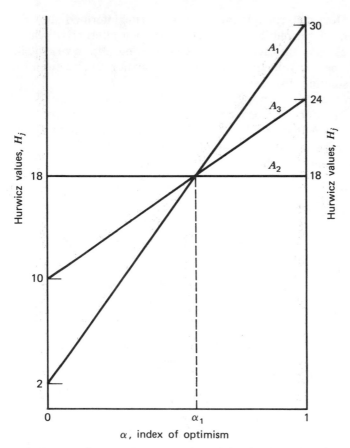

FIGURE 9.2. Graphical interpretation of the Hurwicz principle.

alternative and the minimum of these would be selected—the pessimistic mini-max criterion. If $\alpha = 1$, the optimistic minimin principle would be executed.

### 9.5.5 Savage Principle (Minimax Regret)

This principle, proposed by L. J. Savage, introduces and defines a quantity termed *regret*. A matrix consisting of regret values is first developed. Then the maximum regret value for each alternative $A_j$ is determined, and the alternative associated with the minimum regret value is chosen from the set of maximum regret values.

If the original decision matrix values are *gains,* the procedure for determining the regret matrix is as follows.

1. For a given state $S_l$, search the matrix column values over all alternatives and determine the largest gain. Assign this gain a zero regret value.
2. For all other gain values *under* the given state $S_l$, subtract these from the

largest gain value from step 1. The difference is interpreted as units of regret for a particular alternative $A_j$, given that state $S_l$ occurs.

3. Repeat steps 1 and 2 for each state $S_k$, $k \neq l$, until the regret matrix is completed.

Once the regret matrix is developed, examine each alternative over all states and select the maximum regret value. Then *choose the alternative that minimizes the maximum regret*. This principle of choice is conservative and is very similar to the maximin principle. It is emphasized, however, that the Savage principle deals in units of regret, not gain values in the original matrix; therefore, if both the maximin principle and the minimax regret principle are applied to a common decision problem, different alternatives may be selected.

Applying the Savage principle to the present-worth example of Table 9.4, the regret matrix given in Table 9.7 is obtained.

TABLE 9.7. Regret Matrix for the
Minimax Regret Example

|  | $S_1$ | $S_2$ | $S_3$ |
|---|---|---|---|
| $A_1$ | $3,234 | $0 | $0 |
| $A_2$ | 1,861 | 1,028 | 2,693 |
| $A_3$ | 0 | 1,122 | 4,063 |

The maximum regret values are $3234, $2693, and $4063 for alternatives $A_1$, $A_2$, and $A_3$, respectively. Thus, the minimum of these is $2693, and alternative $A_2$ would be preferred.

In a decision problem involving *costs*, the "logic" for creating the regret matrix is the same as the problem involving gains. That is,

1. For a given state $S_l$, search the matrix column values over all alternatives and determine the smallest cost. Assign this cost a zero regret value.
2. For all other cost values *under* the given state $S_l$, subtract the smallest cost value determined in step 1 from these to determine regret values.
3. Repeat steps 1 and 2 for each state $S_k$, $k \neq l$, until the regret matrix is completed.

After the regret matrix is obtained, the procedural steps for selecting an alternative are the same as before.

## 9.6 SEQUENTIAL DECISIONS

In the previous sections, the decision situations happened once. We now wish to consider situations that may require multiple decisions in sequential fashion. The reader can no doubt recall from personal experience or imagine real-world decisions that are sequential. For example, medical decisions are typically sequential. Based on the results of a routine blood analysis, a physician may require additional medical tests. The physician may then perform exploratory surgery and, based on these findings, recommend additional surgery for the patient. Following surgery, the recuperative procedure may also be conditional and involve further sequential decisions on behalf of the physician.

Sequential decisions in the business and industrial world are also common. For example, the number and type of material-handling units to purchase may depend on the forecast of product demand in each of the next 5 years. In the field of quality control, sampling is concerned with sequential decisions. For example, if a sample of five units is taken from a production machine, subsequent actions taken in regard to the production machine might very well depend on whether 0, 1, 2, 3, 4, or 5 defective units were found in the sample of five units.

In this section, sequential decisions under risk will be discussed, the technique of *decision-tree* representation will be used, and the logic of Bayes's theorem will be applied in the solution of the decision-tree problem.

### 9.6.1 Decision Trees

Magee [11, 12] is generally credited with representing sequential decisions in decision-trees and advocating the expectation criterion for solving the problem. Hespos and Strassman [8] developed stochastic decision trees incorporating risk-analysis methods and simulation techniques into the basic decision-tree methodology originated by Magee. Before presenting the graphic technique and a solution procedure, it is necessary to define certain symbolism.

$\Delta$  = a decision point, which will be labeled $D_i$ for the $i$th decision point in the tree

$\Delta$——  = a branch emanating from a decision point; represents an alternative that can be chosen at this point

0  = a fork (or node) in the tree where chance events influence the outcomes of an alternative choice

0——  = a branch representing a probabilistic outcome for a given alternative. The notation $(\theta; p)$ represents an outcome $\theta$, having an associated value $V(\theta; p)$, which occurs with probability $p$. It is assumed that branches emanating from a fork in the tree represent mutually exclusive and collectively exhaustive outcomes such that their probabilities sum to 1.0

V  = a value associated with a particular outcome (branch)

In Figure 9.3, a symbolic decision tree for a simple sequential decision problem is presented. The sequential decision problem depicted involves two decision points, $D_1$ and $D_2$. If alternative $A_1$ is chosen, it can have two outcomes, $\theta_1$ and $\theta_2$. If $\theta_1$ occurs, then a decision, $D_2$, is required. If alternative $A_{11}$ is chosen, there can be the two outcomes $\theta_{111}$ and $\theta_{112}$. If alternative $A_{12}$ is chosen at $D_2$,

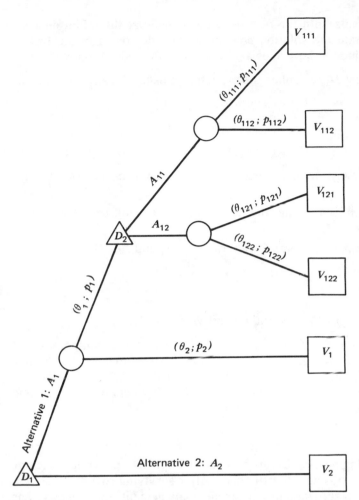

FIGURE 9.3. A symbolic decision tree.

there can be the two outcomes $\theta_{121}$ and $\theta_{122}$. For each branch in the tree there is an associated value. For example, if alternative $A_1$ is chosen *and* outcome $\theta_1$ occurs *and* alternative $A_{11}$ is chosen *and* outcome $\theta_{111}$ occurs, than value $V_{111}$ results with the conditional probability $p_{111}$. The first decision to be made is at decision point $D_1$, and if alternative $A_1$ is chosen, then a second decision is required at decision point $D_2$. The ultimate question to answer is which alternative to choose at $D_1$?

The principle of *maximizing expected gain* or *minimizing expected loss* is adopted as the principle of choice at each decision point. The solution procedure advocated is to reach the best decision at the decision point (or points) most distant from the base (the first decision) of the tree. Then, replace this most distant decision point with the best expected value and work backward through decision points until the best decision is made at the initial decision point, $D_1$. To illustrate the solution procedure, consider the decision tree of Figure 9.3 and assume all the values are *gains*. The procedure follows.

**1.** At decision point $D_2$, calculate the expected gains for $A_{11}$ and $A_{12}$. That is,

$$E(A_{11}) = p_{111}V_{111} + p_{112}V_{112}$$

and

$$E(A_{12}) = p_{121}V_{121} + p_{122}V_{122}$$

If $E(A_{11}) > E(A_{12})$, then alternative $A_{11}$ is chosen as best at $D_2$. Assume this is the case.

**2.** Replace $D_2$ with $E(A_{11})$, and the new reduced decision tree becomes as shown in Figure 9.4.

**3.** Calculate the expected gains for alternatives $A_1$ and $A_2$. That is,

$$E(A_1) = p_1 E(A_{11}) + p_2 V_1$$

and

$$E(A_2) = V_2(\text{a certain event with } p = 1.0).$$

If $E(A_1) > E(A_2)$, alternative $A_1$ is chosen as the best at decision point $D_1$. If $E(A_2) > E(A_1)$, then $A_2$ would, of course, be chosen, and if $E(A_2) = E(A_1)$, then one would be indifferent between the choice of $A_1$ and $A_2$. Given that $E(A_1) > E(A_2)$ from step 3, the optimal sequence of decisions is to choose $A_1$ at $D_1$ and then choose $A_{11}$ later on if outcome $\theta_1$ does occur.

## Example 8.5

The Jax Tool and Engineering Company is a medium-size, job-order machine shop and has been in business for 10 years. After surviving two national economic recessions with shrewd and industrious management, the company has

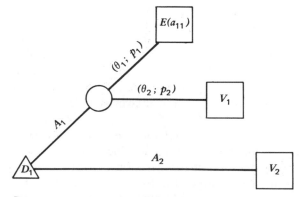

FIGURE 9.4. The reduced decision tree for Figure 9.3.

built a solid reputation for dependability and high-quality work. During the past 2 years, sales have sharply increased, and the company is operating near capacity. A market test has recently been made for a new product—electroplating tanks—and the response has been very favorable. However, production floor space to establish the new product line is limited in the present building. The people in management have three alternatives: (1) do nothing, (2) rearrange an area of the existing floor space, or (3) build an addition to the present plant. Also, there is the consideration of competition from one other job-order machine shop in the local area. If the electroplating tank venture is successful and the market demand cannot be fully met by the Jax Company, there is a good chance that competition will occur.

Management assigns representatives from production engineering and marketing to make a detailed study of the alternatives. Because management desires a report as soon as possible, it is necessary to make rather gross estimates on many items in the analysis. The analysts decide to use a 5-year planning horizon, a minimum attractive rate of return of 30%, and an incremental present worth value for each alternative relative to the "do-nothing" alternative. The results of the analysis now follow.

Rearrangement does not require any new construction, but machinery must be moved, some new machinery purchased, and new storage areas created. Production would be limited to 10 tanks/week on a two-shift operation. If there is competition in the first year from the other local firm, the Jax Company would probably take no further action, and a present worth value of $50,000 is estimated for this outcome. On the other hand, if there is no competition in the first year, the Jax Company will not be able to meet the expected demand. Subjective probabilities of 0.4 and 0.6 are assigned to the outcomes of (1) no competition in the first year, and (2) competition in the first year, respectively. If no competition, Jax management would then face the decision of either expanding the product line by building an addition or not expanding.

Expansion at this point would be expensive; a building addition plus a layout of the existing plant would be involved. However, no competition is expected if the expansion is made now and demand can be met with increased capacity. A present value of $30,000 is estimated for this outcome. If there is no expansion, two outcomes are possible. One possibility is that the other local machine shop can still enter the competition. However, a 1-year advantage in the market will have been gained, and the present worth of this outcome is estimated to be $10,000. It is very unlikely that competition will occur after a 2-year period, and the probability of such an event is assumed to be zero. A subjective probability value of 0.7 is assigned to the outcome of competition in the second year. The other possible outcome is, of course, no competition in the second year, with a probability of 0.3 and estimated present worth of $50,000.

The "build addition initially" alternative has higher first and recurring costs than the "new layout" alternative, but production capacity is estimated at 25 tanks/week, which should satisfy average annual demand over the next 5 years. With adequate production capacity initially, the chance of competition is low and about 20%. The present worth estimate in the event of no competition is $90,000. If there is competition, then the reduced sales and high annual fixed costs for the new building would result in a present worth of −$100,000. A decision-tree representation of this example is given in Figure 9.5 (values are in thousands of dollars).

Using the principle of choice of maximizing expected present worth, a best decision is sought at $D_2$, the decision point most distant from the base of the tree, $D_1$. The expected values at $D_2$ are

$$E(\text{expansion}) = \$30,000$$

and

$$E(\text{no expansion}) = \$10,000(0.7) + \$50,000(0.3)$$
$$= \$22,000$$

Thus, the alternative of expansion is chosen to maximize expected present worth, and this *best* value of $30,000 replaces $D_2$ in the tree, as shown in Figure 9.6.

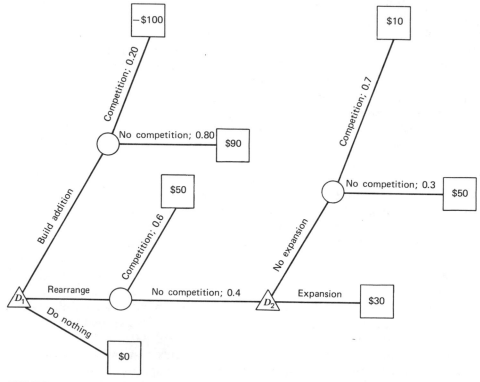

FIGURE 9.5. Jax Company example—original tree.

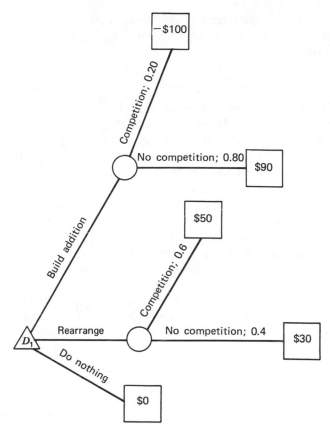

FIGURE 9.6. Jax Company example—reduced tree.

From Figure 9.6, the expected values at $D_1$ are calculated as:

$$E(\text{build addition}) = -\$100,000(0.20) + \$90,000(0.80)$$
$$= \$52,000$$

$$E(\text{rearrange}) = \$50,000(0.6) + \$30,000(0.4)$$
$$= \$42,000$$

$$E(\text{do nothing}) = \$0$$

Thus, the alternative of building an addition should be chosen in order to maximize expected present worth. It is interesting to note from Figure 9.6 that if this alternative is indeed chosen, the Jax Company is exposed to the possibility of a −$100,000 present worth. This event is "smoothed" by the expectation principle, on which most decision-tree analysis is based. The −$100,000 figure is not necessarily a dollar loss, since a *MARR* = 30% was assumed in the calculation of the values. However, management may wish to choose the "re-arrange" alternative instead because of the large negative figure; if so, they would be exercising a maximin philosophy.

## Example 9.6

For further illustration of the decision-tree approach to a problem and exposure to fundamental laws of probability, consider an urn containing five white ($W$) balls and seven black ($B$) balls. Three consecutive, independent, random (without bias in the selection) draws are made from the urn without replacing any ball drawn. The probability of selecting a white or black ball on the first draw is unconditional; that is, it does not depend on a previous draw (event). The probability of selecting a white or black ball on the second draw is, however, conditional on the color of ball selected on the first draw, and the third draw is conditional on both the first and second draws. In this example, we are interested in all the possible sequences of colors for the three draws and their probabilities of occurrence. A decision-tree representation of this problem is helpful and given in Figure 9.7.

All the eight possible sequences of colors are indicated in Figure 9.7. The objective probabilities of occurrence for each sequence will be determined by a relative-frequency argument. That is, for a general event $A$, the probability of $A$, $P(A)$, is calculated as the ratio

$$P(A) = \frac{\text{Number of ways favorable for the event}}{\text{Total number of ways possible}}$$

In the urn example, the probability of the unconditional event "a white ball on the first draw," $W_1$, is given by

$$P(W_1) = \frac{5 \text{ white balls (favorable number)}}{12 \text{ balls (total number)}}$$

Similarly, the probability of the unconditional event "a black ball on the first draw," $B_1$, is

$$P(B_1) = \frac{7}{12}$$

For the second draw, the probability of the event "a white ball on the second draw," $W_2$, is conditional on whether the first draw results in a white or black ball [that is, $P(W_2|W_1)$ or $P(W_2|B_1)$]. The probability that a white ball is selected on the second draw given that a white ball is selected on the first draw is reasoned to be

$$P(W_2|W_1) = \frac{4 \text{ (white balls remaining after first draw)}}{11 \text{ (total balls remaining after first draw)}}$$

Similarly,

$$P(W_2|B_1) = \frac{5}{11}$$

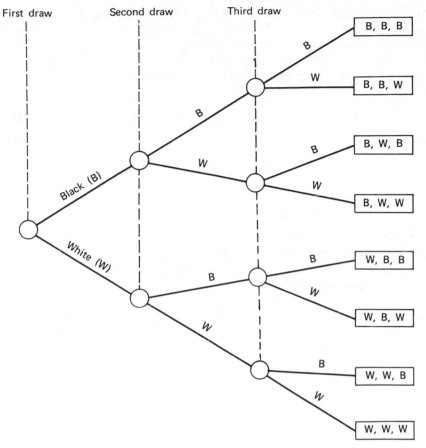

FIGURE 9.7. Tree representation of urn example.

The probability of the event "a black ball on the second draw," $B_2$, is also conditional on whether the first draw results in a white or black ball, and

$$P(B_2|W_1) = \frac{7}{11}$$

$$P(B_2|B_1) = \frac{6}{11}$$

From the above results, note that

$$P(W_2|W_1) + P(B_2|W_1) = 1.0$$

and

$$P(W_2|B_1) + P(B_2|B_1) = 1.0$$

If a white (or black) ball is in fact drawn first, then the second draw can only yield two possible outcomes (collectively exhaustive): either a white or black ball. Since these two outcomes are mutually exclusive (cannot occur together), the individual probabilities of occurrence are additive and sum to 1.0 or certainty. In general, for two mutually exclusive events $A$ and $B$, the additive law of probability states

$$P(\text{either } A \text{ or } B) = P(A \cup B) = P(A) + P(B) \tag{9.4}$$

where the symbol ($\cup$) stands for the union of the events. Thus, in this example,

$$P(W_2|W_1 \cup B_2|W_1) = P(W_2|W_1) + P(B_2|W_1)$$

and

$$P(W_2|B_1 \cup B_2|B_1) = P(W_2|B_1) + P(B_2|B_1)$$

The additive law is easily expanded for multiple, mutually exclusive events.

For the third draw, if the ball selected on each of the first and second draws is white, then the probability of selecting a white ball on the third draw is denoted $P(W_3|W_1, W_2)$ and reasoned, by the relative-frequency logic, to be

$$P(W_3|W_1, W_2) = \frac{3 \text{ (white balls remaining after first and second draw)}}{10 \text{ (total balls remaining after first and second draw)}}$$

The symbolism $(W_3|W_1, W_2)$ is interpreted to mean the event of drawing a white ball on the third draw given that a white ball was selected on the first draw *and* a white ball was selected on the second draw.

Similarly,

$$P(B_3|W_1, W_2) = \frac{7}{10}$$

$$P(W_3|W_1, B_2) = \frac{4}{10}$$

$$P(B_3|W_1, B_2) = \frac{6}{10}$$

$$P(W_3|B_1, W_2) = \frac{4}{10}$$

$$P(B_3|B_1, W_2) = \frac{6}{10}$$

$$P(W_3|B_1, B_2) = \frac{5}{10}$$

$$P(B_3|B_1, B_2) = \frac{5}{10}$$

A summary of all these events and their probabilities is shown in Figure 9.8.

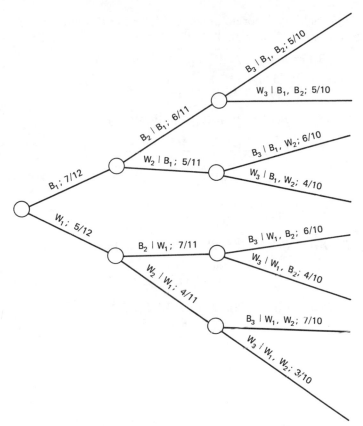

FIGURE 9.8. Probability tree for urn example.

Up to this point the probabilities of each individual outcome (event) of the example have been determined. However, the probability of each sequence (branch of the tree) occurring is yet to be determined (see Figure 9.7 to recall the possible sequences of colors). For instance, it is desired to know the probability of the joint event of a white ball being selected with each of the three draws or $P(W_1, W_2, W_3)$, which may also be written with set notation as $P(W_1 \cap W_2 \cap W_3)$ to mean the joint occurrence of these three events (conditional events in this particular case). This probability may be calculated as

$$P(W_1, W_2, W_3) = P(W_1) \cdot P(W_2|W_1) \cdot P(W_3|W_1, W_2)$$

$$= \left(\frac{5}{12}\right)\left(\frac{4}{11}\right)\left(\frac{3}{10}\right) = \frac{60}{1320}$$

A full justification of the above multiplicative calculation requires a probability theory development that is outside the scope of this textbook. However, the calculation of the joint occurrence of three conditional events is a direct extension of the conditional probability theorem presented later in this chapter. For the other possible sequences of this example,

$$P(W_1, W_2, B_3) = P(W_1) \cdot P(W_2|W_1) \cdot P(B_3|W_1, W_2)$$

$$= \left(\frac{5}{12}\right)\left(\frac{4}{11}\right)\left(\frac{7}{10}\right) = \frac{140}{1320}$$

$$P(W_1, B_2, W_3) = P(W_1) \cdot P(B_2|W_1) \cdot P(W_3|W_1, B_2)$$

$$= \left(\frac{5}{12}\right)\left(\frac{7}{11}\right)\left(\frac{4}{10}\right) = \frac{140}{1320}$$

$$P(W_1, B_2, B_3) = P(W_1) \cdot P(B_2|W_1) \cdot P(B_3|W_1, B_2)$$

$$= \left(\frac{5}{12}\right)\left(\frac{7}{11}\right)\left(\frac{6}{10}\right) = \frac{210}{1320}$$

$$P(B_1, B_2, B_3) = P(B_1) \cdot P(B_2|B_1) \cdot P(B_3|B_1, B_2)$$

$$= \left(\frac{7}{12}\right)\left(\frac{6}{11}\right)\left(\frac{5}{10}\right) = \frac{210}{1320}$$

$$P(B_1, B_2, W_3) = P(B_1) \cdot P(B_2|B_1) \cdot P(W_3|B_1, B_2)$$

$$= \left(\frac{7}{12}\right)\left(\frac{6}{11}\right)\left(\frac{5}{10}\right) = \frac{210}{1320}$$

$$P(B_1, W_2, B_3) = P(B_1) \cdot P(W_2|B_1) \cdot P(B_3|B_1, W_2)$$

$$= \left(\frac{7}{12}\right)\left(\frac{5}{11}\right)\left(\frac{6}{10}\right) = \frac{210}{1320}$$

$$P(B_1, W_2, W_3) = P(B_1) \cdot P(W_2|B_1) \cdot P(W_3|B_1, W_2)$$

$$= \left(\frac{7}{12}\right)\left(\frac{5}{11}\right)\left(\frac{4}{10}\right) = \frac{140}{1320}$$

The sum of the probabilities for each of these eight possible sequences is $1320/1320 = 1.0$, and the complete probability tree is shown in Figure 9.9.

### 9.6.2 The Conditional Probability Theorem

The probability of the joint occurrence of two dependent events, $A$ and $B$, is given by

$$P(A \text{ and } B) = P(A \cap B) = P(A) \cdot P(B|A) \qquad (9.6)$$

where $P(A)$ is an unconditional probability and $P(B|A)$ is the conditional probability of event $B$ given that event $A$ has occurred or will occur. If event $B$ is not conditional on event $A$, then events $A$ and $B$ are *independent* events and Equation 9.6 may be written as

$$P(A \cap B) = P(A) \cdot P(B) \qquad (9.7)$$

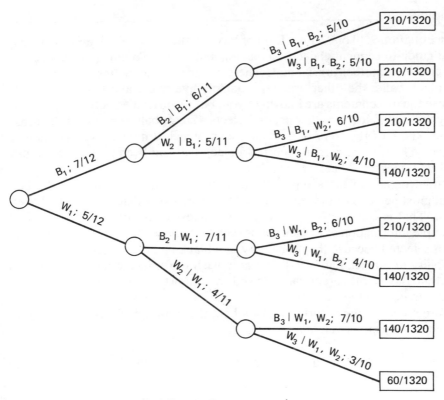

FIGURE 9.9. Complete probability tree for urn example.

An extension of Equation 9.6 for three dependent events is given by

$$P(A \cap B \cap C) = P(A) \cdot P(B|A) \cdot P(C|A, B) \qquad (9.8)$$

which was the relationship used in the previous urn example.

For the purpose of the development to follow, Equation 9.6 is now written in its complete form as

$$P(A \cap B) = P(A) \cdot P(B|A) = P(A|B) \cdot P(B) \qquad (9.9)$$

from which it follows that

$$P(B|A) = \frac{P(A|B)P(B)}{P(A)} \qquad (9.10)$$

or

$$P(A|B) = \frac{P(B|A)P(A)}{P(B)} \qquad (9.11)$$

## Example 9.7

Either Equation 9.10 or 9.11 is a form of Bayes's theorem, which may provide a logical guide to sequential decision making under risk. To illustrate this claim, let us consider another urn example where one urn, $U_1$, contains two white and eight black balls; the other urn, $U_2$, contains five white and five black balls. Suppose the urn contents are known to you, but the urns themselves are hidden from your view. Now, another person selects a single ball from one of the urns. Without knowing the color of the ball drawn, which urn would you guess was selected? Then, if the person reported a black ball had been drawn, which urn would you guess had been selected?

Without benefit of the sample information concerning the color of the ball drawn, most persons would feel that it was equally likely that $U_1$ or $U_2$ had been chosen. That is, since $P(U_1) = P(U_2) = 1/2$, guess either urn. However, after the sample information of a black ball is received, intuition suggests that $U_1$ was the urn selected because the probability of a black ball in $U_1$ is greater than the probability of a black ball in $U_2$. We now wish to formalize this example and show that Bayes's theorem can be used to support the intuitive guess of $U_1$ after receiving the report.

Before receiving the report of a black ball, the prior belief that $U_1$ or $U_2$ was selected is subjectively stated as

$$P(U_1) = \text{the } prior \text{ probability of selecting } U_1$$
$$= 1/2$$

and

$$P(U_2) = \text{the } prior \text{ probability of selecting } U_2$$
$$= 1/2$$

Note that

$$W = \text{the event of a white ball being drawn}$$

and

$$B = \text{the event of a black ball being drawn}$$

Also, before receiving the sample information, the likelihood of a particular report ($W$ or $B$), given that $U_1$ or $U_2$ was selected, can be determined, based on knowledge of the contents of each urn. That is,

$$P(W|U_1) = \text{the conditional probability of a white ball being drawn}$$
$$\text{given that } U_1 \text{ was selected for the draw}$$

$$= \frac{2}{10}$$

Similarly,

$$P(B|U_1) = \frac{8}{10}$$

$$P(W|U_2) = \frac{5}{10}$$

$$P(B|U_2) = \frac{5}{10}$$

In the literature on sequential decision making, these conditional probabilities, either objectively known as above or subjectively assigned, are termed *likelihood statements*.

Ultimately, it is desired to assess the probability of whether $U_1$ or $U_2$ was chosen based on the sample information received (the report of a black ball in this case). Thus, the *posterior probabilities* of $P(U_1|B)$, $P(U_2|B)$, $P(U_1|W)$, and $P(U_2|W)$ need to be determined.

In the calculation of these posterior probabilities, it is convenient to calculate the probability of event $W$ and event $B$ before the ball is drawn. It is reasoned that event $W$ could occur if either $U_1$ or $U_2$ were chosen. The same is true for the event $B$. Thus, $B$ could occur if (1) $U_1$ were selected and $B$ occurred, or (2) $U_2$ were selected and $B$ occurred. That is,

$$P(B) = P(U_1 \text{ and } B \text{ or } U_2 \text{ and } B)$$
$$= P(U_1 \cap B \cup U_2 \cap B)$$
$$= P(U_1 \cap B) + P(U_2 \cap B)$$

Then, using Equation 9.9,

$$P(B) = P(U_1)P(B|U_1) + P(U_2)P(B|U_2)$$

and, from the previous data,

$$P(B) = \left(\frac{1}{2}\right)\left(\frac{8}{10}\right) + \left(\frac{1}{2}\right)\left(\frac{5}{10}\right) = \frac{13}{20}$$

Similarly,

$$P(W) = P(U_1)P(W|U_1) + P(U_2)P(W|U_2)$$
$$= \left(\frac{1}{2}\right)\left(\frac{2}{10}\right) + \left(\frac{1}{2}\right)\left(\frac{5}{10}\right) = \frac{7}{20}$$

With these results, the posterior probabilities may be calculated from Bayes's theorem as

$$P(U_1|B) = \frac{P(B|U_1)P(U_1)}{P(B)} = \frac{(8/10)(1/2)}{13/20} = \frac{8}{13}$$

$$P(U_2|B) = \frac{P(B|U_2)P(U_2)}{P(B)} = \frac{(5/10)(1/2)}{13/20} = \frac{5}{13}$$

$$P(U_1|W) = \frac{P(W|U_1)P(U_1)}{P(W)} = \frac{(2/10)(1/2)}{7/20} = \frac{2}{7}$$

$$P(U_2|W) = \frac{P(W|U_2)P(U_2)}{P(W)} = \frac{(5/10)(1/2)}{7/20} = \frac{5}{7}$$

It is noted from the results of $P(U_1|B) = 8/13$ and $P(U_2|B) = 5/13$ that the event of $B$ should therefore increase the belief that $U_1$ was selected for taking the sample if one is to be consistent in a probabilistic sense. That is, the prior probability of $U_1$ was 1/2 and the posterior probability, given the event $B$, of $U_1$ is 8/13. If, after receiving the information of $B$, the guess had been $U_1$, then the above calculations would support the intuitive guess. If a second sample of a ball is taken from an urn, then the posterior probabilities $P(U_1|B)$ and $P(U_2|B)$ become the prior probabilities $P(U_1)$ and $P(U_2)$ for the second trial.

### 9.6.3 The Value of Perfect Information

In order to use the logic of Bayes's theorem in a more practical way to aid decision making, it is of interest to evaluate the value of the sample information in a monetary sense. Recalling the previous example, suppose the person who drew the ball from the urn had offered to pay a $10 prize if the correct urn were guessed. However, the person will charge for the sample information of whether a white or black ball is drawn. The question then is how much should be paid for the *sample information*. Before answering this particular question, a rationale for determining the *value of perfect information* is considered (i.e., the knowledge of exactly which urn was selected for the draw).

The *prior* probabilities of $P(U_1) = P(U_2) = 1/2$ are now recalled and written with the more suggestive notation of $PR(U_1) = PR(U_2) = 1/2$ [16]. Without any sample information, one should be indifferent as to which urn is guessed and the *prior expected profit* (PEP) would be $5, as determined from the matrix of Table 9.8.

TABLE 9.8. Matrix for the Urn Example

| Urn Guessed / Actual Urn Selected | $PR(U_1) = 1/2$<br>$U_1$ | $PR(U_2) = 1/2$<br>$U_2$ |
|---|---|---|
| $U_1$ | $10 | $ 0 |
| $U_2$ | $ 0 | $10 |

The prior expected profit is $5, as determined from either of the two calculations below.

$$E(\text{profit}|\text{guess } U_1) = \left(\frac{1}{2}\right)(\$10) + \left(\frac{1}{2}\right)(\$0) = \$5$$

$$E(\text{profit}|\text{guess } U_2) = \left(\frac{1}{2}\right)(\$0) + \left(\frac{1}{2}\right)(\$10) = \$5$$

If these expected profits were different, the alternative with the larger expected value would be chosen, and the prior expected profit is the larger value.

Now, given that the person drawing the ball states that $U_1$ was selected and that the person is perfectly reliable, $U_1$ would obviously be guessed and a $10 payoff received. The same is true, of course, for the report of $U_2$. The *expected profit given perfect information (EP|PI)* is thus $10 in this example. It is now reasoned that the maximum amount one should pay for perfect information is $10 − $5 (expected profit without any information) = $5. We can formalize the *expected value of perfect information (EVPI)* as the difference between the *expected profit given perfect information (EP|PI)* and the *prior expected profit (PEP)*, or

$$EVPI = EP|PI - PEP \qquad (9.12)$$

### 9.6.4 The Value of Imperfect Information

In most real-world decision situations, information is incomplete and not perfectly reliable, as in the urn example just discussed. The report that a black ball had been drawn gave additional information to the decision maker, but still did not provide the absolute answer as to which urn was selected. In order to define the *expected value of sample information (EVSI)*, the posterior probability (with new notation) results are repeated below.

For the event of $B$, the posterior probabilities were

$$PO(U_1|B) = \frac{8}{13}$$

and

$$PO(U_2|B) = \frac{5}{13}$$

Thus, since 8/13 > 5/13, the best guess is that $U_1$ was selected. The actual urn chosen could have been $U_1$ or $U_2$, and the expected profit given a guess of $U_1$ is

$$\begin{aligned} E(\text{profit}|\text{guess } U_1) &= (\text{expected profit}|\text{guess } U_1 \text{ and is } U_1)PO(U_1) \\ &\quad + (\text{expected profit}|\text{guess } U_1 \text{ and is } U_2)PO(U_2) \\ &= (\$10)\left(\frac{8}{13}\right) + (\$0)\left(\frac{5}{13}\right) \\ &= \frac{\$80}{13} \end{aligned}$$

For the event of $W$, the posterior probabilities were

$$PO(U_1|W) = \frac{2}{7}$$

$$PO(U_2|W) = \frac{5}{7}$$

Thus, if the report had been $W$, and since $5/7 > 2/7$, the best guess would be $U_2$ with

$$E(\text{profit}|\text{guess } U_2) = (\text{expected profit}|\text{guess } U_2 \text{ and is } U_1)PO(U_1)$$
$$+ (\text{expected profit}|\text{guess } U_2 \text{ and is } U_2)PO(U_2)$$

$$= (\$0) \left(\frac{2}{7}\right) + (\$10) \left(\frac{5}{7}\right)$$

$$= \frac{\$50}{7}$$

However, *before* the sample is taken and the report is given, both events $B$ and $W$ are possible with $P(B) = 13/20$ and $P(W) = 7/20$. The *expected profit given sample information* $(EP|SI)$ is

$$EP|SI = (\text{expected profit}|B)P(B) + (\text{expected profit}|W)P(W)$$

$$= \left(\frac{\$80}{13}\right)\left(\frac{13}{20}\right) + \left(\frac{\$50}{7}\right)\left(\frac{7}{20}\right)$$

$$= \$6.50$$

In the above calculations, both expected profit values (expected profit$|B$) and (expected profit$|W$) assume that the best guess would, in fact, be made for the reports of $B$ and $W$, respectively.

Finally, it is reasoned that the *expected value of sample information* is the difference between the *expected value given sample information* and the *prior expected profit*, or

$$EVSI = EP|SI - PEP$$
$$= \$6.50 - \$5 \tag{9.13}$$
$$= \$1.50$$

which is the maximum amount one should pay for the sample information of this example. This value is, of course, less than the $5 value for perfect information.

A decision-tree representation of this example is given in Figure 9.10 for a *sample information charge* of $1.50.

From Figure 9.10, the appropriate calculations at $D_2$ are

$$E(\text{guess } U_1) = (\$8.50) \left(\frac{8}{13}\right) + (-\$1.50) \left(\frac{5}{13}\right)$$

$$= \frac{\$60.5}{13}$$

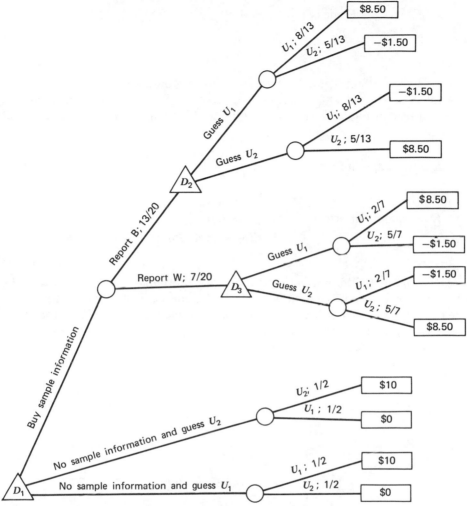

FIGURE 9.10. Decision tree for urn example.

$$E(\text{guess } U_2) = (-\$1.50) \left(\frac{8}{13}\right) + (\$8.50) \left(\frac{5}{13}\right)$$

$$= \frac{\$30.5}{13}$$

Since $\$60.5/13 > \$30.5/13$, the best guess given $B$ would be $U_1$.
    The appropriate calculations at $D_3$ are

$$E(\text{guess } U_1) = (\$8.50) \left(\frac{2}{7}\right) + (-\$1.50) \left(\frac{5}{7}\right)$$

$$= \frac{\$6.5}{7}$$

$$E(\text{guess } U_2) = (-\$1.50)\left(\frac{2}{7}\right) + (\$8.50)\left(\frac{5}{7}\right)$$

$$= \frac{\$39.5}{7}$$

Since $\$39.5/7 > \$6.5/7$, the best guess given $W$ would be $U_2$.

Replacing $D_2$ and $D_3$ with these best expected values yields the reduced decision tree of Figure 9.11. From Figure 9.11, the appropriate calculations at $D_1$ are

$$E(\text{buy sample information}) = \left(\frac{\$60.5}{13}\right)\left(\frac{13}{20}\right) + \left(\frac{\$39.5}{7}\right)\left(\frac{7}{20}\right)$$

$$= \$5$$

$E(\text{no sample information and guess } U_2)$
$$= E(\text{no sample information and guess } U_1)$$
$$= \$5$$

Thus, one is indifferent between the three alternatives at $D_1$ given that the charge for sample information is \$1.50. If the charge is less than this amount, the decision maker should choose the "buy sample information" alternative in order to maximize expected profits.

For a similar sequential decision problem involving only costs, the basic

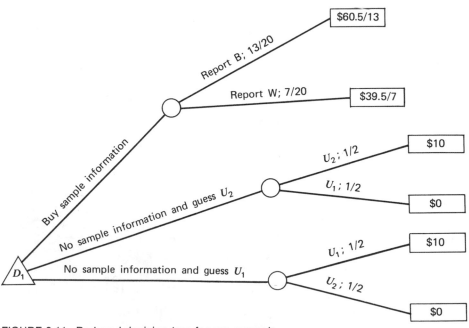

FIGURE 9.11. Reduced decision tree for urn example.

relationships to determine expected value of information become

$EVPI$ = prior expected cost − expected cost given perfect information

$$= PEC - EC|PI \qquad (9.14)$$

$EVSI$ = prior expected cost − expected cost given sample information

$$= PEC - EC|SI \qquad (9.15)$$

## Example 9.3[3]

The editor of the American Book Company is considering a manuscript for a historical novel. The editor estimates the book can be marketed for $5, with the company receiving $4 and the author $1. Since other publishers are interested in the manuscript, the author desires a commitment from the American Book Company in the very near future. American has already invested $300 in a preliminary review of the manuscript and a quick market survey. The editor considers this manuscript competitive with other recently published historical novels and the author, although relatively new, has other published works that have received good reviews.

The editor identifies the immediate alternatives of (1) reject the manuscript, (2) accept the manuscript, or (3) obtain an additional review of the manuscript by an expert. An additional review will cost $400, and the expert's opinion will not perfectly predict the historical novel's success in the market. The editor reasons that at best the expert can only rate the manuscript good, fair, or poor.

If the manuscript is accepted for publication, the editor speculates on the possible market outcomes; there could be (1) a low market demand, or (2) the novel could be a best-seller. Additionally, there is the question of movie rights. There may be no interest in filming the novel, but, on the other hand, a best-seller could yield $30,000 for the movie rights. Even if there is low market demand, a film company might pay $4000 for the rights to make a low-budget, grade B film.

The American Book Company has available historical data on 60 recent manuscripts of similar books, given in Table 9.9.

From available sales figures, the editor can estimate reasonably well the revenues given a particular outcome for a published book. He considers the present worth of after-tax revenues over the effective life of a book using a 10% expected rate of return. For a best-seller, a present worth of $70,000 is estimated, and a low-demand publication yields an estimated $5000 present worth. Offers for movie rights, if any, are usually made 1 year after the manuscript is under contract, and these after-tax revenues should be added to the book sales

---

[3] This example was presented at the Sixth Triennial Symposium, Engineering Economy Division, ASEE, June 19–20, 1971, in a paper entitled "Introduction to Decision Theory" by Barnard E. Smith. The article appears in the publication, *Decision and Risk Analysis: Powerful New Tools For Management,* The Engineering Economist, Stevens Institute of Technology, Hoboken, N.J., and is presented here by the permission of the publisher.

**TABLE 9.9.** Sixty Manuscript Decisions/Outcomes

| Decision/Outcome When Marketed | No Outside Expert Review | Outside Reviewer's Evaluation | | | Total |
|---|---|---|---|---|---|
| | | Good | Fair | Poor | |
| Not published | 5 | 0 | 14 | 12 | 31 |
| Low market demand | 2 | 0 | 4 | 3 | 9 |
| Best-seller | 1 | 2 | 4 | 0 | 7 |
| Best-seller + movie | 1 | 4 | 3 | 0 | 8 |
| Low market + B movie | 1 | 0 | 3 | 1 | 5 |
| Total | 10 | 6 | 28 | 16 | 60 |

(assume that $4000 for a grade B film and $30,000 for a "best-seller film" are appropriate figures). It costs $15,000 to publish a book.

If it is assumed the editor has a linear utility function for dollars and uses a "maximize present worth" principle for choice among alternatives, what decision should be made at the present time?

The editor faces the immediate alternatives of

$A_1$ = reject the manuscript without any additional review by an expert
$A_2$ = accept the manuscript without obtaining an expert's review
$A_3$ = obtain an expert's review

The value of alternative $A_1$ is the $300 cost, which has already been spent on a preliminary review [that is, $V(A_1) = \$300$]. Indeed, the $300 for a preliminary review is a sunk cost and applies whether alternative $A_1$, $A_2$, or $A_3$ is chosen at the present time. This $300 cost can therefore be dropped from consideration, and $V(A_1) = 0$ with this adjustment.

For alternative $A_2$ there could be four possible market outcomes, defined as

$\theta_1$ = low market demand
$\theta_2$ = a best-seller
$\theta_3$ = best-seller plus movie rights
$\theta_4$ = low market demand plus grade B movie

Thus, the value associated with each of these outcomes is determined as

$$V(\theta_1|A_2) = \text{revenues} - \text{publication costs}$$
$$= \$5000 - \$15,000 = -\$10,000$$
$$V(\theta_2|A_2) = \$70,000 - \$15,000 = \$55,000$$
$$V(\theta_3|A_2) = \$70,000 + \$30,000 - \$15,000 = \$85,000$$
$$V(\theta_4|A_2) = \$5000 + \$4000 - \$15,000 = -\$6000$$

For alternative $A_3$, the expert's evaluations could be either

$$Z_1 = \text{Good}$$
$$Z_2 = \text{Fair}$$
$$Z_3 = \text{Poor}$$

For *each* of these evaluations, the editor must make one of two possible decisions, defined as

$$X_1 = \text{reject the manuscript}$$
$$X_2 = \text{accept the manuscript}$$

Given the decision of $X_1$ for *each* of the three possible evaluations ($Z_j$), the cost to the American Book Company would be the same—the additional cost of an expert's review. That is,

$$V(X_1|A_3, Z_j) = -\$400 \qquad \text{for } j = 1, 2, 3$$

Then, for the decision of $X_2$ associated with each of the reviewer's evaluations, there are the four possible market outcomes defined previously. The value for any *given* market outcome *and* decision $X_2$ is the same for all $Z_j$ evaluations. This value is equal to the values for alternative $A_2$ minus $400 for the expert's review. That is,

$$V(\theta_1|A_3, Z, X_2) = -\$10,000 - \$400 = -\$10,400$$

$$V(\theta_2|A_3, Z, X_2) = \$55,000 - \$400 = \$54,600$$

$$V(\theta_3|A_3, Z, X_2) = \$85,000 - \$400 = \$84,600$$

$$V(\theta_4|A_3, Z, X_2) = -\$6000 - \$400 = -\$6400$$

A decision-tree representation of this example, including probability values for the various outcomes, is given by Figure 9.12, where the values are in thousands of dollars.

The probability values for the various outcomes of the decision tree given in Figure 9.12 are computed using the data of Table 9.9. First, consider alternative $A_2$ (accept for publication without further review). It is noted from Table 9.9 that out of the ten manuscripts that were not reviewed by an outside expert, five were published. Two of the five had low market demand, one was a best-seller, one was a best-seller plus movie rights, and another had low market demand plus the rights sold for a grade B movie. Thus, by a relative frequency logic, the following probabilities can be determined:

$$P(\theta_1|A_2) = \frac{2}{5}$$

$$P(\theta_2|A_2) = \frac{1}{5}$$

$$P(\theta_3|A_2) = \frac{1}{5}$$

$$P(\theta_4|A_2) = \frac{1}{5}$$

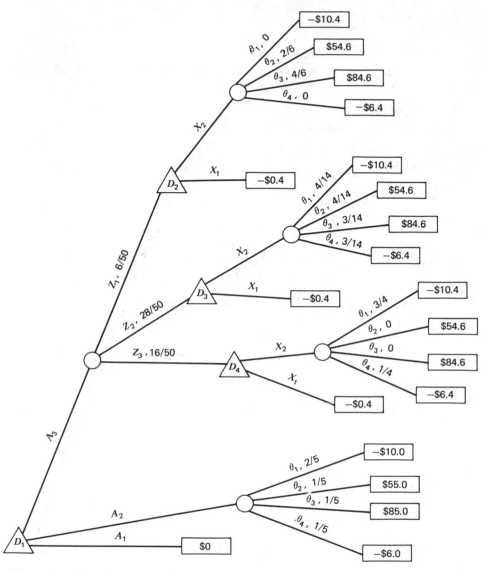

FIGURE 9.12. Decision tree for American Book Company example.

For the 50 manuscripts that were reviewed by an expert, 6 received a good evaluation, 28 received a fair evaluation, and 16 received a poor evaluation. Thus,

$$P(Z_1) = \frac{6}{50}$$

$$P(Z_2) = \frac{28}{50}$$

$$P(Z_3) = \frac{16}{50}$$

Similarly, the following conditional probabilities may be determined from the data of Table 9.9.

$$P(\theta_1|Z_1) = 0, \qquad P(\theta_1|Z_2) = \frac{4}{14}, \qquad P(\theta_1|Z_3) = \frac{3}{4}$$

$$P(\theta_2|Z_1) = \frac{2}{6}, \qquad P(\theta_2|Z_2) = \frac{4}{14}, \qquad P(\theta_2|Z_3) = 0$$

$$P(\theta_3|Z_1) = \frac{4}{6}, \qquad P(\theta_3|Z_2) = \frac{3}{14}, \qquad P(\theta_3|Z_3) = 0$$

$$P(\theta_4|Z_1) = 0, \qquad P(\theta_4|Z_2) = \frac{3}{14}, \qquad P(\theta_4|Z_3) = \frac{1}{4}$$

It is recognized that these probability values are estimates but would seem to be less gross than purely subjective evaluations made without any historical data as a basis.

The solution to the decision tree of Figure 9.12 follows. The appropriate calculations for $D_2$ are

$$E(X_1) = -0.4 = -\$400$$

$$E(X_2) = (-\$10.4)(0) + (\$54.6)\left(\frac{2}{6}\right) + (\$84.6)\left(\frac{4}{4}\right) + (-\$6.4)(0)$$

$$= \$74.6 = \$74,600$$

and the best decision at $D_2$ is therefore $X_2$ (accept the manuscript). The calculations for $D_3$ are

$$E(X_1) = -\$400$$

$$E(X_2) = (-\$10.4)\left(\frac{4}{14}\right) + (\$54.6)\left(\frac{4}{14}\right) + (\$84.6)\left(\frac{3}{14}\right) + (-\$6.4)\left(\frac{3}{14}\right)$$

$$= \$29.386 = \$29,386$$

and the best decision at $D_3$ is therefore $X_2$ (accept the manuscript). The calculations are $D_4$ are

$$E(X_1) = -\$400$$

$$E(X_2) = (-\$10.4)\left(\frac{3}{4}\right) + (\$54.6)(0) + (\$84.6)(0) + (-\$6.4)\left(\frac{1}{4}\right)$$

$$= -\$9.4 = \$9,400$$

and the best decision at $D_4$ is therefore $X_1$ (reject the manuscript).

The reduced decision tree is given by Figure 9.13, and the appropriate calculations for $D_1$ are

$$E(A_3) = (\$74{,}600)\left(\frac{6}{50}\right) + (\$29{,}386)\left(\frac{28}{50}\right) + (-\$400)\left(\frac{16}{50}\right)$$

$$= \$25{,}280$$

$$E(A_2) = (-\$10{,}000)\left(\frac{2}{5}\right) + (\$55{,}000)\left(\frac{1}{5}\right) + (\$85{,}000)\left(\frac{1}{5}\right) + (-\$6000)\left(\frac{1}{5}\right)$$

$$= \$22{,}800$$

$$E(A_1) = \$0$$

and the best decision at $D_1$ in order to maximize the expected present worth is alternative $A_3$ (obtain an expert's opinion). Then, if the evaluation is either good or fair, accept the manuscript for publication; but if the evaluation is poor, reject the manuscript. The expected value of sample information in this example is the difference between $E(A_3)$ and $E(A_2)$ plus the \$400 paid the expert, or \$2880.

$$EVSI = EV|SI - PEP + \$400$$
$$= \$25{,}280 - \$22{,}800 + \$400$$
$$= \$2880$$

Clearly the expert's opinion is well worth the \$400 fee in this example.

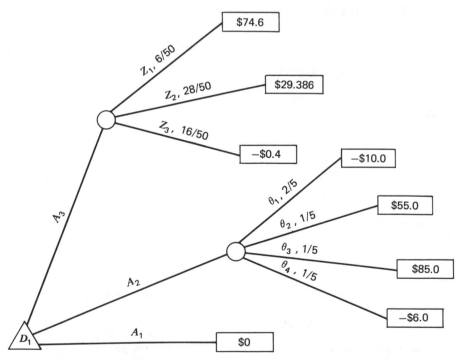

FIGURE 9.13. Reduced decision tree for American Book Company example (value in thousands of dollars).

### 9.6.5 Sequential Decisions—Summary Comments

The decision-tree representation of a sequential-decision situation is, in our opinion, an effective device for visualization. By virtue of developing the tree, serious attention is given to explicitly identifying and describing alternatives, to assessing the outcomes that might occur as a result of the alternative choice, and to making any necessary future decisions. Even if the probabilities for outcomes cannot be adequately assessed or values for the "branches" accurately estimated, considerable insight into the decision situation may be achieved and decision making aided. On the other hand, whether the expectation principle of choice among alternatives is an appropriate criterion or not is a matter of personal preference, and any real-world sequential-decision situation may readily become too large for effective description or convenient solution. As the number of alternatives, the number of outcomes per alternative, and the number of decision points increase, so do the uncertainty of data and the burden of calculation. Furthermore, the expense in time and money to obtain the information required may become prohibitive when compared with the benefits expected from the analysis. An evaluation of such a trade-off is largely, if not primarily, a matter of judgment.

## 9.7 MULTIPLE OBJECTIVES

In order to initiate a discussion of multiple objectives and the confounding effect they have on the decision-making process, quotations from Miller and Starr [13], given in Chapter One, are now paraphrased: "Being unable to describe goals satisfactorily in terms of one objective, people customarily maintain various objectives, and multiple objectives are frequently in conflict with each other." Because organizations are made up of people, this statement would seem to apply to an organizational unit as well as an individual. Accepting this premise, then, managers who are responsible for the organization's behavior and performance usually expect a decision to achieve multiple objectives and evaluate recommendations made to them on this basis.

Establishing the relevant objectives is an important problem for a given decision maker, and the problem is further complicated by conflicting objectives and the difficulty (if not impossibility) of obtaining a measurement of these objectives.

At the outset, we acknowledge that the discussion to follow is somewhat superficial, and we have drawn heavily from Churchman, Ackoff, and Arnoff [3, Chapters 5 and 6], Miller and Starr [13, Chapters 1–3 and pp. 263–241], Kepner and Tregoe [9], and Morris [15, Chapter 7]. The literature on the subject of multiple objectives and the theory of measurement is very extensive, since these factors are involved in all areas of human endeavor. The primary literature reference sources are the general areas of psychology, statistics, mathematics, and management. The previous references have been cited because they are concerned with managerial decision making instead of the psychological motivation stimulating the multiple objectives or the mathematical theory of measurement. For the reader particularly interested in the subject of multiple objectives from a managerial decision-making point of view, the cited references are recommended readings.

Kepner and Tregoe [9] state that ''making the best decision will involve a sequence of procedures based on the following seven concepts:

1. The objectives of a decision must be established first.

2. The objectives are classified as to importance.

3. Alternative actions are developed.

4. The alternatives are evaluated against the established objectives.

5. The choice of the alternative best able to achieve all the objectives represents the tentative decision.

6. The tentative decision is explored for future possible adverse consequences.

7. The effects of the final decision are controlled by taking other actions to prevent possible adverse consequences from becoming problems, and by making sure the actions decided on are carried out.''

For a single objective, Chapters Three to Nine of this text have been concerned primarily with steps 3 to 6 of the above procedure. This particular section of Chapter Nine will be concerned mainly with steps 2 and 4 of the above procedure. Churchman, Ackoff, and Arnoff [3] present an interesting discussion on establishing the objectives of a decision. Morse [17] states that objectives are derived from two general areas: the *results* expected to come from a decision, and the *resources* available for expenditure in carrying out a decision. The results expected may be to achieve certain things, but it may also be desired to maintain certain things. For example, over the next 5 years, management may wish to increase sales by 40% relative to the past year and render better customer service, retaining good employee morale and the company's good financial credit rating. In any case, for the subsequent discussion, it will be assumed that objectives can be precisely stated and are relevant to the decision.

### 9.7.1 Classifying Objectives According to Importance

Assessing the relative importance of multiple objectives, measuring the outcomes of alternatives in terms of these objectives, obtaining a common measure for multiple outcomes having different measures, and combining all these into a single measure of merit form the heart of the multiple-objective problem and the source of much difficulty and controversy. We cannot offer a solution to the problem, but can only discuss approaches to it, citing cautions at appropriate points.

When objectives can be sorted by classes such that each objective in the class is equivalent, then this is nothing more than a classification system or *nominal* system of measurement. However, when objectives can be sorted into classes that can then be ordered, this is termed an *ordinal* system of measurement.

An illustration of a nominal classification of objectives is the concept advanced by Kepner and Tregoe where objectives first are categorized on a "must" or "desirable" basis. For example, a university library must serve the faculty and student population, but perhaps it is only desirable to serve residents of the local community. Once objectives are classed on a "must" versus "desirable" basis, the desirable objectives may be considered only subjectively thereafter, whereas the must objectives are deserving of greater attention and, if possible, quantification. Some of the approaches taken to quantify objectives will now be treated; in the discussion, the term *goal* is used synonymously with *objective* and the symbolism adopted to denote the $k$th goal or objective of interest is $G_k$.

## 9.7.2 Ranking

If a decision maker is capable of stating preferences among objectives, objectives may be ranked in terms of relative importance. A technique to assist the decision maker in making consistent preference statements is the method of paired comparisons. In order to illustrate the method, assume that four goals (objectives) are relevant: $G_1$, $G_2$, $G_3$, and $G_4$. The method of paired comparisons enumerates all possible pairs $[n(n-1)/2]$, makes preference statements, and deduces a ranking for each goal. In this example, the possible pairs are:

$$G_1 \text{ versus } G_2 \qquad G_2 \text{ versus } G_3 \qquad G_3 \text{ versus } G_4$$
$$G_1 \text{ versus } G_3 \qquad G_2 \text{ versus } G_4$$
$$G_1 \text{ versus } G_4$$

If the symbol $>$ represents "preferred to" and the symbol $<$ represents "not preferred to," suppose the results of the decision about the pairs were:

$$G_1 > G_2 \qquad G_2 < G_3 \qquad G_3 > G_4$$
$$G_1 < G_3 \qquad G_2 > G_4$$
$$G_1 > G_4$$

Rewriting so that all comparisons are of the "preferred to" variety yields:

$$G_3 > G_1 \qquad G_1 > G_2 \qquad G_2 > G_4$$
$$G_3 > G_2 \qquad G_1 > G_4$$
$$G_3 > G_4$$

It can then be concluded that $G_3$ is preferred to all others, $G_1$ is preferred to two others, $G_2$ is preferred to one other, and $G_4$ is preferred to zero others and the following ranks should be assigned: $G_3 = $ I, $G_1 = $ II, $G_2 = $ III, and $G_4 = $ IV.

If, in the above procedure, the preference statements had been $G_1 > G_2$, $G_2 > G_4$ and $G_1 < G_4$, the decision maker should reconsider the inconsistent judgments. These results are "intransitive" in a mathematical sense.

The ranking method of classifying objectives according to relative importance is certainly useful; indeed, ranking was implicitly used in the comparison of alternative investments under conditions of assumed certainty in the earlier chapters of this textbook. There are, however, obvious deficiencies in the ranking method. For instance, the ranks do not indicate the *extent* to which one objective is preferred over another. Furthermore, because rankings are an *ordinal scale of measurement,* the mathematical operations of addition, subtraction, multiplication, division, and so forth, cannot be performed. Therefore a higher-order scale of measurement (interval or ratio scale) is generally required to determine choice among alternatives under conditions of risk and uncertainty [15]. Unfortunately, accepted methods for measuring objectives on a ratio scale do not currently exist.

### 9.7.3 Weighting Objectives

Before proceeding with a discussion of certain methods to assign relative weights to multiple objectives, a distinction between objectives and outcomes should be made. This is perhaps most easily done by means of an example where a machine-replacement situation is of interest, and the following two objectives are defined:

$G_1$: minimize the equivalent annual cost over a five-year planning horizon
$G_2$: minimize the number of production personnel transferred to other jobs within the firm

Let

$\theta_{jk}$ = the outcome resulting from taking the $j$th alternative course of action in terms of the $k$th objective or goal
$V(\theta_{jk})$ = the value of the outcome resulting from selecting the $j$th alternative in terms of the $k$th objective

Suppose, in this example, that

$A_1$ = the alternative of keeping the present machining center
$A_2$ = the alternative of replacing the present center with a new machining center

The outcomes for each of these alternatives in terms of the two objectives, $G_1$ and $G_2$, are:

$\theta_{11}$ = an annual cost of \$80,000
$\theta_{12}$ = no persons transferred
$\theta_{21}$ = an annual cost of \$60,000
$\theta_{22}$ = two persons transferred

In order to compare the two alternatives in a quantitative fashion, values must be assigned to each outcome; the values must be weighted in some way to reflect the relative importance of the two objectives, and some type of aggregate number must be determined for each alternative to enable a comparison and final choice. For example, the end result (without regard for validity at this point) might appear as

$$V(A_1) = w_1 V(\theta_{11}) + w_2 V(\theta_{12})$$

$$V(A_2) = w_1 V(\theta_{21}) + w_2 V(\theta_{22})$$

where $V(A_2) > V(A_1)$.

The $w_k$ (or $w_1$, $w_2$) values are weighting factors identifying the relative importance of the objectives $G_k$ (or $G_1$, $G_2$) and hence are applied to the appropriate outcomes ($\theta_{jk}$) of the $j$th alternative. Determining the weighting factors is thus one issue in quantifying multiple objectives. The values $V(\theta_{jk})$ may be in different dimensions and, if so, may have to be transformed to another common dimension. Determining the transformation function is another issue in quantifying multiple objectives. Finally, the values for each alternative, $V(A_j)$, were determined by a linear function. Determining the mathematical functional form of the model is yet another issue in quantifying multiple objectives. Again, all these issues will not be resolved in this chapter, but some approaches that have been taken will be presented.

One method for assigning relative weights to multiple objectives in order to indicate the extent by which one objective is preferred to another is simply to assign weights by judgment. That is, in the case of four objectives ($G_1$, $G_2$, $G_3$, $G_4$), the decision maker may reason that $G_3$ is twice as important as $G_1$, three times as important as $G_2$ and five times as important as $G_4$. Assigning a weighting value of 1.00 to $G_3$, it can be reasoned that the set of weighting values is

$$\text{For } G_3: \quad w_3 = 1.000$$

$$\text{For } G_1: \quad w_1 = 0.500$$

$$\text{For } G_2: \quad w_2 = 0.333$$

$$\text{For } G_4: \quad w_4 = 0.200$$

These weights may be transformed to a scale from 0 to 1.0, or "normalized," by dividing each weight by the sum of all weights; that is,

$$w_3' = \frac{1.000}{2.033} = 0.4918$$

$$w_1' = \frac{0.500}{2.033} = 0.2459$$

$$w_2' = \frac{0.333}{2.033} = 0.1639$$

$$w_4' = \frac{0.200}{2.033} = 0.0984$$

Another method of assigning weights to objectives by judgment, but still keeping consistency, is adapted from Churchman, Ackoff, and Arnoff [3], and the commentary on this method given by Morris [15]. The references state that the method is applicable to weight objectives or outcomes. For a decision involving multiple objectives $G_1, \ldots, G_m$ to be evaluated, this method requires two major assumptions. Paraphrasing Morris, these are:

1. It must be possible for the decision maker to think about and judge the value of any combination of objectives. That is, it must be possible to consider not only the importance or weight value for, say, $G_1$ but also the weighting value for sums of objectives.

2. Values are assumed to be additive. Given the individual weighting values for, say, $G_1$ and $G_2$, it is assumed that the weighting value for both objectives is the sum of their individual weighting values.

The general procedure for the method now follows.

1. Rank the objectives in order of importance, where $G_1$ indicates the most important, $G_2$ the next most important, and so forth, and $G_m$ is the least important.

2. Assign the weighting value of 1.00 to $G_1$ (that is, $w_1 = 1.00$) and weighting values to the other objectives to reflect their importance relative to $G_1$. These two steps actually complete the judgmental process, and the following steps serve to refine the initial judgments and aid consistency.

3. Compare $G_1$ to the linear combination of all other objectives. That is, compare $G_1$ versus $(G_2 + G_3 + \cdots + G_m)$.

   (a) If $G_1 > (G_2 + G_3 + \cdots + G_m)$, adjust (if necessary) the value of $w_1$ such that $w_1 > (w_2 + w_3 + \cdots + w_m)$. Attempt, in all adjustments of the procedural steps, to keep the relative values within the objective *group* invariant. Proceed to step 4.

   (b) If $G_1 = (G_2 + G_3 + \cdots + G_m)$, adjust $w_1$ so that $w_1 = (w_2 + \cdots + w_m)$ and proceed to step 4.

   (c) If $G_1 < (G_2 + G_3 + \cdots + G_m)$, adjust (if necessary) the value of $w_1$ so that $w_1 < (w_2 + \cdots + w_m)$.

      (1) Now compare $G_1$ versus $(G_2 + G_3 + \cdots + G_{m-1})$.

         [a] If $G_1 > (G_2 + G_3 + \cdots + G_{m-1})$, adjust (if necessary) the values so that $w_1 > (G_2 + G_3 + \cdots + G_{m-1})$ and proceed to step 4.

         [b] If $G_1 = (G_2 + G_3 + \cdots + G_{m-1})$, adjust (if necessary) the values such that $w_1 = (w_2 + w_3 + \cdots + w_{m-1})$ and proceed to step 4.

         [c] If $G_1 < (G_2 + G_3 + \cdots + G_{m-1})$, adjust (if necessary) the values such that $w_1 < (w_2 + w_3 + \cdots + w_{m-1})$. Then, compare $G_1$ versus $(G_2 + G_3 + \cdots + G_{m-2})$, and so forth, either until $G_1$ is preferred or equal to the rest, then proceed to step 4, or until the comparison of $G_1$ versus $(G_2 + G_3)$ is completed, then proceed to step 4.

4. Compare $G_2$ versus $(G_3 + G_4 + \cdots + G_m)$ and proceed as in step 3.
5. Continue until the comparison of $G_{m-2}$ versus $(G_{m-1} + G_m)$ is completed.
6. If desired, convert each $w_k$ into a normalized value, dividing $w_k$ by the sum $\Sigma_{k=1}^m w_k$.

## Example 9.9

Assume that it is desired to determine weighting values for four objectives where these have been ranked in order of importance such that $G_1$ is most important and $G_4$ is the least important. Step 1 of the procedure is thus accomplished.

**Step 2.** The objectives are tentatively assigned the weights $w_1 = 1.00$, $w_2 = 0.80$, $w_3 = 0.50$, and $w_4 = 0.20$.

**Step 3.** Assume that $G_1$ is preferred to the linear combination of the other objectives. That is, $G_1 > (G_2 + G_3 + G_4)$. Since $w_1 < (w_2 + w_2 + w_3)$, it is necessary to adjust $w_1$ to be greater than the sum of 1.50—say, $w_1 = 1.75$. Then proceed to step 4.

**Step 4.** Comparing $G_2$ versus $(G_3 + G_4)$, assume that $G_2 < (G_3 + G_4)$, and the step 3 procedure is repeated. Since the weighting values do not reflect this [that is, $w_2 = 0.80$ is not less than $(0.50 + 0.20)$], $w_2$ is adjusted to be 0.65. [The adjustment to decrease $w_2$ does not violate the previous preference of $G_1 > (G_2 + G_3 + G_4)$ with the new value of $w_1 = 1.75$ to reflect this preference.] Since, in this example of only four objectives, the comparison of $G_{m-2}$ versus $(G_{m-1} + G_m)$ or $G_2$ versus $(G_3 + G_4)$ has been completed at this point, step 5 of the procedure has been done.

If it is desired to normalize the results, new weighting values are calculated as

$$
\begin{aligned}
w_1' &= 1.75/(1.75 + 0.65 + 0.50 + 0.20) = 0.5645 \\
w_2' &= 0.65/3.10 &&= 0.2097 \\
w_3' &= 0.50/3.10 &&= 0.1613 \\
w_4' &= 0.20/3.10 &&= \underline{0.0645} \\
&\qquad\qquad\text{Total} &&\ \ 1.0000
\end{aligned}
$$

For a small number of objectives, this method is relatively easy to use, but becomes cumbersome as the number of objectives increases. However, it is believed that many real-world problems reduce to a few primary objectives. It should also be again emphasized that the method still depends on the decision maker's judgment in assigning the relative weights.

### 9.7.4 Determining the Value of Multiple Objectives

In the previous section on weighting objectives, an example of a machine replacement situation was hypothesized to introduce the discussion. This example is now recalled, where the objectives were:

G₁: minimize the equivalent annual cost over a 5-year planning horizon
G₂: minimize the number of production personnel transferred to other jobs within the firm

For the two alternatives of $A_1$ (keep present machining center) and $A_2$ (replace with new machining center), the outcomes were

$A_1$: \$80,000 annual cost $(\theta_{11})$; no persons transferred $(\theta_{12})$
$A_2$: \$60,000 annual cost $(\theta_{21})$; two persons transferred $(\theta_{22})$

Let it be assumed that, by some method, the objectives have been weighted as $w_1 = 0.60$ and $w_2 = 0.40$.

One method to assign values to the outcomes of the two alternatives is to make a second series of judgments as to the degree to which each outcome succeeds in meeting the objective. Suppose these judgments are made in the form of numbers on an arbitrary scale from 0 to 1, where the higher the scale value, the closer to meeting the desired objective. Assume the results of such judgments are as given below.

| Outcome | Value |
|---------|-------|
| $\theta_{11}$ | 0.50 |
| $\theta_{12}$ | 1.00 |
| $\theta_{21}$ | 0.85 |
| $\theta_{22}$ | 0.60 |

Then, assuming that a linear model is appropriate, the values for each alternative are calculated as

$$V(A_1) = w_1 V(\theta_{11}) + w_2 V(\theta_{12})$$
$$= (0.60)(0.50) + (0.40)(1.00) = 0.70$$
$$V(A_2) = w_1 V(\theta_{21}) + w_2 V(\theta_{22})$$
$$= (0.60)(0.85) + (0.40)(0.60) = 0.75$$

Thus, since $V(A_2) > V(A_1)$, alternative $A_2$ would be chosen, and the present machining center would be replaced.

Clearly, these results are sensitive to the weighting values assigned by judgment. Furthermore, the magnitude of the difference in values between $V(A_1)$ and $V(A_2)$ is a function of the arbitrary scale (from 0 to 1) chosen. The linear model assumed is also questionable [13]. All these points are criticisms of this method of quantifying multiple objectives. In the literature on the theory of utility functions, it is argued that if a decision maker's utility function can be established for each objective, then values of outcomes (e.g., \$80,000 annual cost) can be converted to a utility value (e.g., \$80,000 annual cost equals 0.75) from the decision maker's utility curve. Such a utility value could be determined for each objective, and our model for the alternatives above would become

$$U(A_1) = w_1 U(\theta_{11}) + w_2 U(\theta_{12})$$

$$U(A_2) = w_1 U(\theta_{21}) + w_2 U(\theta_{22})$$

Then, the decision maker should choose to maximize the weighted utility value. Whether such utility functions for each objective can be readily obtained for a single decision maker is still questionable in our opinion, and even if this could be accomplished, there is yet another question of whether interaction between multiple objectives would be totally missed. As stated in Chapter One, the study of value measurement, and utility theory in particular, is interesting, but questions remain on the practical application and implementation of utility theory to business and engineering problems. In any event, it is judged that a treatment of utility theory is outside the scope of this textbook.

## 9.8 SUMMARY

In this chapter, some prescriptive or normative models for decisions under risk and uncertainty have been presented. For decisions under risk and uncertainty where the matrix model is appropriate, criteria that are commonly cited in the literature for choosing among mutually exclusive alternatives were discussed. None of these criteria were cited as the best or optimal basis for selecting an alternative; they merely represent observations on how decision makers seem to decide.

The decision-tree model for a sequence of decisions where the outcomes of alternatives are chance events was also presented, and the expectation-variance criterion was suggested as a guideline for choice among alternatives. Some fundamental laws of probability were reviewed, and Bayes's theorem of conditional probability was illustrated as a guideline for revising one's opinion in sequential decisions under risk when only imperfect information is available to the decision maker.

The last section of the chapter was an introductory presentation of some approaches in the literature for quantifying multiple objectives. The discussion offered and examples used were restricted to decisions under assumed certainty. However, the methods illustrated are easily extended to decisions under risk and uncertainty and the matrix model used for analysis.

## BIBLIOGRAPHY

1. Agee, Marvin H., Taylor, Robert E., and Torgersen, Paul E., *Quantitative Analysis For Management Decision,* Prentice-Hall, 1976.
2. Chernoff, Herman, and Moses, Lincoln E., *Elementary Decision Theory,* Wiley, 1959.

3. Churchman, C. West, Ackoff, Russell L., and Arnoff, E. Leonard, *Introduction to Operations Research*, Wiley, 1957.

4. Conrath, David W., "From Statistical Decision Theory to Practice: Some Problems with the Transition," *Management Science,* **19** (8), April 1973. Reprinted in EE Monograph Series No. 2, entitled *Risk and Uncertainty: Non-Deterministic Decision Making in Engineering Economy,* American Institute of Industrial Engineers, 1975, pp. 32–37.

5. de Neufville, Richard, and Stafford, Joseph H., *Systems Analysis for Engineers and Managers,* McGraw-Hill, 1958.

6. Fishburn, Peter C., "Decision Under Uncertainty: An Introductory Exposition," EE Monograph Series No. 2, entitled *Risk and Uncertainty: Non-Deterministic Decision Making in Engineering Economy,* American Institute of Industrial Engineers, Atlanta, Georgia, 1975, pp. 19–31.

7. Goetz, Billy, E., "Perplexing Problems in Decision Theory," *The Engineering Economist,* **14** (3), ASEE, Spring 1969.

8. Hespos, Richard F., and Strassman, Paul A., "Stochastic Decision Trees for the Analysis of Investment Decisions," *Management Science,* **11** (10), August 1965. Reprinted in EE Monograph Series No. 2, entitled *Risk and Uncertainty: Non-Deterministic Decision Making in Engineering Economy,* American Institute of Industrial Engineers, 1975, pp. 48–55.

9. Kepner, Charles H., and Tregoe, Benjamin B., *The Rational Manager,* McGraw-Hill, 1965.

10. Luce, R. Duncan, and Raiffa, Howard, *Games and Decisions: Introduction and Critical Survey,* Wiley, 1958.

11. Magee, John F., "Decision Trees for Decision Making," *Harvard Business Review,* **42** (4), July–August 1964.

12. Magee, John F., "How to Use Decision Trees in Capital Investment," *Harvard Business Review,* **42** (5), September–October 1964.

13. Miller, David W., and Starr, Martin K., *Executive Decisions and Operations Research,* Second Edition, Prentice-Hall, 1969.

14. Miller, Ernest C., *Advanced Techniques for Strategic Planning,* AMA Research Study 104, American Management Association, 1971.

15. Morris, William T., *The Analysis of Management Decisions,* Revised Edition, Richard D. Irwin, 1964.

16. Morris, William T., *Management Science: A Bayesian Introduction,* Prentice-Hall, 1968.

17. Morse, Phillip M., *Library Effectiveness: A Systems Approach,* M.I.T. Press, 1968.

18. Spiegel, Murray R., *Schaum's Outline of Theory and Problems of Statistics,* Schaum Publishing, 1961.

# PROBLEMS

**1.** Assume the following decision matrix where the cell values are units of gain.

|        | $S_1$ | $S_2$ | $S_3$ | $S_4$ |
|--------|-------|-------|-------|-------|
| $A_1$  | $4   | $ 2  | $0    | $1    |
| $A_2$  | - 4   | 10    | 3     | 7     |
| $A_3$  | 4     | 8     | 2     | 3     |
| $A_4$  | 2     | 4     | 9     | 5     |

If a person's index of optimism is estimated as $\alpha = 0.25$, predict the person's choice of alternative by applying the various principles of choice for a decision under certainty. (9.4.1, 9.5)

**2.** Analyze the cost matrix below by the use of the Hurwicz principle. (9.5.4)

|        | $S_1$ | $S_2$ | $S_3$ |
|--------|-------|-------|-------|
| $A_1$  | $6   | $6   | $6   |
| $A_2$  | 12    | 8     | 0     |
| $A_3$  | 4     | 7     | 3     |

**3.** Assume the values in the matrix of Problem 2 are units of gain. Analyze the alternatives by using the Hurwicz principle. (9.5.4).

**4.** For the matrix given in Problem 2, if the probability of $S_1$ occurring is 0.10, what are the probabilities for $S_2$ and $S_3$ which would make the decision maker indifferent between the choice of $A_2$ and $A_3$? Assume the expectation principle of choice. (9.4.2).

**5.** Apply the various decision under uncertainty principles of choice to the matrix below. The values in the matrix are cost units. (9.4.1, 9.5)

|        | $S_1$ | $S_2$ | $S_3$ | $S_4$ |
|--------|-------|-------|-------|-------|
| $A_1$  | $13  | $13  | $5    | $9    |
| $A_2$  | 8     | 8     | 8     | 8     |
| $A_3$  | 0     | 21    | 5     | 5     |
| $A_4$  | 9     | 17    | 5     | 5     |
| $A_5$  | 5     | 7     | 7     | 5     |

**6.** A particular production department supervisor is known to be very conservative in business matters. For the decision model below, where values are equivalent annual profits (in thousands of dollars) for a $MARR = 15\%$ and a 5-year planning horizon, predict the supervisor's choice of an alternative by each of the principles for a decision under uncertainty. (9.4.1, 9.5)

|       | $S_1$ | $S_2$ | $S_3$ | $S_4$ |
|-------|-------|-------|-------|-------|
| $A_1$ | $16  | $16  | $ 8  | $12  |
| $A_2$ | 9     | 20    | 10    | 6     |
| $A_3$ | 15    | 10    | 8     | 9     |
| $A_4$ | 3     | 14    | 6     | 11    |
| $A_5$ | 8     | 16    | 9     | 12    |

**7.** Assume the following decision matrix.

|       | $p_1$ $S_1$ | $p_2$ $S_2$ | $p_3$ $S_3$ |
|-------|-------------|-------------|-------------|
| $A_1$ | −$20       | $100       | $800       |
| $A_2$ | 218         | 218         | 218         |
| $A_3$ | 650         | 120         | 40          |

**(a)** Treat the decision as one under uncertainty and the values in the matrix as costs. Then,

**(1)** If $\alpha$ = an index of optimism, for what value of $\alpha$ is one indifferent between $A_1$ and $A_2$? (9.5.4).

**(2)** Which alternative would be chosen by the Savage principle? (9.5.5).

**(b)** Treat the decision as one under risk where $p_1 = 0.20$, $p_2 = 0.70$, and $p_3 = 0.10$ and the values in the matrix are profits. Which alternative would be chosen by the expectation-variance principle? (9.4.2).

**8.** Given the decision matrix shown below (with *cost* elements), determine the preferred alternative using the following principles of choice. (9.5)
**(a)** Minimax.
**(b)** Minimin.
**(c)** Minimax regret.
**(d)** Expected cost, if each future state is expected to occur with equal probability.
**(e)** Hurwicz (with $\alpha = 0.80$).

|       | $S_1$ | $S_2$  | $S_3$ |
|-------|-------|--------|-------|
| $a_1$ | $120  | $50    | $ 10  |
| $a_2$ | 60    | 60     | 60    |
| $a_3$ | 70    | 50     | 60    |
| $a_4$ | 20    | − 50   | 250   |

9. Given the decision matrix shown below, determine the recommended alternative under the following principles of choice: expectation, most probable future, maximax, maximin, and minimax regret. Entries in the matrix are profits. (9.4, 9.5)

|       | $p_1 = 0.1$ <br> $S_1$ | $p_2 = 0.3$ <br> $S_2$ | $p_3 = 0.5$ <br> $S_3$ | $p_4 = 0.1$ <br> $S_4$ |
|-------|-------|--------|-------|-------|
| $A_1$ | −$ 50 | $100   | $200  | $400  |
| $A_2$ | − 20  | 50     | 500   | 100   |
| $A_3$ | 100   | − 100  | 50    | 100   |
| $A_4$ | 200   | − 50   | 50    | 200   |

10. Shown below is a matrix of costs for three investment alternatives under three future states. Determine the preferred alternative using **(a)** the Laplace principle, **(b)** the Minimax principle, and **(c)** the Minimax Regret principle. (9.5.1, 9.5.2, 9.5.5).

|       | $S_1$ | $S_2$  | $S_3$ |
|-------|-------|--------|-------|
| $A_1$ | $300  | $200   | $100  |
| $A_2$ | 145   | 180    | 200   |
| $A_3$ | 210   | 140    | 175   |

11. An analysis yields a decision under risk given in the matrix below, where the matrix values are annual profits in thousands of dollars.

| $p_k =$ | 0.1 | 0.1 | 0.4 | 0.2 | 0.1 | 0.1 |
|---------|-----|-----|-----|-----|-----|-----|
| $S_k =$ | $S_1$ | $S_2$ | $S_3$ | $S_4$ | $S_5$ | $S_6$ |
| $A_1$ | $12 | $ 5 | -$ 8 | -$3 | $ 6 | $ 9 |
| $A_2$ | 7 | 0 | 1 | 5 | 20 | 7 |
| $A_3$ | 3 | 3 | 7 | 9 | - 5 | 5 |
| $A_4$ | 0 | 12 | 15 | 2 | 8 | - 200 |
| $A_5$ | - 10 | 22 | 9 | 0 | 4 | 12 |

(a) Which alternative should be chosen in order to minimize the probability of a loss? (9.4.4)
(b) Which alternative should be chosen in order to maximize the probability of an annual profit of at least $9000? (9.4.4)
(c) Which alternative should be chosen in order to maximize the probability that annual profits will be be between $3000 and $10,000? (9.4.4)
(d) Would the most probable future principle be a reasonable one to follow in selecting an alternative? (9.4.3)

12. An electronics firm has recently received a United States government contract to produce a certain quantity of expensive electronic guidance systems. The contracted number of units can be produced in 2 years, and the contract terminates then. The firm won the contract with cost estimates based on using present manufacturing equipment. If the firm had more specialized equipment (with almost zero salvage value immediately after purchase), the unit cost of production would be reduced, and profit per unit would thus be increased. However, this would be true only if the contract were for at least 4 years instead of 2 years. The firm feels there is a 50% chance for an additional 2-year contract, a 30% chance for an additional 4-year contract, and a 20% chance of no contract renewal.
Using only present equipment, total profit on this job for a 2-year, 4-year, and 6-year contract is estimated as $40,000, $80,000 and $120,000, respectively. If the specialized equipment is purchased, total profit on this job for a 2-year, 4-year, and 6-year contract is estimated as -$100,000, $90,000, and $180,000, respectively. If an expectation principle is used, should the specialized equipment be purchased? Does the most probable future criterion seem reasonable in this decision? (9.4.1, 9.4.2, 9.4.3).

13. A highway construction firm is considering the purchase of a used mobile crane. Two such cranes are available, $A_1$ and $A_2$. If either is purchased, the firm will use a 5-year depreciation schedule. The cranes differ in capacity, age, and mechanical condition, but both are presently in operating condition and each has the capacity to do the handling jobs expected. The firm expects that a major overhaul of each crane will eventually be required, but when the overhaul will be necessary is uncertain. Estimates of the net annual operating expenses (excluding labor) for each crane are given below:

|        | $A_1$ | | $A_2$ | |
| Year | Expense Schedule 1 | Expense Schedule 2 | Expense Schedule 1 | Expense Schedule 2 |
| --- | --- | --- | --- | --- |
| 1 | $ 6,500 | $ 6,500 | $15,000 | $ 8,000 |
| 2 | 7,500 | 12,500 | 8,000 | 8,500 |
| 3 | 8,500 | 5,500 | 8,000 | 18,000 |
| 4 | 14,000 | 6,000 | 8,000 | 6,000 |
| 5 | 5,500 | 6,500 | 8,000 | 6,500 |

Other data for the two used cranes are:

|  | $A_1$ | $A_2$ |
| --- | --- | --- |
| First cost | $40,000 | $27,500 |
| Salvage value at the end of 5 years | 12,000 | 8,500 |

If the firm uses a before-tax $MARR = 30\%$ and a present worth method of comparing investment alternatives, which crane would be purchased if the Laplace principle were used? (9.4.1, 9.5.1).

14. Suppose a coin is biased such that, when the coin is tossed, the probability of a head (H) showing is 0.60. Now assume that the coin is tossed three consecutive times.
   (a) Sketch a probability tree for the three tosses and determine the probability values for each of the eight possible outcomes: HHH, HHT, and so on. For the three tosses, the following states are now defined:

   $S_1$ = the event of three heads (3H)
   $S_2$ = the event of two heads, one tail—occurring in any order (2H, T)
   $S_3$ = the event of two tails, one head—occurring in any order (2T, H)
   $S_4$ = the event of three tails (3T) (9.6.1, 9.6.2)

   (b) Create a decision matrix where four alternatives are to guess the events defined above. Now assume that before tossing the coin three times, a person offers a $4 prize if the (3H) event is guessed correctly, a $3 prize if the (2H, T), event is guessed correctly, a $4 prize if the (2T, H) event is guessed correctly, and a $5 prize if the (3T) event is guessed correctly. However, there is a $1 charge to play the game. If the game is played, would the expectation principle and the most probable future principle select the same alternative? (9.2, 9.4.2, 9.4.3)

15.* A vehicle manufacturer plans to develop and operate a public bus system for a community. The manufacturer's purpose is to demonstrate the profitability of the bus system and then sell the system to the community. The manufacturer's most serious concern is whether the Public Utilities Commission will authorize a compet-

* Problem 15 is taken from deNeufville and Stafford [5] by permission of the publisher.

ing bus system. It is rationalized that there is a small chance that such an event will happen; therefore, the manufacturer subjectively assigns a probability of 0.2 to the event of competition.

A detailed analysis by the manufacturer results in an estimate of 50 vehicles needed in the event of no competition and 25 vehicles needed if competition results. Furthermore, it is reasoned that operating either a 50-vehicle or a 25-vehicle system will not affect the probability of the Public Utilities Commission authorizing or not authorizing competition.

If the manufacturer initially develops a 50-vehicle system and there is no competition, it is estimated the system will yield a $250,000 profit. In the event of competition, a $120,000 loss is projected.

If a 25-vehicle system is initially developed and competition occurs, the manufacturer plans to sell the system as quickly as possible and estimates a $25,000 profit in this instance. On the other hand, if there is no competition, the 25-vehicle system will be inadequate. Poor service would therefore result, and the value of the demonstration would be reduced, thereby adversely affecting the sales price. Thus, in the event of no competition, the manufacturer reasons that another decision must be faced: whether to expand from 25 vehicles to 50 vehicles or not to expand. If there is no expansion, the manufacturer would sell soon to avoid the bad publicity for expected poor service, estimating a $35,000 profit in this case. If, on the other hand, the system is expanded and the total system is sold as soon as possible, two outcomes are predicted. One outcome is that poor service could result during the expansion and a net $10,000 loss would result (a probability of 0.10 is subjectively assigned to this event). The second outcome is that poor service would not result, and a net $140,000 profit is expected in this instance. Create a decision-tree model of this situation and solve by use of the expectation principle. (9.6.1, 9.6.2, 9.4.2).

16. The Promax Company purchases a steel casting from four suppliers (A, B, C, and D) on a random basis throughout a production year. The steel casting from each supplier may be of either low (L) or high (H) nickel alloy content. Historical records reveal that the Promax Company has purchased castings from the four suppliers in the proportions shown in the following table. (9.6.1, 9.6.2)

| Supplier | Proportion of Total Castings | Proportion by Alloy Content | |
|---|---|---|---|
| | | L | H |
| A | 0.45 | 0.20 | 0.80 |
| B | 0.10 | 0.65 | 0.35 |
| C | 0.30 | 0.55 | 0.45 |
| D | 0.15 | 0.40 | 0.60 |

(a) If a steel casting is selected at random from the shop floor of the Promax Company, estimate the probability that the casting has a high nickel content.
(b) If a casting that is selected from the shop floor of the Promax Company has a low nickel content, estimate the probability that it came from supplier C.

**17.** Let it be supposed that, out of your sight, an experimenter rolls a green, red, and white die. The experimenter covers two of these. Data concerning the dice are that (1) the green die has five surfaces marked H, one surface marked T, (2) the red die has two surfaces marked H, four surfaces marked T, and (3) the white die has three surfaces marked H, three surfaces marked T. Also

> $A$ = the event that the green die is uncovered.
> $B$ = the event that the red die is uncovered.
> $C$ = the event that the white die is uncovered.
> $h$ = the event that a surface marked H is showing on the uppermost surface of the uncovered die.
> $t$ = the event that a surface marked T is showing.

Now, the experimenter reports that the letter H is showing on the uppermost surface of the uncovered die.

**(a)** Using the logic of Bayes's theorem, what is the best guess concerning the color of the uncovered die? (9.6.1, 9.6.2)

**(b)** Assume that the experimenter, after the report but before the guess, offers to pay $6 if the green die is guessed correctly, $15 if the red die is guessed correctly, and $15 if the white die is guessed correctly. However, there is a $3 charge to play the game. Given that the game is played, model this situation as a decision under risk matrix and determine the best guess by the expectation principle. (9.2, 9.4.2)

**18.** Work problem 17 above, with the following two modifications:

**(a)** The experimenter reports that the letter T is showing on the uppermost surface of the uncovered die.

**(b)** The experimenter offers to pay $9, $12, and $15 if the green, red and white die, respectively, is guessed correctly. There is a $5 charge to play the game.

**19*.** After receiving somewhat unreliable information concerning enemy troop movement along a supply route, a military commander of an artillery battalion faces the following combat decision (assume the matrix values are loss units, arbitrarily chosen).

|  | $p_1 = 0.7$<br>$S_1$ = Troop<br>Movement | $p_2 = 0.3$<br>$S_2$ = No Troop<br>Movement |
|---|---|---|
| $A_1$ = bombard supply route | 0 | 4 |
| $A_2$ = not bombard supply route | 12 | 0 |

The commander can choose between $A_1$ and $A_2$ without additional information, or he can send out a reconnaissance plane for additional information. If the plane is

---

* Problem 19 taken from Agee, Taylor, and Torgersen [1] by permission of the publisher.

sent out, he reasons the mutually exclusive and collectively exhaustive outcomes of the flight are:

Outcome $O_1$: the plane is shot down, with a loss of 3 units.
Outcome $O_2$: the plane returns safely with negative information about troop movement, with a loss of 0.2 units.
Outcome $O_3$: the plane returns safely with a report of suspicious activity, with a loss of 0.2 units.

The commander assigns the following subjective conditional probabilities to these outcomes:

$$P(O_1|S_1) = 0.4 \qquad P(O_1|S_2) = 0.2$$
$$P(O_2|S_1) = 0.1 \qquad P(O_2|S_2) = 0.8$$
$$P(O_3|S_1) = 0.5 \qquad P(O_3|S_2) = 0.0$$

(a) Determine the posterior probabilities. (9.6.2)
(b) Create a decision tree for this problem. (9.6.1)
(c) Using the minimizing expected loss principle of choice, what action should the commander take? (9.4.2)
(d) What is the expected value of sample information in this problem? (9.6.3, 9.6.4)

20. A manufacturer is considering the possibility of introducing a new product and the advisability of a test marketing prior to making the final decision. The alternatives are

$$a_1 = \text{market the product}$$

$$a_2 = \text{do not market the product}$$

For simplicity, only three possible futures are considered; they are shown below together with the prior probabilities associated with them.

|  | Profit | Prior Probability |
|---|---|---|
| $S_1$: the product captures 10% of the market | $10,000,000 | 0.70 |
| $S_2$: the product captures 3% of the market | 1,000,000 | 0.10 |
| $S_3$: the product captures less than 1% of the market | − 5,000,000 | 0.20 |

If the test marketing is made, three possible results are considered.

$Z_1$: test sales of more than 10% of the market.
$Z_2$: test sales of 5 to 10% of the market.
$Z_3$: test sales of less than 5% of the market.

The conditional probabilities of the test results are

|           | $Z_1$ | $Z_2$ | $Z_3$ |
|-----------|-------|-------|-------|
| $P(Z|S_1)$ | 0.6   | 0.3   | 0.1   |
| $P(Z|S_2)$ | 0.3   | 0.6   | 0.1   |
| $P(Z|S_3)$ | 0.1   | 0.1   | 0.8   |

(a) Determine the prior expected profit. (9.6.1, 9.6.2, 9.6.3)
(b) Determine *EVPI*. (9.6.3)
(c) Determine *EVSI*. (9.6.4)
(d) If the test marketing costs $250,000, what would be the expected net gain from the sample information? (9.6.4)

21. Work problem 20 above, with the following two modifications:
(a) The prior probabilities of states $S_1$, $S_2$, and $S_3$ are: 0.50, 0.30, and 0.20, respectively.
(b) Test marketing costs are $200,000.

22. A textile firm in a rural town owns two trucks that transport raw materials from and finished goods to a nearby city. One of the trucks (tractor and trailer) is 8 years old, and annual maintenance expenses are increasing significantly. Management is considering the purchase of a new truck but, because of a mild national economic recession, textile sales have declined during the past two years, and management is reluctant to make a large capital expenditure at present. Although in the midst of a recession, inflation is also occurring; if the decision to purchase is delayed for a year, the cost of a new truck will probably increase 10% or more.
Adopting a 2-year planning horizon, the following estimates and judgments are made.

*Present Truck*
The present market value is approximately $9000. If kept one more year, operating (excluding labor and depreciation) and maintenance expenses are estimated as either $6000, $10,000, or $12,000 with subjective probabilities of 0.25, 0.50, and 0.25, respectively. The salvage value at the end of the year is about $8000. If kept one more year, a new decision to keep an additional year or buy a new truck will be made.
If the truck is kept two additional years, operating and maintenance expenses (excluding labor and depreciation) for the second year are estimated to be either $7000, $10,000, or $15,000 with subjective probabilities of 0.10, 0.70, and 0.20, respectively. The salvage value at the end of the second year is about $7500.

*New Truck*
A new truck can be purchased now for $30,000. Average annual operating, maintenance, and depreciation expenses for each of the first 2 years are estimated to be $15,000 with reasonable certainty.
If the purchase of the truck is delayed for 1 year, an increase in purchase price is

virtually certain. It is judged there is a 50–50 chance that the purchase price will be $32,000 or $34,000. If it is $32,000, total operating, maintenance, and depreciation expenses will be $16,000; if it is $34,000, total expenses will be $17,500. For simplicity, assume $MARR = 0\%$ and (a) create a decision-tree model for this situation, and then (b) determine the best present decision by the expectation principle. (9.6.1, 9.4.2)

23. The Jax Tool and Engineering Company is experiencing considerable problems with in-process material handling and storage of finished goods. A new layout of machinery is possible, but this alternative is subjectively judged by management to be prohibitively expensive, and the alternative is discarded. They wish, therefore, to consider a new material handling system and identify the following objectives.

$G_1$: minimize annual costs over a 10-year planning horizon.
$G_2$: minimize the disruption of production during the installation of the new system.
$G_3$: install material handling equipment that has flexibility to meet different handling requirements.
$G_4$: the material handling equipment should have a high index of repairability.
$G_5$: material handling personnel should require little training in order to operate the equipment.

Management ranks these goals in the following order of importance: $G_1 = $ I, $G_3 = $ II, $G_2 = $ III, $G_5 = $ IV, and $G_4 = $ V. Using the Churchman, Ackoff, and Arnoff method of weighting objectives, assign weights to these five objectives. (9.7.4)

24. Suppose a decision may be modeled as

|  | $p_1 = 0.6$<br>$S_1$<br>Contract Renewed | $p_2 = 0.4$<br>$S_2$<br>Contract Not Renewed |
|---|---|---|
| $A_1$: use present equipment | $V(\theta_{11})$ | $V(\theta_{12})$ |
| $A_2$: purchase new equipment | $V(\theta_{21})$ | $V(\theta_{22})$ |

Assume that analysis produces the following descriptions of the outcomes:

$\theta_{11}$ = no new capital required, 20 people required for at least 4 years, average annual profit units are 85.
$\theta_{12}$ = no new capital required, 20 people required for 2 years and then laid off, average annual profit units are 67.
$\theta_{21}$ = two uinits of new capital required, 8 people required for at least 4 years, average annual profit units are 150.
$\theta_{22}$ = two units of new capital required, 8 people required for 2 years and then laid off, average annual profit units are 400.

Presume the relevant values for objectives are

$V$(profit units)                = $+0.003$ (profit units) $+ 1.0$

$V$(capital units required) $= e^{-u}$, where $u = $ the number of capital units required

$$V(\text{work-years required}) = -0.0145 \ (\text{work-years required}) + 1.0$$

Furthermore, the goals of "capital units required" and "annual profit units" are considered equally important, but the labor consideration is only half as important as either of the other two.

On the basis of this information, which alternative would you recommend? (9.2, 9.7, 9.7.2)

25. Work Problem 24 above, with the following two modifications.
    (a) The probabilities of states $S_1$ and $S_2$ are 0.3 and 0.7, respectively.
    (b) The goal of "capital units required" is double the importance of the "annual profit units" goal and four times the importance of the labor factor.

# APPENDIX A
## DISCRETE COMPOUNDING

## SECTION I—DISCRETE COMPOUND INTEREST FACTORS

## SECTION II—GEOMETRIC SERIES FACTORS

# SECTION I—DISCRETE COMPOUND INTEREST FACTORS

| | Single payment | | Uniform series | | | | Gradient series | |
|---|---|---|---|---|---|---|---|---|
| | Compound amount factor | Present worth factor | Compound amount factor | Sinking fund factor | Present worth factor | Capital recovery factor | Uniform series factor | Present worth factor |
| $n$ | To find $F$ given $P$ $F\|P\,i,n$ | To find $P$ given $F$ $P\|F\,i,n$ | To find $F$ given $A$ $F\|A\,i,n$ | To find $A$ given $F$ $A\|F\,i,n$ | To find $P$ given $A$ $P\|A\,i,n$ | To find $A$ given $P$ $A\|P\,i,n$ | To find $A$ given $G$ $A\|G\,i,n$ | To find $P$ given $G$ $P\|G\,i,n$ |
| 1 | 1.0050 | 0.9950 | 1.0000 | 1.0000 | 0.9950 | 1.0050 | 0.0000 | 0.0000 |
| 2 | 1.0100 | 0.9901 | 2.0050 | 0.4988 | 1.9851 | 0.5038 | 0.4988 | 0.9901 |
| 3 | 1.0151 | 0.9851 | 3.0150 | 0.3317 | 2.9702 | 0.3367 | 0.9967 | 2.9604 |
| 4 | 1.0202 | 0.9802 | 4.0301 | 0.2481 | 3.9505 | 0.2531 | 1.4938 | 5.9011 |
| 5 | 1.0253 | 0.9754 | 5.0503 | 0.1980 | 4.9259 | 0.2030 | 1.9900 | 9.8026 |
| 6 | 1.0304 | 0.9705 | 6.0755 | 0.1646 | 5.8964 | 0.1696 | 2.4855 | 14.6552 |
| 7 | 1.0355 | 0.9657 | 7.1059 | 0.1407 | 6.8621 | 0.1457 | 2.9801 | 20.4493 |
| 8 | 1.0407 | 0.9609 | 8.1414 | 0.1228 | 7.8230 | 0.1278 | 3.4738 | 27.1755 |
| 9 | 1.0459 | 0.9561 | 9.1821 | 0.1089 | 8.7791 | 0.1139 | 3.9668 | 34.8244 |
| 10 | 1.0511 | 0.9513 | 10.2280 | 0.0978 | 9.7304 | 0.1028 | 4.4589 | 43.3865 |
| 11 | 1.0564 | 0.9466 | 11.2792 | 0.0887 | 10.6770 | 0.0937 | 4.9501 | 52.8526 |
| 12 | 1.0617 | 0.9419 | 12.3356 | 0.0811 | 11.6189 | 0.0861 | 5.4406 | 63.2136 |
| 13 | 1.0670 | 0.9372 | 13.3972 | 0.0746 | 12.5562 | 0.0796 | 5.9302 | 74.4602 |
| 14 | 1.0723 | 0.9326 | 14.4642 | 0.0691 | 13.4887 | 0.0741 | 6.4190 | 86.5835 |
| 15 | 1.0777 | 0.9279 | 15.5365 | 0.0644 | 14.4166 | 0.0694 | 6.9069 | 99.5743 |
| 16 | 1.0831 | 0.9233 | 16.6142 | 0.0602 | 15.3399 | 0.0652 | 7.3940 | 113.4238 |
| 17 | 1.0885 | 0.9187 | 17.6973 | 0.0565 | 16.2586 | 0.0615 | 7.8803 | 128.1231 |
| 18 | 1.0939 | 0.9141 | 18.7858 | 0.0532 | 17.1728 | 0.0582 | 8.3658 | 143.6634 |
| 19 | 1.0994 | 0.9096 | 19.8797 | 0.0503 | 18.0824 | 0.0553 | 8.8504 | 160.0360 |
| 20 | 1.1049 | 0.9051 | 20.9791 | 0.0477 | 18.9874 | 0.0527 | 9.3342 | 177.2322 |
| 21 | 1.1104 | 0.9006 | 22.0840 | 0.0453 | 19.8880 | 0.0503 | 9.8172 | 195.2434 |
| 22 | 1.1160 | 0.8961 | 23.1944 | 0.0431 | 20.7841 | 0.0481 | 10.2993 | 214.0610 |
| 23 | 1.1216 | 0.8916 | 24.3104 | 0.0411 | 21.6757 | 0.0461 | 10.7806 | 233.6767 |
| 24 | 1.1272 | 0.8872 | 25.4320 | 0.0393 | 22.5629 | 0.0443 | 11.2611 | 254.0820 |
| 25 | 1.1328 | 0.8828 | 26.5591 | 0.0377 | 23.4456 | 0.0427 | 11.7407 | 275.2683 |
| 26 | 1.1385 | 0.8784 | 27.6919 | 0.0361 | 24.3240 | 0.0411 | 12.2195 | 297.2278 |
| 27 | 1.1442 | 0.8740 | 28.8304 | 0.0347 | 25.1980 | 0.0397 | 12.6975 | 319.9522 |
| 28 | 1.1499 | 0.8697 | 29.9745 | 0.0334 | 26.0677 | 0.0384 | 13.1747 | 343.4329 |
| 29 | 1.1556 | 0.8653 | 31.1244 | 0.0321 | 26.9330 | 0.0371 | 13.6510 | 367.6624 |
| 30 | 1.1614 | 0.8610 | 32.2800 | 0.0310 | 27.7941 | 0.0360 | 14.1265 | 392.6323 |
| 31 | 1.1672 | 0.8567 | 33.4414 | 0.0299 | 28.6508 | 0.0349 | 14.6012 | 418.3347 |
| 32 | 1.1730 | 0.8525 | 34.6086 | 0.0289 | 29.5033 | 0.0339 | 15.0750 | 444.7615 |
| 33 | 1.1789 | 0.8482 | 35.7817 | 0.0279 | 30.3515 | 0.0329 | 15.5480 | 471.9053 |
| 34 | 1.1848 | 0.8440 | 36.9606 | 0.0271 | 31.1955 | 0.0321 | 16.0202 | 499.7581 |
| 35 | 1.1907 | 0.8398 | 38.1454 | 0.0262 | 32.0354 | 0.0312 | 16.4915 | 528.3120 |
| 40 | 1.2208 | 0.8191 | 44.1588 | 0.0226 | 36.1722 | 0.0276 | 18.8359 | 681.3345 |
| 45 | 1.2516 | 0.7990 | 50.3242 | 0.0199 | 40.2072 | 0.0249 | 21.1595 | 850.7629 |
| 50 | 1.2832 | 0.7793 | 56.6452 | 0.0177 | 44.1428 | 0.0227 | 23.4624 | 1035.6965 |
| 55 | 1.3156 | 0.7601 | 63.1258 | 0.0158 | 47.9814 | 0.0208 | 25.7447 | 1235.2686 |
| 60 | 1.3489 | 0.7414 | 69.7700 | 0.0143 | 51.7256 | 0.0193 | 28.0064 | 1448.6458 |
| 65 | 1.3829 | 0.7231 | 76.5821 | 0.0131 | 55.3775 | 0.0181 | 30.2475 | 1675.0271 |
| 70 | 1.4178 | 0.7053 | 83.5661 | 0.0120 | 58.9394 | 0.0170 | 32.4680 | 1913.6414 |
| 75 | 1.4536 | 0.6879 | 90.7265 | 0.0110 | 62.4136 | 0.0160 | 34.6679 | 2163.7520 |
| 80 | 1.4903 | 0.6710 | 98.0677 | 0.0102 | 65.8023 | 0.0152 | 36.8474 | 2424.6448 |
| 85 | 1.5280 | 0.6545 | 105.5942 | 0.0095 | 69.1075 | 0.0145 | 39.0065 | 2695.6382 |
| 90 | 1.5666 | 0.6383 | 113.3109 | 0.0088 | 72.3313 | 0.0138 | 41.1451 | 2976.0767 |
| 95 | 1.6061 | 0.6226 | 121.2224 | 0.0082 | 75.4757 | 0.0132 | 43.2633 | 3265.3296 |
| 100 | 1.6467 | 0.6073 | 129.3336 | 0.0077 | 78.5426 | 0.0127 | 45.3613 | 3562.7935 |

| | Single payment | | Uniform series | | | | Gradient series | |
|---|---|---|---|---|---|---|---|---|
| | Compound amount factor | Present worth factor | Compound amount factor | Sinking fund factor | Present worth factor | Capital recovery factor | Uniform series factor | Present worth factor |
| $n$ | To find $F$ given $P$ $F\|P\,i,n$ | To find $P$ given $F$ $P\|F\,i,n$ | To find $F$ given $A$ $F\|A\,i,n$ | To find $A$ given $F$ $A\|F\,i,n$ | To find $P$ given $A$ $P\|A\,i,n$ | To find $A$ given $P$ $A\|P\,i,n$ | To find $A$ given $G$ $A\|G\,i,n$ | To find $P$ given $G$ $P\|G\,i,n$ |
| 1 | 1.0058 | 0.9942 | 1.0000 | 1.0000 | 0.9942 | 1.0058 | 0.0000 | 0.0000 |
| 2 | 1.0117 | 0.9884 | 2.0058 | 0.4985 | 1.9826 | 0.5044 | 0.4985 | 0.9884 |
| 3 | 1.0176 | 0.9827 | 3.0175 | 0.3314 | 2.9653 | 0.3372 | 0.9961 | 2.9538 |
| 4 | 1.0235 | 0.9770 | 4.0351 | 0.2478 | 3.9423 | 0.2537 | 1.4927 | 5.8848 |
| 5 | 1.0295 | 0.9713 | 5.0587 | 0.1977 | 4.9137 | 0.2035 | 1.9884 | 9.7702 |
| 6 | 1.0355 | 0.9657 | 6.0882 | 0.1643 | 5.8794 | 0.1701 | 2.4830 | 14.5987 |
| 7 | 1.0416 | 0.9601 | 7.1237 | 0.1404 | 6.8395 | 0.1462 | 2.9767 | 20.3593 |
| 8 | 1.0476 | 0.9545 | 8.1653 | 0.1225 | 7.7940 | 0.1283 | 3.4695 | 27.0411 |
| 9 | 1.0537 | 0.9490 | 9.2129 | 0.1085 | 8.7430 | 0.1144 | 3.9612 | 34.6331 |
| 10 | 1.0599 | 0.9435 | 10.2666 | 0.0974 | 9.6865 | 0.1032 | 4.4520 | 43.1245 |
| 11 | 1.0661 | 0.9380 | 11.3265 | 0.0883 | 10.6245 | 0.0941 | 4.9418 | 52.5048 |
| 12 | 1.0723 | 0.9326 | 12.3926 | 0.0807 | 11.5571 | 0.0865 | 5.4307 | 62.7632 |
| 13 | 1.0785 | 0.9272 | 13.4649 | 0.0743 | 12.4843 | 0.0801 | 5.9186 | 73.8893 |
| 14 | 1.0848 | 0.9218 | 14.5434 | 0.0688 | 13.4061 | 0.0746 | 6.4055 | 85.8727 |
| 15 | 1.0912 | 0.9165 | 15.6283 | 0.0640 | 14.3225 | 0.0698 | 6.8914 | 98.7030 |
| 16 | 1.0975 | 0.9111 | 16.7194 | 0.0598 | 15.2337 | 0.0656 | 7.3764 | 112.3700 |
| 17 | 1.1039 | 0.9059 | 17.8170 | 0.0561 | 16.1395 | 0.0620 | 7.8604 | 126.8636 |
| 18 | 1.1104 | 0.9006 | 18.9209 | 0.0529 | 17.0401 | 0.0587 | 8.3435 | 142.1738 |
| 19 | 1.1168 | 0.8954 | 20.0313 | 0.0499 | 17.9355 | 0.0558 | 8.8255 | 158.2906 |
| 20 | 1.1234 | 0.8902 | 21.1481 | 0.0473 | 18.8257 | 0.0531 | 9.3066 | 175.2041 |
| 21 | 1.1299 | 0.8850 | 22.2715 | 0.0449 | 19.7107 | 0.0507 | 9.7868 | 192.9045 |
| 22 | 1.1365 | 0.8799 | 23.4014 | 0.0427 | 20.5906 | 0.0486 | 10.2660 | 211.3821 |
| 23 | 1.1431 | 0.8748 | 24.5379 | 0.0408 | 21.4654 | 0.0466 | 10.7442 | 230.6274 |
| 24 | 1.1498 | 0.8697 | 25.6810 | 0.0389 | 22.3351 | 0.0448 | 11.2214 | 250.6308 |
| 25 | 1.1565 | 0.8647 | 26.8308 | 0.0373 | 23.1998 | 0.0431 | 11.6977 | 271.3826 |
| 26 | 1.1633 | 0.8597 | 27.9874 | 0.0357 | 24.0594 | 0.0416 | 12.1730 | 292.8740 |
| 27 | 1.1700 | 0.8547 | 29.1506 | 0.0343 | 24.9141 | 0.0401 | 12.6473 | 315.0955 |
| 28 | 1.1769 | 0.8497 | 30.3207 | 0.0330 | 25.7638 | 0.0388 | 13.1206 | 338.0376 |
| 29 | 1.1837 | 0.8448 | 31.4975 | 0.0317 | 26.6086 | 0.0376 | 13.5930 | 361.6917 |
| 30 | 1.1906 | 0.8399 | 32.6813 | 0.0306 | 27.4485 | 0.0364 | 14.0645 | 386.0481 |
| 31 | 1.1976 | 0.8350 | 33.8719 | 0.0295 | 28.2835 | 0.0354 | 14.5349 | 411.0986 |
| 32 | 1.2046 | 0.8302 | 35.0695 | 0.0285 | 29.1137 | 0.0343 | 15.0044 | 436.8340 |
| 33 | 1.2116 | 0.8254 | 36.2741 | 0.0276 | 29.9390 | 0.0334 | 15.4730 | 463.2454 |
| 34 | 1.2187 | 0.8206 | 37.4857 | 0.0267 | 30.7596 | 0.0325 | 15.9405 | 490.3240 |
| 35 | 1.2258 | 0.8158 | 38.7043 | 0.0258 | 31.5754 | 0.0317 | 16.4071 | 518.0615 |
| 40 | 1.2619 | 0.7924 | 44.9051 | 0.0223 | 35.5840 | 0.0281 | 18.7257 | 666.3342 |
| 45 | 1.2992 | 0.7697 | 51.2889 | 0.0195 | 39.4777 | 0.0253 | 21.0201 | 829.8254 |
| 50 | 1.3375 | 0.7477 | 57.8611 | 0.0173 | 43.2599 | 0.0231 | 23.2904 | 1007.5413 |
| 55 | 1.3770 | 0.7262 | 64.6271 | 0.0155 | 46.9336 | 0.0213 | 25.5368 | 1198.5317 |
| 60 | 1.4176 | 0.7054 | 71.5929 | 0.0140 | 50.5020 | 0.0198 | 27.7591 | 1401.8899 |
| 65 | 1.4595 | 0.6852 | 78.7642 | 0.0127 | 53.9681 | 0.0185 | 29.9575 | 1616.7498 |
| 70 | 1.5025 | 0.6655 | 86.1471 | 0.0116 | 57.3349 | 0.0174 | 32.1320 | 1842.2847 |
| 75 | 1.5469 | 0.6465 | 93.7479 | 0.0107 | 60.6052 | 0.0165 | 34.2827 | 2077.7070 |
| 80 | 1.5925 | 0.6279 | 101.5730 | 0.0098 | 63.7817 | 0.0157 | 36.4095 | 2322.2639 |
| 85 | 1.6395 | 0.6099 | 109.6289 | 0.0091 | 66.8672 | 0.0150 | 38.5127 | 2575.2383 |
| 90 | 1.6879 | 0.5925 | 117.9226 | 0.0085 | 69.8643 | 0.0143 | 40.5922 | 2835.9470 |
| 95 | 1.7377 | 0.5755 | 126.4611 | 0.0079 | 72.7754 | 0.0137 | 42.6482 | 3103.7390 |
| 100 | 1.7890 | 0.5590 | 135.2515 | 0.0074 | 75.6031 | 0.0132 | 44.6806 | 3377.9934 |

## TABLE A.3 Discrete Compounding: $i = 2/3\%$

| | Single payment | | Uniform series | | | | Gradient series | |
|---|---|---|---|---|---|---|---|---|
| | Compound amount factor | Present worth factor | Compound amount factor | Sinking fund factor | Present worth factor | Capital recovery factor | Uniform series factor | Present worth factor |
| $n$ | To find $F$ given $P$ $F\|P\,i,n$ | To find $P$ given $F$ $P\|F\,i,n$ | To find $F$ given $A$ $F\|A\,i,n$ | To find $A$ given $F$ $A\|F\,i,n$ | To find $P$ given $A$ $P\|A\,i,n$ | To find $A$ given $P$ $A\|P\,i,n$ | To find $A$ given $G$ $A\|G\,i,n$ | To find $P$ given $G$ $P\|G\,i,n$ |
| 1 | 1.0067 | 0.9934 | 1.0000 | 1.0000 | 0.9934 | 1.0067 | 0.0000 | 0.0000 |
| 2 | 1.0134 | 0.9868 | 2.0067 | 0.4983 | 1.9802 | 0.5050 | 0.4983 | 0.9868 |
| 3 | 1.0201 | 0.9803 | 3.0200 | 0.3311 | 2.9604 | 0.3378 | 0.9956 | 2.9473 |
| 4 | 1.0269 | 0.9738 | 4.0402 | 0.2475 | 3.9342 | 0.2542 | 1.4917 | 5.8686 |
| 5 | 1.0338 | 0.9673 | 5.0671 | 0.1974 | 4.9015 | 0.2040 | 1.9867 | 9.7379 |
| 6 | 1.0407 | 0.9609 | 6.1009 | 0.1639 | 5.8625 | 0.1706 | 2.4806 | 14.5425 |
| 7 | 1.0476 | 0.9546 | 7.1416 | 0.1400 | 6.8170 | 0.1467 | 2.9734 | 20.2698 |
| 8 | 1.0546 | 0.9482 | 8.1892 | 0.1221 | 7.7652 | 0.1288 | 3.4651 | 26.9075 |
| 9 | 1.0616 | 0.9420 | 9.2438 | 0.1082 | 8.7072 | 0.1148 | 3.9557 | 34.4431 |
| 10 | 1.0687 | 0.9357 | 10.3054 | 0.0970 | 9.6429 | 0.1037 | 4.4452 | 42.8645 |
| 11 | 1.0758 | 0.9295 | 11.3741 | 0.0879 | 10.5724 | 0.0946 | 4.9336 | 52.1597 |
| 12 | 1.0830 | 0.9234 | 12.4499 | 0.0803 | 11.4958 | 0.0870 | 5.4208 | 62.3167 |
| 13 | 1.0902 | 0.9172 | 13.5329 | 0.0739 | 12.4130 | 0.0806 | 5.9070 | 73.3236 |
| 14 | 1.0975 | 0.9112 | 14.6231 | 0.0684 | 13.3242 | 0.0751 | 6.3920 | 85.1688 |
| 15 | 1.1048 | 0.9051 | 15.7206 | 0.0636 | 14.2293 | 0.0703 | 6.8760 | 97.8408 |
| 16 | 1.1122 | 0.8991 | 16.8254 | 0.0594 | 15.1285 | 0.0661 | 7.3588 | 111.3279 |
| 17 | 1.1196 | 0.8932 | 17.9376 | 0.0557 | 16.0217 | 0.0624 | 7.8406 | 125.6189 |
| 18 | 1.1270 | 0.8873 | 19.0572 | 0.0525 | 16.9089 | 0.0591 | 8.3212 | 140.7026 |
| 19 | 1.1346 | 0.8814 | 20.1842 | 0.0495 | 17.7903 | 0.0562 | 8.8007 | 156.5677 |
| 20 | 1.1421 | 0.8756 | 21.3188 | 0.0469 | 18.6659 | 0.0536 | 9.2791 | 173.2034 |
| 21 | 1.1497 | 0.8698 | 22.4609 | 0.0445 | 19.5357 | 0.0512 | 9.7564 | 190.5986 |
| 22 | 1.1574 | 0.8640 | 23.6107 | 0.0424 | 20.3997 | 0.0490 | 10.2327 | 208.7427 |
| 23 | 1.1651 | 0.8583 | 24.7681 | 0.0404 | 21.2579 | 0.0470 | 10.7078 | 227.6248 |
| 24 | 1.1729 | 0.8526 | 25.9332 | 0.0386 | 22.1105 | 0.0452 | 11.1818 | 247.2345 |
| 25 | 1.1807 | 0.8470 | 27.1061 | 0.0369 | 22.9575 | 0.0436 | 11.6546 | 267.5613 |
| 26 | 1.1886 | 0.8413 | 28.2868 | 0.0354 | 23.7988 | 0.0420 | 12.1264 | 288.5947 |
| 27 | 1.1965 | 0.8358 | 29.4754 | 0.0339 | 24.6346 | 0.0406 | 12.5971 | 310.3247 |
| 28 | 1.2045 | 0.8302 | 30.6719 | 0.0326 | 25.4648 | 0.0393 | 13.0667 | 332.7410 |
| 29 | 1.2125 | 0.8247 | 31.8763 | 0.0314 | 26.2896 | 0.0380 | 13.5352 | 355.8338 |
| 30 | 1.2206 | 0.8193 | 33.0889 | 0.0302 | 27.1089 | 0.0369 | 14.0025 | 379.5928 |
| 31 | 1.2287 | 0.8138 | 34.3094 | 0.0291 | 27.9227 | 0.0358 | 14.4688 | 404.0081 |
| 32 | 1.2369 | 0.8085 | 35.5382 | 0.0281 | 28.7312 | 0.0348 | 14.9340 | 429.0703 |
| 33 | 1.2452 | 0.8031 | 36.7751 | 0.0272 | 29.5343 | 0.0339 | 15.3980 | 454.7698 |
| 34 | 1.2535 | 0.7978 | 38.0203 | 0.0263 | 30.3320 | 0.0330 | 15.8610 | 481.0967 |
| 35 | 1.2618 | 0.7925 | 39.2737 | 0.0255 | 31.1246 | 0.0321 | 16.3229 | 508.0418 |
| 40 | 1.3045 | 0.7666 | 45.6675 | 0.0219 | 35.0090 | 0.0286 | 18.6157 | 651.7161 |
| 45 | 1.3485 | 0.7416 | 52.2773 | 0.0191 | 38.7666 | 0.0258 | 20.8810 | 809.4832 |
| 50 | 1.3941 | 0.7173 | 59.1104 | 0.0169 | 42.4013 | 0.0236 | 23.1188 | 980.2688 |
| 55 | 1.4412 | 0.6939 | 66.1743 | 0.0151 | 45.9173 | 0.0218 | 25.3293 | 1163.0535 |
| 60 | 1.4898 | 0.6712 | 73.4769 | 0.0136 | 49.3184 | 0.0203 | 27.5125 | 1356.8711 |
| 65 | 1.5402 | 0.6493 | 81.0261 | 0.0123 | 52.6084 | 0.0190 | 29.6684 | 1560.8049 |
| 70 | 1.5922 | 0.6281 | 88.8303 | 0.0113 | 55.7909 | 0.0179 | 31.7971 | 1773.9871 |
| 75 | 1.6460 | 0.6075 | 96.8982 | 0.0103 | 58.8693 | 0.0170 | 33.8987 | 1995.5950 |
| 80 | 1.7016 | 0.5877 | 105.2385 | 0.0095 | 61.8472 | 0.0162 | 35.9734 | 2224.8511 |
| 85 | 1.7591 | 0.5685 | 113.8607 | 0.0088 | 64.7277 | 0.0154 | 38.0211 | 2461.0191 |
| 90 | 1.8185 | 0.5499 | 122.7741 | 0.0081 | 67.5142 | 0.0148 | 40.0420 | 2703.4014 |
| 95 | 1.8799 | 0.5319 | 131.9886 | 0.0076 | 70.2096 | 0.0142 | 42.0362 | 2951.3406 |
| 100 | 1.9434 | 0.5146 | 141.5144 | 0.0071 | 72.8169 | 0.0137 | 44.0038 | 3204.2144 |

**TABLE A.4  Discrete Compounding:** $i = 3/4\%$

| | Single payment | | Uniform series | | | | Gradient series | |
|---|---|---|---|---|---|---|---|---|
| | Compound amount factor | Present worth factor | Compound amount factor | Sinking fund factor | Present worth factor | Capital recovery factor | Uniform series factor | Present worth factor |
| $n$ | To find $F$ given $P$ $F\|P\,i,n$ | To find $P$ given $F$ $P\|F\,i,n$ | To find $F$ given $A$ $F\|A\,i,n$ | To find $A$ given $F$ $A\|F\,i,n$ | To find $P$ given $A$ $P\|A\,i,n$ | To find $A$ given $P$ $A\|P\,i,n$ | To find $A$ given $G$ $A\|G\,i,n$ | To find $P$ given $G$ $P\|G\,i,n$ |
| 1 | 1.0075 | 0.9926 | 1.0000 | 1.0000 | 0.9926 | 1.0075 | 0.0000 | 0.0000 |
| 2 | 1.0151 | 0.9852 | 2.0075 | 0.4981 | 1.9777 | 0.5056 | 0.4981 | 0.9852 |
| 3 | 1.0227 | 0.9778 | 3.0226 | 0.3308 | 2.9556 | 0.3383 | 0.9950 | 2.9408 |
| 4 | 1.0303 | 0.9706 | 4.0452 | 0.2472 | 3.9261 | 0.2547 | 1.4907 | 5.8525 |
| 5 | 1.0381 | 0.9633 | 5.0756 | 0.1970 | 4.8894 | 0.2045 | 1.9851 | 9.7058 |
| 6 | 1.0459 | 0.9562 | 6.1136 | 0.1636 | 5.8456 | 0.1711 | 2.4782 | 14.4866 |
| 7 | 1.0537 | 0.9490 | 7.1595 | 0.1397 | 6.7946 | 0.1472 | 2.9701 | 20.1808 |
| 8 | 1.0616 | 0.9420 | 8.2132 | 0.1218 | 7.7366 | 0.1293 | 3.4608 | 26.7747 |
| 9 | 1.0696 | 0.9350 | 9.2748 | 0.1078 | 8.6716 | 0.1153 | 3.9502 | 34.2544 |
| 10 | 1.0776 | 0.9280 | 10.3443 | 0.0967 | 9.5996 | 0.1042 | 4.4384 | 42.6064 |
| 11 | 1.0857 | 0.9211 | 11.4219 | 0.0876 | 10.5207 | 0.0951 | 4.9253 | 51.8174 |
| 12 | 1.0938 | 0.9142 | 12.5076 | 0.0800 | 11.4349 | 0.0875 | 5.4110 | 61.8740 |
| 13 | 1.1020 | 0.9074 | 13.6014 | 0.0735 | 12.3423 | 0.0810 | 5.8954 | 72.7632 |
| 14 | 1.1103 | 0.9007 | 14.7034 | 0.0680 | 13.2430 | 0.0755 | 6.3786 | 84.4720 |
| 15 | 1.1186 | 0.8940 | 15.8137 | 0.0632 | 14.1370 | 0.0707 | 6.8606 | 96.9876 |
| 16 | 1.1270 | 0.8873 | 16.9323 | 0.0591 | 15.0243 | 0.0666 | 7.3413 | 110.2973 |
| 17 | 1.1354 | 0.8807 | 18.0593 | 0.0554 | 15.9050 | 0.0629 | 7.8207 | 124.3887 |
| 18 | 1.1440 | 0.8742 | 19.1947 | 0.0521 | 16.7792 | 0.0596 | 8.2989 | 139.2494 |
| 19 | 1.1525 | 0.8676 | 20.3387 | 0.0492 | 17.6468 | 0.0567 | 8.7759 | 154.8670 |
| 20 | 1.1612 | 0.8612 | 21.4912 | 0.0465 | 18.5080 | 0.0540 | 9.2516 | 171.2296 |
| 21 | 1.1699 | 0.8548 | 22.6524 | 0.0441 | 19.3628 | 0.0516 | 9.7261 | 188.3252 |
| 22 | 1.1787 | 0.8484 | 23.8223 | 0.0420 | 20.2112 | 0.0495 | 10.1994 | 206.1420 |
| 23 | 1.1875 | 0.8421 | 25.0010 | 0.0400 | 21.0533 | 0.0475 | 10.6714 | 224.6682 |
| 24 | 1.1964 | 0.8358 | 26.1885 | 0.0382 | 21.8891 | 0.0457 | 11.1422 | 243.8923 |
| 25 | 1.2054 | 0.8296 | 27.3849 | 0.0365 | 22.7188 | 0.0440 | 11.6117 | 263.8027 |
| 26 | 1.2144 | 0.8234 | 28.5903 | 0.0350 | 23.5422 | 0.0425 | 12.0800 | 284.3887 |
| 27 | 1.2235 | 0.8173 | 29.8047 | 0.0336 | 24.3595 | 0.0411 | 12.5470 | 305.6387 |
| 28 | 1.2327 | 0.8112 | 31.0282 | 0.0322 | 25.1707 | 0.0397 | 13.0128 | 327.5415 |
| 29 | 1.2420 | 0.8052 | 32.2609 | 0.0310 | 25.9759 | 0.0385 | 13.4774 | 350.0867 |
| 30 | 1.2513 | 0.7992 | 33.5029 | 0.0298 | 26.7751 | 0.0373 | 13.9407 | 373.2629 |
| 31 | 1.2607 | 0.7932 | 34.7542 | 0.0288 | 27.5683 | 0.0363 | 14.4028 | 397.0601 |
| 32 | 1.2701 | 0.7873 | 36.0148 | 0.0278 | 28.3557 | 0.0353 | 14.8636 | 421.4673 |
| 33 | 1.2796 | 0.7815 | 37.2849 | 0.0268 | 29.1371 | 0.0343 | 15.3232 | 446.4744 |
| 34 | 1.2892 | 0.7757 | 38.5646 | 0.0259 | 29.9128 | 0.0334 | 15.7816 | 472.0711 |
| 35 | 1.2989 | 0.7699 | 39.8538 | 0.0251 | 30.6827 | 0.0326 | 16.2387 | 498.2471 |
| 40 | 1.3483 | 0.7416 | 46.4465 | 0.0215 | 34.4469 | 0.0290 | 18.5058 | 637.4692 |
| 45 | 1.3997 | 0.7145 | 53.2901 | 0.0188 | 38.0732 | 0.0263 | 20.7421 | 789.7173 |
| 50 | 1.4530 | 0.6883 | 60.3943 | 0.0166 | 41.5665 | 0.0241 | 22.9476 | 953.8486 |
| 55 | 1.5083 | 0.6630 | 67.7688 | 0.0148 | 44.9316 | 0.0223 | 25.1223 | 1128.7869 |
| 60 | 1.5657 | 0.6387 | 75.4241 | 0.0133 | 48.1734 | 0.0208 | 27.2665 | 1313.5188 |
| 65 | 1.6253 | 0.6153 | 83.3708 | 0.0120 | 51.2963 | 0.0195 | 29.3801 | 1507.0911 |
| 70 | 1.6872 | 0.5927 | 91.6201 | 0.0109 | 54.3046 | 0.0184 | 31.4634 | 1708.6062 |
| 75 | 1.7514 | 0.5710 | 100.1833 | 0.0100 | 57.2027 | 0.0175 | 33.5163 | 1917.2222 |
| 80 | 1.8180 | 0.5500 | 109.0725 | 0.0092 | 59.9944 | 0.0167 | 35.5391 | 2132.1470 |
| 85 | 1.8873 | 0.5299 | 118.3001 | 0.0085 | 62.6838 | 0.0160 | 37.5318 | 2352.6367 |
| 90 | 1.9591 | 0.5104 | 127.8789 | 0.0078 | 65.2746 | 0.0153 | 39.4946 | 2577.9951 |
| 95 | 2.0337 | 0.4917 | 137.8224 | 0.0073 | 67.7704 | 0.0148 | 41.4277 | 2807.5696 |
| 100 | 2.1111 | 0.4737 | 148.1444 | 0.0068 | 70.1746 | 0.0143 | 43.3311 | 3040.7454 |

| | Single payment | | Uniform series | | | | Gradient series | |
|---|---|---|---|---|---|---|---|---|
| | Compound amount factor | Present worth factor | Compound amount factor | Sinking fund factor | Present worth factor | Capital recovery factor | Uniform series factor | Present worth factor |
| $n$ | To find $F$ given $P$ $F\|P\,i,n$ | To find $P$ given $F$ $P\|F\,i,n$ | To find $F$ given $A$ $F\|A\,i,n$ | To find $A$ given $F$ $A\|F\,i,n$ | To find $P$ given $A$ $P\|A\,i,n$ | To find $A$ given $P$ $A\|P\,i,n$ | To find $A$ given $G$ $A\|G\,i,n$ | To find $P$ given $G$ $P\|G\,i,n$ |
| 1 | 1.0100 | 0.9901 | 1.0000 | 1.0000 | 0.9901 | 1.0100 | 0.0000 | 0.0000 |
| 2 | 1.0201 | 0.9803 | 2.0100 | 0.4975 | 1.9704 | 0.5075 | 0.4975 | 0.9803 |
| 3 | 1.0303 | 0.9706 | 3.0301 | 0.3300 | 2.9410 | 0.3400 | 0.9934 | 2.9215 |
| 4 | 1.0406 | 0.9610 | 4.0604 | 0.2463 | 3.9020 | 0.2563 | 1.4876 | 5.8044 |
| 5 | 1.0510 | 0.9515 | 5.1010 | 0.1960 | 4.8534 | 0.2060 | 1.9801 | 9.6103 |
| 6 | 1.0615 | 0.9420 | 6.1520 | 0.1625 | 5.7955 | 0.1725 | 2.4710 | 14.3205 |
| 7 | 1.0721 | 0.9327 | 7.2135 | 0.1386 | 6.7282 | 0.1486 | 2.9602 | 19.9168 |
| 8 | 1.0829 | 0.9235 | 8.2857 | 0.1207 | 7.6517 | 0.1307 | 3.4478 | 26.3812 |
| 9 | 1.0937 | 0.9143 | 9.3685 | 0.1067 | 8.5660 | 0.1167 | 3.9337 | 33.6959 |
| 10 | 1.1046 | 0.9053 | 10.4622 | 0.0956 | 9.4713 | 0.1056 | 4.4179 | 41.8435 |
| 11 | 1.1157 | 0.8963 | 11.5668 | 0.0865 | 10.3676 | 0.0965 | 4.9005 | 50.8067 |
| 12 | 1.1268 | 0.8874 | 12.6825 | 0.0788 | 11.2551 | 0.0888 | 5.3815 | 60.5687 |
| 13 | 1.1381 | 0.8787 | 13.8093 | 0.0724 | 12.1337 | 0.0824 | 5.8607 | 71.1126 |
| 14 | 1.1495 | 0.8700 | 14.9474 | 0.0669 | 13.0037 | 0.0769 | 6.3384 | 82.4221 |
| 15 | 1.1610 | 0.8613 | 16.0969 | 0.0621 | 13.8651 | 0.0721 | 6.8143 | 94.4810 |
| 16 | 1.1726 | 0.8528 | 17.2579 | 0.0579 | 14.7179 | 0.0679 | 7.2886 | 107.2734 |
| 17 | 1.1843 | 0.8444 | 18.4304 | 0.0543 | 15.5623 | 0.0643 | 7.7613 | 120.7834 |
| 18 | 1.1961 | 0.8360 | 19.6147 | 0.0510 | 16.3983 | 0.0610 | 8.2323 | 134.9957 |
| 19 | 1.2081 | 0.8277 | 20.8109 | 0.0481 | 17.2260 | 0.0581 | 8.7017 | 149.8950 |
| 20 | 1.2202 | 0.8195 | 22.0190 | 0.0454 | 18.0456 | 0.0554 | 9.1694 | 165.4664 |
| 21 | 1.2324 | 0.8114 | 23.2392 | 0.0430 | 18.8570 | 0.0530 | 9.6354 | 181.6950 |
| 22 | 1.2447 | 0.8034 | 24.4716 | 0.0409 | 19.6604 | 0.0509 | 10.0998 | 198.5663 |
| 23 | 1.2572 | 0.7954 | 25.7163 | 0.0389 | 20.4558 | 0.0489 | 10.5626 | 216.0660 |
| 24 | 1.2697 | 0.7876 | 26.9735 | 0.0371 | 21.2434 | 0.0471 | 11.0237 | 234.1800 |
| 25 | 1.2824 | 0.7798 | 28.2432 | 0.0354 | 22.0232 | 0.0454 | 11.4831 | 252.8945 |
| 26 | 1.2953 | 0.7720 | 29.5256 | 0.0339 | 22.7952 | 0.0439 | 11.9409 | 272.1957 |
| 27 | 1.3082 | 0.7644 | 30.8209 | 0.0324 | 23.5596 | 0.0424 | 12.3971 | 292.0702 |
| 28 | 1.3213 | 0.7568 | 32.1291 | 0.0311 | 24.3164 | 0.0411 | 12.8516 | 312.5047 |
| 29 | 1.3345 | 0.7493 | 33.4504 | 0.0299 | 25.0658 | 0.0399 | 13.3044 | 333.4863 |
| 30 | 1.3478 | 0.7419 | 34.7849 | 0.0287 | 25.8077 | 0.0387 | 13.7557 | 355.0021 |
| 31 | 1.3613 | 0.7346 | 36.1327 | 0.0277 | 26.5423 | 0.0377 | 14.2052 | 377.0394 |
| 32 | 1.3749 | 0.7273 | 37.4941 | 0.0267 | 27.2696 | 0.0367 | 14.6532 | 399.5858 |
| 33 | 1.3887 | 0.7201 | 38.8690 | 0.0257 | 27.9897 | 0.0357 | 15.0995 | 422.6291 |
| 34 | 1.4026 | 0.7130 | 40.2577 | 0.0248 | 28.7027 | 0.0348 | 15.5441 | 446.1572 |
| 35 | 1.4166 | 0.7059 | 41.6603 | 0.0240 | 29.4086 | 0.0340 | 15.9871 | 470.1583 |
| 40 | 1.4889 | 0.6717 | 48.8864 | 0.0205 | 32.8347 | 0.0305 | 18.1776 | 596.8561 |
| 45 | 1.5648 | 0.6391 | 56.4811 | 0.0177 | 36.0945 | 0.0277 | 20.3273 | 733.7037 |
| 50 | 1.6446 | 0.6080 | 64.4632 | 0.0155 | 39.1961 | 0.0255 | 22.4363 | 879.4176 |
| 55 | 1.7285 | 0.5785 | 72.8525 | 0.0137 | 42.1472 | 0.0237 | 24.5049 | 1032.8148 |
| 60 | 1.8167 | 0.5504 | 81.6697 | 0.0122 | 44.9550 | 0.0222 | 26.5333 | 1192.8061 |
| 65 | 1.9094 | 0.5237 | 90.9366 | 0.0110 | 47.6266 | 0.0210 | 28.5217 | 1358.3903 |
| 70 | 2.0068 | 0.4983 | 100.6763 | 0.0099 | 50.1685 | 0.0199 | 30.4703 | 1528.6474 |
| 75 | 2.1091 | 0.4741 | 110.9128 | 0.0090 | 52.5871 | 0.0190 | 32.3793 | 1702.7340 |
| 80 | 2.2167 | 0.4511 | 121.6715 | 0.0082 | 54.8882 | 0.0182 | 34.2492 | 1879.8771 |
| 85 | 2.3298 | 0.4292 | 132.9790 | 0.0075 | 57.0777 | 0.0175 | 36.0801 | 2059.3701 |
| 90 | 2.4486 | 0.4084 | 144.8633 | 0.0069 | 59.1609 | 0.0169 | 37.8724 | 2240.5675 |
| 95 | 2.5735 | 0.3886 | 157.3538 | 0.0064 | 61.1430 | 0.0164 | 39.6265 | 2422.8811 |
| 100 | 2.7048 | 0.3697 | 170.4814 | 0.0059 | 63.0289 | 0.0159 | 41.3426 | 2605.7758 |

## TABLE A.6 Discrete Compounding: $i = 3/2\%$

| | Single payment | | Uniform series | | | | Gradient series | |
|---|---|---|---|---|---|---|---|---|
| | Compound amount factor | Present worth factor | Compound amount factor | Sinking fund factor | Present worth factor | Capital recovery factor | Uniform series factor | Present worth factor |
| $n$ | To find $F$ given $P$ $F\|P\,i,n$ | To find $P$ given $F$ $P\|F\,i,n$ | To find $F$ given $A$ $F\|A\,i,n$ | To find $A$ given $F$ $A\|F\,i,n$ | To find $P$ given $A$ $P\|A\,i,n$ | To find $A$ given $P$ $A\|P\,i,n$ | To find $A$ given $G$ $A\|G\,i,n$ | To find $P$ given $G$ $P\|G\,i,n$ |
| 1 | 1.0150 | 0.9852 | 1.0000 | 1.0000 | 0.9852 | 1.0150 | 0.0000 | 0.0000 |
| 2 | 1.0302 | 0.9707 | 2.0150 | 0.4963 | 1.9559 | 0.5113 | 0.4963 | 0.9707 |
| 3 | 1.0457 | 0.9563 | 3.0452 | 0.3284 | 2.9122 | 0.3434 | 0.9901 | 2.8833 |
| 4 | 1.0614 | 0.9422 | 4.0909 | 0.2444 | 3.8544 | 0.2594 | 1.4814 | 5.7098 |
| 5 | 1.0773 | 0.9283 | 5.1523 | 0.1941 | 4.7826 | 0.2091 | 1.9702 | 9.4229 |
| 6 | 1.0934 | 0.9145 | 6.2296 | 0.1605 | 5.6972 | 0.1755 | 2.4566 | 13.9956 |
| 7 | 1.1098 | 0.9010 | 7.3230 | 0.1366 | 6.5982 | 0.1516 | 2.9405 | 19.4018 |
| 8 | 1.1265 | 0.8877 | 8.4328 | 0.1186 | 7.4859 | 0.1336 | 3.4219 | 25.6157 |
| 9 | 1.1434 | 0.8746 | 9.5593 | 0.1046 | 8.3605 | 0.1196 | 3.9008 | 32.6125 |
| 10 | 1.1605 | 0.8617 | 10.7027 | 0.0934 | 9.2222 | 0.1084 | 4.3772 | 40.3675 |
| 11 | 1.1779 | 0.8489 | 11.8633 | 0.0843 | 10.0711 | 0.0993 | 4.8512 | 48.8568 |
| 12 | 1.1956 | 0.8364 | 13.0412 | 0.0767 | 10.9075 | 0.0917 | 5.3227 | 58.0571 |
| 13 | 1.2136 | 0.8240 | 14.2368 | 0.0702 | 11.7315 | 0.0852 | 5.7917 | 67.9454 |
| 14 | 1.2318 | 0.8118 | 15.4504 | 0.0647 | 12.5434 | 0.0797 | 6.2582 | 78.4994 |
| 15 | 1.2502 | 0.7999 | 16.6821 | 0.0599 | 13.3432 | 0.0749 | 6.7223 | 89.6974 |
| 16 | 1.2690 | 0.7880 | 17.9324 | 0.0558 | 14.1313 | 0.0708 | 7.1839 | 101.5178 |
| 17 | 1.2880 | 0.7764 | 19.2014 | 0.0521 | 14.9076 | 0.0671 | 7.6431 | 113.9399 |
| 18 | 1.3073 | 0.7649 | 20.4894 | 0.0488 | 15.6726 | 0.0638 | 8.0997 | 126.9434 |
| 19 | 1.3270 | 0.7536 | 21.7967 | 0.0459 | 16.4262 | 0.0609 | 8 5539 | 140.5084 |
| 20 | 1.3469 | 0.7425 | 23.1237 | 0.0432 | 17.1686 | 0.0582 | 9.0057 | 154.6153 |
| 21 | 1.3671 | 0.7315 | 24.4705 | 0.0409 | 17.9001 | 0.0559 | 9.4550 | 169.2453 |
| 22 | 1.3876 | 0.7207 | 25.8376 | 0.0387 | 18.6208 | 0.0537 | 9.9018 | 184.3797 |
| 23 | 1.4084 | 0.7100 | 27.2251 | 0.0367 | 19.3309 | 0.0517 | 10.3462 | 200.0005 |
| 24 | 1.4295 | 0.6995 | 28.6335 | 0.0349 | 20.0304 | 0.0499 | 10.7881 | 216.0900 |
| 25 | 1.4509 | 0.6892 | 30.0630 | 0.0333 | 20.7196 | 0.0483 | 11.2276 | 232.6310 |
| 26 | 1.4727 | 0.6790 | 31.5140 | 0.0317 | 21.3986 | 0.0467 | 11.6646 | 249.6065 |
| 27 | 1.4948 | 0.6690 | 32.9867 | 0.0303 | 22.0676 | 0.0453 | 12.0992 | 267.0000 |
| 28 | 1.5172 | 0.6591 | 34.4815 | 0.0290 | 22.7267 | 0.0440 | 12.5313 | 284.7957 |
| 29 | 1.5400 | 0.6494 | 35.9987 | 0.0278 | 23.3761 | 0.0428 | 12.9610 | 302.9778 |
| 30 | 1.5631 | 0.6398 | 37.5387 | 0.0266 | 24.0158 | 0.0416 | 13.3883 | 321.5308 |
| 31 | 1.5865 | 0.6303 | 39.1018 | 0.0256 | 24.6461 | 0.0406 | 13.8131 | 340.4402 |
| 32 | 1.6103 | 0.6210 | 40.6883 | 0.0246 | 25.2671 | 0.0396 | 14.2355 | 359.6909 |
| 33 | 1.6345 | 0.6118 | 42.2986 | 0.0236 | 25.8790 | 0.0386 | 14.6555 | 379.2691 |
| 34 | 1.6590 | 0.6028 | 43.9331 | 0.0228 | 26.4817 | 0.0378 | 15.0731 | 399.1604 |
| 35 | 1.6839 | 0.5939 | 45.5921 | 0.0219 | 27.0756 | 0.0369 | 15.4882 | 419.3521 |
| 40 | 1.8140 | 0.5513 | 54.2679 | 0.0184 | 29.9158 | 0.0334 | 17.5277 | 524.3567 |
| 45 | 1.9542 | 0.5117 | 63.6142 | 0.0157 | 32.5523 | 0.0307 | 19.5074 | 635.0110 |
| 50 | 2.1052 | 0.4750 | 73.6828 | 0.0136 | 34.9997 | 0.0286 | 21.4277 | 749.9634 |
| 55 | 2.2679 | 0.4409 | 84.5296 | 0.0118 | 37.2715 | 0.0268 | 23.2894 | 868.0283 |
| 60 | 2.4432 | 0.4093 | 96.2146 | 0.0104 | 39.3803 | 0.0254 | 25.0930 | 988.1675 |
| 65 | 2.6320 | 0.3799 | 108.8027 | 0.0092 | 41.3378 | 0.0242 | 26.8393 | 1109.4751 |
| 70 | 2.8355 | 0.3527 | 122.3637 | 0.0082 | 43.1549 | 0.0232 | 28.5290 | 1231.1658 |
| 75 | 3.0546 | 0.3274 | 136.9727 | 0.0073 | 44.8416 | 0.0223 | 30.1631 | 1352.5601 |
| 80 | 3.2907 | 0.3039 | 152.7108 | 0.0065 | 46.4073 | 0.0215 | 31.7423 | 1473.0742 |
| 85 | 3.5450 | 0.2821 | 169.6651 | 0.0059 | 47.8607 | 0.0209 | 33.2676 | 1592.2095 |
| 90 | 3.8189 | 0.2619 | 187.9298 | 0.0053 | 49.2099 | 0.0203 | 34.7399 | 1709.5440 |
| 95 | 4.1141 | 0.2431 | 207.6061 | 0.0048 | 50.4622 | 0.0198 | 36.1602 | 1824.7214 |
| 100 | 4.4320 | 0.2256 | 228.8029 | 0.0044 | 51.6247 | 0.0194 | 37.5295 | 1937.4495 |

## TABLE A.7  Discrete Compounding: $i = 2\%$

| | Single payment | | Uniform series | | | | Gradient series | |
|---|---|---|---|---|---|---|---|---|
| | Compound amount factor | Present worth factor | Compound amount factor | Sinking fund factor | Present worth factor | Capital recovery factor | Uniform series factor | Present worth factor |
| $n$ | To find $F$ given $P$ $F\|P\ i,n$ | To find $P$ given $F$ $P\|F\ i,n$ | To find $F$ given $A$ $F\|A\ i,n$ | To find $A$ given $F$ $A\|F\ i,n$ | To find $P$ given $A$ $P\|A\ i,n$ | To find $A$ given $P$ $A\|P\ i,n$ | To find $A$ given $G$ $A\|G\ i,n$ | To find $P$ given $G$ $P\|G\ i,n$ |
| 1 | 1.0200 | 0.9804 | 1.0000 | 1.0000 | 0.9804 | 1.0200 | 0.0000 | 0.0000 |
| 2 | 1.0404 | 0.9612 | 2.0200 | 0.4950 | 1.9416 | 0.5150 | 0.4950 | 0.9612 |
| 3 | 1.0612 | 0.9423 | 3.0604 | 0.3268 | 2.8839 | 0.3468 | 0.9868 | 2.8458 |
| 4 | 1.0824 | 0.9238 | 4.1216 | 0.2426 | 3.8077 | 0.2626 | 1.4752 | 5.6173 |
| 5 | 1.1041 | 0.9057 | 5.2040 | 0.1922 | 4.7135 | 0.2122 | 1.9604 | 9.2403 |
| 6 | 1.1262 | 0.8880 | 6.3081 | 0.1585 | 5.6014 | 0.1785 | 2.4423 | 13.6801 |
| 7 | 1.1487 | 0.8706 | 7.4343 | 0.1345 | 6.4720 | 0.1545 | 2.9208 | 18.9035 |
| 8 | 1.1717 | 0.8535 | 8.5830 | 0.1165 | 7.3255 | 0.1365 | 3.3961 | 24.8779 |
| 9 | 1.1951 | 0.8368 | 9.7546 | 0.1025 | 8.1622 | 0.1225 | 3.8681 | 31.5720 |
| 10 | 1.2190 | 0.8203 | 10.9497 | 0.0913 | 8.9826 | 0.1113 | 4.3367 | 38.9551 |
| 11 | 1.2434 | 0.8043 | 12.1687 | 0.0822 | 9.7868 | 0.1022 | 4.8021 | 46.9977 |
| 12 | 1.2682 | 0.7885 | 13.4121 | 0.0746 | 10.5753 | 0.0946 | 5.2642 | 55.6712 |
| 13 | 1.2936 | 0.7730 | 14.6803 | 0.0681 | 11.3484 | 0.0881 | 5.7231 | 64.9475 |
| 14 | 1.3195 | 0.7579 | 15.9739 | 0.0626 | 12.1062 | 0.0826 | 6.1786 | 74.7999 |
| 15 | 1.3459 | 0.7430 | 17.2934 | 0.0578 | 12.8493 | 0.0778 | 6.6309 | 85.2021 |
| 16 | 1.3728 | 0.7284 | 18.6393 | 0.0537 | 13.5777 | 0.0737 | 7.0799 | 96.1288 |
| 17 | 1.4002 | 0.7142 | 20.0121 | 0.0500 | 14.2919 | 0.0700 | 7.5256 | 107.5554 |
| 18 | 1.4282 | 0.7002 | 21.4123 | 0.0467 | 14.9920 | 0.0667 | 7.9681 | 119.4581 |
| 19 | 1.4568 | 0.6864 | 22.8406 | 0.0438 | 15.6785 | 0.0638 | 8.4073 | 131.8139 |
| 20 | 1.4859 | 0.6730 | 24.2974 | 0.0412 | 16.3514 | 0.0612 | 8.8433 | 144.6003 |
| 21 | 1.5157 | 0.6598 | 25.7833 | 0.0388 | 17.0112 | 0.0588 | 9.2760 | 157.7959 |
| 22 | 1.5460 | 0.6468 | 27.2990 | 0.0366 | 17.6580 | 0.0566 | 9.7055 | 171.3795 |
| 23 | 1.5769 | 0.6342 | 28.8450 | 0.0347 | 18.2922 | 0.0547 | 10.1317 | 185.3309 |
| 24 | 1.6084 | 0.6217 | 30.4219 | 0.0329 | 18.9139 | 0.0529 | 10.5547 | 199.6305 |
| 25 | 1.6406 | 0.6095 | 32.0303 | 0.0312 | 19.5235 | 0.0512 | 10.9745 | 214.2592 |
| 26 | 1.6734 | 0.5976 | 33.6709 | 0.0297 | 20.1210 | 0.0497 | 11.3910 | 229.1987 |
| 27 | 1.7069 | 0.5859 | 35.3443 | 0.0283 | 20.7069 | 0.0483 | 11.8043 | 244.4311 |
| 28 | 1.7410 | 0.5744 | 37.0512 | 0.0270 | 21.2813 | 0.0470 | 12.2145 | 259.9392 |
| 29 | 1.7758 | 0.5631 | 38.7922 | 0.0258 | 21.8444 | 0.0458 | 12.6214 | 275.7064 |
| 30 | 1.8114 | 0.5521 | 40.5681 | 0.0246 | 22.3965 | 0.0446 | 13.0251 | 291.7164 |
| 31 | 1.8476 | 0.5412 | 42.3794 | 0.0236 | 22.9377 | 0.0436 | 13.4257 | 307.9538 |
| 32 | 1.8845 | 0.5306 | 44.2270 | 0.0226 | 23.4683 | 0.0426 | 13.8230 | 324.4035 |
| 33 | 1.9222 | 0.5202 | 46.1116 | 0.0217 | 23.9886 | 0.0417 | 14.2171 | 341.0508 |
| 34 | 1.9607 | 0.5100 | 48.0338 | 0.0208 | 24.4986 | 0.0408 | 14.6083 | 357.8817 |
| 35 | 1.9999 | 0.5000 | 49.9945 | 0.0200 | 24.9986 | 0.0400 | 14.9961 | 374.8826 |
| 40 | 2.2080 | 0.4529 | 60.4020 | 0.0166 | 27.3555 | 0.0366 | 16.8885 | 461.9931 |
| 45 | 2.4379 | 0.4102 | 71.8927 | 0.0139 | 29.4902 | 0.0339 | 18.7034 | 551.5652 |
| 50 | 2.6916 | 0.3715 | 84.5794 | 0.0118 | 31.4236 | 0.0318 | 20.4420 | 642.3606 |
| 55 | 2.9717 | 0.3365 | 98.5865 | 0.0101 | 33.1748 | 0.0301 | 22.1057 | 733.3527 |
| 60 | 3.2810 | 0.3048 | 114.0515 | 0.0088 | 34.7609 | 0.0288 | 23.6961 | 823.6975 |
| 65 | 3.6225 | 0.2761 | 131.1262 | 0.0076 | 36.1975 | 0.0276 | 25.2147 | 912.7085 |
| 70 | 3.9996 | 0.2500 | 149.9779 | 0.0067 | 37.4986 | 0.0267 | 26.6632 | 999.8343 |
| 75 | 4.4158 | 0.2265 | 170.7918 | 0.0059 | 38.6771 | 0.0259 | 28.0434 | 1084.6393 |
| 80 | 4.8754 | 0.2051 | 193.7720 | 0.0052 | 39.7445 | 0.0252 | 29.3572 | 1166.7868 |
| 85 | 5.3829 | 0.1858 | 219.1439 | 0.0046 | 40.7113 | 0.0246 | 30.6064 | 1246.0241 |
| 90 | 5.9431 | 0.1683 | 247.1567 | 0.0040 | 41.5869 | 0.0240 | 31.7929 | 1322.1701 |
| 95 | 6.5617 | 0.1524 | 278.0850 | 0.0036 | 42.3800 | 0.0236 | 32.9189 | 1395.1033 |
| 100 | 7.2446 | 0.1380 | 312.2323 | 0.0032 | 43.0984 | 0.0232 | 33.9863 | 1464.7527 |

## TABLE A.8  Discrete Compounding: $i = 5/2\%$

| | Single payment | | Uniform series | | | | Gradient series | |
|---|---|---|---|---|---|---|---|---|
| | Compound amount factor | Present worth factor | Compound amount factor | Sinking fund factor | Present worth factor | Capital recovery factor | Uniform series factor | Present worth factor |
| $n$ | To find $F$ given $P$ $F\|P\,i,n$ | To find $P$ given $F$ $P\|F\,i,n$ | To find $F$ given $A$ $F\|A\,i,n$ | To find $A$ given $F$ $A\|F\,i,n$ | To find $P$ given $A$ $P\|A\,i,n$ | To find $A$ given $P$ $A\|P\,i,n$ | To find $A$ given $G$ $A\|G\,i,n$ | To find $P$ given $G$ $P\|G\,i,n$ |
| 1 | 1.0250 | 0.9756 | 1.0000 | 1.0000 | 0.9756 | 1.0250 | 0.0000 | 0.0000 |
| 2 | 1.0506 | 0.9518 | 2.0250 | 0.4938 | 1.9274 | 0.5188 | 0.4938 | 0.9518 |
| 3 | 1.0769 | 0.9286 | 3.0756 | 0.3251 | 2.8560 | 0.3501 | 0.9835 | 2.8090 |
| 4 | 1.1038 | 0.9060 | 4.1525 | 0.2408 | 3.7620 | 0.2658 | 1.4691 | 5.5269 |
| 5 | 1.1314 | 0.8839 | 5.2563 | 0.1902 | 4.6458 | 0.2152 | 1.9506 | 9.0623 |
| 6 | 1.1597 | 0.8623 | 6.3877 | 0.1565 | 5.5081 | 0.1815 | 2.4280 | 13.3738 |
| 7 | 1.1887 | 0.8413 | 7.5474 | 0.1325 | 6.3494 | 0.1575 | 2.9013 | 18.4214 |
| 8 | 1.2184 | 0.8207 | 8.7361 | 0.1145 | 7.1701 | 0.1395 | 3.3704 | 24.1666 |
| 9 | 1.2489 | 0.8007 | 9.9545 | 0.1005 | 7.9709 | 0.1255 | 3.8355 | 30.5724 |
| 10 | 1.2801 | 0.7812 | 11.2034 | 0.0893 | 8.7521 | 0.1143 | 4.2965 | 37.6032 |
| 11 | 1.3121 | 0.7621 | 12.4835 | 0.0801 | 9.5142 | 0.1051 | 4.7534 | 45.2246 |
| 12 | 1.3449 | 0.7436 | 13.7956 | 0.0725 | 10.2578 | 0.0975 | 5.2062 | 53.4038 |
| 13 | 1.3785 | 0.7254 | 15.1404 | 0.0660 | 10.9832 | 0.0910 | 5.6549 | 62.1088 |
| 14 | 1.4130 | 0.7077 | 16.5190 | 0.0605 | 11.6909 | 0.0855 | 6.0995 | 71.3093 |
| 15 | 1.4483 | 0.6905 | 17.9319 | 0.0558 | 12.3814 | 0.0808 | 6.5401 | 80.9758 |
| 16 | 1.4845 | 0.6736 | 19.3802 | 0.0516 | 13.0550 | 0.0766 | 6.9766 | 91.0802 |
| 17 | 1.5216 | 0.6572 | 20.8647 | 0.0479 | 13.7122 | 0.0729 | 7.4091 | 101.5953 |
| 18 | 1.5597 | 0.6412 | 22.3863 | 0.0447 | 14.3534 | 0.0697 | 7.8375 | 112.4950 |
| 19 | 1.5987 | 0.6255 | 23.9460 | 0.0418 | 14 9789 | 0.0668 | 8.2619 | 123.7545 |
| 20 | 1.6386 | 0.6103 | 25.5447 | 0.0391 | 15.5892 | 0.0641 | 8.6823 | 135.3497 |
| 21 | 1.6796 | 0.5954 | 27.1833 | 0.0368 | 16.1845 | 0.0618 | 9.0086 | 147.2574 |
| 22 | 1.7216 | 0.5809 | 28.8629 | 0.0346 | 16.7654 | 0.0596 | 9.5110 | 159.4556 |
| 23 | 1.7646 | 0.5667 | 30.5844 | 0.0327 | 17.3321 | 0.0577 | 9.9193 | 171.9229 |
| 24 | 1.8087 | 0.5529 | 32.3490 | 0.0309 | 17.8850 | 0.0559 | 10.3237 | 184.6390 |
| 25 | 1.8539 | 0.5394 | 34.1578 | 0.0293 | 18.4244 | 0.0543 | 10.7241 | 197.5844 |
| 26 | 1.9003 | 0.5262 | 36.0117 | 0.0278 | 18.9506 | 0.0528 | 11.1205 | 210.7403 |
| 27 | 1.9478 | 0.5134 | 37.9120 | 0.0264 | 19.4640 | 0.0514 | 11.5130 | 224.0887 |
| 28 | 1.9965 | 0.5009 | 39.8598 | 0.0251 | 19.9649 | 0.0501 | 11.9015 | 237.6124 |
| 29 | 2.0464 | 0.4887 | 41.8563 | 0.0239 | 20.4536 | 0.0489 | 12.2861 | 251.2949 |
| 30 | 2.0976 | 0.4767 | 43.9027 | 0.0228 | 20.9303 | 0.0478 | 12.6668 | 265.1204 |
| 31 | 2.1500 | 0.4651 | 46.0003 | 0.0217 | 21.3954 | 0.0467 | 13.0436 | 279.0737 |
| 32 | 2.2038 | 0.4538 | 48.1503 | 0.0208 | 21.8492 | 0.0458 | 13.4166 | 293.1406 |
| 33 | 2.2589 | 0.4427 | 50.3540 | 0.0199 | 22.2919 | 0.0449 | 13.7856 | 307.3071 |
| 34 | 2.3153 | 0.4319 | 52.6129 | 0.0190 | 22.7238 | 0.0440 | 14.1508 | 321.5601 |
| 35 | 2.3732 | 0.4214 | 54.9282 | 0.0182 | 23.1452 | 0.0432 | 14.5122 | 335.8867 |
| 40 | 2.6851 | 0.3724 | 67.4026 | 0.0148 | 25.1028 | 0.0398 | 16.2620 | 408.2219 |
| 45 | 3.0379 | 0.3292 | 81.5161 | 0.0123 | 26.8330 | 0.0373 | 17.9185 | 480.8069 |
| 50 | 3.4371 | 0.2909 | 97.4843 | 0.0103 | 28.3623 | 0.0353 | 19.4839 | 552.6079 |
| 55 | 3.8888 | 0.2572 | 115.5509 | 0.0087 | 29.7140 | 0.0337 | 20.9608 | 622.8279 |
| 60 | 4.3998 | 0.2273 | 135.9915 | 0.0074 | 30.9087 | 0.0324 | 22.3518 | 690.8655 |
| 65 | 4.9780 | 0.2009 | 159.1183 | 0.0063 | 31.9646 | 0.0313 | 23.6600 | 756.2805 |
| 70 | 5.6321 | 0.1776 | 185.2841 | 0.0054 | 32.8979 | 0.0304 | 24.8881 | 818.7642 |
| 75 | 6.3722 | 0.1569 | 214.8882 | 0.0047 | 33.7227 | 0.0297 | 26.0393 | 878.1150 |
| 80 | 7.2096 | 0.1387 | 248.3826 | 0.0040 | 34.4518 | 0.0290 | 27.1167 | 934.2180 |
| 85 | 8.1570 | 0.1226 | 286.2783 | 0.0035 | 35.0962 | 0.0285 | 28.1235 | 987.0269 |
| 90 | 9.2289 | 0.1084 | 329.1541 | 0.0030 | 35.6658 | 0.0280 | 29.0629 | 1036.5498 |
| 95 | 10.4416 | 0.0958 | 377.6641 | 0.0026 | 36.1692 | 0.0276 | 29.9382 | 1082.8379 |
| 100 | 11.8137 | 0.0846 | 432.5486 | 0.0023 | 36.6141 | 0.0273 | 30.7525 | 1125.9746 |

| | Single payment | | Uniform series | | | | Gradient series | |
|---|---|---|---|---|---|---|---|---|
| | Compound amount factor | Present worth factor | Compound amount factor | Sinking fund factor | Present worth factor | Capital recovery factor | Uniform series factor | Present worth factor |
| $n$ | To find $F$ given $P$ $F\|P\ i,n$ | To find $P$ given $F$ $P\|F\ i,n$ | To find $F$ given $A$ $F\|A\ i,n$ | To find $A$ given $F$ $A\|F\ i,n$ | To find $P$ given $A$ $P\|A\ i,n$ | To find $A$ given $P$ $A\|P\ i,n$ | To find $A$ given $G$ $A\|G\ i,n$ | To find $P$ given $G$ $P\|G\ i,n$ |
| 1 | 1.0300 | 0.0907 | 1.0000 | 1.0000 | 0.9709 | 1.0300 | 0.0000 | 0.000 |
| 2 | 1.0609 | 0.9246 | 2.0300 | 0.4926 | 1.9135 | 0.5226 | 0.4926 | 0.9426 |
| 3 | 1.0927 | 0.9151 | 3.0909 | 0.3235 | 2.8286 | 0.3535 | 0.9803 | 2.7729 |
| 4 | 1.1255 | 0.8885 | 4.1836 | 0.2390 | 3.7171 | 0.2690 | 1.4631 | 5.4383 |
| 5 | 1.1593 | 0.8626 | 5.3091 | 0.1884 | 4.5797 | 0.2184 | 1.9409 | 8.8888 |
| 6 | 1.1941 | 0.8375 | 6.4684 | 0.1546 | 5.4172 | 0.1846 | 2.4138 | 13.0762 |
| 7 | 1.2299 | 0.8131 | 7.6625 | 0.1305 | 6.2303 | 0.1605 | 2.8819 | 17.9547 |
| 8 | 1.2668 | 0.7894 | 8.8923 | 0.1125 | 7.0197 | 0.1425 | 3.3450 | 23.4806 |
| 9 | 1.3048 | 0.7664 | 10.1591 | 0.0984 | 7.7861 | 0.1284 | 3.8032 | 29.6119 |
| 10 | 1.3439 | 0.7441 | 11.4639 | 0.0872 | 8.5302 | 0.1172 | 4.2565 | 36.3088 |
| 11 | 1.3842 | 0.7224 | 12.8078 | 0.0781 | 9.2526 | 0.1081 | 4.7049 | 43.5330 |
| 12 | 1.4258 | 0.7014 | 14.1920 | 0.0705 | 9.9540 | 0.1005 | 5.1485 | 51.2482 |
| 13 | 1.4685 | 0.6810 | 15.6178 | 0.0640 | 10.6350 | 0.0940 | 5.5872 | 59.4196 |
| 14 | 1.5126 | 0.6611 | 17.0863 | 0.0585 | 11.2961 | 0.0885 | 6.0210 | 68.0141 |
| 15 | 1.5580 | 0.6419 | 18.5989 | 0.0538 | 11.9379 | 0.0838 | 6.4500 | 77.0002 |
| 16 | 1.6047 | 0.6232 | 20.1569 | 0.0496 | 12.5611 | 0.0796 | 6.8742 | 86.3477 |
| 17 | 1.6528 | 0.6050 | 21.7616 | 0.0460 | 13.1661 | 0.0760 | 7.2936 | 96.0280 |
| 18 | 1.7024 | 0.5874 | 23.4144 | 0.0427 | 13.7535 | 0.0727 | 7.7081 | 106.0137 |
| 19 | 1.7535 | 0.5703 | 25.1169 | 0.0398 | 14.3238 | 0.0698 | 8.1179 | 116.2788 |
| 20 | 1.8061 | 0.5537 | 26.8704 | 0.0372 | 14.8775 | 0.0672 | 8.5229 | 126.7987 |
| 21 | 1.8603 | 0.5375 | 28.6765 | 0.0349 | 15.4150 | 0.0649 | 8.9231 | 137.5496 |
| 22 | 1.9161 | 0.5219 | 30.5368 | 0.0327 | 15.9369 | 0.0627 | 9.3186 | 148.5094 |
| 23 | 1.9736 | 0.5067 | 32.4529 | 0.0308 | 16.4436 | 0.0608 | 9.7093 | 159.6566 |
| 24 | 2.0328 | 0.4919 | 34.4265 | 0.0290 | 16.9355 | 0.0590 | 10.0954 | 170.9711 |
| 25 | 2.0938 | 0.4776 | 36.4593 | 0.0274 | 17.4131 | 0.0574 | 10.4768 | 182.4336 |
| 26 | 2.1566 | 0.4637 | 38.5530 | 0.0259 | 17.8768 | 0.0559 | 10.8535 | 194.0260 |
| 27 | 2.2213 | 0.4502 | 40.7096 | 0.0246 | 18.3270 | 0.0546 | 11.2255 | 205.7309 |
| 28 | 2.2879 | 0.4371 | 42.9309 | 0.0233 | 18.7641 | 0.0533 | 11.5930 | 217.5320 |
| 29 | 2.3566 | 0.4243 | 45.2189 | 0.0221 | 19.1885 | 0.0521 | 11.9558 | 229.4137 |
| 30 | 2.4273 | 0.4120 | 47.5754 | 0.0210 | 19.6004 | 0.0510 | 12.3141 | 229.4137 |
| 31 | 2.5001 | 0.4000 | 50.0027 | 0.0200 | 20.0004 | 0.0500 | 12.6678 | 253.3609 |
| 32 | 2.5751 | 0.3883 | 52.5028 | 0.0190 | 20.3888 | 0.0490 | 13.0169 | 265.3993 |
| 33 | 2.6523 | 0.3770 | 55.0778 | 0.0182 | 20.7658 | 0.0482 | 13.3616 | 277.4642 |
| 34 | 2.7319 | 0.3660 | 57.7302 | 0.0173 | 21.1318 | 0.0473 | 13.7018 | 289.5437 |
| 35 | 2.8139 | 0.3554 | 60.4621 | 0.0165 | 21.4872 | 0.0465 | 14.0375 | 301.6267 |
| 40 | 3.2620 | 0.3066 | 75.4013 | 0.0133 | 23.1148 | 0.0433 | 15.6502 | 361.7499 |
| 45 | 3.7816 | 0.2644 | 92.7199 | 0.0108 | 24.5187 | 0.0408 | 17.1556 | 420.6325 |
| 50 | 4.3839 | 0.2281 | 112.7969 | 0.0089 | 25.7298 | 0.0389 | 18.5575 | 477.4803 |
| 55 | 5.0821 | 0.1968 | 136.0716 | 0.0073 | 26.7744 | 0.0373 | 19.8600 | 531.7411 |
| 60 | 5.8916 | 0.1697 | 163.0534 | 0.0061 | 27.6756 | 0.0361 | 21.0674 | 583.0526 |
| 65 | 6.8300 | 0.1464 | 194.3328 | 0.0051 | 28.4529 | 0.0351 | 22.1841 | 631.2010 |
| 70 | 7.9178 | 0.1263 | 230.5941 | 0.0043 | 29.1234 | 0.0343 | 23.2145 | 676.0869 |
| 75 | 9.1789 | 0.1089 | 272.6309 | 0.0037 | 29.7018 | 0.0337 | 24.1634 | 717.6978 |
| 80 | 10.6409 | 0.0940 | 321.3630 | 0.0031 | 30.2008 | 0.0331 | 25.0353 | 756.0865 |
| 85 | 12.3357 | 0.0811 | 377.8570 | 0.0026 | 30.6312 | 0.0326 | 25.8349 | 791.3529 |
| 90 | 14.3005 | 0.0699 | 443.3489 | 0.0023 | 31.0024 | 0.0323 | 26.5667 | 823.6302 |
| 95 | 16.5782 | 0.0603 | 519.2720 | 0.0019 | 31.3227 | 0.0319 | 27.2351 | 853.0742 |
| 100 | 19.2186 | 0.0520 | 607.2877 | 0.0016 | 31.5989 | 0.0316 | 27.8444 | 879.8540 |

## TABLE A.10  Discrete Compounding: $i = 4\%$

| | Single payment | | Uniform series | | | | Gradient series | |
|---|---|---|---|---|---|---|---|---|
| | Compound amount factor | Present worth factor | Compound amount factor | Sinking fund factor | Present worth factor | Capital recovery factor | Uniform series factor | Present worth factor |
| $n$ | To find $F$ given $P$ $F\|P\ i,n$ | To find $P$ given $F$ $P\|F\ i,n$ | To find $F$ given $A$ $F\|A\ i,n$ | To find $A$ given $F$ $A\|F\ i,n$ | To find $P$ given $A$ $P\|A\ i,n$ | To find $A$ given $P$ $A\|P\ i,n$ | To find $A$ given $G$ $A\|G\ i,n$ | To find $P$ given $G$ $P\|G\ i,n$ |
| 1 | 1.0400 | 0.9615 | 1.0000 | 1.0000 | 0.9615 | 1.0400 | 0.0000 | 0.0000 |
| 2 | 1.0816 | 0.9246 | 2.0400 | 0.4902 | 1.8861 | 0.5302 | 0.4902 | 0.9246 |
| 3 | 1.1249 | 0.8890 | 3.1216 | 0.3203 | 2.7751 | 0.3603 | 0.9739 | 2.7025 |
| 4 | 1.1699 | 0.8548 | 4.2465 | 0.2355 | 3.6299 | 0.2755 | 1.4510 | 5.2670 |
| 5 | 1.2167 | 0.8219 | 5.4163 | 0.1846 | 4.4518 | 0.2246 | 1.9216 | 8.5547 |
| 6 | 1.2653 | 0.7903 | 6.6330 | 0.1508 | 5.2421 | 0.1908 | 2.3857 | 12.5062 |
| 7 | 1.3159 | 0.7599 | 7.8983 | 0.1266 | 6.0021 | 0.1666 | 2.8433 | 17.0657 |
| 8 | 1.3686 | 0.7307 | 9.2142 | 0.1085 | 6.7327 | 0.1485 | 3.2944 | 22.1806 |
| 9 | 1.4233 | 0.7026 | 10.5828 | 0.0945 | 7.4353 | 0.1345 | 3.7391 | 27.8013 |
| 10 | 1.4802 | 0.6756 | 12.0061 | 0.0833 | 8.1109 | 0.1233 | 4.1773 | 33.8814 |
| 11 | 1.5395 | 0.6496 | 13.4864 | 0.0741 | 8.7605 | 0.1141 | 4.6090 | 40.3772 |
| 12 | 1.6010 | 0.6246 | 15.0258 | 0.0666 | 9.3851 | 0.1066 | 5.0343 | 47.2477 |
| 13 | 1.6651 | 0.6006 | 16.6268 | 0.0601 | 9.9856 | 0.1001 | 5.4533 | 54.4546 |
| 14 | 1.7317 | 0.5775 | 18.2919 | 0.0547 | 10.5631 | 0.0947 | 5.8659 | 61.9618 |
| 15 | 1.8009 | 0.5553 | 20.0236 | 0.0499 | 11.1184 | 0.0899 | 6.2721 | 69.7355 |
| 16 | 1.8730 | 0.5339 | 21.8245 | 0.0458 | 11.6523 | 0.0858 | 6.6720 | 77.7441 |
| 17 | 1.9479 | 0.5134 | 23.6975 | 0.0422 | 12.1657 | 0.0822 | 7.0656 | 85.9581 |
| 18 | 2.0258 | 0.4936 | 25.6454 | 0.0390 | 12.6593 | 0.0790 | 7.4530 | 94.3498 |
| 19 | 2.1068 | 0.4746 | 27.6712 | 0.0361 | 13.1339 | 0.0761 | 7.8342 | 102.8933 |
| 20 | 2.1911 | 0.4564 | 29.7781 | 0.0336 | 13.5903 | 0.0736 | 8.2091 | 111.5647 |
| 21 | 2.2788 | 0.4388 | 31.9692 | 0.0313 | 14.0292 | 0.0713 | 8.5779 | 120.3414 |
| 22 | 2.3699 | 0.4220 | 34.2480 | 0.0292 | 14.4511 | 0.0692 | 8.9407 | 129.2024 |
| 23 | 2.4647 | 0.4057 | 36.6179 | 0.0273 | 14.8568 | 0.0673 | 9.2973 | 138.1284 |
| 24 | 2.5633 | 0.3901 | 39.0826 | 0.0256 | 15.2470 | 0.0656 | 9.6479 | 147.1012 |
| 25 | 2.6658 | 0.3751 | 41.6459 | 0.0240 | 15.6221 | 0.0640 | 9.9925 | 156.1040 |
| 26 | 2.7725 | 0.3607 | 44.3117 | 0.0226 | 15.9828 | 0.0626 | 10.3312 | 165.1212 |
| 27 | 2.8834 | 0.3468 | 47.0842 | 0.0212 | 16.3296 | 0.0612 | 10.6640 | 174.1385 |
| 28 | 2.9987 | 0.3335 | 49.9676 | 0.0200 | 16.6631 | 0.0600 | 10.9909 | 183.1424 |
| 29 | 3.1187 | 0.3207 | 52.9663 | 0.0189 | 16.9837 | 0.0589 | 11.3120 | 192.1206 |
| 30 | 3.2434 | 0.3083 | 56.0849 | 0.0178 | 17.2920 | 0.0578 | 11.6274 | 201.0618 |
| 31 | 3.3731 | 0.2965 | 59.3283 | 0.0169 | 17.5885 | 0.0569 | 11.9371 | 209.9556 |
| 32 | 3.5081 | 0.2851 | 62.7015 | 0.0159 | 17.8736 | 0.0559 | 12.2411 | 218.7924 |
| 33 | 3.6484 | 0.2741 | 66.2095 | 0.0151 | 18.1476 | 0.0551 | 12.5396 | 227.5634 |
| 34 | 3.7943 | 0.2636 | 69.8579 | 0.0143 | 18.4112 | 0.0543 | 12.8324 | 236.2607 |
| 35 | 3.9461 | 0.2534 | 73.6522 | 0.0136 | 18.6646 | 0.0536 | 13.1198 | 244.8768 |
| 40 | 4.8010 | 0.2083 | 95.0255 | 0.0105 | 19.7928 | 0.0505 | 14.4765 | 286.5303 |
| 45 | 5.8412 | 0.1712 | 121.0294 | 0.0083 | 20.7200 | 0.0483 | 15.7047 | 325.4028 |
| 50 | 7.1067 | 0.1407 | 152.6671 | 0.0066 | 21.4822 | 0.0466 | 16.8122 | 361.1638 |
| 55 | 8.6464 | 0.1157 | 191.1592 | 0.0052 | 22.1086 | 0.0452 | 17.8070 | 393.6890 |
| 60 | 10.5196 | 0.0951 | 237.9907 | 0.0042 | 22.6235 | 0.0442 | 18.6972 | 422.9966 |
| 65 | 12.7987 | 0.0781 | 294.9684 | 0.0034 | 23.0467 | 0.0434 | 19.4909 | 449.2014 |
| 70 | 15.5716 | 0.0642 | 264.2905 | 0.0027 | 23.3945 | 0.0427 | 20.1961 | 472.4789 |
| 75 | 18.9453 | 0.0528 | 448.6314 | 0.0022 | 23.6804 | 0.0422 | 20.8206 | 493.0408 |
| 80 | 23.0498 | 0.0434 | 551.2450 | 0.0018 | 23.9154 | 0.0418 | 21.3718 | 511.1161 |
| 85 | 28.0436 | 0.0357 | 676.0901 | 0.0015 | 24.1085 | 0.0415 | 21.8569 | 526.9384 |
| 90 | 34.1193 | 0.0293 | 827.9833 | 0.0012 | 24.2673 | 0.0412 | 22.2826 | 540.7369 |
| 95 | 41.5114 | 0.0241 | 1012.7846 | 0.0010 | 24.3978 | 0.0410 | 22.6550 | 552.7307 |
| 100 | 50.5049 | 0.0198 | 1237.6237 | 0.0008 | 24.5050 | 0.0408 | 22.9800 | 563.1249 |

| | Single payment | | Uniform series | | | | Gradient series | |
|---|---|---|---|---|---|---|---|---|
| | Compound amount factor | Present worth factor | Compound amount factor | Sinking fund factor | Present worth factor | Capital recovery factor | Uniform series factor | Present worth factor |
| $n$ | To find $F$ given $P$ $F\|P\ i,n$ | To find $P$ given $F$ $P\|F\ i,n$ | To find $F$ given $A$ $F\|A\ i,n$ | To find $A$ given $F$ $A\|F\ i,n$ | To find $P$ given $A$ $P\|A\ i,n$ | To find $A$ given $P$ $A\|P\ i,n$ | To find $A$ given $G$ $A\|G\ i,n$ | To find $P$ given $G$ $P\|G\ i,n$ |
| 1 | 1.0500 | 0.9524 | 1.0000 | 1.0000 | 0.9524 | 1.0500 | 0.0000 | 0.0000 |
| 2 | 1.1025 | 0.9070 | 2.0500 | 0.4878 | 1.8594 | 0.5378 | 0.4878 | 0.9070 |
| 3 | 1.1576 | 0.8638 | 3.1525 | 0.3172 | 2.7232 | 0.3672 | 0.9675 | 2.6347 |
| 4 | 1.2155 | 0.8227 | 4.3101 | 0.2320 | 3.5460 | 0.2820 | 1.4391 | 5.1028 |
| 5 | 1.2763 | 0.7835 | 5.5256 | 0.1810 | 4.3295 | 0.2310 | 1.9025 | 8.2369 |
| 6 | 1.3401 | 0.7462 | 6.8019 | 0.1470 | 5.0757 | 0.1970 | 2.3579 | 11.9680 |
| 7 | 1.4071 | 0.7107 | 8.1420 | 0.1228 | 5.7864 | 0.1728 | 2.8052 | 16.2321 |
| 8 | 1.4775 | 0.6768 | 9.5491 | 0.1047 | 6.4632 | 0.1547 | 3.2445 | 20.9700 |
| 9 | 1.5513 | 0.6446 | 11.0266 | 0.0907 | 7.1078 | 0.1407 | 3.6758 | 26.1268 |
| 10 | 1.6289 | 0.6139 | 12.5779 | 0.0795 | 7.7217 | 0.1295 | 4.0991 | 31.6520 |
| 11 | 1.7103 | 0.5847 | 14.2068 | 0.0704 | 8.3064 | 0.1204 | 4.5144 | 37.4988 |
| 12 | 1.7959 | 0.5568 | 15.9171 | 0.0628 | 8.8633 | 0.1128 | 4.9219 | 43.6241 |
| 13 | 1.8856 | 0.5303 | 17.7130 | 0.0565 | 9.3936 | 0.1065 | 5.3215 | 49.9879 |
| 14 | 1.9799 | 0.5051 | 19.5986 | 0.0510 | 9.8986 | 0.1010 | 5.7133 | 56.5538 |
| 15 | 2.0789 | 0.4810 | 21.5786 | 0.0463 | 10.3797 | 0.0963 | 6.0973 | 63.2880 |
| 16 | 2.1829 | 0.4581 | 23.6575 | 0.0423 | 10.8378 | 0.0923 | 6.4736 | 70.1597 |
| 17 | 2.2920 | 0.4363 | 25.8404 | 0.0387 | 11.2741 | 0.0887 | 6.8423 | 77.1405 |
| 18 | 2.4066 | 0.4155 | 28.1324 | 0.0355 | 11.6896 | 0.0855 | 7.2034 | 84.2043 |
| 19 | 2.5270 | 0.3957 | 30.5390 | 0.0327 | 12.0853 | 0.0827 | 7.5569 | 91.3275 |
| 20 | 2.6533 | 0.3769 | 33.0660 | 0.0302 | 12.4622 | 0.0802 | 7.9030 | 98.4884 |
| 21 | 2.7860 | 0.3589 | 35.7193 | 0.0280 | 12.8212 | 0.0780 | 8.2416 | 105.6673 |
| 22 | 2.9253 | 0.3418 | 38.5052 | 0.0260 | 13.1630 | 0.0760 | 8.5730 | 112.8461 |
| 23 | 3.0715 | 0.3256 | 41.4305 | 0.0241 | 13.4886 | 0.0741 | 8.8971 | 120.0087 |
| 24 | 3.2251 | 0.3101 | 44.5020 | 0.0225 | 13.7986 | 0.0725 | 9.2140 | 127.1402 |
| 25 | 3.3864 | 0.2953 | 47.7271 | 0.0210 | 14.0939 | 0.0710 | 9.5238 | 134.2275 |
| 26 | 3.5557 | 0.2812 | 51.1135 | 0.0196 | 14.3752 | 0.0696 | 9.8266 | 141.2585 |
| 27 | 3.7335 | 0.2678 | 54.6691 | 0.0183 | 14.6430 | 0.0683 | 10.1224 | 148.2226 |
| 28 | 3.9201 | 0.2551 | 58.4026 | 0.0171 | 14.8981 | 0.0671 | 10.4114 | 155.1101 |
| 29 | 4.1161 | 0.2429 | 62.3227 | 0.0160 | 15.1411 | 0.0660 | 10.6936 | 161.9126 |
| 30 | 4.3219 | 0.2314 | 66.4388 | 0.0151 | 15.3725 | 0.0651 | 10.9691 | 168.6226 |
| 31 | 4.5380 | 0.2204 | 70.7608 | 0.0141 | 15.5928 | 0.0641 | 11.2381 | 175.2333 |
| 32 | 4.7649 | 0.2099 | 75.2988 | 0.0133 | 15.8027 | 0.0633 | 11.5005 | 181.7392 |
| 33 | 5.0032 | 0.1999 | 80.0638 | 0.0125 | 16.0025 | 0.0625 | 11.7566 | 188.1351 |
| 34 | 5.2533 | 0.1904 | 85.0670 | 0.0118 | 16.1929 | 0.0618 | 12.0063 | 194.4168 |
| 35 | 5.5160 | 0.1813 | 90.3203 | 0.0111 | 16.3742 | 0.0611 | 12.2498 | 200.5807 |
| 40 | 7.0400 | 0.1420 | 120.7998 | 0.0083 | 17.1591 | 0.0583 | 13.3775 | 299.5452 |
| 45 | 8.8950 | 0.1113 | 159.7002 | 0.0063 | 17.7741 | 0.0563 | 14.3644 | 255.3145 |
| 50 | 11.4674 | 0.0872 | 209.3480 | 0.0048 | 18.2559 | 0.0548 | 15.2233 | 277.9148 |
| 55 | 14.6356 | 0.0683 | 272.7126 | 0.0037 | 18.6335 | 0.0537 | 15.9664 | 297.5104 |
| 60 | 18.6792 | 0.0535 | 353.5837 | 0.0028 | 18.9293 | 0.0528 | 16.6062 | 314.3432 |
| 65 | 23.8399 | 0.0419 | 456.7980 | 0.0022 | 19.1611 | 0.0522 | 17.1541 | 328.6910 |
| 70 | 30.4264 | 0.0329 | 588.5285 | 0.0017 | 19.3427 | 0.0517 | 17.6212 | 340.8409 |
| 75 | 38.8327 | 0.0258 | 756.6537 | 0.0013 | 19.4850 | 0.0513 | 18.0176 | 351.0721 |
| 80 | 49.5614 | 0.0202 | 971.2288 | 0.0010 | 19.5965 | 0.0510 | 18.3526 | 359.6460 |
| 85 | 63.2544 | 0.0158 | 1245.0871 | 0.0008 | 19.6838 | 0.0508 | 18.6346 | 366.8007 |
| 90 | 80.7304 | 0.0124 | 1594.6073 | 0.0006 | 19.7523 | 0.0506 | 18.8712 | 372.7488 |
| 95 | 103.0347 | 0.0097 | 2040.6935 | 0.0005 | 19.8059 | 0.0505 | 19.0689 | 377.6774 |
| 100 | 131.5013 | 0.0076 | 2610.0252 | 0.0004 | 19.8479 | 0.0504 | 19.2337 | 381.7492 |

## TABLE A.12  Discrete Compounding: $i = 6\%$

| | Single payment | | Uniform series | | | | Gradient series | |
|---|---|---|---|---|---|---|---|---|
| | Compound amount factor | Present worth factor | Compound amount factor | Sinking fund factor | Present worth factor | Capital recovery factor | Uniform series factor | Present worth factor |
| $n$ | To find $F$ given $P$ $F\|P\ i,n$ | To find $P$ given $F$ $P\|F\ i,n$ | To find $F$ given $A$ $F\|A\ i,n$ | To find $A$ given $F$ $A\|F\ i,n$ | To find $P$ given $A$ $P\|A\ i,n$ | To find $A$ given $P$ $A\|P\ i,n$ | To find $A$ given $G$ $A\|G\ i,n$ | To find $P$ given $G$ $P\|G\ i,n$ |
| 1 | 1.0600 | 0.9434 | 1.0000 | 1.0000 | 0.9434 | 1.0600 | 0.0000 | 0.0000 |
| 2 | 1.1236 | 0.8900 | 2.0600 | 0.4854 | 1.8334 | 0.5454 | 0.4854 | 0.8900 |
| 3 | 1.1910 | 0.8396 | 3.1836 | 0.3141 | 2.6730 | 0.3741 | 0.9612 | 2.5692 |
| 4 | 1.2625 | 0.7921 | 4.3746 | 0.2286 | 3.4651 | 0.2886 | 1.4272 | 4.9455 |
| 5 | 1.3382 | 0.7473 | 5.6371 | 0.1774 | 4.2124 | 0.2374 | 1.8836 | 7.9345 |
| 6 | 1.4185 | 0.7050 | 6.9753 | 0.1434 | 4.9173 | 0.2034 | 2.3304 | 11.4594 |
| 7 | 1.5036 | 0.6651 | 8.3938 | 0.1191 | 5.5824 | 0.1791 | 2.7676 | 15.4497 |
| 8 | 1.5938 | 0.6274 | 9.8975 | 0.1010 | 6.2098 | 0.1610 | 3.1952 | 19.8416 |
| 9 | 1.6895 | 0.5919 | 11.4913 | 0.0870 | 6.8017 | 0.1470 | 3.6133 | 24.5768 |
| 10 | 1.7908 | 0.5584 | 13.1808 | 0.0759 | 7.3601 | 0.1359 | 4.0220 | 29.6023 |
| 11 | 1.8983 | 0.5268 | 14.9716 | 0.0668 | 7.8869 | 0.1268 | 4.4213 | 34.8702 |
| 12 | 2.0122 | 0.4970 | 16.8699 | 0.0593 | 8.3838 | 0.1193 | 4.8113 | 40.3369 |
| 13 | 2.1329 | 0.4688 | 18.8821 | 0.0530 | 8.8527 | 0.1130 | 5.1920 | 45.9629 |
| 14 | 2.2609 | 0.4423 | 21.0151 | 0.0476 | 9.2950 | 0.1076 | 5.5635 | 51.7128 |
| 15 | 2.3966 | 0.4173 | 23.2760 | 0.0430 | 9.7122 | 0.1030 | 5.9260 | 57.5546 |
| 16 | 2.5404 | 0.3936 | 25.6725 | 0.0390 | 10.1059 | 0.0990 | 6.2794 | 63.4592 |
| 17 | 2.6928 | 0.3714 | 28.2129 | 0.0354 | 10.4773 | 0.0954 | 6.6240 | 69.4011 |
| 18 | 2.8543 | 0.3503 | 30.9057 | 0.0324 | 10.8276 | 0.0924 | 6.9597 | 75.3569 |
| 19 | 3.0256 | 0.3305 | 33.7600 | 0.0296 | 11.1581 | 0.0896 | 7.2867 | 81.3062 |
| 20 | 3.2071 | 0.3118 | 36.7856 | 0.0272 | 11.4699 | 0.0872 | 7.6051 | 87.2304 |
| 21 | 3.3996 | 0.2942 | 39.9927 | 0.0250 | 11.7641 | 0.0850 | 7.9151 | 93.1136 |
| 22 | 3.6035 | 0.2775 | 43.3923 | 0.0230 | 12.0416 | 0.0830 | 8.2166 | 98.9412 |
| 23 | 3.8197 | 0.2618 | 46.9958 | 0.0213 | 12.3034 | 0.0813 | 8.5099 | 104.7007 |
| 24 | 4.0489 | 0.2470 | 50.8156 | 0.0197 | 12.5504 | 0.0797 | 8.7951 | 110.3812 |
| 25 | 4.2919 | 0.2330 | 43.8645 | 0.0182 | 12.7834 | 0.0782 | 9.0722 | 115.9732 |
| 26 | 4.5494 | 0.2198 | 59.1564 | 0.0169 | 13.0032 | 0.0769 | 9.3414 | 121.4684 |
| 27 | 4.8223 | 0.2074 | 63.7058 | 0.0157 | 13.2105 | 0.0757 | 9.6029 | 126.8600 |
| 28 | 5.1117 | 0.1956 | 68.5281 | 0.0146 | 13.4062 | 0.0746 | 9.8568 | 132.1420 |
| 29 | 5.4184 | 0.1846 | 73.6398 | 0.0136 | 13.5907 | 0.0736 | 10.1032 | 137.3096 |
| 30 | 5.7435 | 0.1741 | 79.0582 | 0.0126 | 13.7648 | 0.0726 | 10.3422 | 142.3588 |
| 31 | 6.0881 | 0.1643 | 84.8017 | 0.0118 | 13.9291 | 0.0718 | 10.5740 | 147.2864 |
| 32 | 6.4534 | 0.1550 | 90.8898 | 0.0110 | 14.0840 | 0.0710 | 10.7988 | 152.0901 |
| 33 | 6.8406 | 0.1462 | 97.3432 | 0.0103 | 14.2302 | 0.0703 | 11.0166 | 156.7681 |
| 34 | 7.2510 | 0.1379 | 104.1838 | 0.0096 | 14.3681 | 0.0696 | 11.2276 | 161.3192 |
| 35 | 7.6861 | 0.1301 | 111.4348 | 0.0090 | 14.4982 | 0.0690 | 11.4319 | 165.7427 |
| 40 | 10.2857 | 0.0972 | 154.7620 | 0.0065 | 15.0463 | 0.0665 | 12.3590 | 185.9568 |
| 45 | 13.7646 | 0.0727 | 212.7435 | 0.0047 | 15.4558 | 0.0647 | 13.1413 | 203.1096 |
| 50 | 18.4202 | 0.0543 | 290.3359 | 0.0034 | 15.7619 | 0.0634 | 13.7964 | 217.4574 |
| 55 | 24.6503 | 0.0406 | 394.1720 | 0.0025 | 15.9905 | 0.0625 | 14.3411 | 229.3222 |
| 60 | 32.9867 | 0.0303 | 533.1282 | 0.0019 | 16.1614 | 0.0619 | 14.7909 | 239.0428 |
| 65 | 44.1450 | 0.0227 | 719.0829 | 0.0014 | 16.2891 | 0.0614 | 15.1601 | 246.9450 |
| 70 | 59.0759 | 0.0169 | 967.9322 | 0.0010 | 16.3845 | 0.0610 | 15.4613 | 253.3271 |
| 75 | 79.0569 | 0.0126 | 1300.9487 | 0.0008 | 16.4558 | 0.0608 | 15.7058 | 258.4527 |
| 80 | 105.7960 | 0.0095 | 1746.5999 | 0.0006 | 16.5091 | 0.0606 | 15.9033 | 262.5493 |
| 85 | 141.5789 | 0.0071 | 2342.9817 | 0.0004 | 16.5489 | 0.0604 | 16.0620 | 265.8096 |
| 90 | 189.4645 | 0.0053 | 3141.0752 | 0.0003 | 16.5787 | 0.0603 | 16.1891 | 268.3946 |
| 95 | 253.5463 | 0.0039 | 4209.1042 | 0.0002 | 16.6009 | 0.0602 | 16.2905 | 270.4375 |
| 100 | 339.3021 | 0.0029 | 5638.3681 | 0.0002 | 16.6175 | 0.0602 | 16.3711 | 272.0471 |

| | Single payment | | Uniform series | | | | Gradient series | |
|---|---|---|---|---|---|---|---|---|
| | Compound amount factor | Present worth factor | Compound amount factor | Sinking fund factor | Present worth factor | Capital recovery factor | Uniform series factor | Present worth factor |
| $n$ | To find $F$ given $P$ $F\mid P\ i,n$ | To find $P$ given $F$ $P\mid F\ i,n$ | To find $F$ given $A$ $F\mid A\ i,n$ | To find $A$ given $F$ $A\mid F\ i,n$ | To find $P$ given $A$ $P\mid A\ i,n$ | To find $A$ given $P$ $A\mid P\ i,n$ | To find $A$ given $G$ $A\mid G\ i,n$ | To find $P$ given $G$ $P\mid G\ i,n$ |
| 1 | 1.0700 | 0.9346 | 1.0000 | 1.0000 | 0.9346 | 1.0700 | 0.0000 | 0.0000 |
| 2 | 1.1449 | 0.8734 | 2.0700 | 0.4831 | 1.8080 | 0.5531 | 0.4831 | 0.8734 |
| 3 | 1.2250 | 0.8163 | 3.2149 | 0.3111 | 2.6243 | 0.3811 | 0.9549 | 2.5060 |
| 4 | 1.3108 | 0.7629 | 4.4399 | 0.2252 | 3.3872 | 0.2952 | 1.4155 | 4.7947 |
| 5 | 1.4026 | 0.7130 | 5.7507 | 0.1739 | 4.1002 | 0.2439 | 1.8650 | 7.6467 |
| 6 | 1.5007 | 0.6663 | 7.1533 | 0.1398 | 4.7665 | 0.2098 | 2.3032 | 10.9784 |
| 7 | 1.6058 | 0.6227 | 8.6540 | 0.1156 | 5.3893 | 0.1856 | 2.7304 | 14.7149 |
| 8 | 1.7182 | 0.5820 | 10.2598 | 0.0975 | 5.9713 | 0.1675 | 3.1465 | 18.7889 |
| 9 | 1.8385 | 0.5439 | 11.9780 | 0.0835 | 6.5152 | 0.1535 | 3.5517 | 23.1404 |
| 10 | 1.9672 | 0.5083 | 13.8164 | 0.0724 | 7.0236 | 0.1424 | 3.9461 | 27.7156 |
| 11 | 2.1049 | 0.4751 | 15.7836 | 0.0634 | 7.4987 | 0.1334 | 4.3296 | 32.4665 |
| 12 | 2.2522 | 0.4440 | 17.8885 | 0.0559 | 7.9427 | 0.1259 | 4.7025 | 37.3506 |
| 13 | 2.4098 | 0.4150 | 20.1406 | 0.0497 | 8.3577 | 0.1197 | 5.0648 | 42.3302 |
| 14 | 2.5785 | 0.3878 | 22.5505 | 0.0443 | 8.7455 | 0.1143 | 5.4167 | 47.3718 |
| 15 | 2.7590 | 0.3624 | 25.1290 | 0.0398 | 9.1079 | 0.1098 | 5.7583 | 52.4461 |
| 16 | 2.9522 | 0.3387 | 27.8881 | 0.0359 | 9.4466 | 0.1059 | 6.0897 | 57.5271 |
| 17 | 3.1588 | 0.3166 | 30.8402 | 0.0324 | 9.7632 | 0.1024 | 6.4110 | 62.5923 |
| 18 | 3.3799 | 0.2959 | 33.9990 | 0.0294 | 10.0591 | 0.0994 | 6.7225 | 67.6219 |
| 19 | 3.6165 | 0.2765 | 37.3790 | 0.0268 | 10.3356 | 0.0968 | 7.0242 | 72.5991 |
| 20 | 3.8697 | 0.2584 | 40.9955 | 0.0244 | 10.5940 | 0.0944 | 7.3163 | 77.5091 |
| 21 | 4.1406 | 0.2415 | 44.8652 | 0.0223 | 10.8355 | 0.0923 | 7.5990 | 82.3393 |
| 22 | 4.4304 | 0.2257 | 49.0057 | 0.0204 | 11.0612 | 0.0904 | 7.8725 | 87.0793 |
| 23 | 4.7405 | 0.2109 | 53.4361 | 0.0187 | 11.2722 | 0.0887 | 8.1369 | 91.7201 |
| 24 | 5.0724 | 0.1971 | 58.1767 | 0.0172 | 11.4693 | 0.0872 | 8.3923 | 96.2545 |
| 25 | 5.4274 | 0.1842 | 63.2490 | 0.0158 | 11.6536 | 0.0858 | 8.6391 | 100.6765 |
| 26 | 5.8074 | 0.1722 | 68.6765 | 0.0146 | 11.8258 | 0.0846 | 8.8773 | 104.9814 |
| 27 | 6.2139 | 0.1609 | 74.4838 | 0.0134 | 11.9867 | 0.0834 | 9.1072 | 109.1656 |
| 28 | 6.6488 | 0.1504 | 80.6977 | 0.0124 | 12.1371 | 0.0824 | 9.3289 | 113.2264 |
| 29 | 7.1143 | 0.1406 | 87.3465 | 0.0114 | 12.2777 | 0.0814 | 9.5427 | 117.1622 |
| 30 | 7.6123 | 0.1314 | 94.4608 | 0.0106 | 12.4090 | 0.0806 | 9.7487 | 120.9718 |
| 31 | 8.1451 | 0.1228 | 102.0730 | 0.0098 | 12.5318 | 0.0798 | 9.9471 | 124.6550 |
| 32 | 8.7153 | 0.1147 | 110.2182 | 0.0091 | 12.6466 | 0.0791 | 10.1381 | 128.2120 |
| 33 | 9.3253 | 0.1072 | 118.9334 | 0.0084 | 12.7538 | 0.0784 | 10.3219 | 131.6435 |
| 34 | 9.9781 | 0.1002 | 128.2588 | 0.0078 | 12.8540 | 0.0778 | 10.4987 | 134.9507 |
| 35 | 10.6766 | 0.0937 | 138.2369 | 0.0072 | 12.9477 | 0.0772 | 10.6687 | 138.1353 |
| 40 | 14.9745 | 0.0668 | 199.6351 | 0.0050 | 13.3317 | 0.0750 | 11.4233 | 152.2928 |
| 45 | 21.0025 | 0.0476 | 285.7493 | 0.0035 | 13.6055 | 0.0735 | 12.0360 | 163.7559 |
| 50 | 29.4570 | 0.0339 | 406.5289 | 0.0025 | 13.8007 | 0.0725 | 12.5287 | 172.9051 |
| 55 | 41.3150 | 0.0242 | 575.9286 | 0.0017 | 13.9399 | 0.0717 | 12.9215 | 180.1243 |
| 60 | 57.9464 | 0.0173 | 813.5240 | 0.0012 | 14.0392 | 0.0712 | 13.2321 | 185.7677 |
| 65 | 81.2729 | 0.0123 | 1146.7552 | 0.0009 | 14.1099 | 0.0709 | 13.4760 | 190.1452 |
| 70 | 113.9894 | 0.0088 | 1614.1342 | 0.0006 | 14.1604 | 0.0706 | 13.6662 | 193.5185 |
| 75 | 159.8760 | 0.0063 | 2269.6574 | 0.0004 | 14.1964 | 0.0704 | 13.8136 | 196.1035 |
| 80 | 224.2344 | 0.0045 | 3189.0627 | 0.0003 | 14.2220 | 0.0703 | 13.9273 | 198.0748 |
| 85 | 314.5003 | 0.0032 | 4478.5761 | 0.0002 | 14.2403 | 0.0702 | 14.0146 | 199.5717 |
| 90 | 441.1030 | 0.0023 | 6287.1854 | 0.0002 | 14.2533 | 0.0702 | 14.0812 | 200.7042 |
| 95 | 618.6697 | 0.0016 | 8823.8535 | 0.0001 | 14.2626 | 0.0701 | 14.1319 | 201.5581 |
| 100 | 867.7163 | 0.0012 | 12381.6618 | 0.0001 | 14.2693 | 0.0701 | 14.1703 | 202.2001 |

## TABLE A.14  Discrete Compounding: $i = 8\%$

| | Single payment | | Uniform series | | | | Gradient series | |
|---|---|---|---|---|---|---|---|---|
| | Compound amount factor | Present worth factor | Compound amount factor | Sinking fund factor | Present worth factor | Capital recovery factor | Uniform series factor | Present worth factor |
| $n$ | To find $F$ given $P$ $F \mid P\ i,n$ | To find $P$ given $F$ $P \mid F\ i,n$ | To find $F$ given $A$ $F \mid A\ i,n$ | To find $A$ given $F$ $A \mid F\ i,n$ | To find $P$ given $A$ $P \mid A\ i,n$ | To find $A$ given $P$ $A \mid P\ i,n$ | To find $A$ given $G$ $A \mid G\ i,n$ | To find $P$ given $G$ $P \mid G\ i,n$ |
| 1 | 1.0800 | 0.9259 | 1.0000 | 1.0000 | 0.9259 | 1.0800 | 0.0000 | 0.0000 |
| 2 | 1.1664 | 0.8573 | 2.0800 | 0.4808 | 1.7833 | 0.5608 | 0.4808 | 0.8573 |
| 3 | 1.2597 | 0.7938 | 3.2464 | 0.3080 | 2.5771 | 0.3880 | 0.9487 | 2.4450 |
| 4 | 1.3605 | 0.7350 | 4.5061 | 0.2219 | 3.3121 | 0.3019 | 1.4040 | 4.6501 |
| 5 | 1.4693 | 0.6806 | 5.8666 | 0.1705 | 3.9927 | 0.2505 | 1.8465 | 7.3724 |
| 6 | 1.5869 | 0.6302 | 7.3359 | 0.1363 | 4.6229 | 0.2163 | 2.2763 | 10.5233 |
| 7 | 1.7138 | 0.5835 | 8.9228 | 0.1121 | 5.2064 | 0.1921 | 2.6937 | 14.0242 |
| 8 | 1.8509 | 0.5403 | 10.6366 | 0.0940 | 5.7466 | 0.1740 | 3.0985 | 17.8061 |
| 9 | 1.9990 | 0.5002 | 12.4876 | 0.0801 | 6.2469 | 0.1601 | 3.4910 | 21.8081 |
| 10 | 2.1589 | 0.4632 | 14.4866 | 0.0690 | 6.7101 | 0.1490 | 3.8713 | 25.9768 |
| 11 | 2.3316 | 0.4289 | 16.6455 | 0.0601 | 7.1390 | 0.1401 | 4.2395 | 30.2657 |
| 12 | 2.5182 | 0.3971 | 18.9771 | 0.0527 | 7.5361 | 0.1327 | 4.5957 | 34.6339 |
| 13 | 2.7196 | 0.3677 | 21.4953 | 0.0465 | 7.9038 | 0.1265 | 4.9402 | 39.0463 |
| 14 | 2.9372 | 0.3405 | 24.2149 | 0.0413 | 8.2442 | 0.1213 | 5.2731 | 43.4723 |
| 15 | 3.1722 | 0.3152 | 27.1521 | 0.0368 | 8.5595 | 0.1168 | 5.5945 | 47.8857 |
| 16 | 3.4259 | 0.2919 | 30.3243 | 0.0330 | 8.8514 | 0.1130 | 5.9046 | 52.2640 |
| 17 | 3.7000 | 0.2703 | 33.7502 | 0.0296 | 9.1216 | 0.1096 | 6.2037 | 56.5883 |
| 18 | 3.9960 | 0.2502 | 37.4502 | 0.0267 | 9.3719 | 0.1067 | 6.4920 | 60.8426 |
| 19 | 4.3157 | 0.2317 | 41.4463 | 0.0241 | 9.6036 | 0.1041 | 6.7697 | 65.0134 |
| 20 | 4.6610 | 0.2145 | 45.7620 | 0.0219 | 9.8181 | 0.1019 | 7.0369 | 69.0898 |
| 21 | 5.0338 | 0.1987 | 50.4229 | 0.0198 | 10.0168 | 0.0998 | 7.2940 | 73.0629 |
| 22 | 5.4365 | 0.1839 | 55.4568 | 0.0180 | 10.2007 | 0.0980 | 7.5412 | 76.9257 |
| 23 | 5.8715 | 0.1703 | 60.8933 | 0.0164 | 10.3711 | 0.0964 | 7.7786 | 80.6726 |
| 24 | 6.3412 | 0.1577 | 66.7648 | 0.0150 | 10.5288 | 0.0950 | 8.0066 | 84.2997 |
| 25 | 6.8485 | 0.1460 | 73.1059 | 0.0137 | 10.6748 | 0.0937 | 8.2254 | 87.8041 |
| 26 | 7.3964 | 0.1352 | 79.9544 | 0.0125 | 10.8100 | 0.0925 | 8.4352 | 91.1842 |
| 27 | 7.9881 | 0.1252 | 87.3508 | 0.0114 | 10.9353 | 0.0914 | 8.6363 | 94.4390 |
| 28 | 8.6271 | 0.1159 | 95.3388 | 0.0105 | 11.0511 | 0.0905 | 8.8289 | 97.5687 |
| 29 | 9.3173 | 0.1073 | 103.9659 | 0.0096 | 11.1584 | 0.0896 | 9.0133 | 100.5738 |
| 30 | 10.0627 | 0.0994 | 113.2832 | 0.0088 | 11.2578 | 0.0888 | 9.1897 | 103.4558 |
| 31 | 10.8677 | 0.0920 | 123.3459 | 0.0081 | 11.3498 | 0.0881 | 9.3584 | 106.2163 |
| 32 | 11.7371 | 0.0852 | 134.2135 | 0.0075 | 11.4350 | 0.0875 | 9.5197 | 108.8575 |
| 33 | 12.6760 | 0.0789 | 145.9506 | 0.0069 | 11.5139 | 0.0869 | 9.6737 | 111.3819 |
| 34 | 13.6901 | 0.0730 | 158.6267 | 0.0063 | 11.5869 | 0.0863 | 9.8208 | 113.7924 |
| 35 | 14.7853 | 0.0676 | 172.3168 | 0.0058 | 11.6546 | 0.0858 | 9.9611 | 116.0920 |
| 40 | 21.7245 | 0.0460 | 259.0565 | 0.0039 | 11.9246 | 0.0839 | 10.5699 | 126.0422 |
| 45 | 31.9204 | 0.0313 | 386.5056 | 0.0026 | 12.1084 | 0.0826 | 11.0447 | 133.7331 |
| 50 | 46.9016 | 0.0213 | 573.7702 | 0.0017 | 12.2335 | 0.0817 | 11.4107 | 139.5928 |
| 55 | 68.9139 | 0.0145 | 848.9232 | 0.0012 | 12.3186 | 0.0812 | 11.6902 | 144.0065 |
| 60 | 101.2571 | 0.0099 | 1253.2133 | 0.0008 | 12.3766 | 0.0808 | 11.9015 | 147.3000 |
| 65 | 148.7798 | 0.0067 | 1847.2481 | 0.0005 | 12.4160 | 0.0805 | 12.0602 | 149.7387 |
| 70 | 218.6064 | 0.0046 | 2720.0801 | 0.0004 | 12.4428 | 0.0804 | 12.1783 | 151.5326 |
| 75 | 321.2045 | 0.0031 | 4002.5566 | 0.0002 | 12.4611 | 0.0802 | 12.2658 | 152.8448 |
| 80 | 471.9548 | 0.0021 | 5886.9354 | 0.0002 | 12.4735 | 0.0802 | 12.3301 | 153.8001 |
| 85 | 693.4565 | 0.0014 | 8655.7061 | 0.0001 | 12.4820 | 0.0801 | 12.3772 | 154.4925 |
| 90 | 1018.9151 | 0.0010 | 12723.9386 | 0.0001 | 12.4877 | 0.0801 | 12.4116 | 154.9925 |
| 95 | 1497.1205 | 0.0007 | 18701.5069 | 0.0001 | 12.4971 | 0.0801 | 12.4365 | 155.3524 |
| 100 | 2199.7613 | 0.0005 | 27484.5157 | 0.0000 | 12.4943 | 0.0800 | 12.4545 | 155.6107 |

| | Single payment | | Uniform series | | | | Gradient series | |
|---|---|---|---|---|---|---|---|---|
| | Compound amount factor | Present worth factor | Compound amount factor | Sinking fund factor | Present worth factor | Capital recovery factor | Uniform series factor | Present worth factor |
| $n$ | To find $F$ given $P$ $F\|P\,i,n$ | To find $P$ given $F$ $P\|F\,i,n$ | To find $F$ given $A$ $F\|A\,i,n$ | To find $A$ given $F$ $A\|F\,i,n$ | To find $P$ given $A$ $P\|A\,i,n$ | To find $A$ given $P$ $A\|P\,i,n$ | To find $A$ given $G$ $A\|G\,i,n$ | To find $P$ given $G$ $P\|G\,i,n$ |
| 1 | 1.0900 | 0.9174 | 1.0000 | 1.0000 | 0.9174 | 1.0900 | 0.0000 | 0.0000 |
| 2 | 1.1881 | 0.8417 | 2.0900 | 0.4785 | 1.7591 | 0.5685 | 0.4785 | 0.8417 |
| 3 | 1.2950 | 0.7722 | 3.2781 | 0.3051 | 2.5313 | 0.3951 | 0.9426 | 2.3860 |
| 4 | 1.4116 | 0.7084 | 4.5731 | 0.2187 | 3.2397 | 0.3087 | 1.3925 | 4.5113 |
| 5 | 1.5386 | 0.6499 | 5.9847 | 0.1671 | 3.8897 | 0.2571 | 1.8282 | 7.1110 |
| 6 | 1.6771 | 0.5963 | 7.5233 | 0.1329 | 4.4859 | 0.2229 | 2.2498 | 10.0924 |
| 7 | 1.8280 | 0.5470 | 9.2004 | 0.1087 | 5.0330 | 0.1987 | 2.6574 | 13.3746 |
| 8 | 1.9926 | 0.5019 | 11.0285 | 0.0907 | 5.5348 | 0.1807 | 3.0512 | 16.8877 |
| 9 | 2.1719 | 0.4604 | 13.0210 | 0.0768 | 5.9952 | 0.1668 | 3.4312 | 20.5711 |
| 10 | 2.3674 | 0.4224 | 15.1929 | 0.0658 | 6.4177 | 0.1558 | 3.7978 | 24.3728 |
| 11 | 2.5804 | 0.3875 | 17.5603 | 0.0569 | 6.8052 | 0.1469 | 4.1510 | 28.2481 |
| 12 | 2.8127 | 0.3555 | 20.1407 | 0.0497 | 7.1607 | 0.1397 | 4.4910 | 32.1590 |
| 13 | 3.0658 | 0.3262 | 22.9534 | 0.0436 | 7.4869 | 0.1336 | 4.8182 | 36.0731 |
| 14 | 3.3417 | 0.2992 | 26.0192 | 0.0384 | 7.7862 | 0.1284 | 5.1326 | 39.9633 |
| 15 | 3.6425 | 0.2745 | 29.3609 | 0.0341 | 8.0607 | 0.1241 | 5.4346 | 43.8069 |
| 16 | 3.9703 | 0.2519 | 33.0034 | 0.0303 | 8.3126 | 0.1203 | 5.7245 | 47.5849 |
| 17 | 4.3276 | 0.2311 | 36.9737 | 0.0270 | 8.5436 | 0.1170 | 6.0024 | 51.2821 |
| 18 | 4.7171 | 0.2120 | 41.3013 | 0.0242 | 8.7556 | 0.1142 | 6.2687 | 54.8860 |
| 19 | 5.1417 | 0.1945 | 46.0185 | 0.0217 | 8.9501 | 0.1117 | 6.5236 | 58.3868 |
| 20 | 5.6044 | 0.1784 | 51.1601 | 0.0195 | 9.1285 | 0.1095 | 6.7674 | 61.7770 |
| 21 | 6.1088 | 0.1637 | 56.7645 | 0.0176 | 9.2922 | 0.1076 | 7.0006 | 65.0509 |
| 22 | 6.6586 | 0.1502 | 62.8733 | 0.0159 | 9.4424 | 0.1059 | 7.2232 | 68.2048 |
| 23 | 7.2579 | 0.1378 | 69.5319 | 0.0144 | 9.5802 | 0.1044 | 7.4357 | 71.2359 |
| 24 | 7.9111 | 0.1264 | 76.7898 | 0.0130 | 9.7066 | 0.1030 | 7.6384 | 74.1433 |
| 25 | 8.6231 | 0.1160 | 84.7009 | 0.0118 | 9.8226 | 0.1018 | 7.8316 | 76.9265 |
| 26 | 9.3992 | 0.1064 | 93.3240 | 0.0107 | 9.9290 | 0.1007 | 8.0156 | 79.5863 |
| 27 | 10.2451 | 0.0976 | 102.7231 | 0.0097 | 10.0266 | 0.0997 | 8.1906 | 82.1241 |
| 28 | 11.1671 | 0.0895 | 112.9682 | 0.0089 | 10.1161 | 0.0989 | 8.3571 | 84.5419 |
| 29 | 12.1722 | 0.0822 | 124.1354 | 0.0081 | 10.1983 | 0.0981 | 8.5154 | 86.8422 |
| 30 | 13.2677 | 0.0754 | 136.3075 | 0.0073 | 10.2737 | 0.0973 | 8.6657 | 89.0280 |
| 31 | 14.4618 | 0.0691 | 149.5752 | 0.0067 | 10.3428 | 0.0967 | 8.8083 | 91.1024 |
| 32 | 15.7633 | 0.0634 | 164.0370 | 0.0061 | 10.4062 | 0.0961 | 8.9436 | 93.0690 |
| 33 | 17.1820 | 0.0582 | 179.8003 | 0.0056 | 10.4644 | 0.0956 | 9.0718 | 94.9314 |
| 34 | 18.7284 | 0.0534 | 196.9823 | 0.0051 | 10.5178 | 0.0951 | 9.1933 | 96.6935 |
| 35 | 20.4140 | 0.0490 | 215.7108 | 0.0046 | 10.5668 | 0.0946 | 9.3083 | 98.3590 |
| 40 | 31.4094 | 0.0318 | 337.8824 | 0.0030 | 10.7574 | 0.0930 | 9.7957 | 105.3762 |
| 45 | 48.3273 | 0.0207 | 525.8587 | 0.0019 | 10.8812 | 0.0919 | 10.1603 | 110.5561 |
| 50 | 74.3575 | 0.0134 | 815.0836 | 0.0012 | 10.9617 | 0.0912 | 10.4295 | 114.3251 |
| 55 | 114.4083 | 0.0087 | 1260.0918 | 0.0008 | 11.0140 | 0.0908 | 10.6261 | 117.0362 |
| 60 | 176.0313 | 0.0057 | 1944.7921 | 0.0005 | 11.0480 | 0.0905 | 10.7683 | 118.9683 |
| 65 | 270.8460 | 0.0037 | 2998.2885 | 0.0003 | 11.0701 | 0.0903 | 10.8702 | 120.3344 |
| 70 | 416.7301 | 0.0024 | 4619.2232 | 0.0002 | 11.0844 | 0.0902 | 10.9427 | 121.2942 |
| 75 | 641.1909 | 0.0016 | 7113.2321 | 0.0001 | 11.0938 | 0.0901 | 10.9940 | 121.9646 |
| 80 | 986.5517 | 0.0010 | 10950.5741 | 0.0001 | 11.0998 | 0.0901 | 11.0299 | 122.4306 |
| 85 | 1517.9320 | 0.0007 | 16854.8003 | 0.0001 | 11.1038 | 0.0901 | 11.0551 | 122.7533 |
| 90 | 2335.5266 | 0.0004 | 25939.1842 | 0.0000 | 11.1064 | 0.0900 | 11.0726 | 122.9758 |
| 95 | 3593.4971 | 0.0003 | 39916.6350 | 0.0000 | 11.1080 | 0.0900 | 11.0847 | 123.1287 |
| 100 | 5529.0408 | 0.0002 | 61422.6755 | 0.0000 | 11.1091 | 0.0900 | 11.0930 | 123.2335 |

| | Single payment | | Uniform series | | | | Gradient series | |
|---|---|---|---|---|---|---|---|---|
| | Compound amount factor | Present worth factor | Compound amount factor | Sinking fund factor | Present worth factor | Capital recovery factor | Uniform series factor | Present worth factor |
| $n$ | To find $F$ given $P$ $F\|P\,i,n$ | To find $P$ given $F$ $P\|F\,i,n$ | To find $F$ given $A$ $F\|A\,i,n$ | To find $A$ given $F$ $A\|F\,i,n$ | To find $P$ given $A$ $P\|A\,i,n$ | To find $A$ given $P$ $A\|P\,i,n$ | To find $A$ given $G$ $A\|G\,i,n$ | To find $P$ given $G$ $P\|G\,i,n$ |
| 1 | 1.1000 | 0.9091 | 1.0000 | 1.0000 | 0.9091 | 1.1000 | 0.0000 | 0.0000 |
| 2 | 1.2100 | 0.8264 | 2.1000 | 0.4762 | 1.7355 | 0.5762 | 0.4762 | 0.8264 |
| 3 | 1.3310 | 0.7513 | 3.3100 | 0.3021 | 2.4869 | 0.4021 | 0.9366 | 2.3291 |
| 4 | 1.4641 | 0.6830 | 4.6410 | 0.2155 | 3.1699 | 0.3155 | 1.3812 | 4.3781 |
| 5 | 1.6105 | 0.6209 | 6.1051 | 0.1638 | 3.7908 | 0.2638 | 1.8101 | 6.8618 |
| 6 | 1.7716 | 0.5645 | 7.7156 | 0.1296 | 4.3553 | 0.2296 | 2.2236 | 9.6842 |
| 7 | 1.9487 | 0.5132 | 9.4872 | 0.1054 | 4.8684 | 0.2054 | 2.6216 | 12.7631 |
| 8 | 2.1436 | 0.4665 | 11.4359 | 0.0874 | 5.3349 | 0.1874 | 3.0045 | 16.0287 |
| 9 | 2.3579 | 0.4241 | 13.5795 | 0.0736 | 5.7590 | 0.1736 | 3.3724 | 19.4215 |
| 10 | 2.5937 | 0.3855 | 15.9347 | 0.0627 | 6.1446 | 0.1627 | 3.7255 | 22.8913 |
| 11 | 2.8531 | 0.3505 | 18.5312 | 0.0540 | 6.4951 | 0.1540 | 4.0641 | 26.3963 |
| 12 | 3.1384 | 0.3186 | 21.3843 | 0.0468 | 6.8137 | 0.1468 | 4.3884 | 29.9012 |
| 13 | 3.4523 | 0.2897 | 24.5227 | 0.0408 | 7.1034 | 0.1408 | 4.6988 | 33.3772 |
| 14 | 3.7975 | 0.2633 | 27.9750 | 0.0357 | 7.3667 | 0.1357 | 4.9955 | 36.8005 |
| 15 | 4.1772 | 0.2394 | 31.7725 | 0.0315 | 7.6061 | 0.1315 | 5.2789 | 40.1520 |
| 16 | 4.5950 | 0.2176 | 35.9497 | 0.0278 | 7.8237 | 0.1278 | 5.5493 | 43.4164 |
| 17 | 5.0545 | 0.1978 | 40.5447 | 0.0247 | 8.0216 | 0.1247 | 5.8071 | 46.5819 |
| 18 | 5.5599 | 0.1799 | 45.5992 | 0.0219 | 8.2014 | 0.1219 | 6.0526 | 49.6395 |
| 19 | 6.1159 | 0.1635 | 51.1591 | 0.0195 | 8.3649 | 0.1195 | 6.2861 | 52.5827 |
| 20 | 6.7275 | 0.1486 | 57.2750 | 0.0175 | 8.5136 | 0.1175 | 6.5081 | 55.4069 |
| 21 | 7.4002 | 0.1351 | 64.0025 | 0.0156 | 8.6487 | 0.1156 | 6.7189 | 58.1095 |
| 22 | 8.1403 | 0.1228 | 71.4027 | 0.0140 | 8.7715 | 0.1140 | 6.9189 | 60.6893 |
| 23 | 8.9543 | 0.1117 | 79.5430 | 0.0126 | 8.8832 | 0.1126 | 7.1085 | 63.1462 |
| 24 | 9.8497 | 0.1015 | 88.4973 | 0.0113 | 8.9847 | 0.1113 | 7.2881 | 65.4813 |
| 25 | 10.8347 | 0.0923 | 98.3471 | 0.0102 | 9.0770 | 0.1102 | 7.4580 | 67.6964 |
| 26 | 11.9182 | 0.0839 | 109.1818 | 0.0092 | 9.1609 | 0.1092 | 7.6186 | 69.7940 |
| 27 | 13.1100 | 0.0763 | 121.0999 | 0.0083 | 9.2372 | 0.1083 | 7.7704 | 71.7773 |
| 28 | 14.4210 | 0.0693 | 134.2099 | 0.0075 | 9.3066 | 0.1075 | 7.9137 | 73.6495 |
| 29 | 14.8631 | 0.0630 | 148.6309 | 0.0067 | 9.3696 | 0.1067 | 8.0489 | 75.4146 |
| 30 | 17.4494 | 0.0573 | 164.4940 | 0.0061 | 9.4269 | 0.1061 | 8.1762 | 77.0766 |
| 31 | 19.1943 | 0.0521 | 181.9434 | 0.0055 | 9.4790 | 0.1055 | 8.2962 | 78.6395 |
| 32 | 21.1138 | 0.0474 | 201.1378 | 0.0050 | 9.5264 | 0.1050 | 8.4091 | 80.1078 |
| 33 | 23.2252 | 0.0431 | 222.2515 | 0.0045 | 9.5694 | 0.1045 | 8.5152 | 81.4856 |
| 34 | 25.5477 | 0.0391 | 245.4767 | 0.0041 | 9.6086 | 0.1041 | 8.6149 | 82.7773 |
| 35 | 28.1024 | 0.0356 | 271.0244 | 0.0037 | 9.6442 | 0.1037 | 8.7086 | 83.9872 |
| 40 | 45.2593 | 0.0221 | 442.5926 | 0.0023 | 9.7791 | 0.1023 | 9.0962 | 88.9525 |
| 45 | 72.8905 | 0.0137 | 718.9048 | 0.0014 | 9.8628 | 0.1014 | 9.3740 | 92.4544 |
| 50 | 117.3909 | 0.0085 | 1163.9085 | 0.0009 | 9.9148 | 0.1009 | 9.5704 | 94.8889 |
| 55 | 189.0591 | 0.0053 | 1880.5914 | 0.0005 | 9.9471 | 0.1005 | 9.7075 | 96.5619 |
| 60 | 304.4816 | 0.0033 | 3034.8164 | 0.0003 | 9.9672 | 0.1003 | 9.8023 | 97.7010 |
| 65 | 490.3707 | 0.0020 | 4893.7073 | 0.0002 | 9.9796 | 0.1002 | 9.8672 | 98.4705 |
| 70 | 789.7470 | 0.0013 | 7887.4696 | 0.0001 | 9.9873 | 0.1001 | 9.9113 | 98.9870 |
| 75 | 1271.8954 | 0.0008 | 12708.9537 | 0.0001 | 9.9921 | 0.1001 | 9.9410 | 99.3317 |
| 80 | 2048.4002 | 0.0005 | 20474.0021 | 0.0000 | 9.9951 | 0.1000 | 9.9609 | 99.5606 |
| 85 | 3298.9690 | 0.0003 | 32979.6903 | 0.0000 | 9.9970 | 0.1000 | 9.9742 | 99.7120 |
| 90 | 5313.0226 | 0.0002 | 53120.2261 | 0.0000 | 9.9981 | 0.1000 | 9.9831 | 99.8118 |
| 95 | 8556.6760 | 0.0001 | 85556.7605 | 0.0000 | 9.9988 | 0.1000 | 9.9889 | 99.8773 |
| 100 | 13780.6123 | 0.0001 | 137796.1234 | 0.0000 | 9.9993 | 0.1000 | 9.9927 | 99.9202 |

| | Single payment | | Uniform series | | | | Gradient series | |
|---|---|---|---|---|---|---|---|---|
| | Compound amount factor | Present worth factor | Compound amount factor | Sinking fund factor | Present worth factor | Capital recovery factor | Uniform series factor | Present worth factor |
| $n$ | To find $F$ given $P$ $F\|P\,i,n$ | To find $P$ given $F$ $P\|F\,i,n$ | To find $F$ given $A$ $F\|A\,i,n$ | To find $A$ given $F$ $A\|F\,i,n$ | To find $P$ given $A$ $P\|A\,i,n$ | To find $A$ given $P$ $A\|P\,i,n$ | To find $A$ given $G$ $A\|G\,i,n$ | To find $P$ given $G$ $P\|G\,i,n$ |
| 1 | 1.1200 | 0.8929 | 1.0000 | 1.0000 | 0.8929 | 1.1200 | 0.0000 | 0.0000 |
| 2 | 1.2544 | 0.7972 | 2.1200 | 0.4717 | 1.6901 | 0.5917 | 0.4717 | 0.7972 |
| 3 | 1.4049 | 0.7118 | 3.3744 | 0.2963 | 2.4018 | 0.4163 | 0.9246 | 2.2208 |
| 4 | 1.5735 | 0.6355 | 4.7793 | 0.2092 | 3.0373 | 0.3292 | 1.3589 | 4.1273 |
| 5 | 1.7623 | 0.5674 | 6.3528 | 0.1574 | 3.6048 | 0.2774 | 1.7746 | 6.3970 |
| 6 | 1.9738 | 0.5066 | 8.1152 | 0.1232 | 4.1114 | 0.2432 | 2.1720 | 8.9302 |
| 7 | 2.2107 | 0.4523 | 10.0890 | 0.0991 | 4.5638 | 0.2191 | 2.5515 | 11.6443 |
| 8 | 2.4760 | 0.4039 | 12.2997 | 0.0813 | 4.9676 | 0.2013 | 2.9131 | 14.4714 |
| 9 | 2.7731 | 0.3606 | 14.7757 | 0.0677 | 5.3282 | 0.1877 | 3.2574 | 17.3563 |
| 10 | 3.1058 | 0.3220 | 17.5487 | 0.0570 | 5.6502 | 0.1770 | 3.5847 | 20.2541 |
| 11 | 3.4785 | 0.2875 | 20.6546 | 0.0484 | 5.9377 | 0.1684 | 3.8953 | 23.1288 |
| 12 | 3.8960 | 0.2567 | 24.1331 | 0.0414 | 6.1944 | 0.1614 | 4.1897 | 25.9523 |
| 13 | 4.3635 | 0.2292 | 28.0291 | 0.0357 | 6.4235 | 0.1557 | 4.4683 | 28.7024 |
| 14 | 4.8871 | 0.2046 | 32.3926 | 0.0309 | 6.6282 | 0.1509 | 4.7317 | 31.3624 |
| 15 | 5.4736 | 0.1827 | 37.2797 | 0.0268 | 6.8109 | 0.1468 | 4.9803 | 33.9202 |
| 16 | 6.1304 | 0.1631 | 42.7533 | 0.0234 | 6.9740 | 0.1434 | 5.2147 | 36.3670 |
| 17 | 6.8660 | 0.1456 | 48.8837 | 0.0205 | 7.1196 | 0.1405 | 5.4353 | 38.6973 |
| 18 | 7.6900 | 0.1300 | 55.7497 | 0.0179 | 7.2497 | 0.1379 | 5.6247 | 40.9080 |
| 19 | 8.6128 | 0.1161 | 63.4397 | 0.0158 | 7.3658 | 0.1358 | 5.8375 | 42.9979 |
| 20 | 9.6463 | 0.1037 | 72.0524 | 0.0139 | 7.4694 | 0.1339 | 6.0202 | 44.9676 |
| 21 | 10.8038 | 0.0926 | 81.6987 | 0.0122 | 7.5620 | 0.1322 | 6.1913 | 46.8188 |
| 22 | 12.1003 | 0.0826 | 92.5026 | 0.0108 | 7.6446 | 0.1308 | 6.3514 | 48.5543 |
| 23 | 13.5523 | 0.0738 | 104.6029 | 0.0096 | 7.7184 | 0.1296 | 6.5010 | 50.1776 |
| 24 | 15.1786 | 0.0659 | 118.1552 | 0.0085 | 7.7843 | 0.1285 | 6.6406 | 51.6929 |
| 25 | 17.0001 | 0.0588 | 133.3339 | 0.0075 | 7.8431 | 0.1275 | 6.7708 | 53.1046 |
| 26 | 19.0401 | 0.0525 | 150.3339 | 0.0067 | 7.8957 | 0.1267 | 6.8921 | 54.4177 |
| 27 | 21.3249 | 0.0469 | 169.3740 | 0.0059 | 7.9426 | 0.1259 | 7.0049 | 55.6369 |
| 28 | 23.8839 | 0.0419 | 190.6989 | 0.0052 | 7.9844 | 0.1252 | 7.1098 | 56.7674 |
| 29 | 26.7499 | 0.0374 | 214.5828 | 0.0047 | 8.0218 | 0.1247 | 7.2071 | 57.8141 |
| 30 | 29.9599 | 0.0334 | 241.3327 | 0.0041 | 8.0552 | 0.1241 | 7.2974 | 58.7821 |
| 31 | 33.5551 | 0.0298 | 271.2926 | 0.0037 | 8.0850 | 0.1237 | 7.3811 | 59.6761 |
| 32 | 37.5817 | 0.0266 | 304.8477 | 0.0033 | 8.1116 | 0.1233 | 7.4586 | 60.5010 |
| 33 | 42.0915 | 0.0238 | 342.4294 | 0.0029 | 8.1354 | 0.1229 | 7.5302 | 61.2612 |
| 34 | 47.1425 | 0.0212 | 384.5210 | 0.0026 | 8.1566 | 0.1226 | 7.5965 | 61.9612 |
| 35 | 52.7996 | 0.0189 | 431.6635 | 0.0023 | 8.1755 | 0.1223 | 7.6577 | 62.6052 |
| 40 | 93.0510 | 0.0107 | 767.0914 | 0.0013 | 8.2438 | 0.1213 | 7.8988 | 65.1159 |
| 45 | 163.9876 | 0.0061 | 1358.2300 | 0.0007 | 8.2825 | 0.1207 | 8.0572 | 66.7342 |
| 50 | 289.0022 | 0.0035 | 2400.0182 | 0.0004 | 8.3045 | 0.1204 | 8.1597 | 67.7624 |

## TABLE A.18  Discrete Compounding: $i = 15\%$

| | Single payment | | Uniform series | | | | Gradient series | |
|---|---|---|---|---|---|---|---|---|
| | Compound amount factor | Present worth factor | Compound amount factor | Sinking fund factor | Present worth factor | Capital recovery factor | Uniform series factor | Present worth factor |
| $n$ | To find $F$ given $P$ $F\|P\,i,n$ | To find $P$ given $F$ $P\|F\,i,n$ | To find $F$ given $A$ $F\|A\,i,n$ | To find $A$ given $F$ $A\|F\,i,n$ | To find $P$ given $A$ $P\|A\,i,n$ | To find $A$ given $P$ $A\|P\,i,n$ | To find $A$ given $G$ $A\|G\,i,n$ | To find $P$ given $G$ $P\|G\,i,n$ |
| 1 | 1.1500 | 0.8696 | 1.0000 | 1.0000 | 0.8696 | 1.1500 | 0.0000 | 0.0000 |
| 2 | 1.3225 | 0.7561 | 2.1500 | 0.4651 | 1.6257 | 0.6151 | 0.4651 | 0.7561 |
| 3 | 1.5209 | 0.6575 | 3.4725 | 0.2880 | 2.2832 | 0.4380 | 0.9071 | 2.0712 |
| 4 | 1.7490 | 0.5718 | 4.9934 | 0.2003 | 2.8550 | 0.3503 | 1.3263 | 3.7864 |
| 5 | 2.0114 | 0.4972 | 6.7242 | 0.1483 | 3.3522 | 0.2983 | 1.7228 | 5.7751 |
| 6 | 2.3131 | 0.4323 | 8.7537 | 0.1142 | 3.7845 | 0.2642 | 2.0972 | 7.9368 |
| 7 | 2.6600 | 0.3759 | 11.0668 | 0.0904 | 4.1604 | 0.2404 | 2.4498 | 10.1924 |
| 8 | 3.0590 | 0.3269 | 13.7268 | 0.0729 | 4.4873 | 0.2229 | 2.7813 | 12.4807 |
| 9 | 3.5179 | 0.2843 | 16.7858 | 0.0596 | 4.7716 | 0.2096 | 3.0922 | 14.7548 |
| 10 | 4.0456 | 0.2472 | 20.3037 | 0.0493 | 5.0188 | 0.1993 | 3.3832 | 16.9795 |
| 11 | 4.6524 | 0.2149 | 24.3493 | 0.0411 | 5.2337 | 0.1911 | 3.6549 | 19.1289 |
| 12 | 5.3503 | 0.1869 | 29.0017 | 0.0345 | 5.4206 | 0.1845 | 3.9082 | 21.1849 |
| 13 | 6.1528 | 0.1625 | 34.3519 | 0.0291 | 5.5831 | 0.1791 | 4.1438 | 23.1352 |
| 14 | 7.0757 | 0.1413 | 40.5047 | 0.0247 | 5.7245 | 0.1747 | 4.3624 | 24.9725 |
| 15 | 8.1371 | 0.1229 | 47.5804 | 0.0210 | 5.8474 | 0.1710 | 4.5650 | 26.6930 |
| 16 | 9.3576 | 0.1069 | 55.7175 | 0.0179 | 5.9542 | 0.1679 | 4.7522 | 28.2960 |
| 17 | 10.7613 | 0.0929 | 65.0751 | 0.0154 | 6.0472 | 0.1654 | 4.9251 | 29.7828 |
| 18 | 12.3755 | 0.0808 | 75.8364 | 0.0132 | 6.1280 | 0.1632 | 5.0843 | 31.1565 |
| 19 | 14.2318 | 0.0703 | 88.2118 | 0.0113 | 6.1982 | 0.1613 | 5.2307 | 32.4213 |
| 20 | 16.3665 | 0.0611 | 102.4436 | 0.0098 | 6.2593 | 0.1598 | 5.3651 | 33.5822 |
| 21 | 18.8215 | 0.0531 | 118.8101 | 0.0084 | 6.3125 | 0.1584 | 5.4883 | 34.6448 |
| 22 | 21.6447 | 0.0462 | 137.6316 | 0.0073 | 6.3587 | 0.1573 | 5.6010 | 35.6150 |
| 23 | 24.8915 | 0.0402 | 159.2764 | 0.0063 | 6.3988 | 0.1563 | 5.7040 | 36.4988 |
| 24 | 28.6252 | 0.0349 | 184.1678 | 0.0054 | 6.4338 | 0.1554 | 5.7979 | 37.3023 |
| 25 | 32.9190 | 0.0304 | 212.7930 | 0.0047 | 6.4641 | 0.1547 | 5.8834 | 38.0314 |
| 26 | 37.8568 | 0.0264 | 245.7120 | 0.0041 | 6.4906 | 0.1541 | 5.9612 | 38.6918 |
| 27 | 43.5353 | 0.0230 | 283.5688 | 0.0035 | 6.5135 | 0.1535 | 6.0319 | 39.2890 |
| 28 | 50.0656 | 0.0200 | 327.1041 | 0.0031 | 6.5335 | 0.1531 | 6.0960 | 39.8283 |
| 29 | 57.5755 | 0.0174 | 377.1697 | 0.0027 | 6.5509 | 0.1527 | 6.1541 | 40.3146 |
| 30 | 66.2118 | 0.0151 | 434.7451 | 0.0023 | 6.5660 | 0.1523 | 6.2066 | 40.7526 |
| 31 | 76.1435 | 0.0131 | 500.9569 | 0.0020 | 6.5791 | 0.1520 | 6.2541 | 41.1466 |
| 32 | 87.5651 | 0.0114 | 577.1005 | 0.0017 | 6.5905 | 0.1517 | 6.2970 | 41.5006 |
| 33 | 100.6998 | 0.0099 | 664.6655 | 0.0015 | 6.6005 | 0.1515 | 6.3357 | 41.8184 |
| 34 | 115.8048 | 0.0086 | 765.3654 | 0.0013 | 6.6091 | 0.1513 | 6.3705 | 42.1033 |
| 35 | 133.1755 | 0.0075 | 881.1702 | 0.0011 | 6.6166 | 0.1511 | 6.4019 | 42.3586 |
| 40 | 267.8635 | 0.0037 | 1779.0903 | 0.0006 | 6.6418 | 0.1506 | 6.5168 | 43.2830 |
| 45 | 538.7693 | 0.0019 | 3585.1285 | 0.0003 | 6.6543 | 0.1503 | 6.5830 | 43.8051 |
| 50 | 1083.6574 | 0.0009 | 7217.7163 | 0.0001 | 6.6605 | 0.1501 | 6.6205 | 44.0958 |

## TABLE A.19 Discrete Compounding: $i = 18\%$

| | Single payment | | Uniform series | | | | Gradient series | |
|---|---|---|---|---|---|---|---|---|
| | Compound amount factor | Present worth factor | Compound amount factor | Sinking fund factor | Present worth factor | Capital recovery factor | Uniform series factor | Present worth factor |
| $n$ | To find $F$ given $P$ $F\|P\,i,n$ | To find $P$ given $F$ $P\|F\,i,n$ | To find $F$ given $A$ $F\|A\,i,n$ | To find $A$ given $F$ $A\|F\,i,n$ | To find $P$ given $A$ $P\|A\,i,n$ | To find $A$ given $P$ $A\|P\,i,n$ | To find $A$ given $G$ $A\|G\,i,n$ | To find $P$ given $G$ $P\|G\,i,n$ |
| 1 | 1.1800 | 0.8475 | 1.0000 | 1.0000 | 0.8475 | 1.1800 | 0.0000 | 0.0000 |
| 2 | 1.3924 | 0.7182 | 2.1800 | 0.4587 | 1.5656 | 0.6387 | 0.4587 | 0.7182 |
| 3 | 1.6430 | 0.6086 | 3.5724 | 0.2799 | 2.1743 | 0.4599 | 0.8902 | 1.9355 |
| 4 | 1.9388 | 0.5158 | 5.2154 | 0.1917 | 2.6901 | 0.3717 | 1.2947 | 3.4828 |
| 5 | 2.2878 | 0.4371 | 7.1542 | 0.1398 | 3.1272 | 0.3198 | 1.6728 | 5.2313 |
| 6 | 2.6996 | 0.3704 | 9.4420 | 0.1059 | 3.4976 | 0.2859 | 2.0252 | 7.0834 |
| 7 | 3.1855 | 0.3139 | 12.1415 | 0.0824 | 3.8115 | 0.2624 | 2.3526 | 8.9670 |
| 8 | 3.7589 | 0.2660 | 15.3270 | 0.0652 | 4.0776 | 0.2452 | 2.6558 | 10.8292 |
| 9 | 4.4355 | 0.2255 | 19.0859 | 0.0524 | 4.3030 | 0.2324 | 2.9358 | 12.6329 |
| 10 | 5.2338 | 0.1911 | 23.5213 | 0.0425 | 4.4941 | 0.2225 | 3.1936 | 14.3525 |
| 11 | 6.1759 | 0.1619 | 28.7552 | 0.0348 | 4.6560 | 0.2148 | 3.4303 | 15.9716 |
| 12 | 7.2876 | 0.1372 | 34.9311 | 0.0286 | 4.7932 | 0.2086 | 3.6470 | 17.4811 |
| 13 | 8.5994 | 0.1163 | 42.2187 | 0.0237 | 4.9095 | 0.2037 | 3.8449 | 18.8765 |
| 14 | 10.1472 | 0.0985 | 50.8180 | 0.0197 | 5.0081 | 0.1997 | 4.0250 | 20.1576 |
| 15 | 11.9738 | 0.0835 | 60.9653 | 0.0164 | 5.0916 | 0.1964 | 4.1887 | 21.3269 |
| 16 | 14.1290 | 0.0708 | 72.9390 | 0.0137 | 5.1624 | 0.1937 | 4.3369 | 22.3885 |
| 17 | 16.6723 | 0.0600 | 87.0680 | 0.0115 | 5.2223 | 0.1915 | 4.4708 | 23.3482 |
| 18 | 19.6733 | 0.0508 | 103.7403 | 0.0096 | 5.2732 | 0.1896 | 4.5916 | 24.2123 |
| 19 | 23.2144 | 0.0431 | 123.4135 | 0.0081 | 5.3162 | 0.1881 | 4.7003 | 24.9877 |
| 20 | 27.3930 | 0.0365 | 146.6280 | 0.0068 | 5.3528 | 0.1868 | 4.7978 | 25.6813 |
| 21 | 31.3238 | 0.0309 | 174.0210 | 0.0057 | 5.3837 | 0.1857 | 4.8851 | 26.3000 |
| 22 | 38.1421 | 0.0262 | 206.3448 | 0.0048 | 5.4099 | 0.1848 | 4.9632 | 26.8506 |
| 23 | 45.0076 | 0.0222 | 244.4869 | 0.0041 | 5.4321 | 0.1841 | 5.0329 | 27.3394 |
| 24 | 53.1090 | 0.0188 | 289.4946 | 0.0035 | 5.4510 | 0.1835 | 5.0950 | 27.7725 |
| 25 | 62.6686 | 0.0160 | 342.6035 | 0.0029 | 5.4669 | 0.1829 | 5.1502 | 28.1555 |
| 26 | 73.9490 | 0.0135 | 405.2721 | 0.0025 | 5.4804 | 0.1825 | 5.1991 | 28.4935 |
| 27 | 87.2598 | 0.0115 | 479.2211 | 0.0021 | 5.4919 | 0.1821 | 5.2426 | 28.7915 |
| 28 | 102.9666 | 0.0097 | 566.4809 | 0.0018 | 5.5016 | 0.1818 | 5.2810 | 29.0537 |
| 29 | 121.5005 | 0.0083 | 669.4475 | 0.0015 | 5.5098 | 0.1815 | 5.3149 | 29.2842 |
| 30 | 143.3706 | 0.0070 | 790.9480 | 0.0013 | 5.5168 | 0.1813 | 5.3448 | 29.4864 |
| 31 | 169.1774 | 0.0059 | 934.3186 | 0.0011 | 5.5227 | 0.1811 | 5.3712 | 29.6638 |
| 32 | 199.6293 | 0.0050 | 1103.4960 | 0.0009 | 5.5277 | 0.1809 | 5.3945 | 29.8190 |
| 33 | 235.5626 | 0.0042 | 1303.1253 | 0.0008 | 5.5320 | 0.1808 | 5.4149 | 29.9549 |
| 34 | 277.9638 | 0.0036 | 1538.6878 | 0.0006 | 5.5356 | 0.1807 | 5.4328 | 30.0736 |
| 35 | 327.9973 | 0.0030 | 1816.6516 | 0.0006 | 5.5386 | 0.1806 | 5.4485 | 30.1773 |
| 40 | 750.3783 | 0.0013 | 4163.2130 | 0.0002 | 5.5482 | 0.1802 | 5.5022 | 30.5269 |
| 45 | 1716.6839 | 0.0006 | 9531.5771 | 0.0001 | 5.5523 | 0.1801 | 5.5293 | 30.7006 |
| 50 | 3927.3569 | 0.0003 | 21813.0940 | 0.0000 | 5.5541 | 0.1800 | 5.5428 | 30.7856 |

| | Single payment | | Uniform series | | | | Gradient series | |
|---|---|---|---|---|---|---|---|---|
| | Compound amount factor | Present worth factor | Compound amount factor | Sinking fund factor | Present worth factor | Capital recovery factor | Uniform series factor | Present worth factor |
| $n$ | To find $F$ given $P$ $F\|P\,i,n$ | To find $P$ given $F$ $P\|F\,i,n$ | To find $F$ given $A$ $F\|A\,i,n$ | To find $A$ given $F$ $A\|F\,i,n$ | To find $P$ given $A$ $P\|A\,i,n$ | To find $A$ given $P$ $A\|P\,i,n$ | To find $A$ given $G$ $A\|G\,i,n$ | To find $P$ given $G$ $P\|G\,i,n$ |
| 1 | 1.2000 | 0.8333 | 1.0000 | 1.0000 | 0.8333 | 1.2000 | 0.0000 | 0.0000 |
| 2 | 1.4400 | 0.6944 | 2.2000 | 0.4545 | 1.5278 | 0.6545 | 0.4545 | 0.6944 |
| 3 | 1.7280 | 0.5787 | 3.6400 | 0.2747 | 2.1065 | 0.4747 | 0.8791 | 1.8519 |
| 4 | 2.0736 | 0.4823 | 5.3680 | 0.1863 | 2.5887 | 0.3863 | 1.2742 | 3.2986 |
| 5 | 2.4883 | 0.4019 | 7.4416 | 0.1344 | 2.9906 | 0.3344 | 1.6405 | 4.9061 |
| 6 | 2.9860 | 0.3349 | 9.9299 | 0.1007 | 3.3255 | 0.3007 | 1.9788 | 6.5806 |
| 7 | 3.5832 | 0.2791 | 12.9159 | 0.0774 | 3.6046 | 0.2774 | 2.2902 | 8.2551 |
| 8 | 4.2998 | 0.2326 | 16.4991 | 0.0606 | 3.8372 | 0.2606 | 2.5756 | 9.8831 |
| 9 | 5.1598 | 0.1938 | 20.7989 | 0.0481 | 4.0310 | 0.2481 | 2.8364 | 11.4335 |
| 10 | 6.1917 | 0.1615 | 25.9587 | 0.0385 | 4.1925 | 0.2385 | 3.0739 | 12.8871 |
| 11 | 7.4301 | 0.1346 | 32.1504 | 0.0311 | 4.3271 | 0.2311 | 3.2893 | 14.2330 |
| 12 | 8.9161 | 0.1122 | 39.5805 | 0.0253 | 4.4392 | 0.2253 | 3.4841 | 15.4667 |
| 13 | 10.6993 | 0.0935 | 48.4966 | 0.0206 | 4.5327 | 0.2206 | 3.6597 | 16.5883 |
| 14 | 12.8392 | 0.0779 | 59.1959 | 0.0169 | 4.6106 | 0.2169 | 3.8175 | 17.6008 |
| 15 | 15.4070 | 0.0649 | 72.0351 | 0.0139 | 4.6755 | 0.2139 | 3.9588 | 18.5095 |
| 16 | 18.4884 | 0.0541 | 87.4421 | 0.0114 | 4.7296 | 0.2114 | 4.0851 | 19.3208 |
| 17 | 22.1861 | 0.0451 | 105.9306 | 0.0094 | 4.7746 | 0.2094 | 4.1976 | 20.0419 |
| 18 | 26.6233 | 0.0376 | 128.1167 | 0.0078 | 4.8122 | 0.2078 | 4.2975 | 20.6805 |
| 19 | 31.9480 | 0.0313 | 154.7400 | 0.0065 | 4.8435 | 0.2065 | 4.3861 | 21.2439 |
| 20 | 38.3376 | 0.0261 | 186.6880 | 0.0054 | 4.8696 | 0.2054 | 4.4643 | 21.7395 |
| 21 | 46.0051 | 0.0217 | 225.0256 | 0.0044 | 4.8913 | 0.2044 | 4.5334 | 22.1742 |
| 22 | 55.2061 | 0.0181 | 271.0307 | 0.0037 | 4.9094 | 0.2037 | 4.5941 | 22.5546 |
| 23 | 66.2474 | 0.0151 | 326.2369 | 0.0031 | 4.9245 | 0.2031 | 4.6475 | 22.8867 |
| 24 | 79.4968 | 0.0126 | 392.4842 | 0.0025 | 4.9371 | 0.2025 | 4.6943 | 23.1760 |
| 25 | 95.3962 | 0.0105 | 471.9811 | 0.0021 | 4.9476 | 0.2021 | 4.7352 | 23.4276 |
| 26 | 114.4755 | 0.0087 | 567.3773 | 0.0018 | 4.9563 | 0.2018 | 4.7709 | 23.6460 |
| 27 | 137.3706 | 0.0073 | 681.8528 | 0.0015 | 4.9636 | 0.2015 | 4.8020 | 23.8353 |
| 28 | 164.8447 | 0.0061 | 819.2233 | 0.0012 | 4.9697 | 0.2012 | 4.8291 | 23.9991 |
| 29 | 197.8136 | 0.0051 | 984.0680 | 0.0010 | 4.9747 | 0.2010 | 4.8527 | 24.1406 |
| 30 | 237.3763 | 0.0042 | 1181.8816 | 0.0008 | 4.9789 | 0.2008 | 4.8731 | 24.2628 |
| 31 | 284.8516 | 0.0035 | 1419.2579 | 0.0007 | 4.9824 | 0.2007 | 4.8908 | 24.3681 |
| 32 | 341.8219 | 0.0029 | 1704.1095 | 0.0006 | 4.9854 | 0.2006 | 2.0961 | 24.4588 |
| 33 | 410.1863 | 0.0024 | 2045.9314 | 0.0005 | 4.9878 | 0.2005 | 4.9194 | 24.5368 |
| 34 | 492.2235 | 0.0020 | 2456.1176 | 0.0004 | 4.9898 | 0.2004 | 4.9308 | 24.6038 |
| 35 | 590.6682 | 0.0017 | 2948.3411 | 0.0003 | 4.9915 | 0.2003 | 4.9406 | 24.6614 |
| 40 | 1469.7716 | 0.0007 | 7343.8578 | 0.0001 | 4.9966 | 0.2001 | 4.9728 | 24.8469 |
| 45 | 3657.2620 | 0.0003 | 18281.3099 | 0.0001 | 4.9986 | 0.2001 | 4.9877 | 24.9316 |
| 50 | 9100.4382 | 0.0001 | 45497.1908 | 0.0000 | 4.9995 | 0.2000 | 4.9945 | 24.9698 |

## TABLE A.21  Discrete Compounding: $i = 25\%$

| | Single payment | | Uniform series | | | | Gradient series | |
|---|---|---|---|---|---|---|---|---|
| | Compound amount factor | Present worth factor | Compound amount factor | Sinking fund factor | Present worth factor | Capital recovery factor | Uniform series factor | Present worth factor |
| $n$ | To find $F$ given $P$ $F\|P\ i,n$ | To find $P$ given $F$ $P\|F\ i,n$ | To find $F$ given $A$ $F\|A\ i,n$ | To find $A$ given $F$ $A\|F\ i,n$ | To find $P$ given $A$ $P\|A\ i,n$ | To find $A$ given $P$ $A\|P\ i,n$ | To find $A$ given $G$ $A\|G\ i,n$ | To find $P$ given $G$ $P\|G\ i,n$ |
| 1 | 1.2500 | 0.8000 | 1.0000 | 1.0000 | 0.8000 | 1.2500 | 0.0000 | 0.0000 |
| 2 | 1.5625 | 0.6400 | 2.2500 | 0.4444 | 1.4400 | 0.6944 | 0.4444 | 0.6400 |
| 3 | 1.9531 | 0.5120 | 3.8125 | 0.2623 | 1.9520 | 0.5123 | 0.8525 | 1.6640 |
| 4 | 2.4414 | 0.4096 | 5.7656 | 0.1734 | 2.3616 | 0.4234 | 1.2249 | 2.8928 |
| 5 | 3.0518 | 0.3277 | 8.2070 | 0.1218 | 2.6893 | 0.3718 | 1.5631 | 4.2035 |
| 6 | 3.8147 | 0.2621 | 11.2588 | 0.0888 | 2.9514 | 0.3388 | 1.8683 | 5.5142 |
| 7 | 4.7684 | 0.2097 | 15.0735 | 0.0663 | 3.1611 | 0.3163 | 2.1424 | 6.7725 |
| 8 | 5.9605 | 0.1678 | 19.8419 | 0.0504 | 3.3289 | 0.3004 | 2.3872 | 7.9469 |
| 9 | 7.4506 | 0.1342 | 25.8023 | 0.0388 | 3.4631 | 0.2888 | 2.6048 | 9.0207 |
| 10 | 9.3132 | 0.1074 | 33.2529 | 0.0301 | 3.5705 | 0.2801 | 2.7971 | 9.9870 |
| 11 | 11.6415 | 0.0859 | 42.5661 | 0.0235 | 3.6564 | 0.2735 | 2.9663 | 10.8460 |
| 12 | 14.5519 | 0.0687 | 54.2077 | 0.0184 | 3.7251 | 0.2684 | 3.1145 | 11.6020 |
| 13 | 18.1899 | 0.0550 | 68.7596 | 0.0145 | 3.7801 | 0.2645 | 3.2437 | 12.2617 |
| 14 | 22.7374 | 0.0440 | 86.9495 | 0.0115 | 3.8241 | 0.2615 | 3.3559 | 12.8334 |
| 15 | 28.4217 | 0.0352 | 109.6868 | 0.0091 | 3.8593 | 0.2591 | 3.4530 | 13.3260 |
| 16 | 35.5271 | 0.0281 | 138.1085 | 0.0072 | 3.8874 | 0.2572 | 3.5366 | 13.7482 |
| 17 | 44.4089 | 0.0225 | 173.6357 | 0.0058 | 3.9099 | 0.2558 | 3.6084 | 14.1085 |
| 18 | 55.5112 | 0.0180 | 218.0446 | 0.0046 | 3.9279 | 0.2546 | 3.6698 | 14.4147 |
| 19 | 69.3889 | 0.0144 | 273.5558 | 0.0037 | 3.9424 | 0.2537 | 3.7222 | 14.6741 |
| 20 | 86.7362 | 0.0115 | 342.9447 | 0.0029 | 3.9539 | 0.2529 | 3.7667 | 14.8932 |
| 21 | 108.4202 | 0.0092 | 429.6809 | 0.0023 | 3.9631 | 0.2523 | 3.8045 | 15.0777 |
| 22 | 135.5253 | 0.0074 | 538.1011 | 0.0019 | 3.9705 | 0.2519 | 3.8365 | 15.2326 |
| 23 | 169.4066 | 0.0059 | 673.6264 | 0.0015 | 3.9764 | 0.2515 | 3.8634 | 15.3625 |
| 24 | 211.7582 | 0.0047 | 843.0329 | 0.0012 | 3.9811 | 0.2512 | 3.8861 | 15.4711 |
| 25 | 264.6978 | 0.0038 | 1054.7912 | 0.0009 | 3.9849 | 0.2509 | 3.9052 | 15.5618 |
| 26 | 330.8722 | 0.0030 | 1319.4890 | 0.0008 | 3.9879 | 0.2508 | 3.9212 | 15.6373 |
| 27 | 413.5903 | 0.0024 | 1650.3612 | 0.0006 | 3.9903 | 0.2506 | 3.9346 | 15.7002 |
| 28 | 516.9879 | 0.0019 | 2063.9515 | 0.0005 | 3.9923 | 0.2505 | 3.9457 | 15.7524 |
| 29 | 646.2349 | 0.0015 | 2580.9394 | 0.0004 | 3.9938 | 0.2504 | 3.9551 | 15.7957 |
| 30 | 807.7936 | 0.0012 | 3227.1743 | 0.0003 | 3.9950 | 0.2503 | 3.9628 | 15.8316 |
| 31 | 1009.7420 | 0.0010 | 4034.9678 | 0.0002 | 3.9960 | 0.2502 | 3.9693 | 15.8614 |
| 32 | 1262.1774 | 0.0008 | 5044.7098 | 0.0002 | 3.9968 | 0.2502 | 3.9746 | 15.8859 |
| 33 | 1577.7218 | 0.0006 | 6306.8872 | 0.0002 | 3.9975 | 0.2502 | 3.9791 | 15.9062 |
| 34 | 1972.1523 | 0.0005 | 7884.6091 | 0.0001 | 3.9980 | 0.2501 | 3.9828 | 15.9229 |
| 35 | 2465.1903 | 0.0004 | 9856.7613 | 0.0001 | 3.9984 | 0.2501 | 3.9858 | 15.9367 |

| | Single payment | | Uniform series | | | | Gradient series | |
|---|---|---|---|---|---|---|---|---|
| | Compound amount factor | Present worth factor | Compound amount factor | Sinking fund factor | Present worth factor | Capital recovery factor | Uniform series factor | Present worth factor |
| $n$ | To find $F$ given $P$ $F\|P\,i,n$ | To find $P$ given $F$ $P\|F\,i,n$ | To find $F$ given $A$ $F\|A\,i,n$ | To find $A$ given $F$ $A\|F\,i,n$ | To find $P$ given $A$ $P\|A\,i,n$ | To find $A$ given $P$ $A\|P\,i,n$ | To find $A$ given $G$ $A\|G\,i,n$ | To find $P$ given $G$ $P\|G\,i,n$ |
| 1 | 1.3000 | 0.7692 | 1.0000 | 1.0000 | 0.7692 | 1.3000 | 0.0000 | 0.0000 |
| 2 | 1.6900 | 0.5917 | 2.3000 | 0.4348 | 1.3609 | 0.7348 | 0.4348 | 0.5917 |
| 3 | 2.1970 | 0.4552 | 3.9900 | 0.2506 | 1.8161 | 0.5506 | 0.8271 | 1.5020 |
| 4 | 2.8561 | 0.3501 | 6.1870 | 0.1616 | 2.1662 | 0.4616 | 1.1783 | 2.5524 |
| 5 | 3.7129 | 0.2693 | 9.0431 | 0.1106 | 2.4356 | 0.4106 | 1.4903 | 3.6297 |
| 6 | 4.8268 | 0.2072 | 12.7560 | 0.0784 | 2.6427 | 0.3784 | 1.7654 | 4.6656 |
| 7 | 6.2749 | 0.1594 | 17.5828 | 0.0569 | 2.8021 | 0.3569 | 2.0063 | 5.6218 |
| 8 | 8.1573 | 0.1226 | 23.8577 | 0.0419 | 2.9247 | 0.3419 | 2.2156 | 6.4800 |
| 9 | 10.6045 | 0.0943 | 32.0150 | 0.0312 | 3.0190 | 0.3312 | 2.3963 | 7.2344 |
| 10 | 13.7858 | 0.0725 | 42.6195 | 0.0235 | 3.0915 | 0.3235 | 2.5512 | 7.8872 |
| 11 | 17.9216 | 0.0558 | 56.4053 | 0.0177 | 3.1473 | 0.3177 | 2.6833 | 8.4452 |
| 12 | 23.2981 | 0.0429 | 74.3269 | 0.0135 | 3.1903 | 0.3135 | 2.7952 | 8.9173 |
| 13 | 30.2875 | 0.0330 | 97.6250 | 0.0102 | 3.2233 | 0.3102 | 2.8895 | 9.3135 |
| 14 | 39.3737 | 0.0254 | 127.9124 | 0.0078 | 3.2487 | 0.3078 | 2.9685 | 9.6437 |
| 15 | 51.1859 | 0.0195 | 167.2862 | 0.0060 | 3.2682 | 0.3060 | 3.0344 | 9.9172 |
| 16 | 66.5416 | 0.0150 | 218.4721 | 0.0046 | 3.2832 | 0.3046 | 3.0892 | 10.1426 |
| 17 | 86.5041 | 0.0116 | 285.0137 | 0.0035 | 3.2948 | 0.3035 | 3.1345 | 10.3276 |
| 18 | 112.4553 | 0.0089 | 371.5176 | 0.0027 | 3.3037 | 0.3027 | 3.1718 | 10.4788 |
| 19 | 146.1919 | 0.0068 | 483.9729 | 0.0021 | 3.3105 | 0.3021 | 3.2025 | 10.6019 |
| 20 | 190.0494 | 0.0053 | 630.1648 | 0.0016 | 3.3158 | 0.3016 | 3.2275 | 10.7019 |
| 21 | 247.0643 | 0.0040 | 820.2144 | 0.0012 | 3.3198 | 0.3012 | 3.2480 | 10.7828 |
| 22 | 321.1836 | 0.0031 | 1067.2788 | 0.0009 | 3.3230 | 0.3009 | 3.2646 | 10.8482 |
| 23 | 417.5386 | 0.0024 | 1388.4624 | 0.0007 | 3.3254 | 0.3007 | 3.2781 | 10.9009 |
| 24 | 542.8001 | 0.0018 | 1806.0000 | 0.0006 | 3.3272 | 0.3006 | 3.2890 | 10.9433 |
| 25 | 705.6401 | 0.0014 | 2348.7998 | 0.0004 | 3.3286 | 0.3004 | 3.2979 | 10.9773 |
| 26 | 917.3323 | 0.0011 | 3054.4414 | 0.0003 | 3.3297 | 0.3003 | 3.3050 | 11.0045 |
| 27 | 1192.5320 | 0.0008 | 3971.7727 | 0.0003 | 3.3305 | 0.3003 | 3.3107 | 11.0263 |
| 28 | 1550.2915 | 0.0006 | 5164.3047 | 0.0002 | 3.3312 | 0.3002 | 3.3153 | 11.0437 |
| 29 | 2015.3775 | 0.0005 | 6714.5938 | 0.0001 | 3.3317 | 0.3001 | 3.3189 | 11.0576 |
| 30 | 2619.9920 | 0.0004 | 8729.9727 | 0.0001 | 3.3321 | 0.3001 | 3.3219 | 11.0687 |
| 31 | 3405.9902 | 0.0003 | 11349.9688 | 0.0001 | 3.3324 | 0.3001 | 3.3242 | 11.0775 |
| 32 | 4427.7852 | 0.0002 | 14755.9570 | 0.0001 | 3.3326 | 0.3001 | 3.3261 | 11.0845 |
| 33 | 5756.1211 | 0.0002 | 19183.7461 | 0.0001 | 3.3328 | 0.3001 | 3.3276 | 11.0901 |
| 34 | 7482.9570 | 0.0001 | 24939.8672 | 0.0000 | 3.3329 | 0.3000 | 3.3288 | 11.0945 |
| 35 | 9727.8438 | 0.0001 | 32422.8086 | 0.0000 | 3.3330 | 0.3000 | 3.3297 | 11.0980 |

| | Single payment | | Uniform series | | | | Gradient series | |
|---|---|---|---|---|---|---|---|---|
| | Compound amount factor | Present worth factor | Compound amount factor | Sinking fund factor | Present worth factor | Capital recovery factor | Uniform series factor | Present worth factor |
| $n$ | To find $F$ given $P$ $F\|P\,i,n$ | To find $P$ given $F$ $P\|F\,i,n$ | To find $F$ given $A$ $F\|A\,i,n$ | To find $A$ given $F$ $A\|F\,i,n$ | To find $P$ given $A$ $P\|A\,i,n$ | To find $A$ given $P$ $A\|P\,i,n$ | To find $A$ given $G$ $A\|G\,i,n$ | To find $P$ given $G$ $P\|G\,i,n$ |
| 1 | 1.4000 | 0.7143 | 1.0000 | 1.0000 | 0.7143 | 1.4000 | 0.0000 | 0.0000 |
| 2 | 1.9600 | 0.5102 | 2.4000 | 0.4167 | 1.2245 | 0.8167 | 0.4167 | 0.5102 |
| 3 | 2.7440 | 0.3644 | 4.3600 | 0.2294 | 1.5889 | 0.6294 | 0.7798 | 1.2391 |
| 4 | 3.8416 | 0.2603 | 7.1040 | 0.1408 | 1.8492 | 0.5408 | 1.0923 | 2.0200 |
| 5 | 5.3782 | 0.1859 | 10.9455 | 0.0914 | 2.0352 | 0.4914 | 1.3580 | 2.7637 |
| 6 | 7.5295 | 0.1328 | 16.3238 | 0.0613 | 2.1680 | 0.4613 | 1.5811 | 3.4278 |
| 7 | 10.5414 | 0.0949 | 23.8534 | 0.0419 | 2.2628 | 0.4419 | 1.7664 | 3.9970 |
| 8 | 14.7579 | 0.0678 | 34.3947 | 0.0291 | 2.3306 | 0.4291 | 1.9185 | 4.4713 |
| 9 | 20.6610 | 0.0484 | 49.1526 | 0.0203 | 2.3790 | 0.4203 | 2.0422 | 4.8585 |
| 10 | 28.9255 | 0.0346 | 69.8137 | 0.0143 | 2.4136 | 0.4143 | 2.1419 | 5.1696 |
| 11 | 40.4957 | 0.0247 | 98.7391 | 0.0101 | 2.4383 | 0.4101 | 2.2215 | 5.4166 |
| 12 | 56.6939 | 0.0176 | 139.2348 | 0.0072 | 2.4559 | 0.4072 | 2.2845 | 5.6106 |
| 13 | 79.3715 | 0.0126 | 195.9287 | 0.0051 | 2.4685 | 0.4051 | 2.3341 | 5.7618 |
| 14 | 111.1201 | 0.0090 | 275.3002 | 0.0036 | 2.4775 | 0.4036 | 2.3729 | 5.8788 |
| 15 | 155.5681 | 0.0064 | 386.4202 | 0.0026 | 2.4839 | 0.4026 | 2.4030 | 5.9688 |
| 16 | 217.7953 | 0.0046 | 541.9883 | 0.0018 | 2.4885 | 0.4018 | 2.4262 | 6.0376 |
| 17 | 304.9135 | 0.0033 | 759.7837 | 0.0013 | 2.4918 | 0.4013 | 2.4441 | 6.0901 |
| 18 | 426.8789 | 0.0023 | 1064.6971 | 0.0009 | 2.4941 | 0.4009 | 2.4577 | 6.1299 |
| 19 | 597.6304 | 0.0017 | 1491.5760 | 0.0007 | 2.4958 | 0.4007 | 2.4682 | 6.1601 |
| 20 | 836.6826 | 0.0012 | 2089.2064 | 0.0005 | 2.4970 | 0.4005 | 2.4761 | 6.1828 |
| 21 | 1171.3556 | 0.0009 | 2925.8889 | 0.0003 | 2.4979 | 0.4003 | 2.4821 | 6.1998 |
| 22 | 1639.8978 | 0.0006 | 4097.2445 | 0.0002 | 2.4985 | 0.4002 | 2.4866 | 6.2127 |
| 23 | 2295.8569 | 0.0004 | 5737.1423 | 0.0002 | 2.4989 | 0.4002 | 2.4900 | 6.2222 |
| 24 | 3214.1997 | 0.0003 | 8032.9992 | 0.0001 | 2.4992 | 0.4001 | 2.4925 | 6.2294 |
| 25 | 4499.8796 | 0.0002 | 11247.1990 | 0.0001 | 2.4994 | 0.4001 | 2.4944 | 6.2347 |
| 26 | 6299.8314 | 0.0002 | 15747.0790 | 0.0001 | 2.4996 | 0.4001 | 2.4959 | 6.2387 |
| 27 | 8819.7640 | 0.0001 | 22046.6910 | 0.0000 | 2.4997 | 0.4000 | 2.4969 | 6.2416 |
| 28 | 12347.6000 | 0.0001 | 30866.6740 | 0.0000 | 2.4998 | 0.4000 | 2.4977 | 6.2438 |
| 29 | 17286.7370 | 0.0001 | 43214.3430 | 0.0000 | 2.4999 | 0.4000 | 2.4983 | 6.2454 |
| 30 | 24201.4320 | 0.0000 | 60501.0810 | 0.0000 | 2.4999 | 0.4000 | 2.4988 | 6.2466 |

| | Single payment | | Uniform series | | | | Gradient series | |
|---|---|---|---|---|---|---|---|---|
| | Compound amount factor | Present worth factor | Compound amount factor | Sinking fund factor | Present worth factor | Capital recovery factor | Uniform series factor | Present worth factor |
| $n$ | To find $F$ given $P$ $F\|P\,i,n$ | To find $P$ given $F$ $P\|F\,i,n$ | To find $F$ given $A$ $F\|A\,i,n$ | To find $A$ given $F$ $A\|F\,i,n$ | To find $P$ given $A$ $P\|A\,i,n$ | To find $A$ given $P$ $A\|P\,i,n$ | To find $A$ given $G$ $A\|G\,i,n$ | To find $P$ given $G$ $P\|G\,i,n$ |
| 1 | 1.5000 | 0.6667 | 1.0000 | 1.0000 | 0.6667 | 1.5000 | 0.0000 | 0.0000 |
| 2 | 2.2500 | 0.4444 | 2.5000 | 0.4000 | 1.1111 | 0.9000 | 0.4000 | 0.4444 |
| 3 | 3.3750 | 0.2963 | 4.7500 | 0.2105 | 1.4074 | 0.7105 | 0.7368 | 1.0370 |
| 4 | 5.0625 | 0.1975 | 8.1250 | 0.1231 | 1.6049 | 0.6231 | 1.0154 | 1.6296 |
| 5 | 7.5938 | 0.1317 | 13.1875 | 0.0758 | 1.7366 | 0.5758 | 1.2417 | 2.1564 |
| 6 | 11.3906 | 0.0878 | 20.7813 | 0.0481 | 1.8244 | 0.5481 | 1.4226 | 2.5953 |
| 7 | 17.0859 | 0.0585 | 32.1719 | 0.0311 | 1.8830 | 0.5311 | 1.5648 | 2.9465 |
| 8 | 25.6289 | 0.0390 | 49.2579 | 0.0203 | 1.9220 | 0.5203 | 1.6752 | 3.2196 |
| 9 | 38.4434 | 0.0260 | 74.8867 | 0.0134 | 1.9480 | 0.5134 | 1.7596 | 3.4277 |
| 10 | 57.6650 | 0.0173 | 113.3301 | 0.0088 | 1.9653 | 0.5088 | 1.8235 | 3.5838 |
| 11 | 86.4976 | 0.0116 | 170.9951 | 0.0058 | 1.9769 | 0.5058 | 1.8713 | 3.6994 |
| 12 | 129.7463 | 0.0077 | 257.4927 | 0.0039 | 1.9846 | 0.5039 | 1.9068 | 3.7842 |
| 13 | 194.6195 | 0.0051 | 387.2390 | 0.0026 | 1.9897 | 0.5026 | 1.9329 | 3.8459 |
| 14 | 291.9293 | 0.0034 | 581.8585 | 0.0017 | 1.9932 | 0.5017 | 1.9519 | 3.8904 |
| 15 | 437.8939 | 0.0023 | 873.7878 | 0.0011 | 1.9954 | 0.5011 | 1.9657 | 3.9224 |
| 16 | 656.8408 | 0.0015 | 1311.6817 | 0.0008 | 1.9970 | 0.5008 | 1.9756 | 3.9452 |
| 17 | 985.2613 | 0.0010 | 1968.5225 | 0.0005 | 1.9980 | 0.5005 | 1.9827 | 3.9614 |
| 18 | 1477.8919 | 0.0007 | 2953.7838 | 0.0003 | 1.9987 | 0.5003 | 1.9878 | 3.9729 |
| 19 | 2216.8378 | 0.0005 | 4431.6756 | 0.0002 | 1.9991 | 0.5002 | 1.9914 | 3.9811 |
| 20 | 3325.2567 | 0.0003 | 6648.5134 | 0.0002 | 1.9994 | 0.5002 | 1.9940 | 3.9868 |
| 21 | 4987.8851 | 0.0002 | 9973.7702 | 0.0001 | 1.9996 | 0.5001 | 1.9958 | 3.9908 |
| 22 | 7481.8276 | 0.0001 | 14961.6550 | 0.0001 | 1.9997 | 0.5001 | 1.9971 | 3.9936 |
| 23 | 11222.7410 | 0.0001 | 22443.4830 | 0.0000 | 1.9998 | 0.5000 | 1.9980 | 3.9955 |
| 24 | 16834.1120 | 0.0001 | 33666.2240 | 0.0000 | 1.9999 | 0.5000 | 1.9986 | 3.9969 |
| 25 | 25251.1680 | 0.0000 | 50500.3370 | 0.0000 | 1.9999 | 0.5000 | 1.9990 | 3.9979 |

# SECTION II—GEOMETRIC SERIES FACTORS

## TABLE A.25  Discrete Compounding: $i = 5\%$

| | Geometric series present worth factor, $(P \mid A_1\ i, j, n)$ | | | | |
|---|---|---|---|---|---|
| $n$ | $j = 4\%$ | $j = 6\%$ | $j = 8\%$ | $j = 10\%$ | $j = 15\%$ |
| 1 | 0.9524 | 0.9524 | 0.9524 | 0.9524 | 0.9524 |
| 2 | 1.8957 | 1.9138 | 1.9320 | 1.9501 | 1.9955 |
| 3 | 2.8300 | 2.8844 | 2.9396 | 2.9954 | 3.1379 |
| 4 | 3.7554 | 3.8643 | 3.9759 | 4.0904 | 4.3891 |
| 5 | 4.6721 | 4.8535 | 5.0419 | 5.2375 | 5.7595 |
| 6 | 5.5799 | 5.8521 | 6.1383 | 6.4393 | 7.2604 |
| 7 | 6.4792 | 6.8602 | 7.2661 | 7.6983 | 8.9043 |
| 8 | 7.3699 | 7.8779 | 8.4261 | 9.0173 | 10.7047 |
| 9 | 8.2521 | 8.9053 | 9.6192 | 10.3991 | 12.6765 |
| 10 | 9.1258 | 9.9425 | 10.8464 | 11.8467 | 14.8362 |
| 11 | 9.9913 | 10.9896 | 12.1087 | 13.3632 | 17.2016 |
| 12 | 10.8485 | 12.0466 | 13.4070 | 14.9519 | 19.7922 |
| 13 | 11.6976 | 13.1137 | 14.7425 | 16.6163 | 22.6295 |
| 14 | 12.5386 | 14.1910 | 16.1161 | 18.3599 | 25.7371 |
| 15 | 13.3715 | 15.2785 | 17.5289 | 20.1866 | 29.1407 |
| 16 | 14.1966 | 16.3764 | 18.9821 | 22.1002 | 32.8683 |
| 17 | 15.0137 | 17.4848 | 20.4769 | 24.1050 | 36.9510 |
| 18 | 15.8231 | 18.6037 | 22.0143 | 26.2052 | 41.4226 |
| 19 | 16.6248 | 19.7332 | 23.5956 | 28.4055 | 46.3200 |
| 20 | 17.4189 | 20.8736 | 25.2222 | 30.7105 | 51.6838 |
| 21 | 18.2054 | 22.0247 | 26.8952 | 33.1253 | 57.5584 |
| 22 | 18.9844 | 23.1869 | 28.6160 | 35.6550 | 63.9925 |
| 23 | 19.7559 | 24.3601 | 30.3860 | 38.3053 | 71.0394 |
| 24 | 20.5202 | 25.5445 | 32.2066 | 41.0817 | 78.7575 |
| 25 | 21.2771 | 26.7401 | 34.0791 | 43.9904 | 87.2106 |
| 26 | 22.0269 | 27.9472 | 36.0052 | 47.0375 | 96.4687 |
| 27 | 22.7695 | 29.1657 | 37.9863 | 50.2298 | 106.6086 |
| 28 | 23.5050 | 30.3959 | 40.0240 | 53.5741 | 117.7142 |
| 29 | 24.2335 | 31.6377 | 42.1199 | 57.0776 | 129.8774 |
| 30 | 24.9551 | 32.8914 | 44.2757 | 60.7480 | 143.1991 |
| 31 | 25.6698 | 34.1571 | 46.4931 | 64.5931 | 157.7895 |
| 32 | 26.3777 | 35.4348 | 48.7739 | 68.6213 | 173.7695 |
| 33 | 27.0789 | 36.7246 | 51.1198 | 72.8414 | 191.2713 |
| 34 | 27.7734 | 38.0267 | 53.5328 | 77.2624 | 210.4400 |
| 35 | 28.4612 | 39.3413 | 56.0146 | 81.8940 | 231.4343 |
| 36 | 29.1426 | 40.6683 | 58.5674 | 86.7461 | 254.4280 |
| 37 | 29.8174 | 42.0080 | 61.1932 | 91.8292 | 279.6116 |
| 38 | 30.4858 | 43.3605 | 63.8939 | 97.1544 | 307.1937 |
| 39 | 31.1478 | 44.7258 | 66.6719 | 102.7332 | 337.4026 |
| 40 | 31.8036 | 46.1042 | 69.5291 | 108.5776 | 370.4886 |
| 41 | 32.4531 | 47.4957 | 72.4681 | 114.7004 | 406.7256 |
| 42 | 33.0964 | 48.9004 | 75.4910 | 121.1147 | 446.4138 |
| 43 | 33.7335 | 50.3185 | 78.6002 | 127.8344 | 489.8817 |
| 44 | 34.3647 | 51.7501 | 81.7983 | 134.8742 | 537.4895 |
| 45 | 34.9898 | 53.1953 | 85.0878 | 142.2491 | 589.6314 |
| 46 | 35.6089 | 54.6543 | 88.4713 | 149.9753 | 646.7391 |
| 47 | 36.2221 | 56.1272 | 91.9514 | 158.0693 | 709.2857 |
| 48 | 36.8296 | 57.6141 | 95.5310 | 166.5488 | 777.7891 |
| 49 | 37.4312 | 59.1152 | 99.2128 | 175.4321 | 852.8167 |
| 50 | 38.0271 | 60.6306 | 102.9998 | 184.7384 | 934.9897 |

| | Geometric series future worth factor, $(F/A_1\ i,j,n)$ | | | | |
|---|---|---|---|---|---|
| $n$ | $j = 4\%$ | $j = 6\%$ | $j = 8\%$ | $j = 10\%$ | $j = 15\%$ |
| 1 | 1.0000 | 1.0000 | 1.0000 | 1.0000 | 1.0000 |
| 2 | 2.0900 | 2.1100 | 2.1300 | 2.1500 | 2.2000 |
| 3 | 3.2761 | 3.3391 | 3.4029 | 2.4675 | 3.6325 |
| 4 | 4.5648 | 4.6971 | 4.8328 | 4.9719 | 5.3350 |
| 5 | 5.9629 | 6.1944 | 6.4349 | 6.6846 | 7.3508 |
| 6 | 7.4777 | 7.8423 | 8.2260 | 8.6293 | 9.7297 |
| 7 | 9.1169 | 9.6530 | 10.2241 | 10.8323 | 12.5292 |
| 8 | 10.8886 | 11.6393 | 12.4492 | 13.3227 | 15.8157 |
| 9 | 12.8016 | 13.8151 | 14.9225 | 16.1324 | 19.6655 |
| 10 | 14.8650 | 16.1953 | 17.6677 | 19.2970 | 24.1666 |
| 11 | 17.0885 | 18.7959 | 20.7100 | 22.8555 | 29.4205 |
| 12 | 19.4824 | 21.6340 | 24.0771 | 26.8514 | 35.5439 |
| 13 | 22.0576 | 24.7279 | 27.7992 | 31.3324 | 42.6714 |
| 14 | 24.8255 | 28.0972 | 31.9087 | 36.3513 | 50.9577 |
| 15 | 27.7985 | 31.7630 | 36.4414 | 41.9664 | 60.5813 |
| 16 | 30.9893 | 35.7477 | 41.4356 | 48.2420 | 71.7475 |
| 17 | 34.4118 | 40.0754 | 46.9333 | 55.2490 | 84.6925 |
| 18 | 38.0803 | 44.7720 | 52.9800 | 63.0660 | 99.6883 |
| 19 | 42.0101 | 49.8649 | 59.6250 | 71.7792 | 117.0482 |
| 20 | 46.2175 | 55.3838 | 66.9220 | 81.4840 | 137.1324 |
| 21 | 50.7195 | 61.3601 | 74.9290 | 92.2857 | 160.3556 |
| 22 | 55.5342 | 67.8277 | 83.7093 | 104.3003 | 187.1948 |
| 23 | 60.6808 | 74.8226 | 93.3313 | 117.6556 | 218.1993 |
| 24 | 66.1796 | 82.3835 | 103.8694 | 132.4927 | 254.0008 |
| 25 | 72.0519 | 90.5516 | 115.4040 | 148.9670 | 295.3260 |
| 26 | 78.3203 | 99.3710 | 128.0227 | 167.2501 | 343.0112 |
| 27 | 85.0088 | 108.8890 | 141.8202 | 187.5308 | 398.0186 |
| 28 | 92.1426 | 119.1558 | 156.8992 | 210.0173 | 461.4548 |
| 29 | 99.7484 | 130.2252 | 173.3713 | 234.9391 | 534.5932 |
| 30 | 107.8545 | 142.1549 | 191.3572 | 262.5492 | 618.8983 |
| 31 | 116.4906 | 155.0061 | 210.9877 | 293.1261 | 716.0550 |
| 32 | 125.6883 | 168.8445 | 232.4047 | 326.9767 | 828.0013 |
| 33 | 135.4807 | 183.7401 | 255.7620 | 364.4393 | 956.9664 |
| 34 | 145.9032 | 199.7677 | 281.2262 | 405.8864 | 1105.5146 |
| 35 | 156.9926 | 217.0071 | 308.9776 | 451.7284 | 1276.5951 |
| 36 | 168.7884 | 235.5436 | 339.2119 | 502.4173 | 1473.6004 |
| 37 | 181.3317 | 255.4680 | 372.1406 | 558.4508 | 1700.4322 |
| 38 | 194.6664 | 276.8775 | 407.9933 | 620.3773 | 1961.5785 |
| 39 | 208.8385 | 299.8756 | 447.0182 | 688.8005 | 2262.2007 |
| 40 | 223.8968 | 324.5729 | 489.4844 | 764.3853 | 2608.2356 |
| 41 | 239.8927 | 351.0873 | 535.6832 | 847.8639 | 3006.5109 |
| 42 | 256.8804 | 379.5445 | 585.9298 | 940.0422 | 3464.8795 |
| 43 | 274.9172 | 410.0788 | 640.5658 | 1041.8080 | 3992.3730 |
| 44 | 294.0635 | 442.8332 | 699.9607 | 1154.1385 | 4599.3787 |
| 45 | 314.3832 | 477.9603 | 764.5147 | 1278.1095 | 5297.8426 |
| 46 | 335.9435 | 515.6229 | 834.6609 | 1414.9055 | 6101.5040 |
| 47 | 358.8155 | 555.9946 | 910.8680 | 1565.8303 | 7026.1639 |
| 48 | 383.0741 | 599.2602 | 993.6434 | 1732.3193 | 8089.9944 |
| 49 | 408.7984 | 645.6171 | 1083.5362 | 1915.9525 | 9313.8949 |
| 50 | 436.0716 | 695.2754 | 1181.1404 | 2118.4691 | 10721.9004 |

## TABLE A.27 Discrete Compounding: $i = 8\%$

| | Geometric series present worth factor, $(P \mid A_1 \; i, j, n)$ | | | | |
|---|---|---|---|---|---|
| $n$ | $j = 4\%$ | $j = 6\%$ | $j = 8\%$ | $j = 10\%$ | $j = 15\%$ |
| 1 | 0.9259 | 0.9259 | 0.9259 | 0.9259 | 0.9259 |
| 2 | 1.8176 | 1.8347 | 1.8519 | 1.8690 | 1.9119 |
| 3 | 2.6762 | 2.7267 | 2.7778 | 2.8295 | 2.9617 |
| 4 | 3.5030 | 3.6021 | 3.7037 | 3.8079 | 4.0796 |
| 5 | 4.2992 | 4.4613 | 4.6296 | 4.8043 | 5.2699 |
| 6 | 5.0659 | 5.3046 | 5.5556 | 5.8192 | 6.5374 |
| 7 | 5.8042 | 6.1323 | 6.4815 | 6.8529 | 7.8871 |
| 8 | 6.5151 | 6.9447 | 7.4074 | 7.9057 | 9.3242 |
| 9 | 7.1997 | 7.7420 | 8.3333 | 8.9780 | 10.8545 |
| 10 | 7.8590 | 8.5246 | 9.2593 | 10.0702 | 12.4839 |
| 11 | 8.4939 | 9.2926 | 10.1852 | 11.1826 | 14.2190 |
| 12 | 9.1052 | 10.0465 | 11.1111 | 12.3157 | 16.0665 |
| 13 | 9.6939 | 10.7863 | 12.0370 | 13.4696 | 18.0338 |
| 14 | 10.2608 | 11.5125 | 12.9630 | 14.6450 | 20.1286 |
| 15 | 10.8067 | 12.2252 | 13.8889 | 15.8421 | 22.3592 |
| 16 | 11.3324 | 12.9248 | 14.8148 | 17.0614 | 24.7343 |
| 17 | 11.8386 | 13.6114 | 15.7407 | 18.3033 | 27.2634 |
| 18 | 12.3260 | 14.2852 | 16.6667 | 19.5682 | 29.9564 |
| 19 | 12.7954 | 14.9466 | 17.5926 | 20.8565 | 32.8239 |
| 20 | 13.2475 | 15.5957 | 18.5185 | 22.1687 | 35.8773 |
| 21 | 13.6827 | 16.2329 | 19.4444 | 23.5051 | 39.1286 |
| 22 | 14.1019 | 16.8582 | 20.3704 | 24.8663 | 42.5906 |
| 23 | 14.5055 | 17.4719 | 21.2963 | 26.2527 | 46.2771 |
| 24 | 14.8942 | 18.0743 | 22.2222 | 27.6648 | 50.2024 |
| 25 | 15.2685 | 18.6655 | 23.1481 | 29.1031 | 54.3822 |
| 26 | 15.6289 | 19.2458 | 24.0741 | 30.5679 | 58.8329 |
| 27 | 15.9760 | 19.8153 | 25.0000 | 32.0599 | 63.5721 |
| 28 | 16.3102 | 20.3743 | 25.9259 | 33.5796 | 68.6184 |
| 29 | 16.6321 | 20.9229 | 26.8519 | 35.1273 | 73.9919 |
| 30 | 16.9420 | 21.4614 | 27.7778 | 36.7038 | 79.7136 |
| 31 | 17.2404 | 21.9899 | 28.7037 | 38.3094 | 85.8061 |
| 32 | 17.5278 | 22.5086 | 29.6296 | 39.9477 | 92.2935 |
| 33 | 17.8046 | 23.0177 | 30.5556 | 41.6104 | 99.2015 |
| 34 | 18.0711 | 23.5173 | 31.4815 | 43.3069 | 106.5571 |
| 35 | 18.3277 | 24.0078 | 32.4074 | 45.0348 | 114.3895 |
| 36 | 18.5748 | 24.4891 | 33.3333 | 46.7947 | 122.7296 |
| 37 | 18.8128 | 24.9615 | 34.2593 | 48.5872 | 131.6102 |
| 38 | 19.0419 | 25.4252 | 35.1852 | 50.4129 | 141.0664 |
| 39 | 19.2626 | 25.8803 | 36.1111 | 52.2724 | 151.1355 |
| 40 | 19.4751 | 26.3269 | 37.0370 | 54.1663 | 161.8573 |
| 41 | 19.6797 | 26.7653 | 37.9630 | 56.0953 | 173.2739 |
| 42 | 19.8768 | 27.1956 | 38.8889 | 58.0600 | 185.4306 |
| 43 | 20.0665 | 27.6179 | 39.8148 | 60.0611 | 198.3752 |
| 44 | 20.2493 | 28.0324 | 40.7407 | 62.0993 | 212.1587 |
| 45 | 20.4252 | 28.4392 | 41.6667 | 64.1752 | 226.8357 |
| 46 | 20.5946 | 28.8385 | 42.5926 | 66.2896 | 242.4639 |
| 47 | 20.7578 | 29.2304 | 43.5185 | 68.4431 | 259.1051 |
| 48 | 20.9149 | 29.6150 | 44.4444 | 70.6365 | 276.8249 |
| 49 | 21.0662 | 29.9925 | 45.3704 | 72.8705 | 2.95.6932 |
| 50 | 21.2119 | 30.3630 | 46.2963 | 75.1459 | 315.7844 |

| | Geometric series future worth factor, $(F/A_1\ i,j,n)$ | | | | |
|---|---|---|---|---|---|
| $n$ | $j = 4\%$ | $j = 6\%$ | $j = 8\%$ | $j = 10\%$ | $j = 15\%$ |
| 1 | 1.0000 | 1.0000 | 1.0000 | 1.0000 | 1.0000 |
| 2 | 2.1200 | 2.1400 | 2.1600 | 2.1800 | 2.2300 |
| 3 | 3.3712 | 3.4348 | 3.4992 | 3.5644 | 3.7309 |
| 4 | 4.7658 | 4.9006 | 5.0388 | 5.1806 | 5.5502 |
| 5 | 6.3169 | 6.5551 | 6.8024 | 7.0591 | 7.7433 |
| 6 | 8.0389 | 8.4178 | 8.8160 | 9.2343 | 10.3741 |
| 7 | 9.9473 | 10.5097 | 11.1081 | 11.7446 | 13.5171 |
| 8 | 12.0590 | 12.8541 | 13.7106 | 14.6329 | 17.2585 |
| 9 | 14.3923 | 15.4763 | 16.6584 | 17.9472 | 21.6982 |
| 10 | 16.9670 | 18.4039 | 19.9900 | 21.7409 | 26.9519 |
| 11 | 19.8046 | 21.6670 | 23.7482 | 26.0739 | 33.1536 |
| 12 | 22.9284 | 25.2987 | 27.9797 | 31.0129 | 40.4583 |
| 13 | 26.3638 | 29.3348 | 32.7362 | 36.6324 | 49.0452 |
| 14 | 30.1379 | 33.8145 | 38.0747 | 43.0152 | 59.1216 |
| 15 | 34.2806 | 38.7805 | 44.0579 | 50.2540 | 70.9270 |
| 16 | 38.8240 | 44.2795 | 50.7547 | 58.4515 | 84.7383 |
| 17 | 43.8029 | 50.3623 | 58.2410 | 67.7226 | 100.8749 |
| 18 | 49.2551 | 57.0840 | 66.6003 | 78.1949 | 119.7062 |
| 19 | 55.2213 | 64.5051 | 75.9244 | 90.0104 | 141.6582 |
| 20 | 61.7459 | 72.6911 | 86.3140 | 103.3271 | 167.2226 |
| 21 | 68.8766 | 81.7135 | 97.8801 | 118.3208 | 196.9669 |
| 22 | 76.6655 | 91.6501 | 110.7443 | 135.1867 | 231.5458 |
| 23 | 85.1687 | 102.5857 | 125.0404 | 154.1419 | 271.7142 |
| 24 | 94.4469 | 114.6123 | 140.9151 | 175.4276 | 318.3428 |
| 25 | 104.5660 | 127.8302 | 158.5295 | 199.3115 | 372.4354 |
| 26 | 115.5971 | 142.3485 | 178.0604 | 226.0912 | 435.1492 |
| 27 | 127.6173 | 158.2858 | 199.7015 | 256.0966 | 507.8179 |
| 28 | 140.7101 | 175.7710 | 223.6657 | 289.6944 | 591.9787 |
| 29 | 154.9656 | 194.9443 | 250.1861 | 327.2909 | 689.4026 |
| 30 | 170.4815 | 215.9583 | 279.5182 | 369.3373 | 802.1302 |
| 31 | 187.3634 | 238.9784 | 311.9424 | 416.3337 | 932.5124 |
| 32 | 205.7256 | 264.1848 | 347.7654 | 468.8347 | 1083.2569 |
| 33 | 225.6917 | 291.7730 | 387.3237 | 527.4552 | 1257.4826 |
| 34 | 247.3954 | 321.9554 | 430.9857 | 592.8768 | 1458.7810 |
| 35 | 270.9814 | 354.9629 | 479.1547 | 665.8546 | 1691.2883 |
| 36 | 296.6060 | 391.0460 | 532.2724 | 747.2254 | 1959.7669 |
| 37 | 324.4384 | 430.4769 | 590.8224 | 837.9162 | 2269.7001 |
| 38 | 354.6616 | 473.5512 | 655.3338 | 938.9534 | 2627.4007 |
| 39 | 387.4733 | 520.5895 | 726.3857 | 1051.4740 | 3040.1361 |
| 40 | 423.0875 | 571.9402 | 804.6119 | 1176.7367 | 3516.2718 |
| 41 | 461.7355 | 627.9811 | 890.7054 | 1316.1349 | 4065.4371 |
| 42 | 503.6674 | 689.1225 | 985.4243 | 1471.2109 | 4698.7151 |
| 43 | 549.1536 | 755.8093 | 1089.5977 | 1643.6714 | 5428.8619 |
| 44 | 598.4864 | 828.5245 | 1204.1322 | 1835.4052 | 6270.5578 |
| 45 | 651.9818 | 907.7919 | 1330.0187 | 2048.5017 | 7240.6974 |
| 46 | 709.9816 | 994.1799 | 1468.3407 | 2285.2723 | 8358.7225 |
| 47 | 772.8549 | 1088.3048 | 1620.2820 | 2548.2737 | 9647.0049 |
| 48 | 841.0011 | 1190.8351 | 1787.1366 | 2840.3330 | 11131.2877 |
| 49 | 914.8517 | 1302.4957 | 1970.3181 | 3164.5769 | 12841.1914 |
| 50 | 994.8732 | 1424.0729 | 2171.3709 | 3524.4620 | 14810.7976 |

## TABLE A.29  Discrete Compounding: $i = 10\%$

| | Geometric series present worth factor, $(P \mid A_1 \; i, j, n)$ | | | | |
|---|---|---|---|---|---|
| $n$ | $j = 4\%$ | $j = 6\%$ | $j = 8\%$ | $j = 10\%$ | $j = 15\%$ |
| 1 | 0.9091 | 0.9091 | 0.9091 | 0.9091 | 0.9091 |
| 2 | 1.7686 | 1.7851 | 1.8017 | 1.8182 | 1.8595 |
| 3 | 2.5812 | 2.6293 | 2.6780 | 2.7273 | 2.8531 |
| 4 | 3.3495 | 3.4428 | 3.5384 | 3.6364 | 3.8919 |
| 5 | 4.0759 | 4.2267 | 4.3831 | 4.5455 | 4.9779 |
| 6 | 4.7627 | 4.9821 | 5.2125 | 5.4545 | 6.1133 |
| 7 | 5.4120 | 5.7100 | 6.0269 | 6.3636 | 7.3002 |
| 8 | 6.0259 | 6.4115 | 6.8264 | 7.2727 | 8.5411 |
| 9 | 6.6063 | 7.0874 | 7.6113 | 8.1818 | 9.8385 |
| 10 | 7.1550 | 7.7388 | 8.3820 | 9.0909 | 11.1948 |
| 11 | 7.6738 | 8.3664 | 9.1387 | 10.0000 | 12.6127 |
| 12 | 8.1644 | 8.9713 | 9.8817 | 10.9091 | 14.0951 |
| 13 | 8.6281 | 9.5542 | 10.6111 | 11.8182 | 15.6449 |
| 14 | 9.0666 | 10.1158 | 11.3273 | 12.7273 | 17.2651 |
| 15 | 9.4811 | 10.6571 | 12.0304 | 13.6364 | 18.9590 |
| 16 | 9.8731 | 11.1786 | 12.7208 | 14.5455 | 20.7298 |
| 17 | 10.2436 | 11.6812 | 13.3986 | 15.4545 | 22.5812 |
| 18 | 10.5940 | 12.1656 | 14.0640 | 16.3636 | 24.5167 |
| 19 | 10.9252 | 12.6323 | 14.7174 | 17.2727 | 26.5402 |
| 20 | 11.2384 | 13.0820 | 15.3589 | 18.1818 | 28.6556 |
| 21 | 11.5345 | 13.5154 | 15.9888 | 19.0909 | 30.8672 |
| 22 | 11.8144 | 13.9330 | 16.6071 | 20.0000 | 33.1794 |
| 23 | 12.0791 | 14.3354 | 17.2143 | 20.9091 | 35.5966 |
| 24 | 12.3293 | 14.7232 | 17.8104 | 21.8182 | 38.1238 |
| 25 | 12.5659 | 15.0969 | 18.3957 | 22.7273 | 40.7658 |
| 26 | 12.7896 | 15.4570 | 18.9703 | 23.6364 | 43.5278 |
| 27 | 13.0011 | 15.8041 | 19.5345 | 24.5455 | 46.4155 |
| 28 | 13.2010 | 16.1385 | 20.0884 | 25.4545 | 49.4343 |
| 29 | 13.3900 | 16.4607 | 20.6322 | 26.3636 | 52.5905 |
| 30 | 13.5688 | 16.7712 | 21.1662 | 27.2727 | 55.8900 |
| 31 | 13.7377 | 17.0704 | 21.6904 | 28.1818 | 59.3396 |
| 32 | 13.8975 | 17.3588 | 22.2052 | 29.0909 | 62.9459 |
| 33 | 14.0485 | 17.6367 | 22.7105 | 30.0000 | 66.7162 |
| 34 | 14.1913 | 17.9044 | 23.2067 | 30.9091 | 70.6578 |
| 35 | 14.3264 | 18.1624 | 23.6938 | 31.8182 | 74.7786 |
| 36 | 14.4540 | 18.4111 | 24.1721 | 32.7273 | 79.0867 |
| 37 | 14.5747 | 18.6507 | 24.6417 | 33.6364 | 83.5907 |
| 38 | 14.6888 | 18.8816 | 25.1028 | 34.5455 | 88.2994 |
| 39 | 14.7967 | 19.1040 | 25.5555 | 35.4545 | 93.2221 |
| 40 | 14.8987 | 19.3184 | 25.9999 | 36.3636 | 98.3685 |
| 41 | 14.9951 | 19.5250 | 26.4363 | 37.2727 | 103.7489 |
| 42 | 15.0863 | 19.7241 | 26.8647 | 38.1818 | 109.3739 |
| 43 | 15.1725 | 19.9160 | 27.2854 | 39.0909 | 115.2545 |
| 44 | 15.2540 | 20.1009 | 27.6983 | 40.0000 | 121.4024 |
| 45 | 15.3311 | 20.2790 | 28.1038 | 40.9091 | 127.8298 |
| 46 | 15.4039 | 20.4507 | 28.5019 | 41.8182 | 134.5493 |
| 47 | 15.4728 | 20.6161 | 28.8928 | 42.7273 | 141.5743 |
| 48 | 15.5379 | 20.7755 | 29.2766 | 43.6364 | 148.9186 |
| 49 | 15.5995 | 20.9291 | 29.6534 | 44.5455 | 156.5967 |
| 50 | 15.6577 | 21.0772 | 30.0233 | 45.4545 | 164.6238 |

## TABLE A.30   Discrete Compounding: $i = 10\%$

Geometric series future worth factor, $(F/A_1 \; i,j,n)$

| $n$ | $j = 4\%$ | $j = 6\%$ | $j = 8\%$ | $j = 10\%$ | $j = 15\%$ |
|---|---|---|---|---|---|
| 1 | 1.0000 | 1.0000 | 1.0000 | 1.0000 | 1.0000 |
| 2 | 2.1400 | 2.1600 | 2.1800 | 2.2000 | 2.2500 |
| 3 | 3.4356 | 3.4996 | 3.5644 | 3.6300 | 3.7975 |
| 4 | 4.9040 | 5.0406 | 5.1806 | 5.3240 | 5.6981 |
| 5 | 6.5643 | 6.8071 | 7.0591 | 7.3205 | 8.0169 |
| 6 | 8.4374 | 8.8260 | 9.2343 | 9.6631 | 10.8300 |
| 7 | 10.5464 | 11.1272 | 11.7446 | 12.4009 | 14.2261 |
| 8 | 12.9170 | 13.7435 | 14.6329 | 15.5897 | 18.3087 |
| 9 | 15.5773 | 16.7117 | 17.9472 | 19.2923 | 23.1986 |
| 10 | 18.5583 | 20.0724 | 21.7409 | 23.5795 | 29.0363 |
| 11 | 21.8944 | 23.8705 | 26.0739 | 28.5312 | 35.9855 |
| 12 | 25.6233 | 28.1558 | 31.0129 | 34.2374 | 44.2364 |
| 13 | 29.7866 | 32.9836 | 36.6324 | 40.7996 | 54.0103 |
| 14 | 34.4304 | 38.4149 | 43.0152 | 48.3318 | 65.5641 |
| 15 | 39.6051 | 44.5172 | 50.2540 | 56.9625 | 79.1963 |
| 16 | 45.3665 | 51.3655 | 58.4515 | 66.8360 | 95.2530 |
| 17 | 51.7762 | 59.0424 | 67.7226 | 78.1145 | 114.1359 |
| 18 | 58.9017 | 67.6395 | 78.1949 | 90.9805 | 136.3107 |
| 19 | 66.8177 | 77.2577 | 90.0104 | 105.6384 | 162.3173 |
| 20 | 75.6063 | 88.0091 | 103.3271 | 122.3182 | 192.7807 |
| 21 | 85.3580 | 100.0172 | 118.3208 | 141.2775 | 228.4254 |
| 22 | 96.1726 | 113.4184 | 135.1867 | 162.8055 | 270.0894 |
| 23 | 108.1598 | 128.3638 | 154.1419 | 187.2263 | 318.7431 |
| 24 | 121.4405 | 145.0200 | 175.4276 | 214.9033 | 375.5089 |
| 25 | 136.1478 | 163.5709 | 199.3115 | 246.2433 | 441.6849 |
| 26 | 152.4284 | 184.2198 | 226.0912 | 281.7024 | 518.7724 |
| 27 | 170.4438 | 207.1912 | 256.0966 | 321.7908 | 608.5064 |
| 28 | 190.3715 | 232.7327 | 289.6944 | 367.0798 | 712.8924 |
| 29 | 212.4074 | 261.1176 | 327.2909 | 418.2088 | 834.2472 |
| 30 | 236.7667 | 292.6478 | 369.3373 | 475.8928 | 975.2474 |
| 31 | 263.6868 | 327.6560 | 416.3337 | 540.9315 | 1138.9839 |
| 32 | 293.4286 | 366.5098 | 468.8347 | 614.2190 | 1329.0258 |
| 33 | 326.2796 | 409.6141 | 527.4552 | 696.7546 | 1549.4935 |
| 34 | 362.5559 | 457.4161 | 592.8768 | 789.6553 | 1805.1427 |
| 35 | 402.6058 | 510.4088 | 665.8546 | 894.1684 | 2101.4617 |
| 36 | 446.8125 | 569.1357 | 747.2254 | 1011.6877 | 2444.7834 |
| 37 | 495.5976 | 634.1965 | 837.9162 | 1143.7692 | 2842.4136 |
| 38 | 549.4255 | 706.2523 | 938.9534 | 1292.1500 | 3302.7796 |
| 39 | 608.8069 | 786.3018 | 1051.4740 | 1458.7694 | 3835.6009 |
| 40 | 674.3039 | 874.3384 | 1176.7367 | 1645.7911 | 4452.0858 |
| 41 | 746.5353 | 972.0580 | 1316.1349 | 1855.6295 | 5165.1579 |
| 42 | 826.1819 | 1080.1667 | 1471.2109 | 2090.9776 | 5989.7168 |
| 43 | 913.9929 | 1199.7404 | 1643.6714 | 2354.8391 | 6942.9380 |
| 44 | 1010.7927 | 1331.9649 | 1835.4052 | 2650.5630 | 8044.6188 |
| 45 | 1117.4885 | 1478.1468 | 2048.5017 | 2981.8834 | 9317.5757 |
| 46 | 1235.0785 | 1639.7261 | 2285.2723 | 3352.9622 | 10788.1025 |
| 47 | 1364.6612 | 1818.2892 | 2548.2737 | 3768.4380 | 12486.4975 |
| 48 | 1507.4451 | 2015.5841 | 2840.3330 | 4233.4793 | 14447.6696 |
| 49 | 1664.7601 | 2233.5363 | 3164.5769 | 4753.8445 | 16711.8372 |
| 50 | 1838.0695 | 2474.2675 | 3524.4620 | 5335.9479 | 19325.3318 |

## TABLE A.31 Discrete Compounding: $i = 15\%$

| | Geometric series present worth factor, $(P\,|\,A_1\ i, j, n)$ | | | | |
|---|---|---|---|---|---|
| $n$ | $j=4\%$ | $j=6\%$ | $j=8\%$ | $j=10\%$ | $j=15\%$ |
| 1 | 0.8696 | 0.8696 | 0.8696 | 0.8696 | 0.8696 |
| 2 | 1.6560 | 1.6711 | 1.6862 | 1.7013 | 1.7391 |
| 3 | 2.3671 | 2.4099 | 2.4531 | 2.4969 | 2.6087 |
| 4 | 3.0103 | 3.0908 | 3.1734 | 3.2579 | 3.4783 |
| 5 | 3.5919 | 3.7185 | 3.8498 | 3.9858 | 4.3478 |
| 6 | 4.1179 | 4.2971 | 4.4850 | 4.6821 | 5.2174 |
| 7 | 4.5936 | 4.8303 | 5.0816 | 5.3481 | 6.0870 |
| 8 | 5.0237 | 5.3219 | 5.6418 | 5.9851 | 6.9565 |
| 9 | 5.4128 | 5.7749 | 6.1680 | 6.5945 | 7.8261 |
| 10 | 5.7647 | 6.1926 | 6.6621 | 7.1773 | 8.6957 |
| 11 | 6.0828 | 6.5775 | 7.1261 | 7.7348 | 9.5652 |
| 12 | 6.3705 | 6.9323 | 7.5619 | 8.2681 | 10.4348 |
| 13 | 6.6307 | 7.2593 | 7.9712 | 8.7782 | 11.3043 |
| 14 | 6.8660 | 7.5608 | 8.3556 | 9.2661 | 12.1739 |
| 15 | 7.0789 | 7.8386 | 8.7165 | 9.7328 | 13.0435 |
| 16 | 7.2713 | 8.0947 | 9.0555 | 10.1792 | 13.9130 |
| 17 | 7.4454 | 8.3308 | 9.3739 | 10.6062 | 14.7826 |
| 18 | 7.6028 | 8.5484 | 9.6729 | 11.0146 | 15.6522 |
| 19 | 7.7451 | 8.7489 | 9.9537 | 11.4053 | 16.5217 |
| 20 | 7.8738 | 8.9338 | 10.2173 | 11.7790 | 17.3913 |
| 21 | 7.9903 | 9.1042 | 10.4650 | 12.1364 | 18.2609 |
| 22 | 8.0955 | 9.2613 | 10.6976 | 12.4783 | 19.1304 |
| 23 | 8.1907 | 9.4060 | 10.9160 | 12.8053 | 20.0000 |
| 24 | 8.2768 | 9.5395 | 11.1211 | 13.1181 | 20.8696 |
| 25 | 8.3547 | 9.6625 | 11.3137 | 13.4173 | 21.7391 |
| 26 | 8.4251 | 9.7759 | 11.4946 | 13.7035 | 22.6087 |
| 27 | 8.4888 | 9.8803 | 11.6645 | 13.9773 | 23.4783 |
| 28 | 8.5464 | 9.9767 | 11.8241 | 14.2392 | 24.3478 |
| 29 | 8.5985 | 10.0655 | 11.9739 | 14.4896 | 25.2174 |
| 30 | 8.6456 | 10.1473 | 12.1146 | 14.7292 | 26.0870 |
| 31 | 8.6882 | 10.2227 | 12.2468 | 14.9584 | 26.9565 |
| 32 | 8.7267 | 10.2922 | 12.3709 | 15.1776 | 27.8261 |
| 33 | 8.7615 | 10.3563 | 12.4874 | 15.3873 | 28.6957 |
| 34 | 8.7930 | 10.4154 | 12.5969 | 15.5878 | 29.5652 |
| 35 | 8.8215 | 10.4698 | 12.6997 | 15.7796 | 30.4348 |
| 36 | 8.8473 | 10.5200 | 12.7962 | 15.9631 | 31.3043 |
| 37 | 8.8706 | 10.5663 | 12.8869 | 16.1386 | 32.1739 |
| 38 | 8.8917 | 10.6089 | 12.9720 | 16.3065 | 33.0435 |
| 39 | 8.9107 | 10.6482 | 13.0520 | 16.4671 | 33.9130 |
| 40 | 8.9280 | 10.6845 | 13.1271 | 16.6207 | 34.7826 |
| 41 | 8.9436 | 10.7178 | 13.1976 | 16.7676 | 35.6522 |
| 42 | 8.9576 | 10.7486 | 13.2639 | 16.9082 | 36.5217 |
| 43 | 8.9704 | 10.7770 | 13.3261 | 17.0426 | 37.3913 |
| 44 | 8.9819 | 10.8031 | 13.3845 | 17.1712 | 38.2609 |
| 45 | 8.9923 | 10.8272 | 13.4393 | 17.2942 | 39.1304 |
| 46 | 9.0018 | 10.8495 | 13.4908 | 17.4118 | 40.0000 |
| 47 | 9.0103 | 10.8699 | 13.5392 | 17.5244 | 40.8696 |
| 48 | 9.0180 | 10.8888 | 13.5847 | 17.6320 | 41.7391 |
| 49 | 9.0250 | 10.9062 | 13.6273 | 17.7350 | 42.6087 |
| 50 | 9.0313 | 10.9222 | 13.6674 | 17.8334 | 43.4783 |

**TABLE A.32 Discrete Compounding:** $i = 15\%$

| | Geometric series future worth factor, $(F/A_1 \ i,j,n)$ | | | | |
|---|---|---|---|---|---|
| $n$ | $j = 4\%$ | $j = 6\%$ | $j = 8\%$ | $j = 10\%$ | $j = 15\%$ |
| 1 | 1.0000 | 1.0000 | 1.0000 | 1.0000 | 1.0000 |
| 2 | 2.1900 | 2.2100 | 2.2300 | 2.2500 | 2.3000 |
| 3 | 3.6001 | 3.6651 | 3.7309 | 3.7975 | 3.9675 |
| 4 | 5.2650 | 5.4059 | 5.5502 | 5.6981 | 6.0835 |
| 5 | 7.2246 | 7.4792 | 7.7433 | 8.0169 | 8.7450 |
| 6 | 9.5249 | 9.9394 | 10.3741 | 10.8300 | 12.0681 |
| 7 | 12.2190 | 12.8488 | 13.5171 | 14.2261 | 16.1914 |
| 8 | 15.3678 | 16.2797 | 17.2585 | 18.3087 | 21.2802 |
| 9 | 19.0415 | 20.3155 | 21.6982 | 23.1986 | 27.5312 |
| 10 | 23.3210 | 25.0523 | 26.9519 | 29.0363 | 35.1788 |
| 11 | 28.2994 | 30.6010 | 33.1536 | 35.9855 | 44.5011 |
| 12 | 34.0838 | 37.0895 | 40.4583 | 44.2364 | 55.8287 |
| 13 | 40.7974 | 44.6651 | 49.0452 | 54.0103 | 69.5533 |
| 14 | 48.5821 | 53.4978 | 59.1216 | 65.5641 | 86.1390 |
| 15 | 57.6011 | 63.7834 | 70.9270 | 79.1963 | 106.1356 |
| 16 | 68.0422 | 75.7474 | 84.7383 | 95.2530 | 130.1930 |
| 17 | 80.1215 | 89.6499 | 100.8749 | 114.1359 | 159.0796 |
| 18 | 94.0876 | 105.7902 | 119.7062 | 136.3107 | 193.7028 |
| 19 | 110.2266 | 124.5130 | 141.6582 | 162.3173 | 235.1336 |
| 20 | 128.8674 | 146.2156 | 167.2226 | 192.7807 | 284.6354 |
| 21 | 150.3886 | 171.3550 | 196.9669 | 228.4254 | 343.6973 |
| 22 | 175.2257 | 200.4579 | 231.5458 | 270.0894 | 414.0734 |
| 23 | 203.8795 | 234.1301 | 271.7142 | 318.7431 | 497.8292 |
| 24 | 236.9261 | 273.0694 | 318.3428 | 375.5089 | 597.3950 |
| 25 | 275.0283 | 318.0787 | 372.4354 | 441.6849 | 715.6294 |
| 26 | 318.9484 | 370.0824 | 435.1492 | 518.7724 | 855.8928 |
| 27 | 369.5631 | 430.1441 | 507.8179 | 608.5064 | 1022.1335 |
| 28 | 427.8810 | 499.4881 | 591.9787 | 712.8924 | 1218.9888 |
| 29 | 495.0618 | 579.5230 | 689.4026 | 834.2472 | 1451.9027 |
| 30 | 572.4398 | 671.8698 | 802.1302 | 975.2474 | 1727.2636 |
| 31 | 661.5491 | 778.3937 | 932.5124 | 1138.9839 | 2052.5649 |
| 32 | 764.1546 | 901.2409 | 1083.2569 | 1329.0258 | 2436.5932 |
| 33 | 882.2859 | 1042.8804 | 1257.4826 | 1549.4935 | 2889.6473 |
| 34 | 1018.2772 | 1206.1531 | 1458.7810 | 1805.1427 | 3423.7942 |
| 35 | 1174.8130 | 1394.3271 | 1691.2883 | 2101.4617 | 4053.1681 |
| 36 | 1354.9811 | 1611.1622 | 1959.7669 | 2444.7834 | 4794.3188 |
| 37 | 1562.3322 | 1860.9838 | 2269.7001 | 2842.4136 | 5666.6185 |
| 38 | 1800.9501 | 2148.7675 | 2627.4007 | 3302.7796 | 6692.7359 |
| 39 | 2075.5314 | 2480.2368 | 3040.1361 | 3835.6009 | 7899.1896 |
| 40 | 2391.4775 | 2861.9759 | 3516.2718 | 4452.0858 | 9316.9929 |
| 41 | 2755.0002 | 3301.5580 | 4065.4371 | 5165.1579 | 10982.4054 |
| 42 | 3173.2432 | 3807.6945 | 4698.7151 | 5989.7168 | 12937.8093 |
| 43 | 3654.4225 | 4390.4057 | 5428.8619 | 6942.9380 | 15232.7302 |
| 44 | 4207.9864 | 5061.2171 | 6270.5578 | 8044.6188 | 17925.0267 |
| 45 | 4844.8008 | 5833.3851 | 7240.6974 | 9317.5757 | 21082.2757 |
| 46 | 5577.3622 | 6722.1575 | 8358.7225 | 10788.1025 | 24783.3864 |
| 47 | 6420.0413 | 7745.0716 | 9647.0049 | 12486.4975 | 29120.4790 |
| 48 | 7389.3653 | 8922.2982 | 11131.2877 | 14447.6696 | 34201.0732 |
| 49 | 8504.3406 | 10277.0368 | 12841.1914 | 16711.8372 | 40150.6349 |
| 50 | 9786.8251 | 11835.9699 | 14810.7976 | 19325.3318 | 47115.5409 |

# TABLE A.30 Discrete Compounding: $i = 10\%$

| | Geometric series future worth factor $(F\backslash A_1\ i, j, n)$ | | | | |
|---|---|---|---|---|---|
| $n$ | $j = 4\%$ | $j = 6\%$ | $j = 8\%$ | $j = 10\%$ | $j = 15\%$ |
| 1 | 1.0000 | 1.0000 | 1.0000 | 1.0000 | 1.0000 |
| 2 | 2.1400 | 2.1600 | 2.1800 | 2.2000 | 2.2500 |
| 3 | 3.4356 | 3.4996 | 3.5644 | 3.6300 | 3.7975 |
| 4 | 4.9040 | 5.0406 | 5.1806 | 5.3240 | 5.6981 |
| 5 | 6.5643 | 6.8071 | 7.0591 | 7.3205 | 8.0169 |
| 6 | 8.4374 | 8.8260 | 9.2343 | 9.6631 | 10.8300 |
| 7 | 10.5464 | 11.1272 | 11.7446 | 12.4009 | 14.2261 |
| 8 | 12.9170 | 13.7435 | 14.6329 | 15.5897 | 18.3087 |
| 9 | 15.5773 | 16.7117 | 17.9471 | 19.2923 | 23.1986 |
| 10 | 18.5583 | 20.0724 | 21.7409 | 23.5795 | 29.0363 |
| 11 | 21.8944 | 23.8704 | 26.0739 | 28.5312 | 35.9855 |
| 12 | 25.6233 | 28.1558 | 31.0129 | 34.2374 | 44.2364 |
| 13 | 29.7866 | 32.9836 | 36.6324 | 40.7996 | 54.0103 |
| 14 | 34.4304 | 38.4149 | 43.0152 | 48.3318 | 65.5641 |
| 15 | 39.6051 | 44.5172 | 50.2539 | 56.9624 | 79.1962 |
| 16 | 45.3665 | 51.3655 | 58.4515 | 66.8359 | 95.2529 |
| 17 | 51.7761 | 59.0424 | 67.7226 | 78.1145 | 114.1358 |
| 18 | 58.9017 | 67.6394 | 78.1949 | 90.9804 | 136.3106 |
| 19 | 66.8176 | 77.2577 | 90.0104 | 105.6383 | 162.3171 |
| 20 | 75.6063 | 88.0091 | 103.3271 | 122.3181 | 192.7806 |
| 21 | 85.3580 | 100.0171 | 118.3207 | 141.2774 | 228.4252 |
| 22 | 96.1726 | 113.4183 | 135.1866 | 162.8053 | 270.0891 |
| 23 | 108.1597 | 128.3637 | 154.1418 | 187.2261 | 318.7427 |
| 24 | 121.4404 | 145.0198 | 175.4274 | 214.9030 | 375.5086 |
| 25 | 136.1477 | 163.5708 | 199.3114 | 246.2431 | 441.6846 |
| 26 | 152.4283 | 184.2197 | 226.0910 | 281.7019 | 518.7720 |
| 27 | 170.4436 | 207.1911 | 256.0962 | 321.7903 | 608.5059 |
| 28 | 190.3713 | 232.7325 | 289.6941 | 367.0794 | 712.6916 |
| 29 | 212.4072 | 261.1172 | 327.2905 | 418.2083 | 834.2466 |
| 30 | 236.7666 | 292.6475 | 369.3369 | 475.8921 | 975.2466 |
| 31 | 263.6865 | 327.6558 | 416.3333 | 540.9307 | 1138.9829 |
| 32 | 293.4282 | 366.5093 | 468.8342 | 614.2183 | 1329.0247 |
| 33 | 326.2793 | 409.6138 | 527.4546 | 696.7537 | 1549.4922 |
| 34 | 362.5554 | 457.4158 | 592.8762 | 789.6543 | 1805.1406 |
| 35 | 402.6055 | 510.4082 | 665.8538 | 894.1672 | 2101.4590 |
| 36 | 446.8120 | 569.1353 | 747.2246 | 1011.6863 | 2444.7808 |
| 37 | 495.5972 | 634.1958 | 837.9153 | 1143.7676 | 2842.4094 |
| 38 | 549.4248 | 706.2515 | 938.9524 | 1292.1482 | 3302.7759 |
| 39 | 608.8062 | 786.0310 | 1051.4729 | 1458.7673 | 3835.5967 |
| 40 | 674.3032 | 874.3377 | 1176.7354 | 1645.7888 | 4452.0781 |
| 41 | 746.5344 | 972.0569 | 1316.1333 | 1855.6255 | 5165.1484 |
| 42 | 826.1809 | 1080.1655 | 1471.2092 | 2090.9744 | 5989.7070 |
| 43 | 913.9920 | 1199.7393 | 1643.6697 | 2354.8350 | 6942.9258 |
| 44 | 1010.7915 | 1331.9634 | 1835.4031 | 2653.5584 | 8044.6094 |
| 45 | 1117.4871 | 1478.1453 | 2048.4990 | 2981.8782 | 9317.5625 |
| 46 | 1235.0772 | 1639.7244 | 2285.2680 | 3357.9568 | 10788.0859 |
| 47 | 1364.6597 | 1818.2864 | 2548.2703 | 3768.4319 | 12496.4005 |
| 48 | 1507.4434 | 2015.5806 | 2840.3296 | 4233.4688 | 14447.6524 |
| 49 | 1664.7581 | 2233.5327 | 3164.5728 | 4753.8359 | 16711.8125 |
| 50 | 1838.0671 | 2474.2639 | 3524.4575 | 5335.9375 | 19325.3047 |

# APPENDIX B
## CONTINUOUS COMPOUNDING

## SECTION I—CONTINUOUS COMPOUNDING INTEREST FACTORS

## SECTION II—CONTINUOUS COMPOUNDING, CONTINUOUS FLOW INTEREST FACTORS

## SECTION III—CONTINUOUS COMPOUNDING, GEOMETRIC SERIES FACTORS

# SECTION I—CONTINUOUS COMPOUNDING INTEREST FACTORS

## TABLE B.1 Continuous Compounding: $r = 1/2\%$

| | Single payment | | Uniform series | | | | Gradient series | |
|---|---|---|---|---|---|---|---|---|
| | Compound amount factor | Present worth factor | Compound amount factor | Sinking fund factor | Present worth factor | Capital recovery factor | Uniform series factor | Present worth factor |
| $n$ | To find $F$ given $P$ $(F\|P\,r,n)_\infty$ | To find $P$ given $F$ $(P\|F\,r,n)_\infty$ | To find $F$ given $A$ $(F\|A\,r,n)_\infty$ | To find $A$ given $F$ $(A\|F\,r,n)_\infty$ | To find $P$ given $A$ $(P\|A\,r,n)_\infty$ | To find $A$ given $P$ $(A\|P\,r,n)_\infty$ | To find $A$ given $G$ $(A\|G\,r,n)_\infty$ | To find $P$ given $G$ $(P\|G\,r,n)_\infty$ |
| 1 | 1.0050 | 0.9950 | 1.0000 | 1.0000 | 0.9950 | 1.0050 | 0.0000 | 0.0000 |
| 2 | 1.0100 | 0.9901 | 2.0049 | 0.4988 | 1.9850 | 0.5038 | 0.4988 | 0.9771 |
| 3 | 1.0151 | 0.9851 | 3.0150 | 0.3317 | 2.9702 | 0.3367 | 0.9967 | 2.9539 |
| 4 | 1.0202 | 0.9802 | 4.0301 | 0.2481 | 3.9503 | 0.2531 | 1.4938 | 5.8784 |
| 5 | 1.0253 | 0.9753 | 5.0502 | 0.1980 | 4.9255 | 0.2030 | 1.9900 | 9.7732 |
| 6 | 1.0305 | 0.9704 | 6.0755 | 0.1646 | 5.8960 | 0.1696 | 2.4854 | 14.6235 |
| 7 | 1.0356 | 0.9656 | 7.1060 | 0.1407 | 6.8616 | 0.1457 | 2.9800 | 20.4148 |
| 8 | 1.0408 | 0.9608 | 8.1417 | 0.1228 | 7.8225 | 0.1278 | 3.4738 | 27.1690 |
| 9 | 1.0460 | 0.9560 | 9.1825 | 0.1089 | 8.7784 | 0.1139 | 3.9667 | 34.7988 |
| 10 | 1.0513 | 0.9512 | 10.2285 | 0.0978 | 9.7297 | 0.1028 | 4.4588 | 43.3628 |
| 11 | 1.0565 | 0.9465 | 11.2799 | 0.0887 | 10.6762 | 0.0937 | 4.9500 | 52.8465 |
| 12 | 1.0618 | 0.9418 | 12.3364 | 0.0811 | 11.6180 | 0.0861 | 5.4404 | 63.1995 |
| 13 | 1.0672 | 0.9371 | 13.3982 | 0.0746 | 12.5550 | 0.0796 | 5.9300 | 74.4440 |
| 14 | 1.0725 | 0.9324 | 14.4654 | 0.0691 | 13.4874 | 0.0741 | 6.4188 | 86.5656 |
| 15 | 1.0779 | 0.9277 | 15.5379 | 0.0644 | 14.4152 | 0.0694 | 6.9067 | 99.5504 |
| 16 | 1.0833 | 0.9231 | 16.6157 | 0.0602 | 15.3382 | 0.0652 | 7.3938 | 113.3847 |
| 17 | 1.0887 | 0.9185 | 17.6990 | 0.0565 | 16.2568 | 0.0615 | 7.8800 | 128.0895 |
| 18 | 1.0942 | 0.9139 | 18.7877 | 0.0532 | 17.1707 | 0.0582 | 8.3654 | 143.6162 |
| 19 | 1.0997 | 0.9094 | 19.8818 | 0.0503 | 18.0800 | 0.0553 | 8.8500 | 159.9857 |
| 20 | 1.1052 | 0.9048 | 20.9815 | 0.0477 | 18.9849 | 0.0527 | 9.3338 | 177.1844 |
| 21 | 1.1107 | 0.9003 | 22.0867 | 0.0453 | 19.8853 | 0.0503 | 9.8167 | 195.1986 |
| 22 | 1.1163 | 0.8958 | 23.1975 | 0.0431 | 20.7811 | 0.0481 | 10.2988 | 214.0148 |
| 23 | 1.1219 | 0.8914 | 24.3137 | 0.0411 | 21.6725 | 0.0461 | 10.7800 | 233.6195 |
| 24 | 1.1275 | 0.8869 | 25.4357 | 0.0393 | 22.5594 | 0.0443 | 11.2605 | 254.0333 |
| 25 | 1.1331 | 0.8825 | 26.5632 | 0.0376 | 23.4419 | 0.0427 | 11.7401 | 275.2088 |
| 26 | 1.1388 | 0.8781 | 27.6963 | 0.0361 | 24.3200 | 0.0411 | 12.2188 | 297.1670 |
| 27 | 1.1445 | 0.8737 | 28.8350 | 0.0347 | 25.1936 | 0.0397 | 12.6968 | 319.8611 |
| 28 | 1.1503 | 0.8694 | 29.9796 | 0.0334 | 26.0631 | 0.0384 | 13.1739 | 343.3442 |
| 29 | 1.1560 | 0.8650 | 31.1299 | 0.0321 | 26.9281 | 0.0371 | 13.6501 | 367.5703 |
| 30 | 1.1618 | 0.8607 | 32.2859 | 0.0310 | 27.7888 | 0.0360 | 14.1256 | 392.5261 |
| 31 | 1.1677 | 0.8564 | 33.4479 | 0.0299 | 28.6453 | 0.0349 | 14.6002 | 418.2317 |
| 32 | 1.1735 | 0.8521 | 34.6155 | 0.0289 | 29.4974 | 0.0339 | 15.0739 | 444.6414 |
| 33 | 1.1794 | 0.8479 | 35.7890 | 0.0279 | 30.3453 | 0.0330 | 15.5469 | 471.7749 |
| 34 | 1.1853 | 0.8437 | 36.9684 | 0.0271 | 31.1889 | 0.0321 | 16.0190 | 499.6194 |
| 35 | 1.1912 | 0.8395 | 38.1537 | 0.0262 | 32.0284 | 0.0312 | 16.4903 | 528.1621 |
| 40 | 1.2214 | 0.8187 | 44.1699 | 0.0226 | 36.1633 | 0.0277 | 18.8342 | 681.1008 |
| 45 | 1.2523 | 0.7985 | 50.3385 | 0.0199 | 40.1961 | 0.0249 | 21.1574 | 850.4431 |
| 50 | 1.2840 | 0.7788 | 56.6632 | 0.0176 | 44.1294 | 0.0227 | 23.4598 | 1035.2771 |
| 55 | 1.3165 | 0.7596 | 63.1480 | 0.0158 | 47.9655 | 0.0208 | 25.7416 | 1234.7127 |
| 60 | 1.3499 | 0.7408 | 69.7970 | 0.0143 | 51.7069 | 0.0193 | 28.0027 | 1447.9358 |
| 65 | 1.3840 | 0.7225 | 76.6143 | 0.0131 | 55.3560 | 0.0181 | 30.2431 | 1674.1472 |
| 70 | 1.4191 | 0.7047 | 83.6042 | 0.0120 | 58.9149 | 0.0170 | 32.4629 | 1912.5662 |
| 75 | 1.4550 | 0.6873 | 90.7709 | 0.0110 | 62.3859 | 0.0160 | 34.6621 | 2162.4270 |
| 80 | 1.4918 | 0.6703 | 98.1193 | 0.0102 | 65.7713 | 0.0152 | 36.8408 | 2423.0816 |
| 85 | 1.5296 | 0.6538 | 105.6535 | 0.0095 | 69.0731 | 0.0145 | 38.9990 | 2693.7920 |
| 90 | 1.5683 | 0.6376 | 113.3786 | 0.0088 | 72.2934 | 0.0138 | 41.1368 | 2973.9312 |
| 95 | 1.6080 | 0.6219 | 121.2990 | 0.0082 | 75.4341 | 0.0133 | 43.2541 | 3262.8384 |
| 100 | 1.6487 | 0.6065 | 129.4202 | 0.0077 | 78.4974 | 0.0127 | 45.3510 | 3559.9534 |

**TABLE B.2   Continuous Compounding:** $r = 3/4\%$

| | Single payment | | Uniform series | | | | Gradient series | |
|---|---|---|---|---|---|---|---|---|
| | Compound amount factor | Present worth factor | Compound amount factor | Sinking fund factor | Present worth factor | Capital recovery factor | Uniform series factor | Present worth factor |
| $n$ | To find $F$ given $P$ $(F\|P\,r,n)_\infty$ | To find $P$ given $F$ $(P\|F\,r,n)_\infty$ | To find $F$ given $A$ $(F\|A\,r,n)_\infty$ | To find $A$ given $F$ $(A\|F\,r,n)_\infty$ | To find $P$ given $A$ $(P\|A\,r,n)_\infty$ | To find $A$ given $P$ $(A\|P\,r,n)_\infty$ | To find $A$ given $G$ $(A\|G\,r,n)_\infty$ | To find $P$ given $G$ $(P\|G\,r,n)_\infty$ |
| 1 | 1.0075 | 0.9925 | 1.0000 | 1.0000 | 0.9925 | 1.0075 | 0.0000 | 0.0000 |
| 2 | 1.0151 | 0.9851 | 2.0077 | 0.4981 | 1.9778 | 0.5056 | 0.4981 | 1.0114 |
| 3 | 1.0228 | 0.9778 | 3.0229 | 0.3308 | 2.9557 | 0.3383 | 0.9950 | 2.9787 |
| 4 | 1.0305 | 0.9704 | 4.0457 | 0.2472 | 3.9262 | 0.2547 | 1.4906 | 5.8965 |
| 5 | 1.0382 | 0.9632 | 5.0764 | 0.1970 | 4.8896 | 0.2045 | 1.9850 | 9.7757 |
| 6 | 1.0460 | 0.9560 | 6.1147 | 0.1635 | 5.8456 | 0.1711 | 2.4781 | 14.5620 |
| 7 | 1.0539 | 0.9489 | 7.1608 | 0.1396 | 6.7945 | 0.1472 | 2.9700 | 20.2664 |
| 8 | 1.0618 | 0.9418 | 8.2149 | 0.1217 | 7.7365 | 0.1293 | 3.4606 | 26.8834 |
| 9 | 1.0698 | 0.9347 | 9.2768 | 0.1078 | 8.6713 | 0.1153 | 3.9500 | 34.3758 |
| 10 | 1.0779 | 0.9277 | 10.3468 | 0.0966 | 9.5991 | 0.1042 | 4.4381 | 42.7384 |
| 11 | 1.0860 | 0.9208 | 11.4248 | 0.0875 | 10.5201 | 0.0951 | 4.9250 | 51.9661 |
| 12 | 1.0942 | 0.9139 | 12.5108 | 0.0799 | 11.4341 | 0.0875 | 5.4106 | 62.0226 |
| 13 | 1.1024 | 0.9071 | 13.6052 | 0.0735 | 12.3413 | 0.0810 | 5.8950 | 72.9335 |
| 14 | 1.1107 | 0.9003 | 14.7077 | 0.0680 | 13.2417 | 0.0755 | 6.3781 | 84.6479 |
| 15 | 1.1191 | 0.8936 | 15.8186 | 0.0632 | 14.1354 | 0.0707 | 6.8600 | 97.1757 |
| 16 | 1.1275 | 0.8869 | 16.9378 | 0.0590 | 15.0225 | 0.0666 | 7.3407 | 110.4967 |
| 17 | 1.1360 | 0.8803 | 18.0654 | 0.0554 | 15.9028 | 0.0629 | 7.8200 | 124.5913 |
| 18 | 1.1445 | 0.8737 | 19.2014 | 0.0521 | 16.7766 | 0.0596 | 8.2982 | 139.4543 |
| 19 | 1.1532 | 0.8672 | 20.3461 | 0.0491 | 17.6439 | 0.0567 | 8.7751 | 155.0807 |
| 20 | 1.1618 | 0.8607 | 21.4994 | 0.0465 | 18.5047 | 0.0540 | 9.2507 | 171.4510 |
| 21 | 1.1706 | 0.8543 | 22.6613 | 0.0441 | 19.3591 | 0.0517 | 9.7251 | 188.5457 |
| 22 | 1.1794 | 0.8479 | 23.8321 | 0.0420 | 20.2071 | 0.0495 | 10.1983 | 206.3743 |
| 23 | 1.1883 | 0.8416 | 25.0116 | 0.0400 | 21.0488 | 0.0475 | 10.6702 | 224.9033 |
| 24 | 1.1972 | 0.8353 | 26.2000 | 0.0382 | 21.8841 | 0.0457 | 11.1408 | 244.1280 |
| 25 | 1.2062 | 0.8290 | 27.3970 | 0.0365 | 22.7132 | 0.0440 | 11.6102 | 264.0296 |
| 26 | 1.2153 | 0.8228 | 28.6037 | 0.0350 | 23.5361 | 0.0425 | 12.0784 | 284.6172 |
| 27 | 1.2245 | 0.8167 | 29.8192 | 0.0335 | 24.3530 | 0.0411 | 12.5453 | 305.8721 |
| 28 | 1.2337 | 0.8106 | 31.0438 | 0.0322 | 25.1636 | 0.0397 | 13.0110 | 327.7759 |
| 29 | 1.2430 | 0.8045 | 32.2776 | 0.0310 | 25.9682 | 0.0385 | 13.4754 | 350.3103 |
| 30 | 1.2523 | 0.7985 | 33.5207 | 0.0298 | 26.7668 | 0.0374 | 13.9386 | 373.4841 |
| 31 | 1.2618 | 0.7925 | 34.7732 | 0.0288 | 27.5595 | 0.0363 | 14.4005 | 397.2793 |
| 32 | 1.2712 | 0.7866 | 36.0351 | 0.0278 | 28.3462 | 0.0353 | 14.8612 | 421.6777 |
| 33 | 1.2808 | 0.7808 | 37.3065 | 0.0268 | 29.1270 | 0.0343 | 15.3207 | 446.6751 |
| 34 | 1.2905 | 0.7749 | 38.5873 | 0.0259 | 29.9020 | 0.0334 | 15.7789 | 472.2537 |
| 35 | 1.3002 | 0.7691 | 39.8780 | 0.0251 | 30.6712 | 0.0326 | 16.2359 | 498.4214 |
| 40 | 1.3499 | 0.7408 | 46.4783 | 0.0215 | 34.4320 | 0.0290 | 18.5021 | 637.5711 |
| 45 | 1.4014 | 0.7136 | 53.3307 | 0.0188 | 38.0542 | 0.0263 | 20.7374 | 789.7029 |
| 50 | 1.4550 | 0.6873 | 60.4449 | 0.0165 | 41.5432 | 0.0241 | 22.9418 | 953.6831 |
| 55 | 1.5106 | 0.6620 | 67.8311 | 0.0147 | 44.9038 | 0.0223 | 25.1153 | 1128.4336 |
| 60 | 1.5683 | 0.6376 | 75.4995 | 0.0132 | 48.1407 | 0.0208 | 27.2582 | 1312.9395 |
| 65 | 1.6282 | 0.6142 | 83.4609 | 0.0120 | 51.2584 | 0.0195 | 29.3704 | 1506.2349 |
| 70 | 1.6905 | 0.5916 | 91.7266 | 0.0109 | 54.2614 | 0.0184 | 31.4521 | 1707.4382 |
| 75 | 1.7551 | 0.5698 | 100.3080 | 0.0100 | 57.1538 | 0.0175 | 33.5034 | 1915.6895 |
| 80 | 1.8221 | 0.5488 | 109.2174 | 0.0092 | 59.9398 | 0.0167 | 35.5244 | 2130.2127 |
| 85 | 1.8917 | 0.5286 | 118.4672 | 0.0084 | 62.6233 | 0.0160 | 37.5153 | 2350.2559 |
| 90 | 1.9640 | 0.5092 | 128.0705 | 0.0078 | 65.2080 | 0.0153 | 39.4762 | 2575.1262 |
| 95 | 2.0391 | 0.4904 | 138.0407 | 0.0072 | 67.6975 | 0.0148 | 41.4072 | 2804.1648 |
| 100 | 2.1170 | 0.4724 | 148.3919 | 0.0067 | 70.0955 | 0.0143 | 43.3084 | 3036.7615 |

## TABLE B.3  Continuous Compounding: $r = 1\%$

| | Single payment | | Uniform series | | | | Gradient series | |
|---|---|---|---|---|---|---|---|---|
| | Compound amount factor | Present worth factor | Compound amount factor | Sinking fund factor | Present worth factor | Capital recovery factor | Uniform series factor | Present worth factor |
| $n$ | To find $F$ given $P$ $F\|P\,r,n$ | To find $P$ given $F$ $P\|F\,r,n$ | To find $F$ given $A$ $F\|A\,r,n$ | To find $A$ given $F$ $A\|F\,r,n$ | To find $P$ given $A$ $P\|A\,r,n$ | To find $A$ given $P$ $A\|P\,r,n$ | To find $A$ given $G$ $A\|G\,r,n$ | To find $P$ given $G$ $P\|G\,r,n$ |
| 1 | 1.0101 | 0.9900 | 1.0000 | 1.0000 | 0.9900 | 1.0101 | 0.0000 | 0.0000 |
| 2 | 1.0202 | 0.9802 | 2.0101 | 0.4975 | 1.9702 | 0.5076 | 0.4975 | 0.9802 |
| 3 | 1.0305 | 0.9704 | 3.0303 | 0.3300 | 2.9407 | 0.3401 | 0.9933 | 2.9211 |
| 4 | 1.0408 | 0.9608 | 4.0607 | 0.2463 | 3.9015 | 0.2563 | 1.4875 | 5.8035 |
| 5 | 1.0513 | 0.9512 | 5.1015 | 0.1960 | 4.8527 | 0.2061 | 1.9800 | 9.6084 |
| 6 | 1.0618 | 0.9418 | 6.1528 | 0.1625 | 5.7945 | 0.1726 | 2.4708 | 14.3172 |
| 7 | 1.0725 | 0.9324 | 7.2146 | 0.1386 | 6.7269 | 0.1487 | 2.9600 | 19.9116 |
| 8 | 1.0833 | 0.9231 | 8.2871 | 0.1207 | 7.6500 | 0.1307 | 3.4475 | 26.3734 |
| 9 | 1.0942 | 0.9139 | 9.3704 | 0.1067 | 8.5639 | 0.1168 | 3.9333 | 33.6848 |
| 10 | 1.1052 | 0.9048 | 10.4646 | 0.0956 | 9.4688 | 0.1056 | 4.4175 | 41.8284 |
| 11 | 1.1163 | 0.8958 | 11.5698 | 0.0864 | 10.3646 | 0.0965 | 4.9000 | 50.7867 |
| 12 | 1.1275 | 0.8869 | 12.6860 | 0.0788 | 11.2515 | 0.0889 | 5.3809 | 60.5428 |
| 13 | 1.1388 | 0.8781 | 13.8135 | 0.0724 | 12.1296 | 0.0824 | 5.8600 | 71.0800 |
| 14 | 1.1503 | 0.8694 | 14.9524 | 0.0669 | 12.9990 | 0.0769 | 6.3376 | 82.3816 |
| 15 | 1.1618 | 0.8607 | 16.1026 | 0.0621 | 13.8597 | 0.0722 | 6.8134 | 94.4315 |
| 16 | 1.1735 | 0.8521 | 17.2645 | 0.0579 | 14.7118 | 0.0680 | 7.2876 | 107.2137 |
| 17 | 1.1853 | 0.8437 | 18.4380 | 0.0542 | 15.5555 | 0.0643 | 7.7601 | 120.7123 |
| 18 | 1.1972 | 0.8353 | 19.6233 | 0.0510 | 16.3908 | 0.0610 | 8.2310 | 134.9119 |
| 19 | 1.2092 | 0.8270 | 20.8205 | 0.0480 | 17.2177 | 0.0581 | 8.7002 | 149.7972 |
| 20 | 1.2214 | 0.8187 | 22.0298 | 0.0454 | 18.0364 | 0.0554 | 9.1677 | 165.3531 |
| 21 | 1.2337 | 0.8106 | 23.2512 | 0.0430 | 18.8470 | 0.0531 | 9.6336 | 181.5648 |
| 22 | 1.2461 | 0.8025 | 24.4848 | 0.0408 | 19.6495 | 0.0509 | 10.0978 | 198.4177 |
| 23 | 1.2586 | 0.7945 | 25.7309 | 0.0389 | 20.4441 | 0.0489 | 10.5604 | 215.8974 |
| 24 | 1.2712 | 0.7866 | 26.9895 | 0.0371 | 21.2307 | 0.0471 | 11.0213 | 233.9898 |
| 25 | 1.2840 | 0.7788 | 28.2608 | 0.0354 | 22.0095 | 0.0454 | 11.4805 | 252.6811 |
| 26 | 1.2969 | 0.7711 | 29.5448 | 0.0338 | 22.7806 | 0.0439 | 11.9381 | 271.9573 |
| 27 | 1.3100 | 0.7634 | 30.8417 | 0.0324 | 23.5439 | 0.0425 | 12.3941 | 291.8052 |
| 28 | 1.3231 | 0.7558 | 32.1517 | 0.0311 | 24.2997 | 0.0412 | 12.8484 | 312.2114 |
| 29 | 1.3364 | 0.7483 | 33.4748 | 0.0299 | 25.0480 | 0.0399 | 13.3010 | 333.1628 |
| 30 | 1.3499 | 0.7408 | 34.8112 | 0.0287 | 25.7888 | 0.0388 | 13.7520 | 354.6465 |
| 31 | 1.3634 | 0.7334 | 36.1611 | 0.0277 | 26.5222 | 0.0377 | 14.2013 | 376.6499 |
| 32 | 1.3771 | 0.7261 | 37.5245 | 0.0266 | 27.2484 | 0.0367 | 14.6490 | 399.1605 |
| 33 | 1.3910 | 0.7189 | 38.9017 | 0.0257 | 27.9673 | 0.0358 | 15.0950 | 422.1661 |
| 34 | 1.4049 | 0.7118 | 40.2926 | 0.0248 | 28.6791 | 0.0349 | 15.5394 | 445.6545 |
| 35 | 1.4191 | 0.7047 | 41.6976 | 0.0240 | 29.3838 | 0.0340 | 15.9821 | 469.6139 |
| 40 | 1.4918 | 0.6703 | 48.9370 | 0.0204 | 32.8034 | 0.0305 | 18.1710 | 596.0725 |
| 45 | 1.5683 | 0.6376 | 56.5475 | 0.0177 | 36.0563 | 0.0277 | 20.3190 | 732.6280 |
| 50 | 1.6487 | 0.6065 | 64.5483 | 0.0155 | 39.1505 | 0.0255 | 22.4261 | 877.9948 |
| 55 | 1.7333 | 0.5769 | 72.9593 | 0.0137 | 42.0938 | 0.0238 | 24.4926 | 1030.9885 |
| 60 | 1.8221 | 0.5488 | 81.8015 | 0.0122 | 44.8936 | 0.0223 | 26.5187 | 1190.5195 |
| 65 | 1.9155 | 0.5220 | 91.0971 | 0.0110 | 47.5568 | 0.0210 | 28.5045 | 1355.5862 |
| 70 | 2.0138 | 0.4966 | 100.8692 | 0.0099 | 50.0902 | 0.0200 | 30.4505 | 1525.2692 |
| 75 | 2.1170 | 0.4724 | 111.1424 | 0.0090 | 52.5000 | 0.0190 | 32.3567 | 1698.7256 |
| 80 | 2.2255 | 0.4493 | 121.9423 | 0.0082 | 54.7922 | 0.0183 | 34.2235 | 1875.1837 |
| 85 | 2.3396 | 0.4274 | 133.2960 | 0.0075 | 56.9727 | 0.0176 | 36.0513 | 2053.9382 |
| 90 | 2.4596 | 0.4066 | 145.2317 | 0.0069 | 59.0468 | 0.0169 | 37.8402 | 2234.3453 |
| 95 | 2.5857 | 0.3867 | 157.7794 | 0.0063 | 61.0198 | 0.0164 | 39.5907 | 2415.8187 |
| 100 | 2.7183 | 0.3679 | 170.9705 | 0.0058 | 62.8965 | 0.0159 | 41.3032 | 2597.8253 |

| | Single payment | | Uniform series | | | | Gradient series | |
|---|---|---|---|---|---|---|---|---|
| | Compound amount factor | Present worth factor | Compound amount factor | Sinking fund factor | Present worth factor | Capital recovery factor | Uniform series factor | Present worth factor |
| $n$ | To find $F$ given $P$ $(F|Pr,n)_\infty$ | To find $P$ given $F$ $(P|Fr,n)_\infty$ | To find $F$ given $A$ $(F|Ar,n)_\infty$ | To find $A$ given $F$ $(A|Fr,n)_\infty$ | To find $P$ given $A$ $(P|Ar,n)_\infty$ | To find $A$ given $P$ $(A|Pr,n)_\infty$ | To find $A$ given $G$ $(A|Gr,n)_\infty$ | To find $P$ given $G$ $(P|Gr,n)_\infty$ |
| 1 | 1.0151 | 0.9851 | 1.0000 | 1.0000 | 0.9851 | 1.0151 | 0.0000 | 0.0000 |
| 2 | 1.0305 | 0.9704 | 2.0151 | 0.4963 | 1.9555 | 0.5114 | 0.4963 | 0.9684 |
| 3 | 1.0460 | 0.9560 | 3.0456 | 0.3283 | 2.9116 | 0.3435 | 0.9900 | 2.8820 |
| 4 | 1.0618 | 0.9418 | 4.0916 | 0.2444 | 3.8534 | 0.2595 | 1.4813 | 5.7097 |
| 5 | 1.0779 | 0.9277 | 5.1535 | 0.1940 | 4.7811 | 0.2092 | 1.9700 | 9.4210 |
| 6 | 1.0942 | 0.9139 | 6.2313 | 0.1605 | 5.6950 | 0.1756 | 2.4563 | 13.9898 |
| 7 | 1.1107 | 0.9003 | 7.3256 | 0.1365 | 6.5954 | 0.1516 | 2.9400 | 19.3941 |
| 8 | 1.1275 | 0.8869 | 8.4363 | 0.1185 | 7.4823 | 0.1336 | 3.4213 | 25.6047 |
| 9 | 1.1445 | 0.8737 | 9.5638 | 0.1046 | 8.3560 | 0.1197 | 3.9000 | 32.5928 |
| 10 | 1.1618 | 0.8607 | 10.7083 | 0.0934 | 9.2168 | 0.1085 | 4.3763 | 40.3412 |
| 11 | 1.1794 | 0.8479 | 11.8702 | 0.0842 | 10.0647 | 0.0994 | 4.8501 | 48.8216 |
| 12 | 1.1972 | 0.8353 | 13.0496 | 0.0766 | 10.8999 | 0.0917 | 5.3213 | 58.0101 |
| 13 | 1.2153 | 0.8228 | 14.2468 | 0.0702 | 11.7228 | 0.0853 | 5.7901 | 67.8831 |
| 14 | 1.2337 | 0.8106 | 15.4622 | 0.0647 | 12.5334 | 0.0798 | 6.2564 | 78.4240 |
| 15 | 1.2523 | 0.7985 | 16.6958 | 0.0599 | 13.3319 | 0.0750 | 6.7202 | 89.6029 |
| 16 | 1.2712 | 0.7866 | 17.9482 | 0.0557 | 14.1186 | 0.0708 | 7.1816 | 101.4038 |
| 17 | 1.2905 | 0.7749 | 19.2195 | 0.0520 | 14.8935 | 0.0671 | 7.6404 | 113.8041 |
| 18 | 1.3100 | 0.7634 | 20.5099 | 0.0488 | 15.6569 | 0.0639 | 8.0967 | 126.7817 |
| 19 | 1.3298 | 0.7520 | 21.8199 | 0.0458 | 16.4089 | 0.0609 | 8.5506 | 140.3179 |
| 20 | 1.3499 | 0.7408 | 23.1497 | 0.0432 | 17.1497 | 0.0583 | 9.0020 | 154.3943 |
| 21 | 1.3703 | 0.7298 | 24.4995 | 0.0408 | 17.8795 | 0.0559 | 9.4509 | 168.9895 |
| 22 | 1.3910 | 0.7189 | 25.8698 | 0.0387 | 18.5984 | 0.0538 | 9.8973 | 184.0884 |
| 23 | 1.4120 | 0.7082 | 27.2608 | 0.0367 | 19.3067 | 0.0518 | 10.3413 | 199.6702 |
| 24 | 1.4333 | 0.6977 | 28.6728 | 0.0349 | 20.0043 | 0.0500 | 10.7828 | 215.7173 |
| 25 | 1.4550 | 0.6873 | 30.1061 | 0.0332 | 20.6916 | 0.0483 | 11.2218 | 232.2125 |
| 26 | 1.4770 | 0.6771 | 31.5612 | 0.0317 | 21.3687 | 0.0468 | 11.6584 | 249.1408 |
| 27 | 1.4993 | 0.6670 | 33.0382 | 0.0303 | 22.0357 | 0.0454 | 12.0925 | 266.4829 |
| 28 | 1.5220 | 0.6570 | 34.5375 | 0.0290 | 22.6928 | 0.0441 | 12.5241 | 284.2246 |
| 29 | 1.5450 | 0.6473 | 36.0595 | 0.0277 | 23.3400 | 0.0428 | 12.9533 | 302.3491 |
| 30 | 1.5683 | 0.6376 | 37.6045 | 0.0266 | 23.9777 | 0.0417 | 13.3800 | 320.8403 |
| 31 | 1.5920 | 0.6281 | 39.1728 | 0.0255 | 24.6058 | 0.0406 | 13.8043 | 339.6843 |
| 32 | 1.6161 | 0.6188 | 40.7648 | 0.0245 | 25.2246 | 0.0396 | 14.2261 | 358.8677 |
| 33 | 1.6405 | 0.6096 | 42.3809 | 0.0236 | 25.8342 | 0.0387 | 14.6455 | 378.3740 |
| 34 | 1.6653 | 0.6005 | 44.0215 | 0.0227 | 26.4347 | 0.0378 | 15.0625 | 398.1926 |
| 35 | 1.6905 | 0.5916 | 45.6868 | 0.0219 | 27.0262 | 0.0370 | 15.4770 | 418.3059 |
| 40 | 1.8221 | 0.5488 | 54.3985 | 0.0184 | 29.8545 | 0.0335 | 17.5131 | 522.8696 |
| 45 | 1.9640 | 0.5092 | 63.7888 | 0.0157 | 32.4785 | 0.0308 | 19.4890 | 632.9990 |
| 50 | 2.1170 | 0.4724 | 73.9104 | 0.0135 | 34.9128 | 0.0286 | 21.4052 | 747.3428 |
| 55 | 2.2819 | 0.4382 | 84.8203 | 0.0118 | 37.1713 | 0.0269 | 23.2622 | 864.7144 |
| 60 | 2.4596 | 0.4066 | 96.5800 | 0.0104 | 39.2665 | 0.0255 | 25.0609 | 984.0840 |
| 65 | 2.6512 | 0.3772 | 109.2555 | 0.0092 | 41.2104 | 0.0243 | 26.8018 | 1104.5462 |
| 70 | 2.8576 | 0.3499 | 122.9183 | 0.0081 | 43.0138 | 0.0232 | 28.4859 | 1225.3215 |
| 75 | 3.0802 | 0.3247 | 137.6452 | 0.0073 | 44.6869 | 0.0224 | 30.1140 | 1345.7358 |
| 80 | 3.3201 | 0.3012 | 153.5190 | 0.0065 | 46.2391 | 0.0216 | 31.6869 | 1465.2097 |
| 85 | 3.5787 | 0.2794 | 170.6293 | 0.0059 | 47.6791 | 0.0210 | 33.2056 | 1583.2522 |
| 90 | 3.8574 | 0.2592 | 189.0721 | 0.0053 | 49.0151 | 0.0204 | 34.6710 | 1699.4431 |
| 95 | 4.1579 | 0.2405 | 208.9513 | 0.0048 | 50.2546 | 0.0199 | 36.0842 | 1813.4382 |
| 100 | 4.4817 | 0.2231 | 230.3787 | 0.0043 | 51.4045 | 0.0195 | 37.4462 | 1924.9439 |

| | Single payment | | Uniform series | | | | Gradient series | |
|---|---|---|---|---|---|---|---|---|
| | Compound amount factor | Present worth factor | Compound amount factor | Sinking fund factor | Present worth factor | Capital recovery factor | Uniform series factor | Present worth factor |
| $n$ | To find $F$ given $P$ $F\|P\ r,n$ | To find $P$ given $F$ $P\|F\ r,n$ | To find $F$ given $A$ $F\|A\ r,n$ | To find $A$ given $F$ $A\|F\ r,n$ | To find $P$ given $A$ $P\|A\ r,n$ | To find $A$ given $P$ $A\|P\ r,n$ | To find $A$ given $G$ $A\|G\ r,n$ | To find $P$ given $G$ $P\|G\ r,n$ |
| 1 | 1.0202 | 0.9802 | 1.0000 | 1.0000 | 0.9802 | 1.0202 | 0.0000 | 0.0000 |
| 2 | 1.0408 | 0.9608 | 2.0202 | 0.4950 | 1.9410 | 0.5152 | 0.4950 | 0.9608 |
| 3 | 1.0618 | 0.9418 | 3.0610 | 0.3267 | 2.8828 | 0.3469 | 0.9867 | 2.8443 |
| 4 | 1.0833 | 0.9231 | 4.1228 | 0.2426 | 3.8059 | 0.2628 | 1.4750 | 5.6137 |
| 5 | 1.1052 | 0.9048 | 5.2061 | 0.1921 | 4.7107 | 0.2123 | 1.9600 | 9.2330 |
| 6 | 1.1275 | 0.8869 | 6.3113 | 0.1584 | 5.5976 | 0.1786 | 2.4417 | 13.6676 |
| 7 | 1.1503 | 0.8694 | 7.4388 | 0.1344 | 6.4670 | 0.1546 | 2.9200 | 18.8838 |
| 8 | 1.1735 | 0.8521 | 8.5891 | 0.1164 | 7.3191 | 0.1366 | 3.3950 | 24.8488 |
| 9 | 1.1972 | 0.8353 | 9.7626 | 0.1024 | 8.1544 | 0.1226 | 3.8667 | 31.5309 |
| 10 | 1.2214 | 0.8187 | 10.9598 | 0.0912 | 8.9731 | 0.1114 | 4.3351 | 38.8995 |
| 11 | 1.2461 | 0.8025 | 12.1812 | 0.0821 | 9.7756 | 0.1023 | 4.8002 | 46.9247 |
| 12 | 1.2712 | 0.7866 | 13.4273 | 0.0745 | 10.5623 | 0.0947 | 5.2619 | 55.5776 |
| 13 | 1.2969 | 0.7711 | 14.6985 | 0.0680 | 11.3333 | 0.0882 | 5.7203 | 64.8302 |
| 14 | 1.3231 | 0.7558 | 15.9955 | 0.0625 | 12.0891 | 0.0827 | 6.1754 | 74.6554 |
| 15 | 1.3499 | 0.7408 | 17.3186 | 0.0577 | 12.8299 | 0.0779 | 6.6272 | 85.0269 |
| 16 | 1.3771 | 0.7261 | 18.6685 | 0.0536 | 13.5561 | 0.0738 | 7.0757 | 95.9191 |
| 17 | 1.4049 | 0.7118 | 20.0456 | 0.0499 | 14.2678 | 0.0701 | 7.5209 | 107.3074 |
| 18 | 1.4333 | 0.6977 | 21.4505 | 0.0466 | 14.9655 | 0.0668 | 7.9628 | 119.1679 |
| 19 | 1.4623 | 0.6839 | 22.8839 | 0.0437 | 15.6494 | 0.0639 | 8.4014 | 131.4774 |
| 20 | 1.4918 | 0.6703 | 24.3461 | 0.0411 | 16.3197 | 0.0613 | 8.8368 | 144.2135 |
| 21 | 1.5220 | 0.6570 | 25.8380 | 0.0387 | 16.9768 | 0.0589 | 9.2688 | 157.3545 |
| 22 | 1.5527 | 0.6440 | 27.3599 | 0.0365 | 17.6208 | 0.0568 | 9.6976 | 170.8792 |
| 23 | 1.5841 | 0.6313 | 28.9126 | 0.0346 | 18.2521 | 0.0548 | 10.1231 | 184.7675 |
| 24 | 1.6161 | 0.6188 | 30.4967 | 0.0328 | 18.8709 | 0.0530 | 10.5453 | 198.9995 |
| 25 | 1.6487 | 0.6065 | 32.1128 | 0.0311 | 19.4774 | 0.0513 | 10.9643 | 213.5562 |
| 26 | 1.6820 | 0.5945 | 33.7615 | 0.0296 | 20.0719 | 0.0498 | 11.3800 | 228.4192 |
| 27 | 1.7160 | 0.5827 | 35.4435 | 0.0282 | 20.6547 | 0.0484 | 11.7925 | 243.5707 |
| 28 | 1.7507 | 0.5712 | 37.1595 | 0.0269 | 21.2259 | 0.0471 | 12.2018 | 258.9933 |
| 29 | 1.7860 | 0.5599 | 38.9102 | 0.0257 | 21.7858 | 0.0459 | 12.6078 | 274.6705 |
| 30 | 1.8221 | 0.5488 | 40.6963 | 0.0246 | 22.3346 | 0.0448 | 13.0106 | 290.5860 |
| 31 | 1.8589 | 0.5379 | 42.5184 | 0.0235 | 22.8725 | 0.0437 | 13.4102 | 306.7243 |
| 32 | 1.8965 | 0.5273 | 44.3773 | 0.0225 | 23.3998 | 0.0427 | 13.8065 | 323.0704 |
| 33 | 1.9348 | 0.5169 | 46.2738 | 0.0216 | 23.9167 | 0.0418 | 14.1997 | 339.6097 |
| 34 | 1.9739 | 0.5066 | 48.2086 | 0.0207 | 24.4233 | 0.0409 | 14.5897 | 356.3280 |
| 35 | 2.0138 | 0.4966 | 50.1824 | 0.0199 | 24.9199 | 0.0401 | 14.9765 | 373.2119 |
| 40 | 2.2255 | 0.4493 | 60.6663 | 0.0165 | 27.2591 | 0.0367 | 16.8630 | 459.6713 |
| 45 | 2.4596 | 0.4066 | 72.2528 | 0.0138 | 29.3758 | 0.0340 | 18.6714 | 548.4862 |
| 50 | 2.7183 | 0.3679 | 85.0578 | 0.0118 | 31.2910 | 0.0320 | 20.4028 | 638.4254 |
| 55 | 3.0042 | 0.3329 | 99.2096 | 0.0101 | 33.0240 | 0.0303 | 22.0588 | 728.4707 |
| 60 | 3.3201 | 0.3012 | 114.8497 | 0.0087 | 34.5921 | 0.0289 | 23.6409 | 817.7873 |
| 65 | 3.6693 | 0.2725 | 132.1346 | 0.0076 | 36.0109 | 0.0278 | 25.1507 | 905.6984 |
| 70 | 4.0552 | 0.2466 | 151.2375 | 0.0066 | 37.2947 | 0.0268 | 26.5899 | 991.6629 |
| 75 | 4.4817 | 0.2231 | 172.3494 | 0.0058 | 38.4564 | 0.0260 | 27.9604 | 1075.2549 |
| 80 | 4.9530 | 0.2019 | 195.6817 | 0.0051 | 39.5075 | 0.0253 | 29.2640 | 1156.1476 |
| 85 | 5.4739 | 0.1827 | 221.4679 | 0.0045 | 40.4585 | 0.0247 | 30.5028 | 1234.0977 |
| 90 | 6.0496 | 0.1653 | 249.9660 | 0.0040 | 41.3191 | 0.0242 | 31.6786 | 1308.9328 |
| 95 | 6.6859 | 0.1496 | 281.4613 | 0.0036 | 42.0978 | 0.0238 | 32.7937 | 1380.5397 |
| 100 | 7.3891 | 0.1353 | 316.2689 | 0.0032 | 42.8023 | 0.0234 | 33.8499 | 1448.8552 |

| | Single payment | | Uniform series | | | | Gradient series | |
|---|---|---|---|---|---|---|---|---|
| | Compound amount factor | Present worth factor | Compound amount factor | Sinking fund factor | Present worth factor | Capital recovery factor | Uniform series factor | Present worth factor |
| $n$ | To find $F$ given $P$ $F\|P\ r,n$ | To find $P$ given $F$ $P\|F\ r,n$ | To find $F$ given $A$ $F\|A\ r,n$ | To find $A$ given $F$ $A\|F\ r,n$ | To find $P$ given $A$ $P\|A\ r,n$ | To find $A$ given $P$ $A\|P\ r,n$ | To find $A$ given $G$ $A\|G\ r,n$ | To find $P$ given $G$ $P\|G\ r,n$ |
| 1 | 1.0305 | 0.9704 | 1.0000 | 1.0000 | 0.9704 | 1.0305 | 0.0000 | 0.0000 |
| 2 | 1.0618 | 0.9418 | 2.0305 | 0.4925 | 1.9122 | 0.5230 | 0.4925 | 0.9418 |
| 3 | 1.0942 | 0.9139 | 3.0923 | 0.3234 | 2.8261 | 0.3538 | 0.9800 | 2.7696 |
| 4 | 1.1275 | 0.8869 | 4.1865 | 0.2389 | 3.7131 | 0.2693 | 1.4625 | 5.4304 |
| 5 | 1.1618 | 0.8607 | 5.3140 | 0.1882 | 4.5738 | 0.2186 | 1.9400 | 8.8732 |
| 6 | 1.1972 | 0.8353 | 6.4758 | 0.1544 | 5.4090 | 0.1840 | 2.4125 | 13.0496 |
| 7 | 1.2337 | 0.8106 | 7.6730 | 0.1303 | 6.2196 | 0.1608 | 2.8801 | 17.9131 |
| 8 | 1.2712 | 0.7866 | 8.9067 | 0.1123 | 7.0063 | 0.1427 | 3.3427 | 23.4195 |
| 9 | 1.3100 | 0.7634 | 10.1779 | 0.0983 | 7.7696 | 0.1287 | 3.8002 | 29.5265 |
| 10 | 1.3499 | 0.7408 | 11.4879 | 0.0870 | 8.5104 | 0.1175 | 4.2529 | 36.1939 |
| 11 | 1.3910 | 0.7189 | 12.8378 | 0.0779 | 9.2294 | 0.1083 | 4.7005 | 43.3831 |
| 12 | 1.4333 | 0.6977 | 14.2287 | 0.0703 | 9.9270 | 0.1007 | 5.1433 | 51.0575 |
| 13 | 1.4770 | 0.6771 | 15.6621 | 0.0638 | 10.6041 | 0.0943 | 5.5811 | 59.1822 |
| 14 | 1.5220 | 0.6570 | 17.1390 | 0.0583 | 11.2612 | 0.0888 | 6.0139 | 67.7238 |
| 15 | 1.5683 | 0.6376 | 18.6610 | 0.0536 | 11.8988 | 0.0840 | 6.4419 | 76.6506 |
| 16 | 1.6161 | 0.6188 | 20.2293 | 0.0494 | 12.5176 | 0.0799 | 6.8649 | 85.9324 |
| 17 | 1.6653 | 0.6005 | 21.8454 | 0.0458 | 13.1181 | 0.0762 | 7.2831 | 95.5403 |
| 18 | 1.7160 | 0.5827 | 23.5107 | 0.0425 | 13.7008 | 0.0730 | 7.6964 | 105.4470 |
| 19 | 1.7683 | 0.5655 | 25.2267 | 0.0396 | 14.2663 | 0.0701 | 8.1048 | 115.6265 |
| 20 | 1.8221 | 0.5488 | 26.9950 | 0.0370 | 14.8151 | 0.0675 | 8.5084 | 126.0539 |
| 21 | 1.8776 | 0.5326 | 28.8171 | 0.0347 | 15.3477 | 0.0652 | 8.9072 | 136.7057 |
| 22 | 1.9348 | 0.5169 | 30.6947 | 0.0326 | 15.8646 | 0.0630 | 9.3012 | 147.5596 |
| 23 | 1.9937 | 0.5016 | 32.6295 | 0.0306 | 16.3662 | 0.0611 | 9.6904 | 158.5943 |
| 24 | 2.0544 | 0.4868 | 34.6232 | 0.0289 | 16.8529 | 0.0593 | 10.0748 | 169.7896 |
| 25 | 2.1170 | 0.4724 | 36.6776 | 0.0273 | 17.3253 | 0.0577 | 10.4545 | 181.1264 |
| 26 | 2.1815 | 0.4584 | 38.7946 | 0.0258 | 17.7837 | 0.0562 | 10.8294 | 192.5866 |
| 27 | 2.2479 | 0.4449 | 40.9761 | 0.0244 | 18.2285 | 0.0549 | 11.1996 | 204.1529 |
| 28 | 2.3164 | 0.4317 | 43.2240 | 0.0231 | 18.6603 | 0.0536 | 11.5652 | 215.8090 |
| 29 | 2.3869 | 0.4190 | 45.5404 | 0.0220 | 19.0792 | 0.0524 | 11.9261 | 227.5397 |
| 30 | 2.4596 | 0.4066 | 47.9273 | 0.0209 | 19.4858 | 0.0513 | 12.2823 | 239.3302 |
| 31 | 2.5345 | 0.3946 | 50.3869 | 0.0198 | 19.8803 | 0.0503 | 12.6339 | 251.1668 |
| 32 | 2.6117 | 0.3829 | 52.9214 | 0.0189 | 20.2632 | 0.0494 | 12.9810 | 263.0365 |
| 33 | 2.6912 | 0.3716 | 55.5331 | 0.0180 | 20.6348 | 0.0485 | 13.3235 | 274.9270 |
| 34 | 2.7732 | 0.3606 | 58.2243 | 0.0172 | 20.9954 | 0.0476 | 13.6614 | 286.8266 |
| 35 | 2.8577 | 0.3499 | 60.9975 | 0.0164 | 21.3453 | 0.0468 | 13.9948 | 298.7245 |
| 40 | 3.3201 | 0.3012 | 76.1830 | 0.0131 | 22.9459 | 0.0436 | 15.5953 | 357.8483 |
| 45 | 3.8574 | 0.2592 | 93.8259 | 0.0107 | 24.3235 | 0.0411 | 17.0874 | 415.6245 |
| 50 | 4.4817 | 0.2231 | 114.3242 | 0.0087 | 25.5092 | 0.0392 | 18.4750 | 471.2816 |
| 55 | 5.2070 | 0.1920 | 138.1397 | 0.0072 | 26.5297 | 0.0377 | 19.7623 | 524.2887 |
| 60 | 6.0496 | 0.1653 | 165.8094 | 0.0060 | 27.4081 | 0.0365 | 20.9538 | 574.3044 |
| 65 | 7.0287 | 0.1423 | 197.9570 | 0.0051 | 28.1641 | 0.0355 | 22.0541 | 621.1335 |
| 70 | 8.1662 | 0.1225 | 235.3072 | 0.0042 | 28.8149 | 0.0347 | 23.0677 | 664.6933 |
| 75 | 9.4877 | 0.1054 | 278.7019 | 0.0036 | 29.3750 | 0.0340 | 23.9996 | 704.9860 |
| 80 | 11.0232 | 0.0907 | 329.1193 | 0.0030 | 29.8570 | 0.0335 | 24.8543 | 742.0766 |
| 85 | 12.8071 | 0.0781 | 287.6961 | 0.0026 | 30.2720 | 0.0330 | 25.6368 | 776.0754 |
| 90 | 14.8797 | 0.0672 | 455.7526 | 0.0022 | 30.6291 | 0.0326 | 26.3516 | 807.1241 |
| 95 | 17.2878 | 0.0578 | 534.8229 | 0.0019 | 30.9365 | 0.0323 | 27.0032 | 835.3849 |
| 100 | 20.0855 | 0.0498 | 626.6895 | 0.0016 | 31.2010 | 0.0321 | 27.5963 | 861.0319 |

| | Single payment | | Uniform series | | | | Gradient series | |
|---|---|---|---|---|---|---|---|---|
| | Compound amount factor | Present worth factor | Compound amount factor | Sinking fund factor | Present worth factor | Capital recovery factor | Uniform series factor | Present worth factor |
| $n$ | To find $F$ given $P$ $F\|P\ r,n$ | To find $P$ given $F$ $P\|F\ r,n$ | To find $F$ given $A$ $F\|A\ r,n$ | To find $A$ given $F$ $A\|F\ r,n$ | To find $P$ given $A$ $P\|A\ r,n$ | To find $A$ given $P$ $A\|P\ r,n$ | To find $A$ given $G$ $A\|G\ r,n$ | To find $P$ given $G$ $P\|G\ r,n$ |
| 1 | 1.0408 | 0.9608 | 1.0000 | 1.0000 | 0.9608 | 1.0408 | 0.0000 | 0.0000 |
| 2 | 1.0833 | 0.9231 | 2.0408 | 0.4900 | 1.8839 | 0.5308 | 0.4900 | 0.9231 |
| 3 | 1.1275 | 0.8869 | 3.1241 | 0.3201 | 2.7708 | 0.3609 | 0.9733 | 2.6970 |
| 4 | 1.1735 | 0.8521 | 4.2516 | 0.2352 | 3.6230 | 0.2760 | 1.4500 | 5.2534 |
| 5 | 1.2214 | 0.8187 | 5.4251 | 0.1843 | 4.4417 | 0.2251 | 1.9201 | 8.5238 |
| 6 | 1.2712 | 0.7866 | 6.6465 | 0.1505 | 5.2283 | 0.1913 | 2.3834 | 12.4615 |
| 7 | 1.3231 | 0.7558 | 7.9178 | 0.1263 | 5.9841 | 0.1671 | 2.8402 | 16.9962 |
| 8 | 1.3771 | 0.7261 | 9.2409 | 0.1082 | 6.7103 | 0.1490 | 3.2904 | 22.0792 |
| 9 | 1.4333 | 0.6977 | 10.6180 | 0.0942 | 7.4079 | 0.1350 | 3.7339 | 27.6606 |
| 10 | 1.4918 | 0.6703 | 12.0513 | 0.0830 | 8.0783 | 0.1238 | 4.1709 | 33.6935 |
| 11 | 1.5527 | 0.6440 | 13.5432 | 0.0738 | 8.7223 | 0.1146 | 4.6013 | 40.1339 |
| 12 | 1.6161 | 0.6188 | 15.0959 | 0.0662 | 9.3411 | 0.1071 | 5.0252 | 46.9405 |
| 13 | 1.6820 | 0.5945 | 16.7120 | 0.0598 | 9.9356 | 0.1006 | 5.4425 | 54.0747 |
| 14 | 1.7507 | 0.5712 | 18.3940 | 0.0544 | 10.5068 | 0.0952 | 5.8534 | 61.5004 |
| 15 | 1.8221 | 0.5488 | 20.1447 | 0.0496 | 11.0556 | 0.0905 | 6.2578 | 69.1838 |
| 16 | 1.8965 | 0.5273 | 21.9668 | 0.0455 | 11.5829 | 0.0863 | 6.6558 | 77.0932 |
| 17 | 1.9739 | 0.5066 | 23.8633 | 0.0419 | 12.0895 | 0.0827 | 7.0473 | 85.1991 |
| 18 | 2.0544 | 0.4868 | 25.8371 | 0.0387 | 12.5763 | 0.0795 | 7.4326 | 93.4738 |
| 19 | 2.1383 | 0.4677 | 27.8916 | 0.0359 | 13.0439 | 0.0767 | 7.8114 | 101.8918 |
| 20 | 2.2255 | 0.4493 | 30.0298 | 0.0333 | 13.4933 | 0.0741 | 8.1840 | 110.4291 |
| 21 | 2.3164 | 0.4317 | 32.2554 | 0.0310 | 13.9250 | 0.0718 | 8.5503 | 119.0633 |
| 22 | 2.4109 | 0.4148 | 34.5717 | 0.0289 | 14.3398 | 0.0697 | 8.9104 | 127.7737 |
| 23 | 2.5093 | 0.3985 | 36.9826 | 0.0270 | 14.7383 | 0.0679 | 9.2644 | 136.5412 |
| 24 | 2.6117 | 0.3829 | 39.4919 | 0.0253 | 15.1212 | 0.0661 | 9.6122 | 145.3477 |
| 25 | 2.7183 | 0.3679 | 42.1036 | 0.0238 | 15.4891 | 0.0646 | 9.9539 | 154.1768 |
| 26 | 2.8292 | 0.3535 | 44.8219 | 0.0223 | 15.8425 | 0.0631 | 10.2896 | 163.0132 |
| 27 | 2.9447 | 0.3396 | 47.6511 | 0.0210 | 16.1821 | 0.0618 | 10.6193 | 171.8427 |
| 28 | 3.0649 | 0.3263 | 50.5958 | 0.0198 | 16.5084 | 0.0606 | 10.9431 | 180.6522 |
| 29 | 3.1899 | 0.3135 | 53.6607 | 0.0186 | 16.8219 | 0.0594 | 11.2609 | 189.4298 |
| 30 | 3.3201 | 0.3012 | 56.8506 | 0.0176 | 17.1231 | 0.0584 | 11.5730 | 198.1645 |
| 31 | 3.4556 | 0.2894 | 60.1707 | 0.0166 | 17.4125 | 0.0574 | 11.8792 | 206.8460 |
| 32 | 3.5966 | 0.2780 | 63.6263 | 0.0157 | 17.6905 | 0.0565 | 12.1797 | 215.4651 |
| 33 | 3.7434 | 0.2671 | 67.2230 | 0.0149 | 17.9576 | 0.0557 | 12.4746 | 224.0135 |
| 34 | 3.8962 | 0.2567 | 70.9664 | 0.0141 | 18.2143 | 0.0549 | 12.7638 | 232.4833 |
| 35 | 4.0552 | 0.2466 | 74.8626 | 0.0134 | 18.4609 | 0.0542 | 13.0475 | 240.8676 |
| 40 | 4.9530 | 0.2019 | 96.8625 | 0.0103 | 19.5562 | 0.0511 | 14.3845 | 281.3065 |
| 45 | 6.0496 | 0.1653 | 123.7332 | 0.0081 | 20.4530 | 0.0489 | 15.5918 | 318.8989 |
| 50 | 7.3891 | 0.1353 | 156.5532 | 0.0064 | 21.1872 | 0.0472 | 16.6775 | 353.3480 |
| 55 | 9.0250 | 0.1108 | 196.6396 | 0.0051 | 21.7883 | 0.0459 | 17.6498 | 382.5581 |
| 60 | 11.0232 | 0.0907 | 245.6012 | 0.0041 | 22.2804 | 0.0449 | 18.5172 | 412.5715 |
| 65 | 13.4637 | 0.0743 | 305.4031 | 0.0033 | 22.6834 | 0.0441 | 19.2882 | 437.5217 |
| 70 | 16.4446 | 0.0608 | 378.4453 | 0.0026 | 23.0133 | 0.0435 | 19.9710 | 459.5987 |
| 75 | 20.0855 | 0.0498 | 467.6593 | 0.0021 | 23.2834 | 0.0429 | 20.5737 | 479.0243 |
| 80 | 24.5325 | 0.0408 | 576.6254 | 0.0017 | 23.5045 | 0.0425 | 21.1038 | 496.0344 |
| 85 | 29.9641 | 0.0334 | 709.7170 | 0.0014 | 23.6856 | 0.0422 | 21.5687 | 510.8663 |
| 90 | 36.5982 | 0.0273 | 872.2754 | 0.0011 | 23.8338 | 0.0420 | 21.9751 | 523.7508 |
| 95 | 44.7012 | 0.0224 | 1070.8247 | 0.0009 | 23.9552 | 0.0417 | 22.3295 | 534.9066 |
| 100 | 54.5982 | 0.0183 | 1313.3333 | 0.0008 | 24.0545 | 0.0416 | 22.6376 | 544.5370 |

## TABLE B.8   Continuous Compounding: $r = 5\%$

| | Single payment | | Uniform series | | | | Gradient series | |
|---|---|---|---|---|---|---|---|---|
| | Compound amount factor | Present worth factor | Compound amount factor | Sinking fund factor | Present worth factor | Capital recovery factor | Uniform series factor | Present worth factor |
| $n$ | To find $F$ given $P$ $F\|P\ r,n$ | To find $P$ given $F$ $P\|F\ r,n$ | To find $F$ given $A$ $F\|A\ r,n$ | To find $A$ given $F$ $A\|F\ r,n$ | To find $P$ given $A$ $P\|A\ r,n$ | To find $A$ given $P$ $A\|P\ r,n$ | To find $A$ given $G$ $A\|G\ r,n$ | To find $P$ given $G$ $P\|G\ r,n$ |
| 1 | 1.0513 | 0.9512 | 1.0000 | 1.0000 | 0.9512 | 1.0513 | 0.0000 | 0.0000 |
| 2 | 1.1052 | 0.9048 | 2.0513 | 0.4875 | 1.8561 | 0.5388 | 0.4875 | 0.9048 |
| 3 | 1.1618 | 0.8607 | 3.1564 | 0.3168 | 2.7168 | 0.3681 | 0.9667 | 2.6263 |
| 4 | 1.2214 | 0.8187 | 4.3183 | 0.2316 | 3.5355 | 0.2828 | 1.4375 | 5.0824 |
| 5 | 1.2840 | 0.7788 | 5.5397 | 0.1805 | 4.3143 | 0.2318 | 1.9001 | 8.1976 |
| 6 | 1.3499 | 0.7408 | 6.8237 | 0.1465 | 5.0551 | 0.1978 | 2.3544 | 11.9017 |
| 7 | 1.4191 | 0.7047 | 8.1736 | 0.1223 | 5.7598 | 0.1736 | 2.8004 | 16.1299 |
| 8 | 1.4918 | 0.6703 | 9.5926 | 0.1042 | 6.4301 | 0.1555 | 3.2382 | 20.8221 |
| 9 | 1.5683 | 0.6376 | 11.0845 | 0.0902 | 7.0678 | 0.1415 | 3.6678 | 25.9231 |
| 10 | 1.6487 | 0.6065 | 12.6528 | 0.0790 | 7.6743 | 0.1303 | 4.0892 | 31.3819 |
| 11 | 1.7333 | 0.5769 | 14.3015 | 0.0699 | 8.2512 | 0.1212 | 4.5025 | 37.1514 |
| 12 | 1.8221 | 0.5488 | 16.0347 | 0.0624 | 8.8001 | 0.1136 | 4.9077 | 43.1883 |
| 13 | 1.9155 | 0.5220 | 17.8569 | 0.0560 | 9.3221 | 0.1073 | 5.3049 | 49.4529 |
| 14 | 2.0138 | 0.4966 | 19.7724 | 0.0506 | 9.8187 | 0.1018 | 5.6941 | 55.9085 |
| 15 | 2.1170 | 0.4724 | 21.7862 | 0.0459 | 10.2911 | 0.0972 | 6.0753 | 62.5216 |
| 16 | 2.2255 | 0.4493 | 23.9032 | 0.0418 | 10.7404 | 0.0931 | 6.4487 | 69.2616 |
| 17 | 2.3396 | 0.4274 | 26.1287 | 0.0383 | 11.1678 | 0.0895 | 6.8143 | 76.1002 |
| 18 | 2.4596 | 0.4066 | 28.4683 | 0.0351 | 11.5744 | 0.0864 | 7.1720 | 83.0119 |
| 19 | 2.5857 | 0.3867 | 30.9279 | 0.0323 | 11.9611 | 0.0836 | 7.5221 | 89.9732 |
| 20 | 2.7183 | 0.3679 | 33.5137 | 0.0298 | 12.3290 | 0.0811 | 7.8646 | 96.9629 |
| 21 | 2.8577 | 0.3499 | 36.2319 | 0.0276 | 12.6789 | 0.0789 | 8.1996 | 103.9617 |
| 22 | 3.0042 | 0.3329 | 39.0896 | 0.0256 | 13.0118 | 0.0769 | 8.5270 | 110.9520 |
| 23 | 3.1582 | 0.3166 | 42.0938 | 0.0238 | 13.3284 | 0.0750 | 8.8471 | 117.9180 |
| 24 | 3.3201 | 0.3012 | 45.2519 | 0.0221 | 13.6296 | 0.0734 | 9.1599 | 124.8455 |
| 25 | 3.4903 | 0.2865 | 48.5721 | 0.0206 | 13.9161 | 0.0719 | 9.4654 | 131.7216 |
| 26 | 3.6693 | 0.2725 | 52.0624 | 0.0192 | 14.1887 | 0.0705 | 9.7638 | 138.5349 |
| 27 | 3.8574 | 0.2592 | 55.7317 | 0.0179 | 14.4479 | 0.0692 | 10.0551 | 145.2751 |
| 28 | 4.0552 | 0.2466 | 59.5891 | 0.0168 | 14.6945 | 0.0681 | 10.3395 | 151.9332 |
| 29 | 4.2631 | 0.2346 | 63.6443 | 0.0157 | 14.9291 | 0.0670 | 10.6170 | 158.5012 |
| 30 | 4.4817 | 0.2231 | 67.9074 | 0.0147 | 15.1522 | 0.0660 | 10.8877 | 164.9720 |
| 31 | 4.7115 | 0.2122 | 72.3891 | 0.0138 | 15.3644 | 0.0651 | 11.1517 | 171.3394 |
| 32 | 4.9530 | 0.2019 | 77.1006 | 0.0130 | 15.5663 | 0.0642 | 11.4091 | 177.5982 |
| 33 | 5.2070 | 0.1920 | 82.0536 | 0.0122 | 15.7584 | 0.0635 | 11.6601 | 183.7438 |
| 34 | 5.4739 | 0.1827 | 87.2606 | 0.0115 | 15.9411 | 0.0627 | 11.9046 | 189.7724 |
| 35 | 5.7546 | 0.1738 | 92.7346 | 0.0108 | 16.1149 | 0.0621 | 12.1429 | 195.6807 |
| 40 | 7.3891 | 0.1353 | 124.6132 | 0.0081 | 16.8646 | 0.0593 | 13.2435 | 223.3452 |
| 45 | 9.4877 | 0.1054 | 165.5462 | 0.0060 | 17.4484 | 0.0573 | 14.2024 | 247.8097 |
| 50 | 12.1825 | 0.0821 | 218.1052 | 0.0046 | 17.9032 | 0.0559 | 15.0329 | 269.1364 |
| 55 | 15.6426 | 0.0639 | 285.5923 | 0.0035 | 18.2573 | 0.0548 | 15.7480 | 287.5163 |
| 60 | 20.0855 | 0.0498 | 372.2475 | 0.0027 | 18.5331 | 0.0540 | 16.3604 | 303.2096 |
| 65 | 25.7903 | 0.0388 | 483.5149 | 0.0021 | 18.7479 | 0.0533 | 16.8822 | 316.5055 |
| 70 | 33.1155 | 0.0302 | 626.3851 | 0.0016 | 18.9152 | 0.0529 | 17.3245 | 327.6968 |
| 75 | 42.5211 | 0.0235 | 809.8341 | 0.0012 | 19.0455 | 0.0525 | 17.6979 | 337.0640 |
| 80 | 54.5982 | 0.0183 | 1045.3872 | 0.0010 | 19.1469 | 0.0522 | 18.0116 | 344.8665 |
| 85 | 70.1054 | 0.0143 | 1347.8435 | 0.0007 | 19.2260 | 0.0520 | 18.2742 | 351.3382 |
| 90 | 90.0171 | 0.0111 | 1736.2049 | 0.0006 | 19.2875 | 0.0518 | 18.4931 | 356.6861 |
| 95 | 115.5843 | 0.0087 | 2234.8710 | 0.0004 | 19.3354 | 0.0517 | 18.6751 | 361.0906 |
| 100 | 148.4132 | 0.0067 | 2875.1708 | 0.0003 | 19.3727 | 0.0516 | 18.8258 | 364.7075 |

**TABLE B.9** Continuous Compounding: $r = 6\%$

| | Single payment | | Uniform series | | | | Gradient series | |
|---|---|---|---|---|---|---|---|---|
| | Compound amount factor | Present worth factor | Compound amount factor | Sinking fund factor | Present worth factor | Capital recovery factor | Uniform series factor | Present worth factor |
| $n$ | To find $F$ given $P$ $F\|P\,r,n$ | To find $P$ given $F$ $P\|F\,r,n$ | To find $F$ given $A$ $F\|A\,r,n$ | To find $A$ given $F$ $A\|F\,r,n$ | To find $P$ given $A$ $P\|A\,r,n$ | To find $A$ given $P$ $A\|P\,r,n$ | To find $A$ given $G$ $A\|G\,r,n$ | To find $P$ given $G$ $P\|G\,r,n$ |
| 1 | 1.0618 | 0.9418 | 1.0000 | 1.0000 | 0.9148 | 1.0618 | 0.0000 | 0.0000 |
| 2 | 1.1275 | 0.8869 | 2.0618 | 0.4850 | 1.8287 | 0.5468 | 0.4850 | 0.8869 |
| 3 | 1.1972 | 0.8353 | 3.1893 | 0.3135 | 2.6640 | 0.3754 | 0.9600 | 2.5575 |
| 4 | 1.2712 | 0.7866 | 4.3866 | 0.2280 | 3.4506 | 0.2898 | 1.4251 | 4.9173 |
| 5 | 1.3499 | 0.7408 | 5.6578 | 0.1767 | 4.1914 | 0.2386 | 1.8802 | 7.8806 |
| 6 | 1.4333 | 0.6977 | 7.0077 | 0.1427 | 4.8891 | 0.2045 | 2.3254 | 11.3690 |
| 7 | 1.5220 | 0.6570 | 8.4410 | 0.1185 | 5.5461 | 0.1803 | 2.7607 | 15.3113 |
| 8 | 1.6161 | 0.6188 | 9.9629 | 0.1004 | 6.1649 | 0.1622 | 3.1862 | 19.6428 |
| 9 | 1.7160 | 0.5827 | 11.5790 | 0.0864 | 6.7477 | 0.1482 | 3.6020 | 24.3047 |
| 10 | 1.8221 | 0.5488 | 13.2950 | 0.0752 | 7.2965 | 0.1371 | 4.0080 | 29.2441 |
| 11 | 1.9348 | 0.5169 | 15.1171 | 0.0662 | 7.8133 | 0.1280 | 4.4043 | 34.4126 |
| 12 | 2.0544 | 0.4868 | 17.0519 | 0.0586 | 8.3001 | 0.1205 | 4.7911 | 39.7668 |
| 13 | 2.1815 | 0.4584 | 19.1064 | 0.0523 | 8.7585 | 0.1142 | 5.1684 | 45.2677 |
| 14 | 2.3164 | 0.4317 | 21.2878 | 0.0470 | 9.1902 | 0.1088 | 5.5363 | 50.8800 |
| 15 | 2.4596 | 0.4066 | 23.6042 | 0.0424 | 9.5968 | 0.1042 | 5.8949 | 56.5719 |
| 16 | 2.6117 | 0.3829 | 26.0638 | 0.0384 | 9.9797 | 0.1002 | 6.2442 | 62.3153 |
| 17 | 2.7732 | 0.3606 | 28.6755 | 0.0349 | 10.3402 | 0.0967 | 6.5845 | 68.0848 |
| 18 | 2.9447 | 0.3396 | 31.4487 | 0.0318 | 10.6798 | 0.0936 | 6.9156 | 73.8580 |
| 19 | 3.1268 | 0.3198 | 34.3934 | 0.0291 | 10.9997 | 0.0909 | 7.2379 | 79.6147 |
| 20 | 3.3201 | 0.3012 | 37.5202 | 0.0267 | 11.3009 | 0.0885 | 7.5514 | 85.3374 |
| 21 | 3.5254 | 0.2837 | 40.8403 | 0.0245 | 11.5845 | 0.0863 | 7.8562 | 91.0105 |
| 22 | 3.7434 | 0.2671 | 44.3657 | 0.0225 | 11.8516 | 0.0844 | 8.1525 | 96.6203 |
| 23 | 3.9749 | 0.2516 | 48.1091 | 0.0208 | 12.1032 | 0.0826 | 8.4403 | 102.1550 |
| 24 | 4.2207 | 0.2369 | 52.0840 | 0.0192 | 12.3401 | 0.0810 | 8.7199 | 107.6044 |
| 25 | 4.4817 | 0.2231 | 56.3047 | 0.0178 | 12.5633 | 0.0796 | 8.9912 | 112.9595 |
| 26 | 4.7588 | 0.2101 | 60.7864 | 0.0165 | 12.7734 | 0.0783 | 9.2546 | 118.2129 |
| 27 | 5.0531 | 0.1979 | 65.5452 | 0.0153 | 12.9713 | 0.0771 | 9.5101 | 123.3583 |
| 28 | 5.3656 | 0.1864 | 70.5983 | 0.0142 | 13.1577 | 0.0760 | 9.7578 | 128.3904 |
| 29 | 5.6973 | 0.1755 | 75.9639 | 0.0132 | 13.3332 | 0.0750 | 9.9980 | 133.3049 |
| 30 | 6.0496 | 0.1653 | 81.6612 | 0.0122 | 13.4985 | 0.0741 | 10.2307 | 138.0986 |
| 31 | 6.4237 | 0.1557 | 87.7109 | 0.0114 | 13.6542 | 0.0732 | 10.4560 | 142.7688 |
| 32 | 6.8210 | 0.1466 | 94.1346 | 0.0106 | 13.8008 | 0.0725 | 10.6743 | 147.3136 |
| 33 | 7.2427 | 0.1381 | 100.9556 | 0.0099 | 13.9389 | 0.0717 | 10.8855 | 151.7318 |
| 34 | 7.6906 | 0.1300 | 108.1983 | 0.0092 | 14.0689 | 0.0711 | 11.0899 | 156.0228 |
| 35 | 8.1662 | 0.1225 | 115.8889 | 0.0086 | 14.1913 | 0.0705 | 11.2876 | 160.1863 |
| 40 | 11.0232 | 0.0907 | 162.0915 | 0.0062 | 14.7046 | 0.0680 | 12.1809 | 179.1156 |
| 45 | 14.8797 | 0.0672 | 224.4584 | 0.0045 | 15.0848 | 0.0663 | 12.9295 | 195.0399 |
| 50 | 20.0855 | 0.0498 | 308.6449 | 0.0032 | 15.3665 | 0.0651 | 13.5519 | 208.2453 |
| 55 | 27.1126 | 0.0369 | 422.2849 | 0.0024 | 15.5752 | 0.0642 | 14.0654 | 219.0716 |
| 60 | 36.5982 | 0.0273 | 575.6828 | 0.0017 | 15.7298 | 0.0636 | 14.4862 | 227.8648 |
| 65 | 49.4024 | 0.0202 | 782.7483 | 0.0013 | 15.8443 | 0.0631 | 14.8288 | 234.9516 |
| 70 | 66.6863 | 0.0150 | 1062.2574 | 0.0009 | 15.9292 | 0.0628 | 15.1060 | 240.6259 |
| 75 | 90.0171 | 0.0111 | 1439.5553 | 0.0007 | 15.9920 | 0.0625 | 15.3291 | 245.1437 |
| 80 | 121.5104 | 0.0082 | 1948.8543 | 0.0005 | 16.0386 | 0.0623 | 15.5078 | 248.7234 |
| 85 | 164.0219 | 0.0061 | 2636.3359 | 0.0004 | 16.0731 | 0.0622 | 15.6503 | 251.5478 |
| 90 | 221.4064 | 0.0045 | 3564.3390 | 0.0003 | 16.0986 | 0.0621 | 15.7633 | 253.7679 |
| 95 | 298.8674 | 0.0033 | 4817.0122 | 0.0002 | 16.1176 | 0.0620 | 15.8527 | 255.5073 |
| 100 | 403.4288 | 0.0025 | 6507.9442 | 0.0002 | 16.1316 | 0.0620 | 15.9232 | 256.8660 |

| | Single payment | | Uniform series | | | | Gradient series | |
|---|---|---|---|---|---|---|---|---|
| | Compound amount factor | Present worth factor | Compound amount factor | Sinking fund factor | Present worth factor | Capital recovery factor | Uniform series factor | Present worth factor |
| $n$ | To find $F$ given $P$ $F\|P\,r,n$ | To find $P$ given $F$ $P\|F\,r,n$ | To find $F$ given $A$ $F\|A\,r,n$ | To find $A$ given $F$ $A\|F\,r,n$ | To find $P$ given $A$ $P\|A\,r,n$ | To find $A$ given $P$ $A\|P\,r,n$ | To find $A$ given $G$ $A\|G\,r,n$ | To find $P$ given $G$ $P\|G\,r,n$ |
| 1 | 1.0725 | 0.9324 | 1.0000 | 1.0000 | 0.9324 | 1.0725 | 0.0000 | 0.0000 |
| 2 | 1.1503 | 0.8694 | 2.0725 | 0.4825 | 1.8018 | 0.5550 | 0.4825 | 0.8694 |
| 3 | 1.2337 | 0.8106 | 3.2228 | 0.3103 | 2.6123 | 0.3828 | 0.9534 | 2.4905 |
| 4 | 1.3231 | 0.7558 | 4.4565 | 0.2244 | 3.3681 | 0.2969 | 1.4126 | 4.7579 |
| 5 | 1.4191 | 0.7047 | 5.7796 | 0.1730 | 4.0728 | 0.2455 | 1.8603 | 7.5766 |
| 6 | 1.5220 | 0.6570 | 7.1987 | 0.1389 | 4.7299 | 0.2114 | 2.2964 | 10.8619 |
| 7 | 1.6323 | 0.6126 | 8.7206 | 0.1147 | 5.3425 | 0.1872 | 2.7211 | 14.5376 |
| 8 | 1.7507 | 0.5712 | 10.3529 | 0.0966 | 5.9137 | 0.1691 | 3.1344 | 18.5361 |
| 9 | 1.8776 | 0.5326 | 12.1036 | 0.0826 | 6.4463 | 0.1551 | 3.5364 | 22.7968 |
| 10 | 2.0138 | 0.4966 | 13.9812 | 0.0715 | 6.9429 | 0.1440 | 3.9272 | 27.2661 |
| 11 | 2.1598 | 0.4630 | 15.9950 | 0.0625 | 7.4059 | 0.1350 | 4.3069 | 31.8962 |
| 12 | 2.3164 | 0.4317 | 18.1547 | 0.0551 | 7.8376 | 0.1276 | 4.6755 | 36.6450 |
| 13 | 2.4843 | 0.4025 | 20.4711 | 0.0488 | 8.2401 | 0.1214 | 5.0333 | 41.4753 |
| 14 | 2.6645 | 0.3753 | 22.9554 | 0.0436 | 8.6154 | 0.1161 | 5.3804 | 46.3544 |
| 15 | 2.8577 | 0.3499 | 26.6199 | 0.0390 | 8.9654 | 0.1115 | 5.7168 | 51.2535 |
| 16 | 3.0649 | 0.3263 | 28.4775 | 0.0351 | 9.2916 | 0.1076 | 6.0428 | 56.1477 |
| 17 | 3.2871 | 0.3042 | 31.5424 | 0.0317 | 9.5959 | 0.1042 | 6.3585 | 61.0152 |
| 18 | 3.5254 | 0.2837 | 34.8295 | 0.0287 | 9.8795 | 0.1012 | 6.6640 | 65.8374 |
| 19 | 3.7810 | 0.2645 | 38.3549 | 0.0261 | 10.1440 | 0.0986 | 6.9596 | 70.5979 |
| 20 | 4.0552 | 0.2466 | 42.1359 | 0.0237 | 10.3906 | 0.0962 | 7.2453 | 75.2833 |
| 21 | 4.3492 | 0.2299 | 46.1911 | 0.0216 | 10.6205 | 0.0942 | 7.5215 | 79.8818 |
| 22 | 4.6646 | 0.2144 | 50.5404 | 0.0198 | 10.8349 | 0.0923 | 7.7881 | 84.3838 |
| 23 | 5.0028 | 0.1999 | 55.2050 | 0.0181 | 11.0348 | 0.0906 | 8.0456 | 88.7813 |
| 24 | 5.3656 | 0.1864 | 60.2078 | 0.0166 | 11.2212 | 0.0891 | 8.2940 | 93.0679 |
| 25 | 5.7546 | 0.1738 | 65.5733 | 0.0153 | 11.3949 | 0.0878 | 8.5335 | 97.2385 |
| 26 | 6.1719 | 0.1620 | 71.3279 | 0.0140 | 11.5570 | 0.0865 | 8.7643 | 101.2891 |
| 27 | 6.6194 | 0.1511 | 77.4998 | 0.0129 | 11.7080 | 0.0854 | 8.9867 | 105.2170 |
| 28 | 7.0993 | 0.1409 | 84.1192 | 0.0119 | 11.8489 | 0.0844 | 9.2009 | 109.0202 |
| 29 | 7.6141 | 0.1313 | 91.2185 | 0.0110 | 11.9802 | 0.0835 | 9.4070 | 112.6976 |
| 30 | 8.1662 | 0.1225 | 98.8326 | 0.0101 | 12.1027 | 0.0826 | 9.6052 | 116.2488 |
| 31 | 8.7583 | 0.1142 | 106.9987 | 0.0093 | 12.2169 | 0.0819 | 9.7958 | 119.6742 |
| 32 | 9.3933 | 0.1065 | 115.7570 | 0.0086 | 12.3233 | 0.0811 | 9.9790 | 122.9744 |
| 33 | 10.0744 | 0.0993 | 125.1504 | 0.0080 | 12.4226 | 0.0805 | 10.1550 | 126.1507 |
| 34 | 10.8049 | 0.0926 | 135.2248 | 0.0074 | 12.5151 | 0.0799 | 10.3239 | 129.2049 |
| 35 | 11.5883 | 0.0863 | 146.0297 | 0.0068 | 12.6014 | 0.0794 | 10.4860 | 132.1389 |
| 40 | 16.4446 | 0.0608 | 213.0056 | 0.0047 | 12.9529 | 0.0772 | 11.2017 | 145.0937 |
| 45 | 23.3361 | 0.0429 | 308.0489 | 0.0032 | 13.2006 | 0.0758 | 11.7769 | 155.4611 |
| 50 | 33.1155 | 0.0302 | 442.9218 | 0.0023 | 13.3751 | 0.0748 | 12.2347 | 163.6396 |
| 55 | 46.9931 | 0.0213 | 634.3155 | 0.0016 | 13.4981 | 0.0741 | 12.5957 | 170.0178 |
| 60 | 66.6863 | 0.0150 | 905.9161 | 0.0011 | 13.5847 | 0.0736 | 12.8781 | 174.9458 |
| 65 | 94.6324 | 0.0106 | 1291.3358 | 0.0008 | 13.6458 | 0.0733 | 13.0973 | 178.7238 |
| 70 | 134.2898 | 0.0074 | 1838.2723 | 0.0005 | 13.6888 | 0.0731 | 13.2664 | 181.6014 |
| 75 | 190.5663 | 0.0052 | 2614.4121 | 0.0004 | 13.7192 | 0.0729 | 13.3959 | 183.7808 |
| 80 | 270.4264 | 0.0037 | 3715.8070 | 0.0003 | 13.7405 | 0.0728 | 13.4946 | 185.4235 |
| 85 | 383.7533 | 0.0026 | 5278.7607 | 0.0002 | 13.7556 | 0.0727 | 13.5695 | 186.6563 |
| 90 | 544.5719 | 0.0018 | 7496.6976 | 0.0001 | 13.7662 | 0.0726 | 13.6260 | 187.5782 |
| 95 | 772.7843 | 0.0013 | 10644.0999 | 0.0001 | 13.7737 | 0.0726 | 13.6685 | 188.2652 |
| 100 | 1096.6332 | 0.0009 | 15110.4764 | 0.0001 | 13.7790 | 0.0726 | 13.7003 | 188.7757 |

| | Single payment | | Uniform series | | | | Gradient series | |
|---|---|---|---|---|---|---|---|---|
| | Compound amount factor | Present worth factor | Compound amount factor | Sinking fund factor | Present worth factor | Capital recovery factor | Uniform series factor | Present worth factor |
| $n$ | To find $F$ given $P$ $F\|P\ r,n$ | To find $P$ given $F$ $P\|F\ r,n$ | To find $F$ given $A$ $F\|A\ r,n$ | To find $A$ given $F$ $A\|F\ r,n$ | To find $P$ given $A$ $P\|A\ r,n$ | To find $A$ given $P$ $A\|P\ r,n$ | To find $A$ given $G$ $A\|G\ r,n$ | To find $P$ given $G$ $P\|G\ r,n$ |
| 1 | 1.0833 | 0.9231 | 1.0000 | 1.0000 | 0.9231 | 1.0833 | 0.0000 | 0.0000 |
| 2 | 1.1735 | 0.8521 | 2.0833 | 0.4800 | 1.7753 | 0.5633 | 0.4800 | 0.8521 |
| 3 | 1.2712 | 0.7866 | 3.2568 | 0.3071 | 2.5619 | 0.3903 | 0.9467 | 2.4254 |
| 4 | 1.3771 | 0.7261 | 4.5280 | 0.2208 | 3.2880 | 0.3041 | 1.4002 | 4.6038 |
| 5 | 1.4918 | 0.6703 | 5.9052 | 0.1693 | 3.9584 | 0.2526 | 1.8404 | 7.2851 |
| 6 | 1.6161 | 0.6188 | 7.3970 | 0.1352 | 4.5771 | 0.2185 | 2.2676 | 10.3790 |
| 7 | 1.7507 | 0.5712 | 9.0131 | 0.1109 | 5.1483 | 0.1942 | 2.6817 | 13.8063 |
| 8 | 1.8965 | 0.5273 | 10.7637 | 0.0929 | 5.6756 | 0.1762 | 3.0829 | 17.4973 |
| 9 | 2.0544 | 0.4868 | 12.6602 | 0.0790 | 6.1624 | 0.1623 | 3.4713 | 21.3914 |
| 10 | 2.2225 | 0.4493 | 14.7147 | 0.0680 | 6.6117 | 0.1512 | 3.8470 | 25.4353 |
| 11 | 2.4109 | 0.4148 | 16.9402 | 0.0590 | 7.0265 | 0.1423 | 4.2102 | 29.5832 |
| 12 | 2.6117 | 0.3829 | 19.3511 | 0.0517 | 7.4094 | 0.1350 | 4.5611 | 33.7950 |
| 13 | 2.8292 | 0.3535 | 21.9628 | 0.0455 | 7.7629 | 0.1288 | 4.8998 | 38.0364 |
| 14 | 3.0649 | 0.3263 | 24.7920 | 0.0403 | 8.0891 | 0.1236 | 5.2265 | 42.2781 |
| 15 | 3.3201 | 0.3012 | 27.8569 | 0.0359 | 8.3903 | 0.1192 | 5.5415 | 46.4948 |
| 16 | 3.5966 | 0.2780 | 31.1770 | 0.0321 | 8.6684 | 0.1154 | 5.8449 | 50.6653 |
| 17 | 3.8962 | 0.2567 | 34.7736 | 0.0288 | 8.9250 | 0.1120 | 6.1369 | 54.7719 |
| 18 | 4.2207 | 0.2369 | 38.6698 | 0.0259 | 9.1620 | 0.1091 | 6.4178 | 58.7997 |
| 19 | 4.5722 | 0.2187 | 42.8905 | 0.0233 | 9.3807 | 0.1066 | 6.6879 | 62.7365 |
| 20 | 4.9530 | 0.2019 | 47.4627 | 0.0211 | 9.5826 | 0.1044 | 6.9473 | 66.5725 |
| 21 | 5.3656 | 0.1864 | 52.4158 | 0.0191 | 9.7689 | 0.1024 | 7.1963 | 70.3000 |
| 22 | 5.8124 | 0.1720 | 57.7813 | 0.0173 | 9.9410 | 0.1006 | 7.4352 | 73.9130 |
| 23 | 6.2965 | 0.1588 | 63.5938 | 0.0157 | 10.0998 | 0.0990 | 7.6642 | 77.4069 |
| 24 | 6.8210 | 0.1466 | 69.8903 | 0.0143 | 10.2464 | 0.0976 | 7.8836 | 80.7789 |
| 25 | 7.3891 | 0.1353 | 76.7113 | 0.0130 | 10.3817 | 0.0963 | 8.0937 | 84.0270 |
| 26 | 8.0045 | 0.1249 | 84.1003 | 0.0119 | 10.5067 | 0.0952 | 8.2947 | 87.1502 |
| 27 | 8.6711 | 0.1153 | 92.1048 | 0.0109 | 10.6220 | 0.0941 | 8.4870 | 90.1487 |
| 28 | 9.3933 | 0.1065 | 100.7759 | 0.0099 | 10.7285 | 0.0932 | 8.6707 | 93.0230 |
| 29 | 10.1757 | 0.0983 | 110.1693 | 0.0091 | 10.8267 | 0.0924 | 8.8461 | 95.7747 |
| 30 | 11.0232 | 0.0907 | 120.3449 | 0.0083 | 10.9174 | 0.0916 | 9.0136 | 98.4055 |
| 31 | 11.9413 | 0.0837 | 131.3681 | 0.0076 | 11.0012 | 0.0909 | 9.1734 | 100.9178 |
| 32 | 12.9358 | 0.0773 | 143.3094 | 0.0070 | 11.0785 | 0.0903 | 9.3257 | 103.3143 |
| 33 | 14.0132 | 0.0714 | 156.2452 | 0.0064 | 11.1499 | 0.0897 | 9.4708 | 105.5978 |
| 34 | 15.1803 | 0.0659 | 170.3584 | 0.0059 | 11.2157 | 0.0892 | 9.6090 | 107.7717 |
| 35 | 16.4446 | 0.0608 | 185.4387 | 0.0054 | 11.2765 | 0.0887 | 9.7405 | 109.8392 |
| 40 | 24.5325 | 0.0408 | 282.5472 | 0.0035 | 11.5172 | 0.0868 | 10.3069 | 118.7070 |
| 45 | 36.5982 | 0.0273 | 427.4161 | 0.0023 | 11.6786 | 0.0856 | 10.7426 | 125.4580 |
| 50 | 54.5982 | 0.0183 | 643.5351 | 0.0016 | 11.7868 | 0.0848 | 11.0738 | 130.5242 |
| 55 | 81.4509 | 0.0123 | 965.9467 | 0.0010 | 11.8593 | 0.0843 | 11.3230 | 134.2826 |
| 60 | 121.5104 | 0.0082 | 1446.9283 | 0.0007 | 11.9079 | 0.0840 | 11.5088 | 137.0449 |
| 65 | 181.2722 | 0.0055 | 2164.4686 | 0.0005 | 11.9404 | 0.0837 | 11.6461 | 139.0594 |
| 70 | 270.4264 | 0.0037 | 3234.9129 | 0.0003 | 11.9623 | 0.0836 | 11.7469 | 140.5190 |
| 75 | 403.4288 | 0.0025 | 4831.8281 | 0.0002 | 11.9769 | 0.0835 | 11.8203 | 141.5706 |
| 80 | 601.8450 | 0.0017 | 7214.1457 | 0.0001 | 11.9867 | 0.0834 | 11.8735 | 142.3245 |
| 85 | 897.8473 | 0.0011 | 10768.1458 | 0.0001 | 11.9933 | 0.0834 | 11.9119 | 142.8628 |
| 90 | 1339.4308 | 0.0007 | 16070.0911 | 0.0001 | 11.9977 | 0.0833 | 11.9394 | 143.2456 |
| 95 | 1998.1959 | 0.0005 | 23979.6640 | 0.0000 | 12.0007 | 0.0833 | 11.9591 | 143.5171 |
| 100 | 2980.9580 | 0.0003 | 35779.3601 | 0.0000 | 12.0026 | 0.0833 | 11.9731 | 143.7089 |

| | Single payment | | Uniform series | | | | Gradient series | |
|---|---|---|---|---|---|---|---|---|
| | Compound amount factor | Present worth factor | Compound amount factor | Sinking fund factor | Present worth factor | Capital recovery factor | Uniform series factor | Present worth factor |
| $n$ | To find $F$ given $P$ $F\mid P\ r,n$ | To find $P$ given $F$ $P\mid F\ r,n$ | To find $F$ given $A$ $F\mid A\ r,n$ | To find $A$ given $F$ $A\mid F\ r,n$ | To find $P$ given $A$ $P\mid A\ r,n$ | To find $A$ given $P$ $A\mid P\ r,n$ | To find $A$ given $G$ $A\mid G\ r,n$ | To find $P$ given $G$ $P\mid G\ r,n$ |
| 1 | 1.0942 | 0.9139 | 1.0000 | 1.0000 | 0.9139 | 1.0942 | 0.0000 | 0.0000 |
| 2 | 1.1972 | 0.8353 | 2.0942 | 0.4775 | 1.7492 | 0.5717 | 0.4775 | 0.8353 |
| 3 | 1.3100 | 0.7634 | 3.2914 | 0.3038 | 2.5126 | 0.3980 | 0.9401 | 2.3620 |
| 4 | 1.4333 | 0.6977 | 4.6014 | 0.2173 | 3.2103 | 0.3115 | 1.3878 | 4.4551 |
| 5 | 1.5683 | 0.6376 | 6.0347 | 0.1657 | 3.8479 | 0.2599 | 1.8206 | 7.0056 |
| 6 | 1.7160 | 0.5827 | 7.6030 | 0.1315 | 4.4306 | 0.2257 | 2.2388 | 9.9193 |
| 7 | 1.8776 | 0.5326 | 9.3190 | 0.1073 | 4.9632 | 0.2015 | 2.6424 | 13.1149 |
| 8 | 2.0544 | 0.4868 | 11.1966 | 0.0893 | 5.4500 | 0.1835 | 3.0316 | 16.5221 |
| 9 | 2.2479 | 0.4449 | 13.2510 | 0.0755 | 5.8948 | 0.1696 | 3.4065 | 20.0810 |
| 10 | 2.4596 | 0.4066 | 15.4490 | 0.0645 | 6.3014 | 0.1587 | 3.7674 | 23.7401 |
| 11 | 2.6912 | 0.3716 | 17.9586 | 0.0557 | 6.6730 | 0.1499 | 4.1145 | 27.4559 |
| 12 | 2.9447 | 0.3396 | 20.6498 | 0.0484 | 7.0126 | 0.1426 | 4.4479 | 31.1914 |
| 13 | 3.2220 | 0.3104 | 23.5945 | 0.0424 | 7.3229 | 0.1366 | 4.7680 | 34.9158 |
| 14 | 3.5254 | 0.2837 | 26.8165 | 0.0373 | 7.6066 | 0.1315 | 5.0750 | 38.6033 |
| 15 | 3.8574 | 0.2592 | 30.3419 | 0.0330 | 7.8658 | 0.1271 | 5.3691 | 42.2327 |
| 16 | 4.2207 | 0.2369 | 34.1993 | 0.0292 | 8.1028 | 0.1234 | 5.6507 | 45.7866 |
| 17 | 4.6182 | 0.2165 | 38.4200 | 0.0260 | 8.3193 | 0.1202 | 5.9201 | 49.2512 |
| 18 | 5.0531 | 0.1979 | 43.0382 | 0.0232 | 8.5172 | 0.1174 | 6.1776 | 52.6155 |
| 19 | 5.5290 | 0.1809 | 48.0913 | 0.0208 | 8.6981 | 0.1150 | 6.4234 | 55.8711 |
| 20 | 6.0496 | 0.1653 | 53.6202 | 0.0186 | 8.8634 | 0.1128 | 6.6579 | 59.0117 |
| 21 | 6.6194 | 0.1511 | 59.6699 | 0.0168 | 9.0144 | 0.1109 | 6.8815 | 62.0332 |
| 22 | 7.2427 | 0.1381 | 66.2893 | 0.0151 | 9.1525 | 0.1093 | 7.0945 | 64.9326 |
| 23 | 7.9248 | 0.1262 | 73.5320 | 0.0136 | 9.2787 | 0.1078 | 7.2971 | 67.7087 |
| 24 | 8.6711 | 0.1153 | 81.4568 | 0.0123 | 9.3940 | 0.1065 | 7.4900 | 70.3612 |
| 25 | 9.4877 | 0.1054 | 90.1280 | 0.0111 | 9.4994 | 0.1053 | 7.6732 | 72.8908 |
| 26 | 10.3812 | 0.0963 | 99.6157 | 0.0100 | 9.5957 | 0.1042 | 7.8471 | 75.2990 |
| 27 | 11.3589 | 0.0880 | 109.9969 | 0.0091 | 9.6838 | 0.1033 | 8.0122 | 77.5879 |
| 28 | 12.4286 | 0.0805 | 121.3558 | 0.0082 | 9.7642 | 0.1024 | 8.1686 | 79.7603 |
| 29 | 13.5991 | 0.0735 | 133.7844 | 0.0075 | 9.8378 | 0.1016 | 8.3168 | 81.8193 |
| 30 | 14.8797 | 0.0672 | 147.3835 | 0.0068 | 9.9050 | 0.1010 | 8.4572 | 83.7683 |
| 31 | 16.2810 | 0.0614 | 162.2632 | 0.0062 | 9.9664 | 0.1003 | 8.5899 | 85.6109 |
| 32 | 17.8143 | 0.0561 | 178.5442 | 0.0056 | 10.0225 | 0.0998 | 8.7155 | 87.3511 |
| 33 | 19.4919 | 0.0513 | 196.3585 | 0.0051 | 10.0738 | 0.0993 | 8.8340 | 88.9928 |
| 34 | 21.3276 | 0.0469 | 215.8504 | 0.0046 | 10.1207 | 0.0988 | 8.9460 | 90.5401 |
| 35 | 23.3361 | 0.0429 | 237.1780 | 0.0042 | 10.1636 | 0.0984 | 9.0516 | 91.9970 |
| 40 | 36.5982 | 0.0273 | 378.0038 | 0.0026 | 10.3285 | 0.0968 | 9.4950 | 98.0684 |
| 45 | 57.3975 | 0.0174 | 598.8626 | 0.0017 | 10.4336 | 0.0958 | 9.8207 | 102.4654 |
| 50 | 90.0171 | 0.0111 | 945.2382 | 0.0011 | 10.5006 | 0.0952 | 10.0569 | 105.6042 |
| 55 | 141.1750 | 0.0071 | 1488.4633 | 0.0007 | 10.5434 | 0.0948 | 10.2262 | 107.8193 |
| 60 | 221.4064 | 0.0045 | 2340.4098 | 0.0004 | 10.5707 | 0.0946 | 10.3464 | 109.3680 |
| 65 | 347.2344 | 0.0029 | 3676.5279 | 0.0003 | 10.5880 | 0.0944 | 10.4309 | 110.4424 |
| 70 | 544.5719 | 0.0018 | 5771.9782 | 0.0002 | 10.5991 | 0.0943 | 10.4898 | 111.1829 |
| 75 | 854.0588 | 0.0012 | 9058.2984 | 0.0001 | 10.6062 | 0.0943 | 10.5307 | 111.6904 |
| 80 | 1339.4308 | 0.0007 | 14212.2744 | 0.0001 | 10.6107 | 0.0942 | 10.5588 | 112.0365 |
| 85 | 2100.6456 | 0.0005 | 22295.3179 | 0.0000 | 10.6136 | 0.0942 | 10.5781 | 112.2715 |
| 90 | 3294.4681 | 0.0003 | 34972.0534 | 0.0000 | 10.6154 | 0.0942 | 10.5913 | 112.4306 |
| 95 | 5166.7544 | 0.0002 | 54853.1321 | 0.0000 | 10.6166 | 0.0942 | 10.6002 | 112.5378 |
| 100 | 8103.0839 | 0.0001 | 86032.8702 | 0.0000 | 10.6173 | 0.0942 | 10.6063 | 112.6099 |

| | Single payment | | Uniform series | | | | Gradient series | |
|---|---|---|---|---|---|---|---|---|
| | Compound amount factor | Present worth factor | Compound amount factor | Sinking fund factor | Present worth factor | Capital recovery factor | Uniform series factor | Present worth factor |
| $n$ | To find $F$ given $P$ $F\|P\,r,n$ | To find $P$ given $F$ $P\|F\,r,n$ | To find $F$ given $A$ $F\|A\,r,n$ | To find $A$ given $F$ $A\|F\,r,n$ | To find $P$ given $A$ $P\|A\,r,n$ | To find $A$ given $P$ $A\|P\,r,n$ | To find $A$ given $G$ $A\|G\,r,n$ | To find $P$ given $G$ $P\|G\,r,n$ |
| 1 | 1.1052 | 0.9048 | 1.0000 | 1.0000 | 0.9048 | 1.1052 | 0.0000 | 0.0000 |
| 2 | 1.2214 | 0.8187 | 2.1052 | 0.4750 | 1.7236 | 0.5802 | 0.4750 | 0.8187 |
| 3 | 1.3499 | 0.7408 | 3.3266 | 0.3006 | 2.4644 | 0.4058 | 0.9334 | 2.3004 |
| 4 | 1.4918 | 0.6703 | 4.6764 | 0.2138 | 3.1347 | 0.3190 | 1.3754 | 4.3113 |
| 5 | 1.6487 | 0.6065 | 6.1683 | 0.1621 | 3.7412 | 0.2673 | 1.8009 | 6.7374 |
| 6 | 1.8221 | 0.5488 | 7.8170 | 0.1279 | 4.2900 | 0.2331 | 2.2101 | 9.4815 |
| 7 | 2.0138 | 0.4966 | 9.6391 | 0.1037 | 4.7866 | 0.2089 | 2.6033 | 12.4610 |
| 8 | 2.2255 | 0.4493 | 11.6528 | 0.0858 | 5.2360 | 0.1910 | 2.9806 | 15.6063 |
| 9 | 2.4596 | 0.4066 | 13.8784 | 0.0721 | 5.6425 | 0.1772 | 3.3423 | 18.8589 |
| 10 | 2.7183 | 0.3679 | 16.3380 | 0.0612 | 6.0104 | 0.1664 | 3.6886 | 22.1698 |
| 11 | 3.0042 | 0.3329 | 19.0563 | 0.0525 | 6.3433 | 0.1576 | 4.0198 | 25.4985 |
| 12 | 3.3201 | 0.3012 | 22.0604 | 0.0453 | 6.6445 | 0.1505 | 4.3362 | 28.8116 |
| 13 | 3.6693 | 0.2725 | 25.3806 | 0.0394 | 6.9170 | 0.1446 | 4.6381 | 32.0820 |
| 14 | 4.0552 | 0.2466 | 29.0499 | 0.0344 | 7.1636 | 0.1396 | 4.9260 | 35.2878 |
| 15 | 4.4817 | 0.2231 | 33.1051 | 0.0302 | 7.3867 | 0.1354 | 5.2001 | 38.4116 |
| 16 | 4.9530 | 0.2019 | 37.5867 | 0.0266 | 7.5886 | 0.1318 | 5.4608 | 41.4401 |
| 17 | 5.4739 | 0.1827 | 42.5398 | 0.0235 | 7.7713 | 0.1287 | 5.7086 | 44.3630 |
| 18 | 6.0496 | 0.1653 | 48.0137 | 0.0208 | 7.9366 | 0.1260 | 5.9437 | 47.1731 |
| 19 | 6.6859 | 0.1496 | 54.0634 | 0.0185 | 8.0862 | 0.1237 | 6.1667 | 49.8653 |
| 20 | 7.3891 | 0.1353 | 60.7483 | 0.0165 | 8.2215 | 0.1216 | 6.3780 | 52.4367 |
| 21 | 8.1662 | 0.1225 | 68.1383 | 0.0147 | 8.3440 | 0.1198 | 6.5779 | 54.8858 |
| 22 | 9.0250 | 0.1108 | 76.3045 | 0.0131 | 8.4548 | 0.1183 | 6.7669 | 57.2127 |
| 23 | 9.9742 | 0.1003 | 85.3295 | 0.0117 | 8.5550 | 0.1169 | 6.9454 | 59.4184 |
| 24 | 11.0232 | 0.0907 | 95.3037 | 0.0105 | 8.6458 | 0.1157 | 7.1139 | 61.5049 |
| 25 | 12.1825 | 0.0821 | 106.3269 | 0.0094 | 8.7278 | 0.1146 | 7.2727 | 63.4749 |
| 26 | 13.4637 | 0.0743 | 118.5094 | 0.0084 | 8.8021 | 0.1136 | 7.4223 | 65.3318 |
| 27 | 14.8797 | 0.0672 | 131.9731 | 0.0076 | 8.8693 | 0.1127 | 7.5630 | 67.0791 |
| 28 | 16.4446 | 0.0608 | 146.8528 | 0.0068 | 8.9301 | 0.1120 | 7.6954 | 68.7210 |
| 29 | 18.1741 | 0.0550 | 163.2975 | 0.0061 | 8.9852 | 0.1113 | 7.8197 | 70.2616 |
| 30 | 20.0855 | 0.0498 | 181.4716 | 0.0055 | 9.0349 | 0.1107 | 7.9365 | 71.7054 |
| 31 | 22.1980 | 0.0450 | 201.5572 | 0.0050 | 9.0800 | 0.1101 | 8.0459 | 73.0569 |
| 32 | 24.5325 | 0.0408 | 223.7551 | 0.0045 | 9.1208 | 0.1096 | 8.1485 | 74.3206 |
| 33 | 27.1126 | 0.0369 | 248.2876 | 0.0040 | 9.1576 | 0.1092 | 8.2446 | 75.5008 |
| 34 | 29.9641 | 0.0334 | 275.4003 | 0.0036 | 9.1910 | 0.1088 | 8.3345 | 76.6021 |
| 35 | 33.1155 | 0.0302 | 305.3644 | 0.0033 | 9.2212 | 0.1084 | 8.4185 | 77.6288 |
| 40 | 54.5982 | 0.0183 | 509.6290 | 0.0020 | 9.3342 | 0.1071 | 8.7620 | 81.7864 |
| 45 | 90.0171 | 0.0111 | 846.4044 | 0.0012 | 9.4027 | 0.1064 | 9.0028 | 84.6508 |
| 50 | 148.4132 | 0.0067 | 1401.6532 | 0.0007 | 9.4443 | 0.1059 | 9.1691 | 86.5959 |
| 55 | 244.6919 | 0.0041 | 2317.1038 | 0.0004 | 9.4695 | 0.1056 | 9.2826 | 87.9017 |
| 60 | 403.4288 | 0.0025 | 3826.4266 | 0.0003 | 9.4848 | 0.1054 | 9.3592 | 88.7701 |
| 65 | 665.1416 | 0.0015 | 6314.8791 | 0.0002 | 9.4940 | 0.1053 | 9.4105 | 89.3433 |
| 70 | 1096.6332 | 0.0009 | 10417.6438 | 0.0001 | 9.4997 | 0.1053 | 9.4444 | 89.7190 |
| 75 | 1808.0424 | 0.0006 | 17181.9591 | 0.0001 | 9.5031 | 0.1052 | 9.4668 | 89.9640 |
| 80 | 2980.9580 | 0.0003 | 28334.4297 | 0.0000 | 9.5051 | 0.1052 | 9.4815 | 90.1229 |
| 85 | 4914.7688 | 0.0002 | 46721.7452 | 0.0000 | 9.5064 | 0.1052 | 9.4910 | 90.2255 |
| 90 | 8103.0839 | 0.0001 | 77037.3034 | 0.0000 | 9.5072 | 0.1052 | 9.4972 | 90.2916 |
| 95 | 13359.7268 | 0.0001 | 127019.2091 | 0.0000 | 9.5076 | 0.1052 | 9.5012 | 90.3340 |
| 100 | 22026.4658 | 0.0000 | 209425.4400 | 0.0000 | 9.5079 | 0.1052 | 9.5038 | 90.3611 |

## TABLE B.14  Continuous Compounding: $r = 12\%$

| | Single payment | | Uniform series | | | | Gradient series | |
|---|---|---|---|---|---|---|---|---|
| | Compound amount factor | Present worth factor | Compound amount factor | Sinking fund factor | Present worth factor | Capital recovery factor | Uniform series factor | Present worth factor |
| $n$ | To find $F$ given $P$ $F\|P\ r,n$ | To find $P$ given $F$ $P\|F\ r,n$ | To find $F$ given $A$ $F\|A\ r,n$ | To find $A$ given $F$ $A\|F\ r,n$ | To find $P$ given $A$ $P\|A\ r,n$ | To find $A$ given $P$ $A\|P\ r,n$ | To find $A$ given $G$ $A\|G\ r,n$ | To find $P$ given $G$ $P\|G\ r,n$ |
| 1 | 1.1275 | 0.8869 | 1.0000 | 1.0000 | 0.8869 | 1.1275 | 0.0000 | 0.0000 |
| 2 | 1.2712 | 0.7866 | 2.1275 | 0.4700 | 1.6735 | 0.5975 | 0.4700 | 0.7866 |
| 3 | 1.4333 | 0.6977 | 3.3987 | 0.2942 | 2.3712 | 0.4217 | 0.9202 | 2.1820 |
| 4 | 1.6161 | 0.6188 | 4.8321 | 0.2070 | 2.9900 | 0.3344 | 1.3506 | 4.0383 |
| 5 | 1.8221 | 0.5488 | 6.4481 | 0.1551 | 3.5388 | 0.2826 | 1.7615 | 6.2336 |
| 6 | 2.0544 | 0.4868 | 8.2703 | 0.1209 | 4.0256 | 0.2484 | 2.1531 | 8.6673 |
| 7 | 2.3164 | 0.4317 | 10.3247 | 0.0969 | 4.4573 | 0.2244 | 2.5257 | 11.2576 |
| 8 | 2.6117 | 0.3829 | 12.6411 | 0.0791 | 4.8402 | 0.2066 | 2.8796 | 13.9379 |
| 9 | 2.9447 | 0.3396 | 15.2528 | 0.0656 | 5.1798 | 0.1931 | 3.2153 | 16.6546 |
| 10 | 3.3201 | 0.3012 | 18.1974 | 0.0550 | 5.4810 | 0.1824 | 3.5332 | 19.3654 |
| 11 | 3.7434 | 0.2671 | 21.5176 | 0.0465 | 5.7481 | 0.1740 | 3.8337 | 22.0367 |
| 12 | 4.2207 | 0.2369 | 25.2610 | 0.0396 | 5.9850 | 0.1671 | 4.1174 | 24.6429 |
| 13 | 4.7588 | 0.2101 | 29.4817 | 0.0339 | 6.1952 | 0.1614 | 4.3848 | 27.1646 |
| 14 | 5.3656 | 0.1864 | 34.2405 | 0.0292 | 6.3815 | 0.1567 | 4.6364 | 29.5874 |
| 15 | 6.0496 | 0.1653 | 39.6061 | 0.0252 | 6.5468 | 0.1527 | 4.8728 | 31.9016 |
| 16 | 6.8210 | 0.1466 | 45.6557 | 0.0219 | 6.6934 | 0.1494 | 5.0946 | 34.1007 |
| 17 | 7.6906 | 0.1300 | 52.4767 | 0.0191 | 6.8235 | 0.1466 | 5.3025 | 36.1812 |
| 18 | 8.6711 | 0.1153 | 60.1673 | 0.0166 | 6.9388 | 0.1441 | 5.4969 | 38.1417 |
| 19 | 9.7767 | 0.1023 | 68.8384 | 0.0145 | 7.0411 | 0.1420 | 5.6785 | 39.9828 |
| 20 | 11.0232 | 0.0907 | 78.6151 | 0.0127 | 7.1318 | 0.1402 | 5.8480 | 41.7064 |
| 21 | 12.4286 | 0.0805 | 89.6383 | 0.0112 | 7.2123 | 0.1387 | 6.0058 | 43.3156 |
| 22 | 14.0132 | 0.0714 | 102.0669 | 0.0098 | 7.2836 | 0.1373 | 6.1527 | 44.8142 |
| 23 | 15.7998 | 0.0633 | 116.0801 | 0.0086 | 7.3469 | 0.1361 | 6.2893 | 46.2066 |
| 24 | 17.8143 | 0.0561 | 131.8799 | 0.0076 | 7.4030 | 0.1351 | 6.4160 | 47.4977 |
| 25 | 20.0855 | 0.0498 | 149.6942 | 0.0067 | 7.4528 | 0.1342 | 6.5334 | 48.6926 |
| 26 | 22.6464 | 0.0442 | 169.7797 | 0.0059 | 7.4970 | 0.1334 | 6.6422 | 49.7966 |
| 27 | 25.5337 | 0.0392 | 192.4261 | 0.0052 | 7.5362 | 0.1327 | 6.7428 | 50.8148 |
| 28 | 28.7892 | 0.0347 | 217.9598 | 0.0046 | 7.5709 | 0.1321 | 6.8357 | 51.7527 |
| 29 | 32.4597 | 0.0308 | 246.7490 | 0.0041 | 7.6017 | 0.1315 | 6.9215 | 52.6153 |
| 30 | 36.5982 | 0.0273 | 279.2087 | 0.0036 | 7.6290 | 0.1311 | 7.0006 | 53.4077 |
| 31 | 41.2644 | 0.0242 | 315.8070 | 0.0032 | 7.6533 | 0.1307 | 7.0734 | 54.1347 |
| 32 | 46.5255 | 0.0215 | 357.0714 | 0.0028 | 7.6747 | 0.1303 | 7.1404 | 54.8010 |
| 33 | 52.4573 | 0.0191 | 403.5968 | 0.0025 | 7.6938 | 0.1300 | 7.2020 | 55.4110 |
| 34 | 59.1455 | 0.0169 | 456.0542 | 0.0022 | 7.7107 | 0.1297 | 7.2586 | 55.9690 |
| 35 | 66.6863 | 0.0150 | 515.1996 | 0.0019 | 7.7257 | 0.1294 | 7.3105 | 56.4788 |
| 40 | 121.5104 | 0.0082 | 945.2031 | 0.0011 | 7.7788 | 0.1286 | 7.5114 | 58.4296 |
| 45 | 221.4064 | 0.0045 | 1728.7205 | 0.0006 | 7.8079 | 0.1281 | 7.6392 | 59.6459 |
| 50 | 403.4288 | 0.0025 | 3156.3822 | 0.0003 | 7.8239 | 0.1278 | 7.7191 | 60.3933 |

| | Single payment | | Uniform series | | | | Gradient series | |
|---|---|---|---|---|---|---|---|---|
| | Compound amount factor | Present worth factor | Compound amount factor | Sinking fund factor | Present worth factor | Capital recovery factor | Uniform series factor | Present worth factor |
| $n$ | To find $F$ given $P$ $F\|P\ r,n$ | To find $P$ given $F$ $P\|F\ r,n$ | To find $F$ given $A$ $F\|A\ r,n$ | To find $A$ given $F$ $A\|F\ r,n$ | To find $P$ given $A$ $P\|A\ r,n$ | To find $A$ given $P$ $A\|P\ r,n$ | To find $A$ given $G$ $A\|G\ r,n$ | To find $P$ given $G$ $P\|G\ r,n$ |
| 1 | 1.1618 | 0.8607 | 1.0000 | 1.0000 | 0.8607 | 1.1618 | 0.0000 | 0.0000 |
| 2 | 1.3499 | 0.7408 | 2.1618 | 0.4626 | 1.6015 | 0.6244 | 0.4626 | 0.7408 |
| 3 | 1.5683 | 0.6376 | 3.5117 | 0.2848 | 2.2392 | 0.4466 | 0.9004 | 2.0161 |
| 4 | 1.8221 | 0.5488 | 5.0800 | 0.1969 | 2.7880 | 0.3587 | 1.3137 | 3.6625 |
| 5 | 2.1170 | 0.4724 | 6.9021 | 0.1449 | 3.2603 | 0.3067 | 1.7029 | 5.5520 |
| 6 | 2.4596 | 0.4066 | 9.0191 | 0.1109 | 3.6669 | 0.2727 | 2.0685 | 7.5848 |
| 7 | 2.8577 | 0.3499 | 11.4787 | 0.0871 | 4.0168 | 0.2490 | 2.4110 | 9.6845 |
| 8 | 3.3201 | 0.3012 | 14.3364 | 0.0698 | 4.3180 | 0.2316 | 2.7311 | 11.7928 |
| 9 | 3.8574 | 0.2592 | 17.6565 | 0.0566 | 4.5773 | 0.2185 | 3.0295 | 13.8667 |
| 10 | 4.4817 | 0.2231 | 21.5139 | 0.0465 | 4.8004 | 0.2083 | 3.3070 | 15.8749 |
| 11 | 5.2070 | 0.1920 | 25.9956 | 0.0385 | 4.9925 | 0.2003 | 3.5645 | 17.7954 |
| 12 | 6.0496 | 0.1653 | 31.2026 | 0.0320 | 5.1578 | 0.1939 | 3.8028 | 19.6137 |
| 13 | 7.0287 | 0.1423 | 37.2522 | 0.0268 | 5.3000 | 0.1887 | 4.0228 | 21.3210 |
| 14 | 8.1662 | 0.1225 | 44.2809 | 0.0226 | 5.4225 | 0.1844 | 4.2255 | 22.9129 |
| 15 | 9.4877 | 0.1054 | 52.4471 | 0.0191 | 5.5279 | 0.1809 | 4.4119 | 24.3885 |
| 16 | 11.0232 | 0.0907 | 61.9348 | 0.0161 | 5.6186 | 0.1780 | 4.5829 | 25.7493 |
| 17 | 12.8071 | 0.0781 | 72.9580 | 0.0137 | 5.6967 | 0.1755 | 4.7394 | 26.9986 |
| 18 | 14.8797 | 0.0672 | 85.7651 | 0.0117 | 5.7639 | 0.1735 | 4.8823 | 28.1411 |
| 19 | 17.2878 | 0.0578 | 100.6448 | 0.0099 | 5.8217 | 0.1718 | 5.0126 | 29.1823 |
| 20 | 20.0855 | 0.0498 | 117.9326 | 0.0085 | 5.8715 | 0.1703 | 5.1312 | 30.1282 |
| 21 | 23.3361 | 0.0429 | 138.0182 | 0.0072 | 5.9144 | 0.1691 | 5.2390 | 30.9853 |
| 22 | 27.1126 | 0.0369 | 161.3542 | 0.0062 | 5.9513 | 0.1680 | 5.3367 | 31.7598 |
| 23 | 31.5004 | 0.0317 | 188.4669 | 0.0053 | 5.9830 | 0.1671 | 5.4251 | 32.4582 |
| 24 | 36.5982 | 0.0273 | 219.9673 | 0.0045 | 6.0103 | 0.1664 | 5.5050 | 33.0867 |
| 25 | 42.5211 | 0.0235 | 256.5655 | 0.0039 | 6.0338 | 0.1657 | 5.5771 | 33.6511 |
| 26 | 49.4024 | 0.0202 | 299.0866 | 0.0033 | 6.0541 | 0.1652 | 5.6420 | 34.1571 |
| 27 | 57.3975 | 0.0174 | 348.4890 | 0.0029 | 6.0715 | 0.1647 | 5.7004 | 34.6101 |
| 28 | 66.6863 | 0.0150 | 405.8865 | 0.0025 | 6.0865 | 0.1643 | 5.7529 | 35.0150 |
| 29 | 77.4785 | 0.0129 | 472.5728 | 0.0021 | 6.0994 | 0.1640 | 5.8000 | 35.3764 |
| 30 | 90.0171 | 0.0111 | 550.0513 | 0.0018 | 6.1105 | 0.1637 | 5.8421 | 35.6985 |
| 31 | 104.5850 | 0.0096 | 640.0684 | 0.0016 | 6.1201 | 0.1634 | 5.8799 | 35.9854 |
| 32 | 121.5104 | 0.0082 | 744.6534 | 0.0013 | 6.1283 | 0.1632 | 5.9136 | 36.2405 |
| 33 | 141.1750 | 0.0071 | 866.1638 | 0.0012 | 6.1354 | 0.1630 | 5.9437 | 36.4672 |
| 34 | 164.0219 | 0.0061 | 1007.3388 | 0.0010 | 6.1415 | 0.1628 | 5.9706 | 36.6684 |
| 35 | 190.5663 | 0.0052 | 1171.3607 | 0.0009 | 6.1467 | 0.1627 | 5.9945 | 36.8468 |
| 40 | 403.4288 | 0.0025 | 2486.6727 | 0.0004 | 6.1638 | 0.1622 | 6.0798 | 37.4747 |
| 45 | 854.0588 | 0.0012 | 5271.1883 | 0.0002 | 6.1719 | 0.1620 | 6.1264 | 37.8118 |
| 50 | 1808.0424 | 0.0006 | 11166.0078 | 0.0001 | 6.1757 | 0.1619 | 6.1515 | 37.9900 |

## TABLE B.16  Continuous Compounding: $r = 18\%$

| $n$ | Single payment Compound amount factor<br>To find $F$ given $P$<br>$F\|P\,r,n$ | Single payment Present worth factor<br>To find $P$ given $F$<br>$P\|F\,r,n$ | Uniform series Compound amount factor<br>To find $F$ given $A$<br>$F\|A\,r,n$ | Uniform series Sinking fund factor<br>To find $A$ given $F$<br>$A\|F\,r,n$ | Uniform series Present worth factor<br>To find $P$ given $A$<br>$P\|A\,r,n$ | Uniform series Capital recovery factor<br>To find $A$ given $P$<br>$A\|P\,r,n$ | Gradient series Uniform series factor<br>To find $A$ given $G$<br>$A\|G\,r,n$ | Gradient series Present worth factor<br>To find $P$ given $G$<br>$P\|G\,r,n$ |
|---|---|---|---|---|---|---|---|---|
| 1 | 1.1972 | 0.8353 | 1.0000 | 1.0000 | 0.8353 | 1.1972 | 0.0000 | 0.0000 |
| 2 | 1.4333 | 0.6977 | 2.1972 | 0.4551 | 1.5330 | 0.6523 | 0.4551 | 0.6977 |
| 3 | 1.7160 | 0.5827 | 3.6305 | 0.2754 | 2.1157 | 0.4727 | 0.8806 | 1.8623 |
| 4 | 2.0544 | 0.4868 | 5.3466 | 0.1870 | 2.6025 | 0.3843 | 1.2770 | 3.3234 |
| 5 | 2.4596 | 0.4066 | 7.4010 | 0.1351 | 3.0090 | 0.3323 | 1.6450 | 4.9497 |
| 6 | 2.9447 | 0.3396 | 9.8606 | 0.1014 | 3.3486 | 0.2986 | 1.9852 | 6.6477 |
| 7 | 3.5254 | 0.2837 | 12.8053 | 0.0781 | 3.6323 | 0.2753 | 2.2987 | 8.3496 |
| 8 | 4.2207 | 0.2369 | 16.3307 | 0.0612 | 3.8692 | 0.2585 | 2.5866 | 10.0081 |
| 9 | 5.0531 | 0.1979 | 20.5514 | 0.0487 | 4.0671 | 0.2459 | 2.8500 | 11.5913 |
| 10 | 6.0497 | 0.1653 | 25.6045 | 0.0391 | 4.2324 | 0.2363 | 3.0902 | 13.0790 |
| 11 | 7.2427 | 0.1381 | 31.6541 | 0.0316 | 4.3705 | 0.2288 | 3.3085 | 14.4596 |
| 12 | 8.6711 | 0.1153 | 38.8968 | 0.0257 | 4.4858 | 0.2229 | 3.5062 | 15.7282 |
| 13 | 10.3812 | 0.0963 | 47.5680 | 0.0210 | 4.5821 | 0.2182 | 3.6848 | 16.8842 |
| 14 | 12.4286 | 0.0805 | 57.9492 | 0.0173 | 4.6626 | 0.2145 | 3.8456 | 17.9301 |
| 15 | 14.8797 | 0.0672 | 70.3778 | 0.1421 | 4.7298 | 0.2114 | 3.9898 | 18.8710 |
| 16 | 17.8143 | 0.0561 | 85.2575 | 0.1173 | 4.7859 | 0.2089 | 4.1190 | 19.7130 |
| 17 | 21.3276 | 0.0469 | 103.0719 | 0.0970 | 4.8328 | 0.2069 | 4.2342 | 20.4632 |
| 18 | 25.5337 | 0.0392 | 124.3994 | 0.0804 | 4.8720 | 0.2053 | 4.3369 | 21.1290 |
| 19 | 30.5694 | 0.0327 | 149.9332 | 0.0667 | 4.9047 | 0.2039 | 4.4280 | 21.7178 |
| 20 | 36.5982 | 0.0273 | 180.5026 | 0.0554 | 4.9320 | 0.2028 | 4.5087 | 22.2370 |
| 21 | 43.8160 | 0.0228 | 217.1008 | 0.0461 | 4.9548 | 0.2018 | 4.5801 | 22.6934 |
| 22 | 52.4573 | 0.0191 | 260.9169 | 0.0383 | 4.9739 | 0.2011 | 4.6430 | 23.0938 |
| 23 | 62.8028 | 0.0159 | 313.3742 | 0.0319 | 4.9898 | 0.2004 | 4.6984 | 23.4441 |
| 24 | 75.1886 | 0.0133 | 376.1771 | 0.0266 | 5.0031 | 0.1999 | 4.7471 | 23.7500 |
| 25 | 90.0171 | 0.0111 | 451.3657 | 0.0222 | 5.0142 | 0.1994 | 4.7897 | 24.0166 |
| 26 | 107.7701 | 0.0093 | 541.3829 | 0.0185 | 5.0235 | 0.1991 | 4.8270 | 24.2486 |
| 27 | 129.0242 | 0.0078 | 649.1530 | 0.0154 | 5.0313 | 0.1988 | 4.8597 | 24.4501 |
| 28 | 154.4700 | 0.0065 | 778.1772 | 0.0129 | 5.0377 | 0.1985 | 4.8881 | 24.6249 |
| 29 | 184.9342 | 0.0054 | 932.6473 | 0.0107 | 5.0431 | 0.1983 | 4.9129 | 24.7763 |
| 30 | 221.4064 | 0.0045 | 1117.5816 | 0.0089 | 5.0476 | 0.1981 | 4.9344 | 24.9072 |
| 31 | 265.0716 | 0.0038 | 1338.9880 | 0.0075 | 5.0514 | 0.1980 | 4.9532 | 25.0204 |
| 32 | 317.3483 | 0.0032 | 1604.0597 | 0.0062 | 5.0546 | 0.1978 | 4.9694 | 25.1181 |
| 33 | 379.9349 | 0.0026 | 1921.4082 | 0.0052 | 5.0572 | 0.1977 | 4.9835 | 25.2023 |
| 34 | 454.8647 | 0.0022 | 2306.4138 | 0.0043 | 5.0594 | 0.1977 | 4.9956 | 25.2749 |
| 35 | 544.5719 | 0.0018 | 2756.2080 | 0.0036 | 5.0612 | 0.1976 | 5.0062 | 25.3373 |
| 40 | 1339.4308 | 0.0007 | 6786.5789 | 0.0015 | 5.0668 | 0.1974 | 5.0407 | 25.5398 |
| 45 | 3294.4681 | 0.0003 | 16699.6920 | 0.0006 | 5.0690 | 0.1973 | 5.0569 | 25.6333 |
| 50 | 8103.0839 | 0.0001 | 41082.0140 | 0.0002 | 5.0699 | 0.1972 | 5.0644 | 25.6760 |

| | Single payment | | Uniform series | | | | Gradient series | |
|---|---|---|---|---|---|---|---|---|
| | Compound amount factor | Present worth factor | Compound amount factor | Sinking fund factor | Present worth factor | Capital recovery factor | Uniform series factor | Present worth factor |
| $n$ | To find $F$ given $P$ $F\|P\ r,n$ | To find $P$ given $F$ $P\|F\ r,n$ | To find $F$ given $A$ $F\|A\ r,n$ | To find $A$ given $F$ $A\|F\ r,n$ | To find $P$ given $A$ $P\|A\ r,n$ | To find $A$ given $P$ $A\|P\ r,n$ | To find $A$ given $G$ $A\|G\ r,n$ | To find $P$ given $G$ $P\|G\ r,n$ |
| 1 | 1.2214 | 0.8187 | 1.0000 | 1.0000 | 0.8187 | 1.2214 | 0.0000 | 0.0000 |
| 2 | 1.4918 | 0.6703 | 2.2214 | 0.4502 | 1.4891 | 0.6716 | 0.4502 | 0.6703 |
| 3 | 1.8221 | 0.5488 | 3.7132 | 0.2693 | 2.0379 | 0.4907 | 0.8675 | 1.7679 |
| 4 | 2.2255 | 0.4493 | 5.5353 | 0.1807 | 2.4872 | 0.4021 | 1.2528 | 3.1159 |
| 5 | 2.7183 | 0.3679 | 7.7609 | 0.1289 | 2.8551 | 0.3503 | 1.6068 | 4.5874 |
| 6 | 3.3201 | 0.3012 | 10.4792 | 0.0954 | 3.1563 | 0.3168 | 1.9306 | 6.0934 |
| 7 | 4.0552 | 0.2466 | 13.7993 | 0.0725 | 3.4029 | 0.2939 | 2.2255 | 7.5730 |
| 8 | 4.9530 | 0.2019 | 17.8545 | 0.0560 | 3.6048 | 0.2774 | 2.4929 | 8.9863 |
| 9 | 6.0496 | 0.1653 | 22.8075 | 0.0438 | 3.7701 | 0.2652 | 2.7344 | 10.3087 |
| 10 | 7.3891 | 0.1353 | 28.8572 | 0.0347 | 3.9054 | 0.2561 | 2.9515 | 11.5267 |
| 11 | 9.0250 | 0.1108 | 36.2462 | 0.0276 | 4.0162 | 0.2490 | 3.1459 | 12.6347 |
| 12 | 11.0232 | 0.0907 | 45.2712 | 0.0221 | 4.1069 | 0.2435 | 3.3194 | 13.6326 |
| 13 | 13.4637 | 0.0743 | 56.2944 | 0.0178 | 4.1812 | 0.2392 | 3.4736 | 14.5239 |
| 14 | 16.4446 | 0.0608 | 69.7581 | 0.0143 | 4.2420 | 0.2357 | 3.6102 | 15.3144 |
| 15 | 20.0855 | 0.0498 | 86.2028 | 0.0116 | 4.2918 | 0.2330 | 3.7307 | 16.0114 |
| 16 | 24.5325 | 0.0408 | 106.2883 | 0.0094 | 4.3325 | 0.2308 | 3.8367 | 16.6229 |
| 17 | 29.9641 | 0.0334 | 130.8209 | 0.0076 | 4.3659 | 0.2290 | 3.9297 | 17.1569 |
| 18 | 36.5982 | 0.0273 | 160.7850 | 0.0062 | 4.3932 | 0.2276 | 4.0110 | 17.6214 |
| 19 | 44.7012 | 0.0224 | 197.3832 | 0.0051 | 4.4156 | 0.2265 | 4.0819 | 18.0240 |
| 20 | 54.5982 | 0.0183 | 242.0844 | 0.0041 | 4.4339 | 0.2255 | 4.1435 | 18.3720 |
| 21 | 66.6863 | 0.0150 | 296.6825 | 0.0034 | 4.4489 | 0.2248 | 4.1970 | 18.6719 |
| 22 | 81.4509 | 0.0123 | 363.3689 | 0.0028 | 4.4612 | 0.2242 | 4.2432 | 18.9298 |
| 23 | 99.4843 | 0.0101 | 444.8197 | 0.0022 | 4.4713 | 0.2237 | 4.2831 | 19.1509 |
| 24 | 121.5104 | 0.0082 | 544.3040 | 0.0018 | 4.4795 | 0.2232 | 4.3175 | 19.3402 |
| 25 | 148.4132 | 0.0067 | 665.8145 | 0.0015 | 4.4862 | 0.2229 | 4.3471 | 19.5019 |
| 26 | 181.2722 | 0.0055 | 814.2276 | 0.0012 | 4.4917 | 0.2226 | 4.3724 | 19.6398 |
| 27 | 221.4064 | 0.0045 | 995.4999 | 0.0010 | 4.4963 | 0.2224 | 4.3942 | 19.7572 |
| 28 | 270.4264 | 0.0037 | 1216.9063 | 0.0008 | 4.5000 | 0.2222 | 4.4127 | 19.8571 |
| 29 | 330.2996 | 0.0030 | 1487.3327 | 0.0007 | 4.5030 | 0.2221 | 4.4286 | 19.9419 |
| 30 | 403.4288 | 0.0025 | 1817.6323 | 0.0006 | 4.5055 | 0.2220 | 4.4421 | 20.0137 |
| 31 | 492.7490 | 0.0020 | 2221.0610 | 0.0005 | 4.5075 | 0.2219 | 4.4536 | 20.0746 |
| 32 | 601.8450 | 0.0017 | 2713.8101 | 0.0004 | 4.5092 | 0.2218 | 4.4634 | 20.1261 |
| 33 | 735.0952 | 0.0014 | 3315.6551 | 0.0003 | 4.5105 | 0.2217 | 4.4717 | 20.1697 |
| 34 | 897.8473 | 0.0011 | 4050.7503 | 0.0002 | 4.5116 | 0.2216 | 4.4787 | 20.2064 |
| 35 | 1096.6332 | 0.0009 | 4948.5976 | 0.0002 | 4.5125 | 0.2216 | 4.4847 | 20.2374 |
| 40 | 2980.9580 | 0.0003 | 13459.4438 | 0.0001 | 4.5151 | 0.2215 | 4.5032 | 20.3327 |
| 45 | 8103.0839 | 0.0001 | 36594.3225 | 0.0000 | 4.5161 | 0.2214 | 4.5111 | 20.3726 |
| 50 | 22026.4658 | 0.0000 | 99481.4427 | 0.0000 | 4.5165 | 0.2214 | 4.5144 | 20.3890 |

| | Single payment | | Uniform series | | | | Gradient series | |
|---|---|---|---|---|---|---|---|---|
| | Compound amount factor | Present worth factor | Compound amount factor | Sinking fund factor | Present worth factor | Capital recovery factor | Uniform series factor | Present worth factor |
| $n$ | To find $F$ given $P$ $F\|P\ r,n$ | To find $P$ given $F$ $P\|F\ r,n$ | To find $F$ given $A$ $F\|A\ r,n$ | To find $A$ given $F$ $A\|F\ r,n$ | To find $P$ given $A$ $P\|A\ r,n$ | To find $A$ given $P$ $A\|P\ r,n$ | To find $A$ given $G$ $A\|G\ r,n$ | To find $P$ given $G$ $P\|G\ r,n$ |
| 1 | 1.2840 | 0.7788 | 1.0000 | 1.0000 | 0.7788 | 1.2840 | 0.0000 | 0.0000 |
| 2 | 1.6487 | 0.6065 | 2.2840 | 0.4378 | 1.3853 | 0.7218 | 0.4378 | 0.6065 |
| 3 | 2.1170 | 0.4724 | 3.9327 | 0.2543 | 1.8577 | 0.5383 | 0.8350 | 1.5513 |
| 4 | 2.7183 | 0.3679 | 6.0497 | 0.1653 | 2.2256 | 0.4493 | 1.1929 | 2.6549 |
| 5 | 3.4903 | 0.2865 | 8.7680 | 0.1141 | 2.5121 | 0.3981 | 1.5131 | 3.8009 |
| 6 | 4.4817 | 0.2231 | 12.2584 | 0.0816 | 2.7352 | 0.3656 | 1.7975 | 4.9166 |
| 7 | 5.7546 | 0.1738 | 16.7401 | 0.0597 | 2.9090 | 0.3438 | 2.0486 | 5.9592 |
| 8 | 7.3891 | 0.1353 | 22.4947 | 0.0445 | 3.0443 | 0.3285 | 2.2687 | 6.9066 |
| 9 | 9.4877 | 0.1054 | 29.8837 | 0.0335 | 3.1497 | 0.3175 | 2.4605 | 7.7498 |
| 10 | 12.1825 | 0.0821 | 39.3715 | 0.0254 | 3.2318 | 0.3094 | 2.6266 | 8.4885 |
| 11 | 15.6426 | 0.0639 | 51.5539 | 0.0194 | 3.2957 | 0.3034 | 2.7696 | 9.1278 |
| 12 | 20.0855 | 0.0498 | 67.1966 | 0.0149 | 3.3455 | 0.2989 | 2.8921 | 9.6755 |
| 13 | 25.7903 | 0.0388 | 87.2821 | 0.0115 | 3.3843 | 0.2955 | 2.9964 | 10.1407 |
| 14 | 33.1155 | 0.0302 | 113.0725 | 0.0088 | 3.4145 | 0.2929 | 3.0849 | 10.5333 |
| 15 | 42.5211 | 0.0235 | 146.1879 | 0.0068 | 3.4380 | 0.2909 | 3.1595 | 10.8626 |
| 16 | 54.5982 | 0.0183 | 188.7090 | 0.0053 | 3.4563 | 0.2893 | 3.2223 | 11.1373 |
| 17 | 70.1054 | 0.0143 | 243.3071 | 0.0041 | 3.4706 | 0.2881 | 3.2748 | 11.3655 |
| 18 | 90.0171 | 0.0111 | 313.4126 | 0.0032 | 3.4817 | 0.2872 | 3.3186 | 11.5544 |
| 19 | 115.5843 | 0.0087 | 403.4297 | 0.0025 | 3.4904 | 0.2865 | 3.3550 | 11.7101 |
| 20 | 148.4132 | 0.0067 | 519.0140 | 0.0019 | 3.4971 | 0.2860 | 3.3851 | 11.8381 |
| 21 | 190.5663 | 0.0052 | 667.4271 | 0.0015 | 3.5023 | 0.2855 | 3.4100 | 11.9431 |
| 22 | 244.6919 | 0.0041 | 857.9934 | 0.0012 | 3.5064 | 0.2852 | 3.4305 | 12.0289 |
| 23 | 314.1907 | 0.0032 | 1102.6853 | 0.0009 | 3.5096 | 0.2849 | 3.4474 | 12.0989 |
| 24 | 403.4288 | 0.0025 | 1416.8760 | 0.0007 | 3.5121 | 0.2847 | 3.4612 | 12.1559 |
| 25 | 518.0128 | 0.0019 | 1820.3048 | 0.0005 | 3.5140 | 0.2846 | 3.4725 | 12.2023 |
| 26 | 665.1416 | 0.0015 | 2338.3176 | 0.0004 | 3.5155 | 0.2845 | 3.4817 | 12.2399 |
| 27 | 854.0588 | 0.0012 | 3003.4592 | 0.0003 | 3.5167 | 0.2844 | 3.4892 | 12.2703 |
| 28 | 1096.6332 | 0.0009 | 3857.5180 | 0.0003 | 3.5176 | 0.2843 | 3.4953 | 12.2949 |
| 29 | 1408.1048 | 0.0007 | 4954.1512 | 0.0002 | 3.5183 | 0.2842 | 3.5002 | 12.3148 |
| 30 | 1808.0424 | 0.0006 | 6362.2560 | 0.0002 | 3.5189 | 0.2842 | 3.5042 | 12.3308 |
| 31 | 2321.5724 | 0.0004 | 8170.2984 | 0.0001 | 3.5193 | 0.2841 | 3.5075 | 12.3438 |
| 32 | 2980.9580 | 0.0003 | 10491.8708 | 0.0001 | 3.5196 | 0.2841 | 3.5101 | 12.3542 |
| 33 | 3827.6258 | 0.0003 | 13472.8288 | 0.0001 | 3.5199 | 0.2841 | 3.5122 | 12.3625 |
| 34 | 4914.7688 | 0.0002 | 17300.4546 | 0.0001 | 3.5201 | 0.2841 | 3.5139 | 12.3692 |
| 35 | 6310.6881 | 0.0002 | 22215.2235 | 0.0000 | 3.5203 | 0.2841 | 3.5153 | 12.3746 |

| | Single payment | | Uniform series | | | | Gradient series | |
|---|---|---|---|---|---|---|---|---|
| | Compound amount factor | Present worth factor | Compound amount factor | Sinking fund factor | Present worth factor | Capital recovery factor | Uniform series factor | Present worth factor |
| $n$ | To find $F$ given $P$ $(F|P\,r,n)_x$ | To find $P$ given $F$ $(P|F\,r,n)_x$ | To find $F$ given $A$ $(F|A\,r,n)_x$ | To find $A$ given $F$ $(A|F\,r,n)_x$ | To find $P$ given $A$ $(P|A\,r,n)_x$ | To find $A$ given $P$ $(A|P\,r,n)_x$ | To find $A$ given $G$ $(A|G\,r,n)_x$ | To find $P$ given $G$ $(P|G\,r,n)_x$ |
| 1 | 1.3499 | 0.7408 | 1.0000 | 1.0000 | 0.7408 | 1.3449 | 0.0000 | 0.0000 |
| 2 | 1.8221 | 0.5488 | 2.3499 | 0.4256 | 1.2896 | 0.7754 | 0.4256 | 0.5488 |
| 3 | 2.4596 | 0.4066 | 4.1720 | 0.2397 | 1.6962 | 0.5896 | 0.8029 | 1.3620 |
| 4 | 3.3201 | 0.3012 | 6.6316 | 0.1508 | 1.9974 | 0.5007 | 1.1342 | 2.2655 |
| 5 | 4.4817 | 0.2231 | 9.9517 | 0.1005 | 2.2205 | 0.4503 | 1.4222 | 3.1581 |
| 6 | 6.0496 | 0.1653 | 14.4334 | 0.0693 | 2.3858 | 0.4191 | 1.6701 | 3.9846 |
| 7 | 8.1662 | 0.1225 | 20.4830 | 0.0488 | 2.5083 | 0.3987 | 1.8815 | 4.7193 |
| 8 | 11.0232 | 0.0907 | 28.6492 | 0.0349 | 2.5990 | 0.3848 | 2.0601 | 5.3543 |
| 9 | 14.8797 | 0.0672 | 39.6724 | 0.0252 | 2.6662 | 0.3751 | 2.2099 | 5.8920 |
| 10 | 20.0855 | 0.0498 | 54.5522 | 0.0183 | 2.7160 | 0.3682 | 2.3343 | 6.3401 |
| 11 | 27.1126 | 0.0369 | 74.6377 | 0.0134 | 2.7529 | 0.3633 | 2.4370 | 6.7089 |
| 12 | 36.5982 | 0.0273 | 101.7504 | 0.0098 | 2.7802 | 0.3597 | 2.5212 | 7.0095 |
| 13 | 49.4024 | 0.0202 | 138.3486 | 0.0072 | 2.8004 | 0.3571 | 2.5897 | 7.2524 |
| 14 | 66.6863 | 0.0150 | 187.7510 | 0.0053 | 2.8154 | 0.3552 | 2.6452 | 7.4473 |
| 15 | 90.0170 | 0.0111 | 254.4374 | 0.0039 | 2.8265 | 0.3538 | 2.6898 | 7.6028 |
| 16 | 121.5102 | 0.0082 | 344.4544 | 0.0029 | 2.8348 | 0.3528 | 2.7255 | 7.7263 |
| 17 | 164.0217 | 0.0061 | 465.9646 | 0.0021 | 2.8409 | 0.3520 | 2.7540 | 7.8238 |
| 18 | 221.4061 | 0.0045 | 629.9866 | 0.0016 | 2.8454 | 0.3514 | 2.7766 | 7.9006 |
| 19 | 298.8667 | 0.0033 | 851.3928 | 0.0012 | 2.8487 | 0.3510 | 2.7945 | 7.9608 |
| 20 | 403.4280 | 0.0025 | 1150.2603 | 0.0009 | 2.8512 | 0.3507 | 2.8086 | 8.0079 |
| 21 | 544.5708 | 0.0018 | 1553.6890 | 0.0006 | 2.8531 | 0.3505 | 2.8197 | 8.0447 |
| 22 | 735.0940 | 0.0014 | 2098.2608 | 0.0005 | 2.8544 | 0.3503 | 2.8283 | 8.0732 |
| 23 | 992.2730 | 0.0010 | 2833.3550 | 0.0004 | 2.8554 | 0.3502 | 2.8351 | 8.0954 |
| 24 | 1339.4285 | 0.0007 | 3825.6304 | 0.0003 | 2.8562 | 0.3501 | 2.8404 | 8.1126 |
| 25 | 1808.0383 | 0.0006 | 5165.0547 | 0.0002 | 2.8567 | 0.3501 | 2.8445 | 8.1258 |
| 26 | 2440.5967 | 0.0004 | 6973.0977 | 0.0001 | 2.8571 | 0.3500 | 2.8476 | 8.1361 |
| 27 | 3294.4622 | 0.0003 | 9413.6992 | 0.0001 | 2.8574 | 0.3500 | 2.8501 | 8.1440 |
| 28 | 4447.0547 | 0.0002 | 12708.1641 | 0.0001 | 2.8577 | 0.3499 | 2.8520 | 8.1500 |
| 29 | 6002.8945 | 0.0002 | 17155.2227 | 0.0001 | 2.8578 | 0.3499 | 2.8535 | 8.1547 |
| 30 | 8103.0625 | 0.0001 | 23158.1328 | 0.0000 | 2.8579 | 0.3499 | 2.8546 | 8.1583 |
| 31 | 10937.9922 | 0.0001 | 31261.2070 | 0.0000 | 2.8580 | 0.3499 | 2.8555 | 8.1610 |
| 32 | 14764.7500 | 0.0001 | 42199.2188 | 0.0000 | 2.8581 | 0.3499 | 2.8561 | 8.1631 |
| 33 | 19930.3281 | 0.0001 | 56963.9922 | 0.0000 | 2.8582 | 0.3499 | 2.8566 | 8.1647 |
| 34 | 26903.1133 | 0.0000 | 76894.2500 | 0.0000 | 2.8582 | 0.3499 | 2.8570 | 8.1660 |
| 35 | 36315.4141 | 0.0000 | 103797.4375 | 0.0000 | 2.8582 | 0.3499 | 2.8573 | 8.1669 |

## TABLE B.20  Continuous Compounding: $r = 40\%$

| | Single payment | | Uniform series | | | | Gradient series | |
|---|---|---|---|---|---|---|---|---|
| | Compound amount factor | Present worth factor | Compound amount factor | Sinking fund factor | Present worth factor | Capital recovery factor | Uniform series factor | Present worth factor |
| $n$ | To find $F$ given $P$ $F\|P\,r,n$ | To find $P$ given $F$ $P\|F\,r,n$ | To find $F$ given $A$ $F\|A\,r,n$ | To find $A$ given $F$ $A\|F\,r,n$ | To find $P$ given $A$ $P\|A\,r,n$ | To find $A$ given $P$ $A\|P\,r,n$ | To find $A$ given $G$ $A\|G\,r,n$ | To find $P$ given $G$ $P\|G\,r,n$ |
| 1 | 1.4918 | 0.6703 | 1.0000 | 1.0000 | 0.6703 | 1.4918 | 0.0000 | 0.0000 |
| 2 | 2.2255 | 0.4493 | 2.4918 | 0.4013 | 1.1197 | 0.8931 | 0.4013 | 0.4493 |
| 3 | 3.3201 | 0.3012 | 4.7174 | 0.2120 | 1.4208 | 0.7038 | 0.7402 | 1.0517 |
| 4 | 4.9530 | 0.2019 | 8.0375 | 0.1244 | 1.6227 | 0.6162 | 1.0214 | 1.6574 |
| 5 | 7.3891 | 0.1353 | 12.9905 | 0.0770 | 1.7581 | 0.5688 | 1.2507 | 2.1988 |
| 6 | 11.0232 | 0.0907 | 20.3796 | 0.0491 | 1.8488 | 0.5409 | 1.4346 | 2.6523 |
| 7 | 16.4446 | 0.0608 | 31.4027 | 0.0318 | 1.9096 | 0.5237 | 1.5800 | 3.0172 |
| 8 | 24.5325 | 0.0408 | 47.8474 | 0.0209 | 1.9504 | 0.5127 | 1.6933 | 3.3025 |
| 9 | 36.5982 | 0.0273 | 72.3799 | 0.0138 | 1.9777 | 0.5056 | 1.7804 | 3.5211 |
| 10 | 54.5982 | 0.0183 | 108.9782 | 0.0092 | 1.9960 | 0.5010 | 1.8467 | 3.6860 |
| 11 | 81.4509 | 0.0123 | 163.5763 | 0.0061 | 2.0083 | 0.4979 | 1.8965 | 3.8087 |
| 12 | 121.5104 | 0.0082 | 245.0272 | 0.0041 | 2.0165 | 0.4959 | 1.9337 | 3.8993 |
| 13 | 181.2722 | 0.0055 | 366.5376 | 0.0027 | 2.0220 | 0.4946 | 1.9611 | 3.9655 |
| 14 | 270.4264 | 0.0037 | 547.8100 | 0.0018 | 2.0257 | 0.4937 | 1.9813 | 4.0135 |
| 15 | 403.4288 | 0.0025 | 818.2362 | 0.0012 | 2.0282 | 0.4930 | 1.9960 | 4.0482 |
| 16 | 601.8450 | 0.0017 | 1221.6650 | 0.0008 | 2.0299 | 0.4926 | 2.0066 | 4.0732 |
| 17 | 897.8473 | 0.0011 | 1823.5101 | 0.0005 | 2.0310 | 0.4924 | 2.0143 | 4.0910 |
| 18 | 1339.4308 | 0.0007 | 2721.3574 | 0.0004 | 2.0317 | 0.4922 | 2.0198 | 4.1037 |
| 19 | 1998.1959 | 0.0005 | 4060.7881 | 0.0002 | 2.0322 | 0.4921 | 2.0237 | 4.1127 |
| 20 | 2980.9580 | 0.0003 | 6058.9840 | 0.0002 | 2.0326 | 0.4920 | 2.0265 | 4.1191 |
| 21 | 4447.0667 | 0.0002 | 9039.9420 | 0.0001 | 2.0328 | 0.4919 | 2.0285 | 4.1236 |
| 22 | 6634.2440 | 0.0002 | 13487.0090 | 0.0001 | 2.0329 | 0.4919 | 2.0299 | 4.1267 |
| 23 | 9897.1291 | 0.0001 | 20121.2530 | 0.0000 | 2.0330 | 0.4919 | 2.0309 | 4.1289 |
| 24 | 14764.7820 | 0.0001 | 30018.3820 | 0.0000 | 2.0331 | 0.4919 | 2.0316 | 4.1305 |
| 25 | 22026.4660 | 0.0000 | 44783.1630 | 0.0000 | 2.0332 | 0.4918 | 2.0321 | 4.1316 |

| | Single payment | | Uniform series | | | | Gradient series | |
|---|---|---|---|---|---|---|---|---|
| | Compound amount factor | Present worth factor | Compound amount factor | Sinking fund factor | Present worth factor | Capital recovery factor | Uniform series factor | Present worth factor |
| $n$ | To find $F$ given $P$ $F\|P\,r,n$ | To find $P$ given $F$ $P\|F\,r,n$ | To find $F$ given $A$ $F\|A\,r,n$ | To find $A$ given $F$ $A\|F\,r,n$ | To find $P$ given $A$ $P\|A\,r,n$ | To find $A$ given $P$ $A\|P\,r,n$ | To find $A$ given $G$ $A\|G\,r,n$ | To find $P$ given $G$ $P\|G\,r,n$ |
| 1 | 1.6487 | 0.6065 | 1.0000 | 0.1000 | 0.6065 | 1.6487 | 0.0000 | 0.0000 |
| 2 | 2.7183 | 0.3679 | 2.6487 | 0.3775 | 0.9744 | 1.0263 | 0.3775 | 0.3679 |
| 3 | 4.4817 | 0.2231 | 5.3670 | 0.1863 | 1.1975 | 0.8350 | 0.6798 | 0.8141 |
| 4 | 7.3891 | 0.1353 | 9.8487 | 0.1015 | 1.3329 | 0.7503 | 0.9154 | 1.2202 |
| 5 | 12.1825 | 0.0821 | 17.2378 | 0.0580 | 1.4150 | 0.7067 | 1.0944 | 1.5485 |
| 6 | 20.0855 | 0.0498 | 29.4202 | 0.0340 | 1.4648 | 0.6827 | 1.2271 | 1.7974 |
| 7 | 33.1155 | 0.0302 | 49.5058 | 0.0202 | 1.4950 | 0.6689 | 1.3235 | 1.9786 |
| 8 | 54.5982 | 0.0183 | 82.6212 | 0.0121 | 1.5133 | 0.6608 | 1.3922 | 2.1068 |
| 9 | 90.0171 | 0.0111 | 137.2194 | 0.0073 | 1.5244 | 0.6560 | 1.4404 | 2.1957 |
| 10 | 148.4132 | 0.0067 | 227.2365 | 0.0044 | 1.5311 | 0.6531 | 1.4737 | 2.2563 |
| 11 | 244.6919 | 0.0041 | 375.6497 | 0.0027 | 1.5352 | 0.6514 | 1.4964 | 2.2972 |
| 12 | 403.4288 | 0.0025 | 620.3417 | 0.0016 | 1.5377 | 0.6503 | 1.5117 | 2.3245 |
| 13 | 665.1416 | 0.0015 | 1023.7705 | 0.0010 | 1.5392 | 0.6497 | 1.5219 | 2.3425 |
| 14 | 1096.6332 | 0.0009 | 1688.9122 | 0.0006 | 1.5401 | 0.6493 | 1.5287 | 2.3543 |
| 15 | 1808.0424 | 0.0006 | 2785.5455 | 0.0004 | 1.5406 | 0.6491 | 1.5332 | 2.3621 |
| 16 | 2980.9580 | 0.0003 | 4593.5881 | 0.0002 | 1.5410 | 0.6489 | 1.5361 | 2.3671 |
| 17 | 4914.7688 | 0.0002 | 7574.5464 | 0.0001 | 1.5412 | 0.6489 | 1.5380 | 2.3704 |
| 18 | 8103.0839 | 0.0001 | 12489.3160 | 0.0001 | 1.5413 | 0.6488 | 1.5393 | 2.3725 |
| 19 | 13359.7270 | 0.0001 | 20592.4010 | 0.0000 | 1.5414 | 0.6488 | 1.5401 | 2.3738 |
| 20 | 22026.4660 | 0.0000 | 33952.1290 | 0.0000 | 1.5414 | 0.6488 | 1.5406 | 2.3747 |
| 21 | 36315.5030 | 0.0000 | 55978.5970 | 0.0000 | 1.5415 | 0.6487 | 1.5409 | 2.3753 |
| 22 | 59874.1420 | 0.0000 | 92294.1040 | 0.0000 | 1.5415 | 0.6487 | 1.5411 | 2.3756 |
| 23 | 98715.7710 | 0.0000 | 152168.2500 | 0.0000 | 1.5415 | 0.6487 | 1.5413 | 2.3758 |
| 24 | 162754.7900 | 0.0000 | 250884.0300 | 0.0000 | 1.5415 | 0.6487 | 1.5414 | 2.3760 |
| 25 | 268337.2900 | 0.0000 | 413638.8400 | 0.0000 | 1.5415 | 0.6487 | 1.5414 | 2.3761 |

# SECTION II—CONTINUOUS COMPOUNDING, CONTINUOUS FLOW INTEREST FACTORS

| | Continuous flow, uniform series | | | |
|---|---|---|---|---|
| | Present worth factor | Capital recovery factor | Compound amount factor | Sinking fund factor |
| $n$ | To find $P$ given $\overline{A}$ $P\|\overline{A}\ r,n$ | To find $\overline{A}$ given $P$ $\overline{A}\|P\ r,n$ | To find $F$ given $\overline{A}$ $F\|\overline{A}\ r,n$ | To find $\overline{A}$ given $F$ $\overline{A}\|F\ r,n$ |
| 1 | 0.9803 | 1.0201 | 1.0203 | 0.9801 |
| 2 | 1.9221 | 0.5203 | 2.0822 | 0.4803 |
| 3 | 2.8270 | 0.3537 | 3.1874 | 0.3137 |
| 4 | 3.6964 | 0.2705 | 4.3378 | 0.2305 |
| 5 | 4.5317 | 0.2207 | 5.5351 | 0.1807 |
| 6 | 5.3343 | 0.1875 | 6.7812 | 0.1475 |
| 7 | 6.1054 | 0.1638 | 8.0782 | 0.1238 |
| 8 | 6.8463 | 0.1461 | 9.4282 | 0.1061 |
| 9 | 7.5581 | 0.1323 | 10.8332 | 0.0923 |
| 10 | 8.2420 | 0.1213 | 12.2956 | 0.0813 |
| 11 | 8.8991 | 0.1124 | 13.8177 | 0.0724 |
| 12 | 9.5304 | 0.1049 | 15.4019 | 0.0649 |
| 13 | 10.1370 | 0.0986 | 17.0507 | 0.0586 |
| 14 | 10.7198 | 0.0933 | 18.7668 | 0.0533 |
| 15 | 11.2797 | 0.0887 | 20.5530 | 0.0487 |
| 16 | 11.8177 | 0.0846 | 22.4120 | 0.0446 |
| 17 | 12.3346 | 0.0811 | 24.3469 | 0.0411 |
| 18 | 12.8312 | 0.0779 | 26.3608 | 0.0379 |
| 19 | 13.3083 | 0.0751 | 28.4569 | 0.0351 |
| 20 | 13.7668 | 0.0726 | 30.6385 | 0.0326 |
| 21 | 14.2072 | 0.0704 | 32.9092 | 0.0304 |
| 22 | 14.6304 | 0.0684 | 35.2725 | 0.0284 |
| 23 | 15.0370 | 0.0665 | 37.7323 | 0.0265 |
| 24 | 15.4277 | 0.0648 | 40.2924 | 0.0248 |
| 25 | 15.8030 | 0.0633 | 42.9570 | 0.0233 |
| 26 | 16.1636 | 0.0619 | 45.7304 | 0.0219 |
| 27 | 16.5101 | 0.0606 | 48.6170 | 0.0206 |
| 28 | 16.8430 | 0.0594 | 51.6214 | 0.0194 |
| 29 | 17.1628 | 0.0583 | 54.7483 | 0.0183 |
| 30 | 17.4701 | 0.0572 | 58.0029 | 0.0172 |
| 31 | 17.7654 | 0.0563 | 61.3903 | 0.0163 |
| 32 | 18.0491 | 0.0554 | 64.9160 | 0.0154 |
| 33 | 18.3216 | 0.0546 | 68.5855 | 0.0146 |
| 34 | 18.5835 | 0.0538 | 72.4048 | 0.0138 |
| 35 | 18.8351 | 0.0531 | 76.3800 | 0.0131 |
| 40 | 19.9526 | 0.0501 | 98.8258 | 0.0101 |
| 45 | 20.8675 | 0.0479 | 126.2412 | 0.0079 |
| 50 | 21.6166 | 0.0463 | 159.7264 | 0.0063 |
| 55 | 22.2299 | 0.0450 | 200.6253 | 0.0050 |
| 60 | 22.7321 | 0.0440 | 250.5794 | 0.0040 |
| 65 | 23.1432 | 0.0432 | 311.5935 | 0.0032 |
| 70 | 23.4797 | 0.0426 | 386.1162 | 0.0026 |
| 75 | 23.7553 | 0.0421 | 477.1384 | 0.0021 |
| 80 | 23.9809 | 0.0417 | 588.3133 | 0.0017 |
| 85 | 24.1657 | 0.0414 | 724.1025 | 0.0014 |
| 90 | 24.3169 | 0.0411 | 889.9559 | 0.0011 |
| 95 | 24.4407 | 0.0409 | 1092.5296 | 0.0009 |
| 100 | 24.5421 | 0.0407 | 1339.9538 | 0.0007 |

## TABLE B.23  Continuous Compounding: $r = 5\%$

| | Continuous flow, uniform series | | | |
|---|---|---|---|---|
| | Present worth factor | Capital recovery factor | Compound amount factor | Sinking fund factor |
| $n$ | To find $P$ given $\overline{A}$ $P\|\overline{A}\ r,n$ | To find $\overline{A}$ given $P$ $\overline{A}\|P\ r,n$ | To find $F$ given $\overline{A}$ $F\|\overline{A}\ r,n$ | To find $\overline{A}$ given $F$ $\overline{A}\|F\ r,n$ |
| 1 | 0.9754 | 1.0252 | 1.0254 | 0.9752 |
| 2 | 1.9033 | 0.5254 | 2.1034 | 0.4754 |
| 3 | 2.7858 | 0.3590 | 3.2367 | 0.3090 |
| 4 | 3.6254 | 0.2758 | 4.4281 | 0.2258 |
| 5 | 4.4240 | 0.2260 | 5.6805 | 0.1760 |
| 6 | 5.1836 | 0.1929 | 6.9972 | 0.1429 |
| 7 | 5.9062 | 0.1693 | 8.3814 | 0.1193 |
| 8 | 6.5936 | 0.1517 | 9.8365 | 0.1017 |
| 9 | 7.2474 | 0.1380 | 11.3662 | 0.0880 |
| 10 | 7.8694 | 0.1271 | 12.9744 | 0.0771 |
| 11 | 8.4610 | 0.1182 | 14.6651 | 0.0682 |
| 12 | 9.0238 | 0.1108 | 16.4424 | 0.0608 |
| 13 | 9.5591 | 0.1046 | 18.3108 | 0.0546 |
| 14 | 10.0683 | 0.0993 | 20.2751 | 0.0493 |
| 15 | 10.5527 | 0.0948 | 22.3400 | 0.0448 |
| 16 | 11.0134 | 0.0908 | 24.5108 | 0.0408 |
| 17 | 11.4517 | 0.0873 | 26.7929 | 0.0373 |
| 18 | 11.8686 | 0.0843 | 29.1921 | 0.0343 |
| 19 | 12.2652 | 0.0815 | 31.7142 | 0.0315 |
| 20 | 12.6424 | 0.0791 | 34.3656 | 0.0291 |
| 21 | 13.0012 | 0.0769 | 37.1530 | 0.0269 |
| 22 | 13.3426 | 0.0749 | 40.0833 | 0.0249 |
| 23 | 13.6673 | 0.0732 | 43.1639 | 0.0232 |
| 24 | 13.9761 | 0.0716 | 46.4023 | 0.0216 |
| 25 | 14.2699 | 0.0701 | 49.8069 | 0.0201 |
| 26 | 14.5494 | 0.0687 | 53.3859 | 0.0187 |
| 27 | 14.8152 | 0.0675 | 57.1485 | 0.0175 |
| 28 | 15.0681 | 0.0664 | 61.1040 | 0.0164 |
| 29 | 15.3086 | 0.0653 | 65.2623 | 0.0153 |
| 30 | 15.5374 | 0.0644 | 69.6338 | 0.0144 |
| 31 | 15.7550 | 0.0635 | 74.2294 | 0.0135 |
| 32 | 15.9621 | 0.0626 | 79.0606 | 0.0126 |
| 33 | 16.1590 | 0.0619 | 84.1396 | 0.0119 |
| 34 | 16.3463 | 0.0612 | 89.4789 | 0.0112 |
| 35 | 16.5245 | 0.0605 | 95.0921 | 0.0105 |
| 40 | 17.2933 | 0.0578 | 127.7811 | 0.0078 |
| 45 | 17.8920 | 0.0559 | 169.7547 | 0.0059 |
| 50 | 18.3583 | 0.0545 | 223.6499 | 0.0045 |
| 55 | 18.7214 | 0.0534 | 292.8526 | 0.0034 |
| 60 | 19.0043 | 0.0526 | 381.7107 | 0.0026 |
| 65 | 19.2245 | 0.0520 | 495.8068 | 0.0020 |
| 70 | 19.3961 | 0.0516 | 642.3090 | 0.0016 |
| 75 | 19.5296 | 0.0512 | 830.4216 | 0.0012 |
| 80 | 19.6337 | 0.0509 | 1071.9630 | 0.0009 |
| 85 | 19.7147 | 0.0507 | 1382.1082 | 0.0007 |
| 90 | 19.7778 | 0.0506 | 1780.3426 | 0.0006 |
| 95 | 19.8270 | 0.0504 | 2291.6857 | 0.0004 |
| 100 | 19.8652 | 0.0503 | 2948.2632 | 0.0003 |

| | Continuous flow, uniform series | | | |
|---|---|---|---|---|
| | Present worth factor | Capital recovery factor | Compound amount factor | Sinking fund factor |
| $n$ | To find $P$ given $\overline{A}$ $P\|\overline{A}\ r,n$ | To find $\overline{A}$ given $P$ $\overline{A}\|P\ r,n$ | To find $F$ given $\overline{A}$ $F\|\overline{A}\ r,n$ | To find $\overline{A}$ given $F$ $\overline{A}\|F\ r,n$ |
| 1 | 0.9706 | 1.0303 | 1.0306 | 0.9703 |
| 2 | 1.8847 | 0.5306 | 2.1249 | 0.4706 |
| 3 | 2.7455 | 0.3642 | 3.2870 | 0.3042 |
| 4 | 3.5562 | 0.2812 | 4.5208 | 0.2212 |
| 5 | 4.3197 | 0.2315 | 5.8310 | 0.1715 |
| 6 | 5.0387 | 0.1985 | 7.2222 | 0.1385 |
| 7 | 5.7159 | 0.1750 | 8.6994 | 0.1150 |
| 8 | 6.3536 | 0.1574 | 10.2679 | 0.0974 |
| 9 | 6.9542 | 0.1438 | 11.9334 | 0.0838 |
| 10 | 7.5198 | 0.1330 | 13.7020 | 0.0730 |
| 11 | 8.0525 | 0.1242 | 15.5799 | 0.0642 |
| 12 | 8.5541 | 0.1169 | 17.5739 | 0.0569 |
| 13 | 9.0266 | 0.1108 | 19.6912 | 0.0508 |
| 14 | 9.4715 | 0.1056 | 21.9394 | 0.0456 |
| 15 | 9.8905 | 0.1011 | 24.3267 | 0.0411 |
| 16 | 10.2851 | 0.0972 | 26.8616 | 0.0372 |
| 17 | 10.6568 | 0.0938 | 29.5532 | 0.0338 |
| 18 | 11.0067 | 0.0909 | 32.4113 | 0.0309 |
| 19 | 11.3363 | 0.0882 | 35.4461 | 0.0282 |
| 20 | 11.6468 | 0.0859 | 38.6686 | 0.0259 |
| 21 | 11.9391 | 0.0838 | 42.0904 | 0.0238 |
| 22 | 12.2144 | 0.0819 | 45.7237 | 0.0219 |
| 23 | 12.4737 | 0.0802 | 49.5817 | 0.0202 |
| 24 | 12.7179 | 0.0786 | 53.6783 | 0.0186 |
| 25 | 12.9478 | 0.0772 | 58.0282 | 0.0172 |
| 26 | 13.1644 | 0.0760 | 62.6470 | 0.0160 |
| 27 | 13.3684 | 0.0748 | 67.5515 | 0.0148 |
| 28 | 13.5604 | 0.0737 | 72.7593 | 0.0137 |
| 29 | 13.7413 | 0.0728 | 78.2891 | 0.0128 |
| 30 | 13.9117 | 0.0719 | 84.1608 | 0.0119 |
| 31 | 14.0721 | 0.0711 | 90.3956 | 0.0111 |
| 32 | 14.2232 | 0.0703 | 97.0160 | 0.0103 |
| 33 | 14.3655 | 0.0696 | 104.0457 | 0.0096 |
| 34 | 14.4995 | 0.0690 | 111.5102 | 0.0090 |
| 35 | 14.6257 | 0.0684 | 119.4362 | 0.0084 |
| 40 | 15.1547 | 0.0660 | 167.0529 | 0.0060 |
| 45 | 15.5466 | 0.0643 | 231.3289 | 0.0043 |
| 50 | 15.8369 | 0.0631 | 318.0923 | 0.0031 |
| 55 | 16.0519 | 0.0623 | 435.2106 | 0.0023 |
| 60 | 16.2113 | 0.0617 | 593.3039 | 0.0017 |
| 65 | 16.3293 | 0.0612 | 806.7075 | 0.0012 |
| 70 | 16.4167 | 0.0609 | 1094.7722 | 0.0009 |
| 75 | 16.4815 | 0.0607 | 1483.6189 | 0.0007 |
| 80 | 16.5295 | 0.0605 | 2008.5070 | 0.0005 |
| 85 | 16.5651 | 0.0604 | 2717.0318 | 0.0004 |
| 90 | 16.5914 | 0.0603 | 3673.4403 | 0.0003 |
| 95 | 16.6109 | 0.0602 | 4964.4567 | 0.0002 |
| 100 | 16.6254 | 0.0601 | 6707.1466 | 0.0001 |

## TABLE B.25   Continuous Compounding: $r = 8\%$

| | Continuous flow, uniform series | | | |
|---|---|---|---|---|
| | Present worth factor | Capital recovery factor | Compound amount factor | Sinking fund factor |
| $n$ | To find $P$ given $\overline{A}$ $P\|\overline{A}\ r,n$ | To find $\overline{A}$ given $P$ $\overline{A}\|P\ r,n$ | To find $F$ given $\overline{A}$ $F\|\overline{A}\ r,n$ | To find $\overline{A}$ given $F$ $\overline{A}\|F\ r,n$ |
| 1 | 0.9610 | 1.0405 | 1.0411 | 0.9605 |
| 2 | 1.8482 | 0.5411 | 2.1689 | 0.4611 |
| 3 | 2.6672 | 0.3749 | 3.3906 | 0.2949 |
| 4 | 3.4231 | 0.2921 | 4.7141 | 0.2121 |
| 5 | 4.1210 | 0.2427 | 6.1478 | 0.1627 |
| 6 | 4.7652 | 0.2099 | 7.7009 | 0.1299 |
| 7 | 5.3599 | 0.1866 | 9.3834 | 0.1066 |
| 8 | 5.9088 | 0.1692 | 11.2060 | 0.0892 |
| 9 | 6.4156 | 0.1559 | 13.1804 | 0.0759 |
| 10 | 6.8834 | 0.1453 | 15.3193 | 0.0653 |
| 11 | 7.3152 | 0.1367 | 17.6362 | 0.0567 |
| 12 | 7.7138 | 0.1296 | 20.1462 | 0.0496 |
| 13 | 8.0818 | 0.1237 | 22.8652 | 0.0437 |
| 14 | 8.4215 | 0.1187 | 25.8107 | 0.0387 |
| 15 | 8.7351 | 0.1145 | 29.0015 | 0.0345 |
| 16 | 9.0245 | 0.1108 | 32.4580 | 0.0308 |
| 17 | 9.2917 | 0.1076 | 36.2024 | 0.0276 |
| 18 | 9.5384 | 0.1048 | 40.2587 | 0.0248 |
| 19 | 9.7661 | 0.1024 | 44.6528 | 0.0224 |
| 20 | 9.9763 | 0.1002 | 49.4129 | 0.0202 |
| 21 | 10.1703 | 0.0983 | 54.5694 | 0.0183 |
| 22 | 10.3494 | 0.0966 | 60.1555 | 0.0166 |
| 23 | 10.5148 | 0.0951 | 66.2067 | 0.0151 |
| 24 | 10.6674 | 0.0937 | 72.7620 | 0.0137 |
| 25 | 10.8083 | 0.0925 | 79.8632 | 0.0125 |
| 26 | 10.9384 | 0.0914 | 87.5559 | 0.0114 |
| 27 | 11.0584 | 0.0904 | 95.8892 | 0.0104 |
| 28 | 11.1693 | 0.0895 | 104.9166 | 0.0095 |
| 29 | 11.2716 | 0.0887 | 114.6959 | 0.0087 |
| 30 | 11.3660 | 0.0880 | 125.2897 | 0.0080 |
| 31 | 11.4532 | 0.0873 | 136.7658 | 0.0073 |
| 32 | 11.5337 | 0.0867 | 149.1977 | 0.0067 |
| 33 | 11.6080 | 0.0861 | 162.6650 | 0.0061 |
| 34 | 11.6766 | 0.0856 | 177.2540 | 0.0056 |
| 35 | 11.7399 | 0.0852 | 193.0581 | 0.0052 |
| 40 | 11.9905 | 0.0834 | 294.1566 | 0.0034 |
| 45 | 12.1585 | 0.0822 | 444.9779 | 0.0022 |
| 50 | 12.2711 | 0.0815 | 669.9769 | 0.0015 |
| 55 | 12.3465 | 0.0810 | 1005.6359 | 0.0010 |
| 60 | 12.3971 | 0.0807 | 1506.3802 | 0.0007 |
| 65 | 12.4310 | 0.0804 | 2253.4030 | 0.0004 |
| 70 | 12.4538 | 0.0803 | 3367.8301 | 0.0003 |
| 75 | 12.4690 | 0.0802 | 5030.3599 | 0.0002 |
| 80 | 12.4792 | 0.0801 | 7510.5630 | 0.0001 |
| 85 | 12.4861 | 0.0801 | 11210.5911 | 0.0001 |
| 90 | 12.4907 | 0.0801 | 16730.3846 | 0.0001 |
| 95 | 12.4937 | 0.0800 | 24964.9487 | 0.0000 |
| 100 | 12.4958 | 0.0800 | 37249.4748 | 0.0000 |

| | Continuous flow, uniform series | | | |
|---|---|---|---|---|
| | Present worth factor | Capital recovery factor | Compound amount factor | Sinking fund factor |
| $n$ | To find $P$ given $\overline{A}$ $P\|\overline{A}\ r,n$ | To find $\overline{A}$ given $P$ $\overline{A}\|P\ r,n$ | To find $F$ given $\overline{A}$ $F\|\overline{A}\ r,n$ | To find $\overline{A}$ given $F$ $\overline{A}\|F\ r,n$ |
| 1 | 0.9516 | 1.0508 | 1.0517 | 0.9508 |
| 2 | 1.8127 | 0.5517 | 2.2140 | 0.4517 |
| 3 | 2.5918 | 0.3858 | 3.4986 | 0.2858 |
| 4 | 3.2968 | 0.3033 | 4.9182 | 0.2033 |
| 5 | 3.9347 | 0.2541 | 6.4872 | 0.1541 |
| 6 | 4.5119 | 0.2216 | 8.2212 | 0.1216 |
| 7 | 5.0341 | 0.1986 | 10.1375 | 0.0986 |
| 8 | 5.5067 | 0.1816 | 12.2554 | 0.0816 |
| 9 | 5.9343 | 0.1685 | 14.5960 | 0.0685 |
| 10 | 6.3212 | 0.1582 | 17.1828 | 0.0582 |
| 11 | 6.6713 | 0.1499 | 20.0417 | 0.0499 |
| 12 | 6.9881 | 0.1431 | 23.2021 | 0.0431 |
| 13 | 7.2747 | 0.1375 | 26.6930 | 0.0375 |
| 14 | 7.5340 | 0.1327 | 30.5520 | 0.0327 |
| 15 | 7.7687 | 0.1287 | 34.8169 | 0.0287 |
| 16 | 7.9810 | 0.1253 | 39.5303 | 0.0253 |
| 17 | 8.1732 | 0.1224 | 44.7395 | 0.0224 |
| 18 | 8.3470 | 0.1198 | 50.4965 | 0.0198 |
| 19 | 8.5043 | 0.1176 | 56.8589 | 0.0176 |
| 20 | 8.6466 | 0.1157 | 63.8906 | 0.0157 |
| 21 | 8.7754 | 0.1140 | 71.6617 | 0.0140 |
| 22 | 8.8920 | 0.1125 | 80.2501 | 0.0125 |
| 23 | 8.9974 | 0.1111 | 89.7418 | 0.0111 |
| 24 | 9.0928 | 0.1100 | 100.2318 | 0.0100 |
| 25 | 9.1792 | 0.1089 | 111.8249 | 0.0089 |
| 26 | 9.2573 | 0.1080 | 124.6374 | 0.0080 |
| 27 | 9.3279 | 0.1072 | 138.7973 | 0.0072 |
| 28 | 9.3919 | 0.1065 | 154.4465 | 0.0065 |
| 29 | 9.4498 | 0.1058 | 171.7415 | 0.0058 |
| 30 | 9.5021 | 0.1052 | 190.8554 | 0.0052 |
| 31 | 9.5495 | 0.1047 | 211.9795 | 0.0047 |
| 32 | 9.5924 | 0.1042 | 235.3253 | 0.0042 |
| 33 | 9.6312 | 0.1038 | 261.1264 | 0.0038 |
| 34 | 9.6663 | 0.1035 | 289.6410 | 0.0035 |
| 35 | 9.6980 | 0.1031 | 321.1545 | 0.0031 |
| 40 | 9.8168 | 0.1019 | 535.9815 | 0.0019 |
| 45 | 9.8889 | 0.1011 | 890.1713 | 0.0011 |
| 50 | 9.9326 | 0.1007 | 1474.1316 | 0.0007 |
| 55 | 9.9591 | 0.1004 | 2436.9193 | 0.0004 |
| 60 | 9.9752 | 0.1002 | 4024.2879 | 0.0002 |
| 65 | 9.9850 | 0.1002 | 6641.4163 | 0.0002 |
| 70 | 9.9909 | 0.1001 | 10956.3316 | 0.0001 |
| 75 | 9.9945 | 0.1001 | 18070.4241 | 0.0001 |
| 80 | 9.9966 | 0.1000 | 29799.5799 | 0.0000 |
| 85 | 9.9980 | 0.1000 | 49137.6884 | 0.0000 |
| 90 | 9.9988 | 0.1000 | 81020.8398 | 0.0000 |
| 95 | 9.9993 | 0.1000 | 133587.2683 | 0.0000 |
| 100 | 9.995 | 0.1000 | 220254.6579 | 0.0000 |

| | Continuous flow, uniform series | | | |
|---|---|---|---|---|
| | Present worth factor | Capital recovery factor | Compound amount factor | Sinking fund factor |
| $n$ | To find $P$ given $\overline{A}$ $P\mid\overline{A}\ r,n$ | To find $\overline{A}$ given $P$ $\overline{A}\mid P\ r,n$ | To find $F$ given $\overline{A}$ $F\mid\overline{A}\ r,n$ | To find $\overline{A}$ given $F$ $\overline{A}\mid F\ r,n$ |
| 1 | 0.9286 | 1.0769 | 1.0789 | 0.9269 |
| 2 | 1.7279 | 0.5787 | 2.3324 | 0.4287 |
| 3 | 2.4158 | 0.4139 | 3.7887 | 0.2639 |
| 4 | 3.0079 | 0.3325 | 5.4808 | 0.1825 |
| 5 | 3.5176 | 0.2843 | 7.4467 | 0.1343 |
| 6 | 3.9562 | 0.2528 | 9.7307 | 0.1028 |
| 7 | 4.3337 | 0.2307 | 12.3843 | 0.0807 |
| 8 | 4.6587 | 0.2147 | 15.4674 | 0.0647 |
| 9 | 4.9384 | 0.2025 | 19.0495 | 0.0525 |
| 10 | 5.1791 | 0.1931 | 23.2113 | 0.0431 |
| 11 | 5.3863 | 0.1857 | 28.0465 | 0.0357 |
| 12 | 5.5647 | 0.1797 | 33.6643 | 0.0297 |
| 13 | 5.7182 | 0.1749 | 40.1913 | 0.0249 |
| 14 | 5.8503 | 0.1709 | 47.7745 | 0.0209 |
| 15 | 5.9640 | 0.1677 | 56.5849 | 0.0177 |
| 16 | 6.0619 | 0.1650 | 66.8212 | 0.0150 |
| 17 | 6.1461 | 0.1627 | 78.7140 | 0.0127 |
| 18 | 6.2186 | 0.1608 | 92.5315 | 0.0108 |
| 19 | 6.2810 | 0.1592 | 108.5852 | 0.0092 |
| 20 | 6.3348 | 0.1579 | 127.2369 | 0.0079 |
| 21 | 6.3810 | 0.1567 | 148.9071 | 0.0067 |
| 22 | 6.4208 | 0.1557 | 174.0843 | 0.0057 |
| 23 | 6.4550 | 0.1549 | 203.3359 | 0.0049 |
| 24 | 6.4845 | 0.1542 | 237.3216 | 0.0042 |
| 25 | 6.5099 | 0.1536 | 276.8072 | 0.0036 |
| 26 | 6.5317 | 0.1532 | 322.6830 | 0.0031 |
| 27 | 6.5505 | 0.1527 | 375.9830 | 0.0027 |
| 28 | 6.5667 | 0.1523 | 437.9089 | 0.0023 |
| 29 | 6.5806 | 0.1520 | 509.8564 | 0.0020 |
| 30 | 6.5926 | 0.1517 | 593.4475 | 0.0017 |

| | Continuous flow, uniform series | | | |
|---|---|---|---|---|
| | Present worth factor | Capital recovery factor | Compound amount factor | Sinking fund factor |
| $n$ | To find $P$ given $\overline{A}$ $P\vert\overline{A}\ r,n$ | To find $\overline{A}$ given $P$ $\overline{A}\vert P\ r,n$ | To find $F$ given $\overline{A}$ $F\vert\overline{A}\ r,n$ | To find $\overline{A}$ given $F$ $\overline{A}\vert F\ r,n$ |
| 1 | 0.9063 | 1.1033 | 1.1070 | 0.9033 |
| 2 | 1.6484 | 0.6066 | 2.4591 | 0.4066 |
| 3 | 2.2559 | 0.4433 | 4.1106 | 0.2433 |
| 4 | 2.7534 | 0.3632 | 6.1277 | 0.1632 |
| 5 | 3.1606 | 0.3164 | 8.5914 | 0.1164 |
| 6 | 3.4940 | 0.2862 | 11.6006 | 0.0862 |
| 7 | 3.7670 | 0.2655 | 15.2760 | 0.0655 |
| 8 | 3.9905 | 0.2506 | 19.7652 | 0.0506 |
| 9 | 4.1735 | 0.2396 | 25.2482 | 0.0396 |
| 10 | 4.3233 | 0.2313 | 31.9453 | 0.0313 |
| 11 | 4.4460 | 0.2249 | 40.1251 | 0.0249 |
| 12 | 4.5464 | 0.2200 | 50.1159 | 0.0200 |
| 13 | 4.6286 | 0.2160 | 62.3187 | 0.0160 |
| 14 | 4.6959 | 0.2129 | 77.2332 | 0.0129 |
| 15 | 4.7511 | 0.2105 | 95.4277 | 0.0105 |
| 16 | 4.7962 | 0.2085 | 117.6627 | 0.0085 |
| 17. | 4.8331 | 0.2069 | 144.8205 | 0.0069 |
| 18 | 4.8634 | 0.2056 | 177.9912 | 0.0056 |
| 19 | 4.8881 | 0.2046 | 218.5059 | 0.0046 |
| 20 | 4.9084 | 0.2037 | 267.9908 | 0.0037 |
| 22 | 4.9386 | 0.2025 | 402.2543 | 0.0025 |
| 23 | 4.9497 | 0.2020 | 492.4216 | 0.0020 |
| 24 | 4.9589 | 0.2017 | 602.5521 | 0.0017 |
| 25 | 4.9663 | 0.2014 | 737.0658 | 0.0014 |
| 26 | 4.9724 | 0.2011 | 901.3612 | 0.0011 |
| 27 | 4.9774 | 0.2009 | 1102.0321 | 0.0009 |
| 28 | 4.9815 | 0.2007 | 1347.1320 | 0.0007 |
| 29 | 4.9849 | 0.2006 | 1646.4978 | 0.0006 |
| 30 | 4.9876 | 0.2005 | 2012.1440 | 0.0005 |

## TABLE B.29  Continuous Compounding: $r = 25\%$

| | Continuous flow, uniform series | | | |
|---|---|---|---|---|
| | Present worth factor | Capital recovery factor | Compound amount factor | Sinking fund factor |
| $n$ | To find $P$ given $\overline{A}$ $P \mid \overline{A}\ r,n$ | To find $\overline{A}$ given $P$ $\overline{A} \mid P\ r,n$ | To find $F$ given $\overline{A}$ $F \mid \overline{A}\ r,n$ | To find $\overline{A}$ given $F$ $\overline{A} \mid F\ r,n$ |
| 1 | 0.8848 | 1.1302 | 1.1361 | 0.8802 |
| 2 | 1.5739 | 0.6354 | 2.5949 | 0.3854 |
| 3 | 2.1105 | 0.4738 | 4.4680 | 0.2238 |
| 4 | 2.5285 | 0.3955 | 6.8731 | 0.1455 |
| 5 | 2.8540 | 0.3504 | 9.9614 | 0.1004 |
| 6 | 3.1075 | 0.3218 | 13.9268 | 0.0718 |
| 7 | 3.3049 | 0.3026 | 19.0184 | 0.0526 |
| 8 | 3.4587 | 0.2891 | 25.5562 | 0.0391 |
| 9 | 3.5784 | 0.2795 | 33.9509 | 0.0295 |
| 10 | 3.6717 | 0.2724 | 44.7300 | 0.0224 |
| 11 | 3.7443 | 0.2671 | 58.5705 | 0.0171 |
| 12 | 3.8009 | 0.2631 | 76.3421 | 0.0131 |
| 13 | 3.8449 | 0.2601 | 99.1614 | 0.0101 |
| 14 | 3.8792 | 0.2578 | 128.4618 | 0.0078 |
| 15 | 3.9059 | 0.2560 | 166.0843 | 0.0060 |
| 16 | 3.9267 | 0.2547 | 214.3926 | 0.0047 |
| 17 | 3.9429 | 0.2536 | 276.4216 | 0.0036 |
| 18 | 3.9556 | 0.2528 | 356.0685 | 0.0028 |
| 19 | 3.9654 | 0.2522 | 458.3371 | 0.0022 |
| 20 | 3.9730 | 0.2517 | 389.6526 | 0.0017 |
| 21 | 3.9790 | 0.2513 | 758.2651 | 0.0013 |
| 22 | 3.9837 | 0.2510 | 974.7677 | 0.0010 |
| 23 | 3.9873 | 0.2508 | 1252.7626 | 0.0008 |
| 24 | 3.9901 | 0.2506 | 1609.7152 | 0.0006 |
| 25 | 3.9923 | 0.2505 | 2068.0513 | 0.0005 |
| 26 | 3.9940 | 0.2504 | 2656.5665 | 0.0004 |
| 27 | 3.9953 | 0.2503 | 3412.2351 | 0.0003 |
| 28 | 3.9964 | 0.2502 | 4382.5326 | 0.0002 |
| 29 | 3.9972 | 0.2502 | 5628.4194 | 0.0002 |
| 30 | 3.9978 | 0.2501 | 7228.1697 | 0.0001 |

# SECTION III—CONTINUOUS COMPOUNDING, GEOMETRIC SERIES FACTORS

**Continuous Compounding:** $r = 5\%$

| | Geometric series present worth factor, $(P\,|\,A_1\ r, c, n)_\infty$ | | | | |
|---|---|---|---|---|---|
| $n$ | $c = 4\%$ | $c = 6\%$ | $c = 8\%$ | $c = 10\%$ | $c = 15\%$ |
| 1 | 0.9512 | 0.9512 | 0.9512 | 0.9512 | 0.9512 |
| 2 | 1.8930 | 1.9120 | 1.9314 | 1.9512 | 2.0025 |
| 3 | 2.8254 | 2.8825 | 2.9415 | 3.0025 | 3.1643 |
| 4 | 3.7485 | 3.8627 | 3.9823 | 4.1077 | 4.4484 |
| 5 | 4.6624 | 4.8527 | 5.0548 | 5.2695 | 5.8674 |
| 6 | 5.5673 | 5.8527 | 6.1600 | 6.4909 | 7.4357 |
| 7 | 6.4631 | 6.8628 | 7.2988 | 7.7749 | 9.1690 |
| 8 | 7.3500 | 7.8830 | 8.4723 | 9.1248 | 11.0845 |
| 9 | 8.2281 | 8.9134 | 9.6816 | 10.5439 | 13.2015 |
| 10 | 9.0975 | 9.9542 | 10.9276 | 12.0357 | 15.5412 |
| 11 | 9.9582 | 11.0055 | 12.2117 | 13.6040 | 18.1269 |
| 12 | 10.8103 | 12.0673 | 13.5348 | 15.2527 | 20.9845 |
| 13 | 11.6540 | 13.1398 | 14.8982 | 16.9860 | 24.1427 |
| 14 | 12.4893 | 14.2231 | 16.3032 | 18.8081 | 27.6331 |
| 15 | 13.3162 | 15.3173 | 17.7509 | 20.7236 | 31.4905 |
| 16 | 14.1350 | 16.4225 | 19.2427 | 22.7374 | 35.7536 |
| 17 | 14.9455 | 17.5388 | 20.7800 | 24.8544 | 40.4651 |
| 18 | 15.7481 | 18.6663 | 22.3641 | 27.0799 | 45.6721 |
| 19 | 16.5426 | 19.8051 | 23.9964 | 29.4196 | 51.4267 |
| 20 | 17.3292 | 20.9554 | 25.6784 | 31.8792 | 57.7865 |
| 21 | 18.1080 | 22.1172 | 27.4116 | 34.4649 | 64.8152 |
| 22 | 18.8791 | 23.2907 | 29.1977 | 37.1832 | 72.5831 |
| 23 | 19.6425 | 24.4760 | 31.0381 | 40.0408 | 81.1679 |
| 24 | 20.3982 | 25.6732 | 32.9346 | 43.0450 | 90.6557 |
| 25 | 21.1465 | 26.8825 | 34.8888 | 46.2032 | 101.1412 |
| 26 | 21.8873 | 28.1039 | 36.9026 | 49.5233 | 112.7296 |
| 27 | 22.6208 | 29.3376 | 38.9777 | 53.0136 | 125.5367 |
| 28 | 23.3469 | 30.5836 | 41.1159 | 56.6829 | 139.6907 |
| 29 | 24.0658 | 31.8422 | 43.3193 | 60.5404 | 155.3334 |
| 30 | 24.7776 | 33.1135 | 45.5898 | 64.5956 | 172.6211 |
| 31 | 25.4823 | 34.3975 | 47.9295 | 68.8587 | 191.7271 |
| 32 | 26.1800 | 35.6944 | 50.3404 | 73.3404 | 212.8424 |
| 33 | 26.8707 | 37.0044 | 52.8247 | 78.0518 | 236.1785 |
| 34 | 27.5546 | 38.3275 | 55.3847 | 83.0049 | 261.9688 |
| 35 | 28.2316 | 39.6640 | 58.0226 | 88.2118 | 290.4716 |
| 36 | 28.9019 | 41.0138 | 60.7409 | 93.6858 | 321.9720 |
| 37 | 29.5656 | 43.3772 | 63.5420 | 99.4404 | 356.7853 |
| 38 | 30.2226 | 43.7544 | 66.4284 | 105.4900 | 395.2600 |
| 39 | 30.8732 | 45.1453 | 69.4026 | 111.8499 | 437.7810 |
| 40 | 31.5172 | 46.5503 | 72.4675 | 118.5358 | 484.7741 |
| 41 | 32.1548 | 47.9694 | 75.6257 | 125.5644 | 536.7095 |
| 42 | 32.7861 | 49.4027 | 78.8801 | 132.9535 | 594.1069 |
| 43 | 33.4111 | 50.8504 | 82.2335 | 140.7214 | 657.5409 |
| 44 | 34.0299 | 52.3127 | 85.6892 | 148.8876 | 727.6463 |
| 45 | 34.6425 | 53.7897 | 89.2500 | 157.4724 | 805.1248 |
| 46 | 35.2490 | 55.2815 | 92.9193 | 166.4974 | 890.7517 |
| 47 | 35.8495 | 56.7883 | 96.7003 | 175.9852 | 985.3842 |
| 48 | 36.4441 | 58.3103 | 100.5965 | 185.9594 | 1089.9691 |
| 49 | 37.0327 | 59.8475 | 104.6114 | 196.4449 | 1205.5534 |
| 50 | 37.6154 | 61.4002 | 108.7485 | 207.4681 | 1333.2938 |

Geometric series future worth factor, $(F|A_1\ r, c, n)_\infty$

| $n$ | $c = 4\%$ | $c = 6\%$ | $c = 8\%$ | $c = 10\%$ | $c = 15\%$ |
|---|---|---|---|---|---|
| 1 | 1.0000 | 1.0000 | 1.0000 | 1.0000 | 1.0000 |
| 2 | 2.0921 | 2.1131 | 2.1346 | 2.1564 | 2.2131 |
| 3 | 3.2826 | 3.3489 | 3.4175 | 3.4884 | 3.6764 |
| 4 | 4.5784 | 4.7179 | 4.8640 | 5.0171 | 5.4332 |
| 5 | 5.9867 | 6.2310 | 6.4905 | 6.7662 | 7.5339 |
| 6 | 7.5150 | 7.9003 | 8.3151 | 8.7618 | 10.0372 |
| 7 | 9.1716 | 9.7387 | 10.3575 | 11.0332 | 13.0114 |
| 8 | 10.9650 | 11.7600 | 12.6392 | 13.6126 | 16.5362 |
| 9 | 12.9043 | 13.9790 | 15.1837 | 16.5361 | 20.7041 |
| 10 | 14.9992 | 16.4118 | 18.0166 | 19.8435 | 25.6231 |
| 11 | 17.2601 | 19.0753 | 21.1659 | 23.5792 | 31.4185 |
| 12 | 19.6977 | 21.9881 | 24.6620 | 27.7923 | 38.2363 |
| 13 | 22.3237 | 25.1699 | 28.5381 | 32.5373 | 46.2464 |
| 14 | 25.1503 | 28.6419 | 32.8305 | 37.8748 | 55.6462 |
| 15 | 28.1905 | 32.4267 | 37.5786 | 43.8719 | 66.6654 |
| 16 | 31.4579 | 36.5489 | 42.8255 | 50.6030 | 79.5711 |
| 17 | 34.9673 | 41.0345 | 48.6178 | 58.1505 | 94.6740 |
| 18 | 38.7340 | 45.9116 | 55.0067 | 66.6059 | 112.3352 |
| 19 | 42.7743 | 51.2102 | 62.0476 | 76.0705 | 132.9744 |
| 20 | 47.1057 | 56.9626 | 69.8011 | 86.6566 | 157.0800 |
| 21 | 51.7464 | 63.2032 | 78.3329 | 98.4886 | 185.2191 |
| 22 | 56.7159 | 69.9691 | 87.7147 | 111.7044 | 218.0516 |
| 23 | 62.0347 | 77.2999 | 98.0244 | 126.4566 | 256.3440 |
| 24 | 67.7245 | 85.2381 | 109.3467 | 142.9144 | 300.9874 |
| 25 | 73.8085 | 93.8290 | 121.7740 | 161.2649 | 353.0176 |
| 26 | 80.3111 | 103.1215 | 135.4066 | 181.7157 | 413.6383 |
| 27 | 87.2579 | 113.1674 | 150.3535 | 204.4962 | 484.2484 |
| 28 | 94.6764 | 124.0227 | 166.7334 | 229.8606 | 566.4738 |
| 29 | 102.5954 | 135.7471 | 184.6753 | 258.0905 | 662.2039 |
| 30 | 111.0455 | 148.4043 | 204.3195 | 289.4972 | 773.6343 |
| 31 | 120.0591 | 162.0628 | 225.8184 | 324.4256 | 903.3165 |
| 32 | 129.6703 | 176.7957 | 249.3376 | 363.2572 | 1054.2155 |
| 33 | 139.9152 | 192.6812 | 275.0572 | 406.4143 | 1229.7767 |
| 34 | 150.8323 | 209.8029 | 303.1729 | 454.3643 | 1434.0037 |
| 35 | 162.4618 | 228.2503 | 333.8972 | 507.6241 | 1671.5485 |
| 36 | 174.8466 | 248.1191 | 367.4612 | 566.7660 | 1947.8169 |
| 37 | 188.0319 | 269.5116 | 404.1156 | 632.4230 | 2269.0901 |
| 38 | 202.0654 | 292.5371 | 444.1330 | 705.2953 | 2642.6664 |
| 39 | 216.9977 | 317.3125 | 487.8094 | 786.1577 | 3077.0262 |
| 40 | 232.8823 | 343.9627 | 535.4663 | 875.8673 | 3582.0230 |
| 41 | 249.7754 | 372.6212 | 587.4528 | 975.3722 | 4169.1061 |
| 42 | 267.7369 | 403.4307 | 644.1479 | 1085.7209 | 4851.5781 |
| 43 | 286.8296 | 436.5436 | 705.9633 | 1208.0733 | 5644.8957 |
| 44 | 307.1202 | 472.1228 | 773.3457 | 1343.7123 | 6567.0108 |
| 45 | 328.6790 | 510.3423 | 846.7805 | 1494.0568 | 7638.8114 |
| 46 | 351.5804 | 551.3878 | 926.7940 | 1660.6759 | 8884.5204 |
| 47 | 375.9028 | 595.4579 | 1013.9582 | 1845.3049 | 10332.3143 |
| 48 | 401.7293 | 642.7645 | 1108.8934 | 2049.8628 | 12014.9221 |
| 49 | 429.1474 | 693.5341 | 1212.2730 | 2276.4720 | 13970.3711 |
| 50 | 458.2495 | 748.0082 | 1324.8280 | 2527.4790 | 16242.8438 |

Geometric series present worth factor, $(P \mid A_1 \ r, c, n)_\infty$

| $n$ | $c = 4\%$ | $c = 6\%$ | $c = 8\%$ | $c = 10\%$ | $c = 15\%$ |
|---|---|---|---|---|---|
| 1 | 0.9231 | 0.9231 | 0.9231 | 0.9231 | 0.9231 |
| 2 | 1.8100 | 1.8280 | 1.8462 | 1.8649 | 1.9132 |
| 3 | 2.6622 | 2.7149 | 2.7693 | 2.8257 | 2.9750 |
| 4 | 3.4809 | 3.5842 | 3.6925 | 3.8059 | 4.1138 |
| 5 | 4.2675 | 4.4364 | 4.6156 | 4.8059 | 5.3352 |
| 6 | 5.0233 | 5.2716 | 5.5387 | 5.8261 | 6.6452 |
| 7 | 5.7495 | 6.0904 | 6.4618 | 6.8669 | 8.0501 |
| 8 | 6.4471 | 6.8929 | 7.3849 | 7.9287 | 9.5570 |
| 9 | 7.1175 | 7.6795 | 8.3080 | 9.0120 | 11.1730 |
| 10 | 7.7615 | 8.4506 | 9.2312 | 10.1172 | 12.9063 |
| 11 | 8.3803 | 9.2064 | 10.1543 | 11.2447 | 14.7652 |
| 12 | 8.9748 | 9.9472 | 11.0774 | 12.3949 | 16.7589 |
| 13 | 9.5460 | 10.6733 | 12.0005 | 13.5685 | 18.8972 |
| 14 | 10.0948 | 11.3851 | 12.9236 | 14.7657 | 21.1905 |
| 15 | 10.6221 | 12.0828 | 13.8467 | 15.9871 | 23.6501 |
| 16 | 11.1287 | 12.7666 | 14.7699 | 17.2332 | 26.2881 |
| 17 | 11.6155 | 13.4370 | 15.6930 | 18.5044 | 29.1173 |
| 18 | 12.0832 | 14.0940 | 16.6161 | 19.8013 | 32.1517 |
| 19 | 12.5325 | 14.7380 | 17.5392 | 21.1245 | 35.4060 |
| 20 | 12.9642 | 15.3693 | 18.4623 | 22.4743 | 38.8964 |
| 21 | 13.3790 | 15.9881 | 19.3854 | 23.8514 | 42.6398 |
| 22 | 13.7775 | 16.5946 | 20.3086 | 25.2564 | 46.6546 |
| 23 | 14.1604 | 17.1892 | 21.2317 | 26.6897 | 50.9606 |
| 24 | 14.5283 | 17.7719 | 22.1548 | 28.1520 | 55.5788 |
| 25 | 14.8817 | 18.3431 | 23.0779 | 29.6438 | 60.5318 |
| 26 | 15.2213 | 18.9030 | 24.0010 | 32.1658 | 65.8440 |
| 27 | 15.5476 | 19.4518 | 24.9241 | 32.7185 | 71.5413 |
| 28 | 15.8611 | 19.9898 | 25.8473 | 34.3026 | 77.6518 |
| 29 | 16.1623 | 20.5171 | 26.7704 | 35.9187 | 84.2053 |
| 30 | 16.4517 | 21.0339 | 27.6935 | 37.5674 | 91.2340 |
| 31 | 16.7297 | 21.5404 | 28.6166 | 39.2494 | 98.7723 |
| 32 | 16.9968 | 22.0371 | 29.5397 | 40.9654 | 106.8572 |
| 33 | 17.2535 | 22.5239 | 30.4628 | 42.7161 | 115.5283 |
| 34 | 17.5001 | 23.0010 | 31.3860 | 44.5021 | 124.8282 |
| 35 | 17.7370 | 23.4686 | 32.3091 | 46.3242 | 134.8024 |
| 36 | 17.9647 | 23.9271 | 33.2322 | 48.1832 | 145.4998 |
| 37 | 18.1834 | 24.3764 | 34.1553 | 50.0796 | 156.9728 |
| 38 | 18.3935 | 24.8168 | 35.0784 | 52.0144 | 169.2778 |
| 39 | 18.5954 | 25.2485 | 36.0015 | 53.9883 | 182.4749 |
| 40 | 18.7894 | 25.6717 | 36.9247 | 56.0021 | 196.6289 |
| 41 | 18.9758 | 26.0865 | 37.8478 | 58.0565 | 211.8092 |
| 42 | 19.1548 | 26.4930 | 38.7709 | 60.1524 | 228.0903 |
| 43 | 19.3269 | 26.8916 | 39.6940 | 62.2907 | 245.5518 |
| 44 | 19.4922 | 27.2822 | 40.6171 | 64.4722 | 264.2794 |
| 45 | 19.6510 | 27.6651 | 41.5402 | 66.6977 | 284.3650 |
| 46 | 19.8036 | 28.0404 | 42.4634 | 68.9682 | 305.9069 |
| 47 | 19.9502 | 28.4083 | 43.3865 | 71.2846 | 329.0107 |
| 48 | 20.0910 | 28.7689 | 44.3096 | 73.6478 | 353.7898 |
| 49 | 20.2264 | 29.1223 | 45.2327 | 76.0587 | 380.3656 |
| 50 | 20.3564 | 29.4688 | 46.1558 | 78.5183 | 408.8683 |

| | Geometric series future worth factor, $(F|A_1\ r, c, n)_\infty$ | | | | |
|---|---|---|---|---|---|
| $n$ | $c = 4\%$ | $c = 6\%$ | $c = 8\%$ | $c = 10\%$ | $c = 15\%$ |
| 1 | 1.0000 | 1.0000 | 1.0000 | 1.0000 | 1.0000 |
| 2 | 2.1460 | 2.1670 | 2.1885 | 2.2103 | 2.2670 |
| 3 | 3.4550 | 3.5224 | 3.5921 | 3.6642 | 3.8553 |
| 4 | 4.9458 | 5.0901 | 5.2412 | 5.3994 | 5.8291 |
| 5 | 6.6395 | 6.8967 | 7.1695 | 7.4591 | 8.2642 |
| 6 | 8.5592 | 8.9718 | 9.4154 | 9.8923 | 11.2504 |
| 7 | 10.7306 | 11.3488 | 12.0217 | 12.7548 | 14.8932 |
| 8 | 13.1823 | 14.0643 | 15.0367 | 16.1100 | 19.3172 |
| 9 | 15.9458 | 17.1595 | 18.5146 | 20.0299 | 24.6689 |
| 10 | 19.0562 | 20.6802 | 22.5162 | 24.5960 | 31.1208 |
| 11 | 22.5521 | 24.6773 | 27.1098 | 29.9011 | 38.8755 |
| 12 | 26.4767 | 29.2074 | 32.3718 | 36.0500 | 48.1710 |
| 13 | 30.8773 | 34.3336 | 38.3881 | 43.1615 | 59.2869 |
| 14 | 35.8067 | 40.1260 | 45.2546 | 51.3702 | 72.5508 |
| 15 | 41.3232 | 46.6624 | 53.0790 | 60.8280 | 88.3472 |
| 16 | 47.4914 | 54.0295 | 61.9814 | 71.7070 | 107.1265 |
| 17 | 54.3826 | 62.3236 | 72.0967 | 84.2016 | 129.4163 |
| 18 | 62.0759 | 71.6514 | 83.5754 | 98.5311 | 155.8342 |
| 19 | 70.6589 | 82.1317 | 96.5858 | 114.9433 | 187.1032 |
| 20 | 80.2285 | 93.8963 | 111.3160 | 133.7179 | 224.0688 |
| 21 | 90.8917 | 107.0916 | 127.9763 | 155.1702 | 267.7198 |
| 22 | 102.7672 | 121.8800 | 146.8012 | 179.6557 | 319.2122 |
| 23 | 115.9863 | 138.4416 | 168.0529 | 207.5753 | 379.8967 |
| 24 | 130.6939 | 156.9766 | 192.0237 | 239.3804 | 451.3512 |
| 25 | 147.0508 | 177.7066 | 219.0400 | 275.5794 | 535.4184 |
| 26 | 165.2346 | 200.8779 | 249.4657 | 316.7448 | 634.2500 |
| 27 | 185.4417 | 226.7632 | 283.7067 | 363.5209 | 750.3571 |
| 28 | 207.8894 | 255.6652 | 322.2155 | 416.6325 | 886.6703 |
| 29 | 232.8182 | 287.9193 | 365.4965 | 476.8948 | 1046.6085 |
| 30 | 260.4938 | 323.8974 | 414.1118 | 545.2244 | 1234.1597 |
| 31 | 291.2103 | 364.0116 | 468.6875 | 622.6516 | 1453.9746 |
| 32 | 325.2928 | 408.7188 | 529.9210 | 710.3344 | 1711.4754 |
| 33 | 363.1008 | 458.5251 | 598.5891 | 809.5735 | 2012.9833 |
| 34 | 405.0318 | 513.9913 | 675.5565 | 921.8297 | 2365.8655 |
| 35 | 451.5256 | 575.7389 | 761.7857 | 1048.7435 | 2778.7077 |
| 36 | 503.0682 | 644.4560 | 858.3481 | 1192.1563 | 3261.5132 |
| 37 | 560.1970 | 720.9052 | 966.4356 | 1354.1347 | 3825.9360 |
| 38 | 623.5064 | 805.9308 | 1087.3845 | 1536.9976 | 4485.5507 |
| 39 | 693.6533 | 900.4679 | 1222.6399 | 1743.3462 | 5256.1676 |
| 40 | 771.3643 | 1005.5522 | 1373.8725 | 1976.0980 | 6156.1980 |
| 41 | 857.4424 | 1122.3303 | 1542.8964 | 2238.5242 | 7207.0797 |
| 42 | 952.7756 | 1252.0716 | 1731.7400 | 2534.2921 | 8433.7723 |
| 43 | 1058.3454 | 1396.1817 | 1942.6579 | 2867.5122 | 9865.3318 |
| 44 | 1175.2371 | 1556.2165 | 2178.1560 | 3242.7909 | 11535.5801 |
| 45 | 1304.6503 | 1733.8984 | 2441.0191 | 3665.2891 | 13483.8828 |
| 46 | 1447.9113 | 1931.1339 | 2734.3416 | 4140.7880 | 15756.0539 |
| 47 | 1606.4860 | 2150.0328 | 3061.5612 | 4675.7628 | 18405.4073 |
| 48 | 1781.9951 | 2392.9306 | 3426.4968 | 5277.4643 | 21493.9796 |
| 49 | 1976.2301 | 2662.4116 | 3833.3901 | 5954.0105 | 25093.9520 |
| 50 | 2191.1713 | 2961.3357 | 4286.9517 | 6714.4890 | 29289.3025 |

# TABLE B.34  Continuous Compounding: $r = 10\%$

| | Geometric series present worth factor, $(P \mid A_1 \ r, c, n)_\infty$ | | | | |
|---|---|---|---|---|---|
| $n$ | $c = 4\%$ | $c = 6\%$ | $c = 8\%$ | $c = 10\%$ | $c = 15\%$ |
| 1 | 0.8607 | 0.8607 | 0.8607 | 0.8607 | 0.8607 |
| 2 | 1.6318 | 1.6472 | 1.6632 | 1.6794 | 1.7214 |
| 3 | 2.3225 | 2.3663 | 2.4115 | 2.4582 | 2.5821 |
| 4 | 2.9413 | 3.0233 | 3.1092 | 3.1991 | 3.4428 |
| 5 | 3.4956 | 3.6238 | 3.7597 | 3.9037 | 4.3035 |
| 6 | 3.9922 | 4.1726 | 4.3662 | 4.5741 | 5.1642 |
| 7 | 4.4370 | 4.6742 | 4.9317 | 5.2117 | 6.0250 |
| 8 | 4.8356 | 5.1326 | 5.4590 | 5.8182 | 6.8857 |
| 9 | 5.1926 | 5.5515 | 5.9507 | 6.3952 | 7.7464 |
| 10 | 5.5124 | 5.9344 | 6.4091 | 6.9440 | 8.6071 |
| 11 | 5.7989 | 6.2844 | 6.8365 | 7.4660 | 9.4678 |
| 12 | 6.0556 | 6.6042 | 7.2350 | 7.9626 | 10.3285 |
| 13 | 6.2855 | 6.8965 | 7.6066 | 8.4350 | 11.1892 |
| 14 | 6.4915 | 7.1636 | 7.9530 | 8.8843 | 12.0499 |
| 15 | 6.6760 | 7.4078 | 8.2761 | 9.3117 | 12.9106 |
| 16 | 6.8413 | 7.6309 | 8.5773 | 9.7183 | 13.7713 |
| 17 | 6.9894 | 7.8348 | 8.8581 | 10.1050 | 14.6320 |
| 18 | 7.1220 | 8.0212 | 9.1199 | 10.4729 | 15.4927 |
| 19 | 7.2409 | 8.1915 | 9.3641 | 10.8229 | 16.3535 |
| 20 | 7.3472 | 8.3472 | 9.5917 | 11.1557 | 17.2142 |
| 21 | 7.4427 | 8.4895 | 9.8040 | 11.4724 | 18.0749 |
| 22 | 7.5281 | 8.6195 | 10.0019 | 11.7736 | 18.9356 |
| 23 | 7.6046 | 8.7383 | 10.1864 | 12.0601 | 19.7963 |
| 24 | 7.6732 | 8.8470 | 10.3584 | 12.3326 | 20.6570 |
| 25 | 7.7346 | 8.9462 | 10.5189 | 12.5918 | 21.5177 |
| 26 | 7.7897 | 9.0369 | 10.6684 | 12.8384 | 22.3784 |
| 27 | 7.8389 | 9.1198 | 10.8079 | 13.0730 | 23.2391 |
| 28 | 7.8831 | 9.1956 | 10.9379 | 13.2961 | 24.0998 |
| 29 | 7.9227 | 9.2649 | 11.0591 | 13.5084 | 24.9605 |
| 30 | 7.9581 | 9.3282 | 11.1722 | 13.7103 | 25.8212 |
| 31 | 7.9898 | 9.3860 | 11.2776 | 13.9023 | 26.6819 |
| 32 | 8.0183 | 9.4389 | 11.3759 | 14.0850 | 27.5427 |
| 33 | 8.0438 | 9.4872 | 11.4675 | 14.2588 | 28.4034 |
| 34 | 8.0666 | 9.5313 | 11.5529 | 14.4241 | 29.2641 |
| 35 | 8.0870 | 9.5717 | 11.6326 | 14.5813 | 30.1248 |
| 36 | 8.1053 | 9.6086 | 11.7069 | 14.7309 | 30.9855 |
| 37 | 8.1218 | 9.6423 | 11.7761 | 14.8732 | 31.8462 |
| 38 | 8.1365 | 9.6731 | 11.8407 | 15.0085 | 32.7069 |
| 39 | 8.1496 | 9.7013 | 11.9009 | 15.1372 | 33.5676 |
| 40 | 8.1614 | 9.7270 | 11.9570 | 15.2597 | 34.4283 |
| 41 | 8.1720 | 9.7505 | 12.0094 | 15.3762 | 35.2890 |
| 42 | 8.1814 | 9.7720 | 12.0582 | 15.4870 | 36.1497 |
| 43 | 8.1899 | 9.7916 | 12.1037 | 15.5924 | 37.0104 |
| 44 | 8.1975 | 9.8096 | 12.1461 | 15.6926 | 37.8712 |
| 45 | 8.2043 | 9.8260 | 12.1856 | 15.7880 | 38.7319 |
| 46 | 8.2104 | 9.8410 | 12.2225 | 15.8787 | 39.5926 |
| 47 | 8.2159 | 9.8547 | 12.2569 | 15.9650 | 40.4533 |
| 48 | 8.2208 | 9.8672 | 12.2890 | 16.0471 | 41.3140 |
| 49 | 8.2252 | 9.8787 | 12.3189 | 16.1252 | 42.1747 |
| 50 | 8.2291 | 9.8891 | 12.3468 | 16.1995 | 43.0354 |

| | | Geometric series future worth factor, $(F|A_1\ r, c, n)_\infty$ | | | |
|---|---|---|---|---|---|
| $n$ | $c = 4\%$ | $c = 6\%$ | $c = 8\%$ | $c = 10\%$ | $c = 15\%$ |
| 1 | 1.0000 | 1.0000 | 1.0000 | 1.0000 | 1.0000 |
| 2 | 2.1241 | 2.145 | 2.1666 | 2.1885 | 2.2451 |
| 3 | 3.3843 | 3.4513 | 3.5205 | 3.5921 | 3.7820 |
| 4 | 4.7937 | 4.9359 | 5.0850 | 5.2412 | 5.6653 |
| 5 | 6.3664 | 6.6183 | 6.8856 | 7.1695 | 7.9592 |
| 6 | 8.1181 | 8.5194 | 8.9509 | 9.4154 | 10.7391 |
| 7 | 10.0654 | 10.6623 | 11.3125 | 12.0217 | 14.0932 |
| 8 | 12.2269 | 13.0722 | 14.0054 | 15.0367 | 18.1246 |
| 9 | 14.6224 | 15.7771 | 17.0683 | 18.5146 | 22.9543 |
| 10 | 17.2735 | 18.8071 | 20.5543 | 22.5162 | 28.7235 |
| 11 | 20.2040 | 22.1956 | 24.4810 | 27.1098 | 35.5975 |
| 12 | 23.4395 | 25.9790 | 28.9308 | 32.3718 | 43.7693 |
| 13 | 27.0078 | 30.1972 | 33.9521 | 38.3881 | 53.4643 |
| 14 | 30.9392 | 34.8937 | 39.6090 | 45.2546 | 64.9459 |
| 15 | 35.2667 | 40.1162 | 45.9728 | 53.0790 | 78.5212 |
| 16 | 40.0261 | 45.9170 | 53.1219 | 61.9814 | 94.5487 |
| 17 | 45.2562 | 52.3530 | 61.1429 | 72.0967 | 113.4466 |
| 18 | 50.9993 | 59.4865 | 70.1315 | 83.5754 | 135.7023 |
| 19 | 57.3014 | 67.3856 | 80.1932 | 96.5858 | 161.8843 |
| 20 | 64.2121 | 76.1247 | 91.4445 | 111.3160 | 192.6550 |
| 21 | 71.7857 | 85.7851 | 104.0137 | 127.9763 | 228.7862 |
| 22 | 80.0809 | 96.4553 | 118.0422 | 146.8012 | 271.1772 |
| 23 | 89.1614 | 108.2322 | 133.6861 | 168.0529 | 320.8754 |
| 24 | 99.0967 | 121.2214 | 151.1169 | 192.0237 | 379.1005 |
| 25 | 109.9619 | 135.5383 | 170.5240 | 219.0400 | 447.2729 |
| 26 | 121.8386 | 151.3086 | 192.1155 | 249.4657 | 527.0461 |
| 27 | 134.8154 | 168.6694 | 216.1207 | 283.7067 | 620.3446 |
| 28 | 148.9884 | 187.7705 | 242.7919 | 322.2155 | 729.4088 |
| 29 | 164.4621 | 208.7749 | 272.4066 | 365.4965 | 856.8454 |
| 30 | 181.3496 | 231.8605 | 305.2702 | 414.1118 | 1005.6880 |
| 31 | 199.7738 | 357.2211 | 341.7185 | 468.6875 | 1179.4660 |
| 32 | 219.8680 | 285.0681 | 382.1205 | 529.9210 | 1382.2852 |
| 33 | 241.7768 | 315.6315 | 426.8820 | 598.5891 | 1618.9221 |
| 34 | 265.6571 | 349.1623 | 476.4489 | 675.5565 | 1894.9324 |
| 35 | 291.6791 | 385.9336 | 531.3113 | 761.7857 | 2216.7776 |
| 36 | 320.0274 | 426.2430 | 592.0073 | 858.3481 | 2591.9728 |
| 37 | 350.9022 | 470.4147 | 659.1281 | 966.4356 | 3029.2571 |
| 38 | 384.5208 | 518.8015 | 733.3229 | 1087.3745 | 3538.7925 |
| 39 | 421.1186 | 571.7877 | 815.3045 | 1222.6399 | 4132.3956 |
| 40 | 460.9512 | 629.7914 | 905.8552 | 1373.8725 | 4823.8051 |
| 41 | 504.2955 | 693.2681 | 1005.8337 | 1542.8964 | 5628.9945 |
| 42 | 551.4520 | 762.7131 | 1116.1825 | 1731.7400 | 6566.5343 |
| 43 | 602.7463 | 838.6659 | 1237.9352 | 1942.6579 | 7658.0136 |
| 44 | 658.5318 | 921.7130 | 1372.2262 | 2178.1560 | 8928.5294 |
| 45 | 791.1914 | 1012.4930 | 1520.2993 | 2441.0191 | 10407.2556 |
| 46 | 785.1404 | 1111.7003 | 1683.5188 | 2734.3416 | 12128.1042 |
| 47 | 856.8290 | 1220.0904 | 1863.3805 | 3061.5612 | 14130.4931 |
| 48 | 934.7453 | 1338.4850 | 2061.5244 | 3426.4968 | 16460.2392 |
| 49 | 1019.4185 | 1467.7778 | 2279.7482 | 3833.3901 | 19170.5950 |
| 50 | 1111.4222 | 1608.9405 | 2520.0222 | 4286.9517 | 22323.4542 |

| | Geometric series present worth factor, $(P\mid A_1\ r, c, n)_\infty$ | | | | |
|---|---|---|---|---|---|
| $n$ | $c = 4\%$ | $c = 6\%$ | $c = 8\%$ | $c = 10\%$ | $c = 15\%$ |
| 1 | 0.9048 | 0.9048 | 0.9048 | 0.9048 | 0.9048 |
| 2 | 1.7570 | 1.7742 | 1.7918 | 1.8097 | 1.8561 |
| 3 | 2.5595 | 2.6095 | 2.6611 | 2.7145 | 2.8561 |
| 4 | 3.3153 | 3.4120 | 3.5133 | 3.6193 | 3.9073 |
| 5 | 4.0271 | 4.1830 | 4.3485 | 4.5242 | 5.0125 |
| 6 | 4.6974 | 4.9239 | 5.1673 | 5.4290 | 6.1743 |
| 7 | 5.3287 | 5.6356 | 5.9698 | 6.3339 | 7.3957 |
| 8 | 5.9232 | 6.3195 | 6.7564 | 7.2387 | 8.6798 |
| 9 | 6.4831 | 6.9765 | 7.5275 | 8.1435 | 10.0296 |
| 10 | 7.0104 | 7.6078 | 8.2832 | 9.0484 | 11.4487 |
| 11 | 7.5070 | 8.2143 | 9.0241 | 9.9532 | 12.9405 |
| 12 | 7.9746 | 8.7971 | 9.7502 | 10.8580 | 14.5088 |
| 13 | 8.4151 | 9.3570 | 10.4620 | 11.7629 | 16.1576 |
| 14 | 8.8298 | 9.8949 | 11.1597 | 12.6677 | 17.8908 |
| 15 | 9.2205 | 10.4118 | 11.8435 | 13.5726 | 19.7129 |
| 16 | 9.5883 | 10.9084 | 12.5138 | 14.4774 | 21.6285 |
| 17 | 9.9348 | 11.3855 | 13.1709 | 15.3822 | 23.6422 |
| 18 | 10.2611 | 11.8439 | 13.8149 | 16.2871 | 25.7592 |
| 19 | 10.5684 | 12.2843 | 14.4462 | 17.1919 | 27.9848 |
| 20 | 10.8577 | 12.7075 | 15.0650 | 18.0967 | 30.3244 |
| 21 | 11.1303 | 13.1141 | 15.6715 | 19.0016 | 32.7840 |
| 22 | 11.3869 | 13.5047 | 16.2660 | 19.9064 | 35.3697 |
| 23 | 11.6286 | 13.8800 | 16.8488 | 20.8113 | 38.0880 |
| 24 | 11.8563 | 14.2406 | 17.4200 | 21.7161 | 40.9457 |
| 25 | 12.0707 | 14.5870 | 17.9799 | 22.6209 | 43.9498 |
| 26 | 12.2726 | 14.9199 | 18.5287 | 23.5258 | 47.1080 |
| 27 | 12.4627 | 15.2397 | 19.0667 | 24.4306 | 50.4281 |
| 28 | 12.6418 | 15.5470 | 19.5939 | 25.3354 | 53.9185 |
| 29 | 12.8104 | 15.8422 | 20.1108 | 26.2403 | 57.5878 |
| 30 | 12.9692 | 16.1259 | 20.6174 | 27.1451 | 61.4452 |
| 31 | 13.1188 | 16.3984 | 21.1140 | 28.0500 | 65.5004 |
| 32 | 13.2597 | 16.6603 | 21.6007 | 28.9548 | 69.7635 |
| 33 | 13.3923 | 16.9119 | 22.0779 | 29.8596 | 74.2452 |
| 34 | 13.5172 | 17.1536 | 22.5455 | 30.7645 | 78.9567 |
| 35 | 13.6349 | 17.3858 | 23.0039 | 31.6693 | 83.9097 |
| 36 | 13.7457 | 17.6089 | 23.4533 | 32.5741 | 89.1167 |
| 37 | 13.8500 | 17.8233 | 23.8937 | 33.4790 | 94.5906 |
| 38 | 13.9483 | 18.0293 | 24.3254 | 34.3838 | 100.3452 |
| 39 | 14.0409 | 18.2272 | 24.7486 | 35.2887 | 106.3949 |
| 40 | 14.1280 | 18.4173 | 25.1634 | 36.1935 | 112.7547 |
| 41 | 14.2101 | 18.6000 | 25.5699 | 37.0983 | 119.4406 |
| 42 | 14.2874 | 18.7755 | 25.9684 | 38.0032 | 126.4693 |
| 43 | 14.3602 | 18.9442 | 26.3591 | 38.9080 | 133.8583 |
| 44 | 14.4288 | 19.1062 | 26.7420 | 39.8128 | 141.6262 |
| 45 | 14.4934 | 19.2619 | 27.1173 | 40.7177 | 149.7924 |
| 46 | 14.5542 | 19.4114 | 27.4852 | 41.6225 | 158.3773 |
| 47 | 14.6114 | 19.5551 | 27.8457 | 42.5274 | 167.4023 |
| 48 | 14.6654 | 19.6932 | 28.1992 | 43.4322 | 176.8900 |
| 49 | 14.7162 | 19.8259 | 28.5457 | 44.3370 | 186.8642 |
| 50 | 14.7640 | 19.9533 | 28.8853 | 45.2419 | 197.3498 |

## TABLE B.37 Continuous Compounding: $r = 15\%$

| | Geometric series future worth factor, $(F|A_1 \ r,c,n)_x$ | | | | |
|---|---|---|---|---|---|
| $n$ | $c = 4\%$ | $c = 6\%$ | $c = 8\%$ | $c = 10\%$ | $c = 15\%$ |
| 1 | 1.0000 | 1.0000 | 1.0000 | 1.0000 | 1.0000 |
| 2 | 2.2026 | 2.2237 | 2.2451 | 2.2670 | 2.3237 |
| 3 | 3.6424 | 3.7110 | 3.7820 | 3.8553 | 4.0496 |
| 4 | 5.3594 | 5.5088 | 5.6653 | 5.8291 | 6.2732 |
| 5 | 7.4002 | 7.6716 | 7.9592 | 8.2642 | 9.1106 |
| 6 | 9.8192 | 10.2630 | 10.7391 | 11.2504 | 12.7020 |
| 7 | 12.6795 | 13.3572 | 14.0932 | 14.8932 | 17.2172 |
| 8 | 16.0546 | 17.0408 | 18.1246 | 19.3172 | 22.8612 |
| 9 | 20.0300 | 21.4147 | 22.9543 | 24.6689 | 29.8810 |
| 10 | 24.7048 | 26.5963 | 28.7235 | 31.1208 | 38.5742 |
| 11 | 30.1947 | 32.7226 | 35.5975 | 38.8755 | 49.2986 |
| 12 | 36.6340 | 39.9531 | 43.7692 | 48.1710 | 62.4837 |
| 13 | 44.1787 | 48.4733 | 53.4643 | 59.2868 | 78.6454 |
| 14 | 53.0103 | 58.4994 | 64.9459 | 72.5508 | 98.4016 |
| 15 | 63.3399 | 70.2829 | 78.5212 | 88.3472 | 122.4924 |
| 16 | 75.4126 | 84.1167 | 94.5487 | 107.1264 | 151.8037 |
| 17 | 89.5134 | 100.3414 | 113.4465 | 129.4162 | 187.3939 |
| 18 | 105.9736 | 119.3532 | 135.7022 | 155.8341 | 230.5277 |
| 19 | 125.1782 | 141.6133 | 161.8842 | 187.1030 | 282.7146 |
| 20 | 147.5745 | 167.6580 | 192.6548 | 224.0686 | 345.7554 |
| 21 | 173.6827 | 198.1109 | 228.7860 | 267.7195 | 421.7959 |
| 22 | 204.1069 | 233.6974 | 271.1768 | 319.2117 | 513.3931 |
| 23 | 239.5493 | 275.2610 | 320.8750 | 379.8963 | 623.5901 |
| 24 | 280.8257 | 323.7825 | 379.1001 | 451.3506 | 756.0088 |
| 25 | 328.8848 | 380.4024 | 447.2725 | 535.4180 | 914.9551 |
| 26 | 384.8279 | 446.4463 | 527.0457 | 634.2493 | 1105.5471 |
| 27 | 449.9353 | 523.4553 | 620.3440 | 750.3562 | 1333.8650 |
| 28 | 525.6948 | 613.2214 | 729.4080 | 886.6692 | 1607.1275 |
| 29 | 613.8352 | 717.8272 | 856.8447 | 1046.6074 | 1933.9006 |
| 30 | 716.3648 | 839.6934 | 1005.6873 | 1234.1585 | 2324.3504 |
| 31 | 835.6172 | 981.6343 | 1179.4649 | 1453.9729 | 2790.5278 |
| 32 | 974.3042 | 1146.9202 | 1382.2839 | 1711.4734 | 3346.7151 |
| 33 | 1135.5767 | 1339.3521 | 1618.9207 | 2012.9807 | 4009.8399 |
| 34 | 1323.0952 | 1563.3477 | 1894.9295 | 2365.8623 | 4799.9414 |
| 35 | 1541.1135 | 1824.0415 | 2216.7742 | 2778.7039 | 5740.7578 |
| 36 | 1794.5728 | 2127.3999 | 2591.9695 | 3261.5088 | 6860.3750 |
| 37 | 2089.2175 | 2480.3567 | 3029.2527 | 3825.9312 | 8192.0274 |
| 38 | 2431.7168 | 2890.9712 | 3538.7886 | 4485.5430 | 9775.0156 |
| 39 | 2829.8240 | 3368.6047 | 4132.3867 | 5256.1563 | 11655.8125 |
| 40 | 3292.5454 | 3924.1423 | 4823.7969 | 6156.1875 | 13889.3555 |
| 41 | 3830.3455 | 4570.2227 | 5628.9844 | 7207.0703 | 16540.5586 |
| 42 | 4455.3789 | 5321.5469 | 6566.5234 | 8433.7578 | 19686.1055 |
| 43 | 5181.7774 | 6195.1836 | 7658.0039 | 9865.3164 | 23416.5625 |
| 44 | 6025.9531 | 7210.9766 | 8928.5156 | 11535.5625 | 27838.8477 |
| 45 | 7006.9727 | 8391.9727 | 10407.2422 | 13483.8633 | 33079.2188 |
| 46 | 8146.9883 | 9764.9609 | 12128.0859 | 15756.0274 | 39286.6289 |
| 47 | 9471.7461 | 11361.0664 | 14130.4727 | 18405.3789 | 46636.8242 |
| 48 | 11011.1524 | 13216.4531 | 16460.2149 | 21493.9453 | 55337.1367 |
| 49 | 12799.9570 | 15373.1406 | 19170.5664 | 25093.9102 | 65632.0000 |
| 50 | 14878.5274 | 17879.9570 | 22323.4219 | 29289.2383 | 77809.6875 |

# APPENDIX C

## Answers to Even-Numbered Problems

### Chapter 1

2. 32" × 32" × 6" pallet, with cost/carton = $0.60
4. 5" × 6" × 100" ingots, monthly cost = $49,535.60
6. Ceramic insert, with V = 247.75 fpm
8. (a) Type B, with $898/mo.
   (b) $0.898/mile
10. Wheel A, unit cost = (10/R) + $(3.46 \times 10^{-5})R$

### Chapter 2

2. 37 units, with OH rates on machining time only
4. 1775 units/year
6. 40%
8. (b) Net profit before taxes = $7,700
   (c) $9,000
10. (a) $32,727
    (b) Total Assets = $283,727. Add Net Worth accounts to achieve "balance".
12. (a) 113.47 lbs./acre
    (b) $21.40/acre
    (c) 79.68 to 147.25 lbs./acre
14. (a) t = 4.375 units/hour
    (b) $2.66/hour
    (c) t = 6.25 units/hour, with profit = $1.25/hour
    (d) Between 1.78 and 6.95 units/hour
16. (a) $22,050
    (b) $6.62/unit
18. $3.78/hr., $4.20/hr., $7.48/hr., and $3.82/hr. for cost centers A, B, C, and D, respectively.

### Chapter 3

2. (a) $570.50, $661.78, $767.66
   (b) $320.00, $228.72, $122.84
4. 10 periods
6. $1,192.60
8. (a) $4,128.80
   (b) $4,459.10
10. $23,326.70
12. $388.00
14. $1,959.11
18. (a) $838.55
    (b) $21,306.59
20. $7,575.19
22. (a) $1,838.10
    (b) $1,671.32
24. $8,977.60
26. $433.94
28. $1,608.00
30. $1,830.06, $1,822.03
32. $2,711.88
34. $411.68
36. $6,419.90
38. 9.545%
40. (a) $1,893.04
    (b) $4,907.46
42. (a) $10,817.80
    (b) $572.91
44. $2,300.10
46. (a) $1,837.42
    (b) $1,557.35
48. (a) $18,372.64
    (b) $47,653.12
    (c) $2,989.23
50. $68,496.00
52. (a) $41,763.22
    (b) $113,524.96
54. Don't pay cash.
56. $3,767.16
58. −$815.16
60. $1,083.46
62. $3,123.75
64. $1,755.94

66. $4,357.22
68. $1,058.70
70. −$1,412.19
72. −$1,212.34
74. $25,037.50
76. X = $1,047.53, Y = $771.41
78. 0.09365
82. 12%
86. $5,786.60
88. $11,195.67
90. $1,330.17
92. $16,909.50
94. $1,041.59
96. $20,474.11
98. $11,337.37,   $2,967.47

## Chapter 4

4.2  (a) 16.64%      (b) 8.99%
4.4  (a) 12.55%      (b) 6.78%
4.6  yes, 10.53%
4.8  8.74%
4.10 (a) $5,652.50   (b)$1,161.02/year
     (d) i = 22.46%   (e)i′ = 20.47%
4.12 (a) $211,350.62
     (b) −37,409.06/year
     (c) −$656,412.75
4.14 (b) $444/year      (d) i = 27.7%
     (f) 1.074     Purchase condenser
4.16 (c) $124,064.30     (e)i′ = 18.77%
     (f) 1.7989   Purchase filter
4.18 (a) PW(15) = $1,003.35
     (c) $3,529.68      (d) 17.51%
4.20 $8,333.33
4.22 $50,240.19
4.24 $28,145.87
4.26 $736,300
4.28 11.34%
4.30 $2,213.15
4.32 $986.06
4.34 $10,222.40/year
4.36 $29,275/year
4.38 At least 9 years
4.40 $50,000, −$2,000, 8, .18, 300,
     A/G

4.42 (a) 1984: 9.02%   (b) 7.95%
     (c) 7.58%
4.44 $9,174.31,   $10,100.16,
     $11,582.75,   $12,397.44
4.46 $1,717.47   $1,066.41
4.48 4.55%
4.50 (a,b) $1,485,277.77
4.52 $3,950.85

## Chapter 5

5.2  None; A; B; B,C
5.4  (a) 24 years   (b) 3 years   (c) 8 years
     (d) 2 years   (e) 6 years   (f) 24 years
     For (d) Alt. 3:−$140,000, $65,000,
     65,000 + $40,000
5.6  (a) Alt. 2:−$80,000, $35,000,
     $45,000 + $20,000
5.8  None; C; B; A; A,D   NCF(A,D):
     (0)−$160,000 (1-5) $62,000
5.10 None; B; A; A,C; A,B
     NCF (A,B):(0)−$1,000,000
     (8)−$70,000 (24) $530,000
5.12 NCF(A,B):−$75,000, $20,000,
     $25,000,$30,000,$35,000,$40,000
5.14 (b)−$5,000, $0, $0, $0, $0, $3,500
5.16 (d) $AW_A(25)$ = $109,862.78
     i. $i_{B-A}$ = 27.9   Select B
5.18 (b) $PW_{B,D}(20)$ = $28,875
     j. $i′_{B,D}$ = 20.45%   Select B,D
5.20 (f) $FW_{2''}(25)$ = $10.41
     (l) $SIR_{2''}$ = 2.80   Select 2″
5.22 (1) 1 2/3 years, (2) 3 2/5 years
     Prefer Alternative 1
5.24 (a) $2813.53, $722.00, $319.50,
     (b) 3 1/8 yr, 1 yr, 2 yr
5.26 (W) 62,943.03, (S) $53,750,
     (C) $52,500   Select concrete
5.28 (b) (U)−$4,730.83/yr,
     (V)−$4,767.20/yr   Select motor U
5.30 (c) (I)−$170,272.26/yr,
     II−$198,033.55/yr   Select I
5.32 (e) $i_{2-1}$ = 15.73%
     (g) $SIR_{2-1}(12)$ = 1.103
     Select 2

5.34 (b) $PW_C(20) = -\$134,160$
$PW_M(20) = -\$115,887.50$
Select machine

5.36 (c) $AW_1(20) = -\$12,409.80/\text{yr}$,
$AW_2(20) = -\$12,241.50/\text{yr}$
Select 2

5.38 (b) $AW_1(15) = -\$64,916/\text{year}$,
$AW_2(15) = -\$58,977.58/\text{year}$
Prefer 2

5.40 (a) $PW_L(12) = -\$41,114$
$PW_P(12) = -\$66,117$
Select lease alternative

5.42 (a,b) $166,666.67/year (factor truncation causes different answers)

5.44 (a) $76,788.52/year
(b) $76.796.41/year

5.46 $PW_{36}(10) = -\$11,676$
$PW_{50}(10) = -\$11,000$
Place 50″ culvert

5.48 (a) $PW_K(20) = -\$20,709.60$
$PW_R = -\$21,308.45$
Keep old

5.50 (b) $AW_K(15) = -\$9,808/\text{year}$
$AW_R(15) = -\$10,431.20/\text{year}$
Keep old

5.52 (a) $PW_K(20) = -\$342,599.72$
$PW_R(20) = -\$313,946.60$
Keep old

5.54 $AW_K(15) = \$4,686.46/\text{year}$
$AW_R(15) = \$2,637.75/\text{year}$
Keep old

5.56 $AW_G(20) = -\$3,960.64/\text{year}$
$AW_B(20) = -\$1,981.37/\text{year}$
Prefer battery-powered

5.58 $AW_K(25) = -\$161,574/\text{yr}$
$AW_2(25) = -\$156,010/\text{yr}$
Prefer replacement

5.60 $PW_X(20) = -\$591,020$
$PW_Y(20) = -\$600,865$
Keep old and buy X

5.62 $AW_K(15) = -\$22,808/\text{year}$
$AW_R(15) = -\$28.020/\text{year}$
Keep old

## Chapter 6

6.2 (a,f,g) Neither (b,d,e) 1245-5 year
(c) 1245-3 year (h) 1250-10 year
(i,j) 1250-15 year (real or p.u.)

6.4 (a) year 2: $4,180  $4,070
(b) year 2: $2,200 $7,700

6.6 (a) $6,000, $8,400, $0
(b) $4,000, $8,000, $0

6.8 (a) year 3:$46,200
(b) year 3: $44,000

6.10 (a) year 4: $14,400, $113,400
(b) year 4: $5,142.86, $161,142.86

6.12 (a) $29,400, $58,800
(b) $28,000, $70,000 (e) $14,577.26,
$51,020.41

6.14 (a) year 4: $21,000, $21,000
(b) year 4: $20,000, $30,000

6.16 AECR + R = $1,528.11/year
for each part

6.18 $2,085.71, $3,128.57, $1,460.00,
$625.71

6.20 (b) $5,730 (d) $21,350 (f) $439,750

6.22 (a) year 5: $12,930
(b) year 5: $12,700
(e) year 5: $9,919.34

6.24 (a) $4,840.33/yr
(b) $4,894.12/yr
(d) $4,964.64/yr

6.26 (a) PW(15) = $7,435.17
(b) $6,912.91
All. Yes, undertake investment

6.28 year 4 ATCF = $8,298
PW(15) = -$2,800.56
Do not invest

6.30 (a) $3,186.70/year
(b) $3,100.59/year

6.32 (a) $71,449.70/year
(b) $7.15/unit

6.34 (a) $13,110.89
(b) $9,206.35

6.36 (a) year 3 ATCF = $23,480.16
(b) year 3 ATCF = $28,616.67

6.38 ATCF only: -$60,000, $15,540,
$18,760, $21,000, $23,700,

$29,100, $8,640, $10,260,
$16,740, $22,140, $14,040,
−$40,000
6.40 $2,967.51/year
6.42 $32.310
6.44 (a) $25,000 (b) $3,680
(c) Back or over (d) $3,080
(e,f) Back or over
6.46 (a) $74,500 Section 1231 loss
(b) $111,250 Section 1231 loss
6.48 (a) $8,000 Section 1231 gain and
$30,000 depreciation recapture
(b) $10,000 depreciation recapture
(c) $60,000 Section 1231 loss
6.50 (a) $521,666.67 Section 1231 gain
(b) $121,666.67 Section 1231 gain
(c) $178,333.33 Section 1231 loss
6.52 −$10,000, $2,850, $3,172,
$2,160 + $3,438
6.54 year 2: $25,092.95
year 5: $16,200 + $9,568.73
6.56 $24,765.27
6.58 Before tax: PW(20) = −$63,675
After-tax:PW (15) = $6,187.80
6.60 Purchase: $135,429.42/year
Lease: $137,167.15/year
6.62 $161,030.85/year
6.64 (a) $233,333.33   (b) $236,250

## Chapter 7

2. (a) 1.170, 1.071, 1.044
(b) A
(c) $23,000, $15,000, $12,000
4. (a) fuel oil
(b) fuel oil
6. Yes
8. A, A
14. Yes

## Chapter 8

2. 8,983.08 hrs/year
4. (a) 31.10%
(b) 50.56%

(c) 73.37%
(d) 98.18%
6. 3 years
8. Z if usage ≤ 2,203.42 hrs/yr
Y if 2,203.42 < usage ≤ 4,115.06
hrs/yr
X if usage > 4,115.06 hrs/yr
10. 67.902 or larger
12. B if volume ≥ 3.674/yr
A otherwise
14. 51.3 < X < 1948.7
16. 203.58%, 41.46%
24. 18.6%
30. −$9,763.60, 664.8012 × $10^4$,
$P(pw ≥ 0) = 0$
32. −$9,763.60, 857.1478 × $10^4$,
$P(pw ≥ 0) = 0.00040$
34. 0.003125

## Chapter 9

2. If $0 ≤ A < 1/4$, choose $A_1$
$A = 1/4$, choose either $A_1$ or $A_3$
$1/4 < A < 5/8$, choose $A_3$
$A = 5/8$, choose either $A_3$ or $A_2$
$5/8 < A ≤ 1.0$, choose $A_2$.
4. $P(S_2) = 0.475$, $P(S_3) = 0.425$
6. Laplace principle: Choose $A_1$
with $E(A_1)$ =$13,000
Maximim principle: Choose either
$A_1$ or $A_5$ with value = $8,000
Maximax prinicple: Choose $A_2$
with value = $20,000
Savage principle: Choose $A_1$ with
minimax regret value = $4,000
Hurwicz principle: Choose $A_1$ or
$A_5$ if $0 ≤ A < 1/3$. Choose either
$A_1, A_2$, or $A_5$ if $A = 1/3$. Choose
$A_2$ if $1/3 < A ≤ 1.0$.
8. Minimax principle: Choose $a_2$ with
value = $60
Minimin principle: Choose $a_4$ with
value = −$50
Minimax regret principle: Choose
either $a_1$ or $a_3$, with value =
$100

Expectation principle: Choose $a_2$,
  with $E(a_2) = \$60$ and Var
  $(a_2) = 0$
Hurwicz principle: Choose $a_4$,
  with $H_4 = 10$
10. (a) Choose $A_2$, with $E(A_2) = \$175$
    and $Var(A_2) = 516.67$.
   (b) Choose $A_2$ with value $= \$200$
   (c) Choose $A_3$ with value $= \$75$

12. Keep present equipment with
    $E(Keep) = \$84,000$. The most
    probable future principle would
    select the purchase alternative,
    ignoring the $\$100,000$ loss prob-
    ability.

14. (a) $p(S_1) = 0.216$; $p(S_2) = 0.432$;
   $p(S_3) = 0.288$; $p(S_4) = 0.064$.
   (b)

|            | 0.216 | 0.432 | 0.288 | 0.064 |          |
|            | $S_1$ | $S_2$ | $S_3$ | $S_4$ | $E(a_i)$ |
|------------|-------|-------|-------|-------|----------|
| Guess $S_1$ | $3   | -$1   | -$1   | -$1   | -$0.136  |
| Guess $S_2$ | -$1   | $2    | -$1   | $1    | $0.296   |
| Guess $S_3$ | -$1   | -$1   | $3    | -$1   | $0.152   |
| Guess $S_4$ | -$1   | -$1   | -$1   | $4    | -$0.680  |

Both the expectation principle and most probable future principle
selects the "Guess $S_2$" alternative.

16. (a) $P(H) = 0.62$
   (b) $P(C|L) = 0.434$
18. (a) $P(t) = 4/9$. Best guess is
    "Red die uncovered" with
    $P(B|t) = 1/2$
   (b) Best guess is "Red die
    uncovered" with E (Guess
    Red) $= \$1.00$.
20. (a) Choose $a_1$ with PEP $=$
    $\$6,100,000$

   (b) EVPI $= \$1,000,000$
   (c) EVSI $= \$89,637$
   (d) $-\$160,363$
22. (b) Keep the present truck (by
    dominance principle). For
    interest, $E$ (Keep Present Truck)
    at $D_1 = -\$21,700$.
24. Choose $A_1$ with $E(A_1) = 0.9078 >$
    $E(A_2) = 0.8799$

# INDEX

# INDEX